MW01097317

Digital Signal Processors
Architectures, Implementations, and Applications

Sen M. Kuo

Northern Illinois University, DeKalb, IL

Woon-Seng Gan

Nanyang Technological University, Singapore

PEARSON

Prentice
Hall

Upper Saddle River, NJ 07458

Library of Congress Cataloging-in-Publication Data

Kuo, Sen M. (Sen-Maw)
  Digital signal processors: architectures, implementations, and applications / Sen M. Kuo
and Woon-Seng Gan.
      p. cm.
  Includes bibliographical references and index.
  ISBN 0-13-035214-4
    1. Signal processing—Digital techniques. I. Gan, Woon-Seng. II. Title.

  TK5102.9K84 2004
  621.382'2—dc22

                                                                            2003064794

Vice President and Editorial Director, ECS: *Marcia J. Horton*
Acquisitions Editor: *Laura Fischer*
Vice President and Director of Production and Manufacturing, ESM: *David W. Riccardi*
Executive Managing Editor: *Vince O'Brien*
Managing Editor: *David A. George*
Production Editor: *Scott Disanno*
Director of Creative Services: *Paul Belfanti*
Creative Director: *Jayne Conte*
Art Editor: *Greg Dulles*
Manufacturing Manager: *Trudy Pisciotti*
Manufacturing Buyer: *Lisa McDowell*
Marketing Manager: *Holly Stark*

© 2005 Pearson Education, Inc.
Pearson Prentice Hall
Pearson Education, Inc.
Upper Saddle River, New Jersey 07458

Printed in the United States of America

10 9 8 7 6 5 4 3 2 1

ISBN 0-13-035214-4

Pearson Education Ltd., *London*
Pearson Education Australia Pty. Ltd., *Sydney*
Pearson Education Singapore, Pte. Ltd.
Pearson Education North Asia Ltd., *Hong Kong*
Pearson Education Canada Inc., *Toronto*
Pearson Educatión de Mexico, S.A. de C.V.
Pearson Education—Japan, *Tokyo*
Pearson Education Malaysia, Pte. Ltd.
Pearson Education, Inc., *Upper Saddle River, New Jersey*

Contents

Preface

Real-time digital signal processing (DSP) using general-purpose digital signal processors is a very hot subject and challenging work in today's engineering fields. Real-time DSP provides an effective way for designing and implementing a variety of DSP algorithms for real-world applications. In fact, many universities and industrial companies are currently engaged in real-time DSP research, education, and development. With DSP penetrating into various applications, the demand for high-performance digital signal processors has expanded rapidly in recent years. It has become increasingly important for today's students and practicing engineers to master not only the theory of DSP, but also the techniques involved in real-time DSP system design and implementation.

Digital Signal Processors: Architectures, Implementations, and Applications offers readers a hands-on approach to understanding the architecture and programming of DSP processors, the design of real-time DSP systems and real-world applications, and the implementation of DSP algorithms using both fixed-point and floating-point processors, including the TMS320C2000, TMS320C54x, TMS320C55x, TMS320C62x, TMS320C64x, TMS320C3x, and TMS320C67x. This book is intended as a text for senior/graduate-level college students. With its emphasis on DSP implementation, experiments, and applications, this book can also serve as a desktop reference for practicing engineers who want to learn DSP concepts and develop real-time DSP applications at work. The minimum prerequisite requirements are a basic understanding of the concepts of signals and systems, the C language, and assembly programming.

This book gives an overview of real-time DSP technologies. To illustrate the hands-on aspects of real-time DSP applications effectively, MATLAB and Simulink are introduced for use in the design, analysis, and implementation of DSP algorithms. In addition, Code Composer Studio (CCS) for the TMS320C54x and TMS320C55x is used for lab experiments, projects, and applications. These useful tools effectively illustrate the concepts of real-time DSP and bridge the gap between theoretical signal processing and real-time implementation. By completing the hands-on exercises and problems, readers can gain practical knowledge of real-time implementation issues. In using the advanced DSP architecture for fast software development and maintenance, the mixing of C and assembly programs is emphasized.

This book uses a two-level approach in the hands-on exercises. First, we introduce the software tools and guide readers step by step in designing, simulating, verifying, and finally developing programs in both floating- and fixed-point formats. Then, we provide additional exercises and problems in order to reinforce readers' understanding of these topics. The MATLAB, Simulink, C, and assembly programs that implement many DSP examples and applications are listed throughout the book and are available on the companion CD and at *www.prenhall.com/kuo* and *www.ntu.edu.sg/home/ewsgan/book.htm*. Several real-world data files for some practical applications introduced in the book can also be found at this website. The use of the Web allows the authors to continuously update the program and data files referenced in the book as the software and/or its interface is updated.

This book is organized into three main sections: application, architecture and implementation. (1) The application section (Chapters 1–3) introduces DSP algorithms and applications, processor architectures, peripheral components, and real-time implementation issues. (2) The architecture section (Chapters 4 and 5) gives detailed descriptions of fixed-point and floating-point processors with a focus on software development. (3) The implementation section (Chapters 6–9) uses a top-down approach in the design, simulation, verification, and final implementation of commonly used DSP algorithms on both fixed- and floating-point processors. Additional hands-on experiments and practical applications using MATLAB, C, and CCS are given in Appendix B. Relevant coding issues and design examples that address connections of peripheral devices with DSP processors for certain applications are introduced in Appendix C.

Chapter 1 gives an overview of real-time DSP systems and applications, processor architectures, software and hardware development issues, and system-design considerations. This chapter lays out the framework for subsequent chapters. The experiment in the last section introduces CCS, an integrated software development tool, in a step-by-step manner. Chapter 2 reviews some fundamental DSP concepts, which are used to explain the topics given in subsequent chapters. This chapter only summarizes key points; as such, readers who are already familiar with DSP theories may skip Chapter 2. However, this chapter does provide numerous useful examples that serve as an introduction to using MATLAB and C to solve DSP problems, and it also introduces some advanced features of CCS that are useful for experiments given in subsequent chapters. Chapter 3 introduces a real-time implementation of DSP algorithms. Several practical topics are explained, such as fixed-point and floating-point number representations and arithmetic, programming issues to satisfy real-time constraints, and peripherals linked to DSP processors. In addition, a hands-on approach using fixed-point C programming in CCS is used to show some important fixed-point implementation issues.

Chapters 4 and 5 introduce architecture, instruction set, programming, and system-design issues for several fixed-point and floating-point DSP processors. These two chapters provide a comprehensive understanding of how to program different DSP processors, and they highlight the strengths and weaknesses of each processor. Examples and experiments demonstrate important differences between fixed-point and floating-point processors. Due to their complexity and length, the instruction sets of the major DSP processors used in this book are not included in the appendices. However, these instruction sets are readily available in the **Help** menu of CCS and on the websites listed in Appendix E.

Chapters 6, 7, 8, and 9 introduce the design, analysis, and software implementation of some commonly used DSP algorithms, which include finite-impulse response (FIR) filtering, infinite-impulse response (IIR) filtering, the fast Fourier transform (FFT), and adaptive filtering using fixed-point and floating-point DSP processors. Implementation of these algorithms on different processors serves to emphasize many of the important characteristics and features involved in programming different processors. It is important to note that a systematic approach is developed to guide readers through the different stages of DSP software development. We begin with floating-point C on general-purpose computers, and we continue with fixed-point C on C5000 processors, fixed-point C using C5000 intrinsics, and assembly programming. Finally, we use fixed-point C to call assembly-optimized routines in C5000 DSP libraries.

A quick reference guide on the use of MATLAB with DSP-related toolboxes and on the use of Simulink with supporting DSP blocksets is given in Appendix A. This appendix helps familiarize readers with these useful MATLAB tools for signal processing. We also introduce two powerful interactive tools, the Signal Processing Tool and the Filter Design and Analysis Tool, which are very effective in the design and evaluation of DSP algorithms for both fixed-point and floating-point implementations. Appendix B provides additional hands-on experiments, practical applications, and in-depth design projects. Appendix C discusses integrated system-design problems involving DSP processors with peripheral components. Appendix D lists and describes all of the program and data files used in this book, and Appendix E lists many websites that provide useful information on DSP.

This book can be used for courses with different emphases. A course on DSP algorithms and applications might begin with Appendix A, proceed with Chapters 1, 3, and end with the materials and experiments in Chapters 6–9. A course on DSP architecture might focus on Chapters 1, 3, 4, and 5 and on Appendix C. The experiments at the end of Chapters 6–9 and at Appendix B also give more insight into the characteristics of DSP processors. Finally, a course that concentrates on real-time DSP system design might start with Chapters 1 and 3, continue with a focus on the C5000 DSP architectures in Chapters 4 and 5, and end with the design and implementation of DSP algorithms using C5000 processors in Chapters 6–9 and in Appendices B and C. The end-of-chapter problems are separated into two groups: Part A includes computer-related problems that use MATLAB, C, and CCS, and part B consists of traditional paper-and-pencil problems that reinforce basic DSP principles.

The intention of this book is to provide wide coverage of real-time DSP in a practical manner without relying too heavily on the use of mathematics. A unique feature of this book is its recurring links between the floating-point world of simulation and the fixed-point world of real-time implementation using C and assembly programs. Readers can select different topics, processors, and experiments and can work toward an understanding of the important concepts of architectures, implementations, and applications in real-time DSP.

Throughout this book, we use several software tools in examples, experiments, and problems. These software tools are commonly used in universities and industrial companies. These tools and their respective versions are listed next in alphabetical order:

- Code Composer Studio™ Version 2.10 from Texas Instruments
- DSP Blockset Version 5.0 from The MathWorks
- Embedded Target for Texas Instruments TMS320C6000™ DSP Platform Version 1.0 from The MathWorks
- Filter Design Toolbox Version 2.2 from The MathWorks
- Fixed-Point Blockset Version 4.0 from The MathWorks
- MATLAB® Version 6.5 from The MathWorks
- MATLAB Link for Code Composer Studio Version 1.0 from The MathWorks
- Microsoft® Visual C++® Version 6.0 from Microsoft
- Real-Time Workshop® Version 5.0 from The MathWorks
- Signal Processing Toolbox Version 6.0 from The MathWorks

- Simulink® Version 5.0 from The MathWorks
- Wavelet Toolbox Version 2.2 from The MathWorks

Different versions of the preceding software may also be used, though there may be some differences in terms of results, options, and settings.

An old Chinese saying states:

I hear, and I forget; I see, and I remember; I do, and I understand

This book emphasizes seeing and doing. In fact, we would like to call the book a real-time DSP workbook, since every opportunity is given to support readers' exploration of DSP using the most effective tools. We hope that readers will enjoy working on these experiments and exercises and that they will experience the power of digital signal processors for real-world applications. A more rigorous mathematical approach to DSP and additional details regarding some subjects can be found in references at the end of each chapter.

This book contains many examples, experiments, and exercise problems that involve MATLAB, Simulink, C, and TMS320Cx assembly programs. Every attempt has been made to ensure the accuracy of all of the programs in this book. We encourage readers to bring to our attention any errors and to share any comments by contacting us at *kuo@ceet.niu.edu* and *ewsgan@ntu.edu.sg*.

ACKNOWLEDGMENTS

We are grateful to Christina Peterson and Gene Frantz at Texas Instruments and to Naomi Fernandes at The MathWorks for providing us with the support needed to write this book. Several individuals at Pearson Prentice Hall also helped to make the book a reality. We wish to thank our editor, Tom Robbins, for his support of this book, as well as field editor William D. Winschief for his encouragement of this project. A special thank you must go to Alice Dworkin, associate editor, who promptly answered our questions and arranged for a number of reviewers to provide constructive comments and suggestions that have improved the book in its finished form. In addition, we also would like to thank our production editor, Scott Disanno, and the staff at Pearson Prentice Hall for the final preparation of the book. We thank Jennifer Y. Kuo for her proofreading efforts. We would also like to thank Kwee-Song Lim, Meng-Tong Wong, Wee-Beng Lee, Francis Kua, and Hong-Swee Lim from Texas Instruments (Singapore) for their valuable input and help during the course of writing this book.

This book is also dedicated to many of our past and present students who have taken our DSP courses and have written M.S. theses and Ph.D. dissertations and completed senior design projects under our guidance at both NIU and NTU. Both institutions have provided us with a stimulating environment for research and teaching, and we appreciate the strong encouragement and support we have received. Finally, we are greatly indebted to our parents and families for their understanding, patience, and encouragement throughout this period.

SEN M. KUO
WOON-SENG GAN

Acronyms
and Abbreviations

3-D	Three-dimensional
ACC	Accumulator
ADC	Analog-to-digital converter
ADPCM	Adaptive-differential pulse-code modulation
AIC	Analog interface circuit
ALE	Adaptive line enhancer
ALU	Arithmetic and logic unit
ANC	Active noise control
AR	Auxiliary register
ARAU	Auxiliary-register arithmetic unit
ASIC	Application-specific integrated circuit
BIOS	Basic input/output system
BSP	Buffered serial port
CCS	Code Composer Studio
CD	Compact disc
CODEC	Coder-decoder
COFF	Common-object file format
CPU	Central processing unit
DAC	Digital-to-analog converter
DAGEN	Data-address generator
DARAM	Dual-access RAM
DC	Direct current
DFT	Discrete Fourier transform
DIF	Decimation-in-frequency
DIT	Decimation-in-time
DMA	Direct memory access
DSK	DSP starter kit
DSP	Digital signal processing
DSPLIB	DSP Library
DTFT	Discrete-time Fourier transform
DTMF	Dual-tone multifrequency
E1	European system for digital multiplexing

EMIF	External-memory interface
EPROM	Erasable-programmable ROM
FDATool	Filter Design and Analysis Tool
FFT	Fast Fourier transform
FIR	Finite-impulse response
GUI	Graphical user interface
HPI	Host port interface
IDE	Integrated development environment
IDFT	Inverse DFT
IEEE	Institute of Electrical and Electronics Engineers
IFFT	Inverse FFT
IIR	Infinite-impulse response
I/O	Input/output
ISR	Interrupt service routine
JTAG	Joint test action group
LMS	Least mean square
LS	Least square
LSB	Least significant bit
MAC	Multiply-add computation
McBSP	Multichannel BSP
MFLOPS	Million floating-point instructions per second
MIPS	Million instructions per second
MMR	Memory-mapped register
MSB	Most significant bit
MSE	Mean square error
OVM	Overflow mode
PC	Program counter
PLL	Phase-locked loop
PSD	Power-spectrum density
QMF	Quadrature-mirror filter
RAM	Random access memory
RLS	Recursive least square
ROM	Read-only memory
SARAM	Single-access RAM
SIMD	Single-instruction multiple-data
SNDR	Signal-to-noise-plus-distortion ratio
SQNR	Signal-to-quantization-noise ratio
SP	Stack pointer
SPTool	Signal Processing Tool
SXM	Sign-extension mode
T1	Primary carrier used in USA for digital transmission
UART	Universal asynchronous receiver-transmitter
UTOPIA	Universal-test-and-operations physical interface for asynchronous transfer mode
VLIW	Very-long-instruction word
VLSI	Very-large-scale integration

1

Introduction to Digital Signal Processing Systems

Digital signal processing (DSP) gained popularity in the 1960s with the introduction of digital technology. It became the method of choice in processing signals as digital hardware increased in speed and became easier to use, less expensive, and more available. In 1979, Intel introduced the first DSP available processor (the Intel 2920), which had an architecture and an instruction set specifically tailored for DSP applications. Today, general-purpose DSP processors are commercially available from Texas Instruments, Motorola, Analog Devices, Agere, DSP Groups, and many other companies. As DSP processors became less expensive and more powerful, real-world DSP applications such as high-speed modems and Internet access, wireless and cellular phones, audio and video players, and digital cameras have exploded onto the marketplace. In this book, we focus on the TMS320 family of DSP processors from Texas Instruments.

1.1 INTRODUCTION TO DIGITAL SIGNAL PROCESSING

A signal is a physical quantity that is usually a function of time, position, pressure, etc. For example, the voltage output from a microphone represents sound pressure as a function of time. Signals that we encounter frequently in our daily life include speech, music, data, images, and video signals. The objectives of signal processing are

1

to transmit or store signals, to enhance desired signal components, and to extract useful information carried by the signals.

In practical applications, many of the signals that we encounter are analog signals, which are present at every moment in time and can be of any amplitude. For example, the musical signals recorded in cassette tapes are analog signals. In order to process these continuous-time signals with DSP processors, signals must be sampled and quantized to produce a sequence of numbers. This discrete-time sequence with discrete-valued amplitudes that are represented with a finite number of bits is called a digital signal. For example, the music stored in a compact disc (CD) is recorded as a digital signal. Digital signals can be stored almost indefinitely without any loss of information on various storage media such as magnetic and optical disks. On the other hand, stored analog signals deteriorate rapidly and cannot be recovered in their original forms.

DSP theories are based on discrete-time signals and systems since they are easier to deal with mathematically. Discrete-time signals can be obtained by sampling (without quantization) analog signals. Discrete-time signals are present at any discrete moment of time and can be of any amplitude. They need an infinite number of bits to represent each sample, which cannot be realized using physical digital hardware. In addition, most DSP algorithms were originally developed using general-purpose computers, for which the speed of modifying and testing the algorithm was more important than the speed of executing the algorithm. To apply these algorithms to real-world applications, the effects of using finite wordlength (see Chapter 3), hardware, execution speed, and algorithm complexity should be considered. Realistic (sampled) data obtained for a given application must be used to refine the algorithm.

As illustrated in Fig. 1.1, DSP systems use a DSP processor or other digital hardware, an analog-to-digital converter (ADC), and a digital-to-analog converter (DAC) to replace analog devices such as amplifiers, modulators, and filters. The analog signal to be processed is sampled and encoded to a digital signal in binary form (see Chapter 3) with the ADC. A DSP processor performs digital operations based on a specific signal-processing algorithm (or computational descriptions) implemented in software to process the digital signals. The output digital signal may be converted back to analog form using the DAC. The DSP processor processes the digital-input signals, and the output digital signals can be transmitted to or processed by other digital systems without using converters. In some applications, a DSP system creates a digital signal internally. For example, we can synthesize a pair of sinewaves (dial tones) using a DSP processor for telephone dialing.

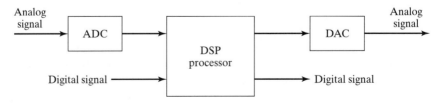

Figure 1.1   A typical DSP system

DSP algorithms can be performed on a wide variety of digital hardware and in many computer languages such as C and C++. Some DSP hardware includes programmable and nonprogrammable logic, general-purpose microprocessors and microcontrollers, and general-purpose digital signal processors. For example, it is possible to integrate highly sophisticated DSP systems such as video-coding and video-decoding algorithms on a single very large-scale integration (VLSI) circuit chip. However, hardware solutions focus on the specific problem. Such solutions are faster and smaller than programmable solutions, but they take longer to design. In addition, they are more expensive to develop and harder to change or upgrade.

The programmable processor can be programmed for a variety of tasks. It is used for systems that are too complicated to implement with nonprogrammable circuits, products that need shorter development time and lower development cost, or systems that need to be upgraded frequently with new algorithms and standards. For a particular task, the processor is slower than the VLSI chip that was custom built for that task.

General-purpose microprocessors or microcontrollers are designed to handle the most general form of processing using a von Neumann architecture with unified memory for both instructions and data. In contrast, a Harvard architecture separates memory space into program memory and data memory so that each may be accessed simultaneously. Most DSP processors employ a modified Harvard architecture, which gives a crossover path between program and data memory. Most processors are also optimized for performing repetitive multiply-and-add operations that access sequentially data stored in consecutive memory locations. In this book, we focus on using DSP processors for implementing a DSP system.

DSP systems are gradually replacing existing analog systems and creating new applications. The main advantage of DSP is its programmability. A DSP system allows the user to change tasks by writing new code to the memory of the system. It also allows the parameters of the system to be updated using algorithms in order to adapt to dynamic changing environments. The DSP system has consistent performance because the memory and processor are fairly independent of temperature and aging in different environments during the life of the product. Furthermore, the same DSP algorithm can be coded in different languages and run on different digital hardware with the same results. The costs of DSP systems are constantly decreasing while performance is increasing. The power consumption of DSP is dropping with each new generation of processors and is well suited for portable devices. In addition, some signal-processing operations are impossible to implement using analog techniques.

In parallel with the advances in digital hardware technology, new research on DSP algorithms and applications has been conducted. This research on DSP has focused on improving algorithm performance and reducing system complexity, which has resulted in the development of new DSP techniques including adaptive filtering (see Chapter 9) for unknown or time-varying environments and fast algorithms such as the fast Fourier transform (FFT) (see Chapter 8) for computing discrete Fourier transforms.

As DSP hardware and algorithm capabilities have advanced, so have processing demands, which has resulted in the development of higher-performance systems with more sophisticated algorithms for the new generation of applications. In today's evolving DSP applications, flexibility and upgradability of design are key factors in longer product cycles. Many industrial standards are in the early stages of development, and some of these standards must maintain compatibility with other standards. A good example is digital cellular phones, which have been upgraded from 2G to 2.5G, 3G, and 4G standards. Programmable DSP processors are especially suitable for designs that require multiple modes of operation and future upgradability.

1.2 DIGITAL SIGNAL PROCESSING SYSTEMS AND APPLICATIONS

DSP systems are often embedded in larger systems to perform specialized DSP operations, thus allowing the overall system to handle general-purpose tasks. For example, a DSP processor in a modem used for data transmission is the embedded DSP system of a computer. Often, this type of DSP system runs only one application and is not programmed by the end user. In fact, the user may not even know that a processor (e.g., a DSP processor in hard drives used for motor-control and head-position control) is embedded in the computer. In this section, we briefly introduce the concepts of DSP system design, the TMS320 family, and some typical real-world applications.

1.2.1 Digital Signal Processing Systems

The research results of DSP are increasingly applied to the development of complete solutions that integrate algorithms, software, and hardware into a system. Software is a program that controls a set of computing hardware resources to implement an algorithm, while hardware consists of particular computing components. Software development can either be processor independent or device specific. Because software development has become a larger expense than hardware development in major DSP systems, processor-independent design has the advantage of porting software on different processors and the ability to migrate to more advanced processors in the future.

In processor-independent design, a high-level language such as C or C++ is preferred and is available for most DSP processors. C programs are easier and faster to write, and they may be ported from one processor to another by simply recompiling the source code for a new processor using the C compiler for that processor. This portable feature is very important if the algorithm will be implemented on several different processors. Processor-dependent software requires the writing of assembly programs that reflect the architecture of a specific processor. While the handcrafted assembly code runs faster and requires less memory, it also takes longer to write and must be rewritten for each different processor. Therefore, C language is the preferred method of implementation if it can run fast enough for the processor used in a given application.

In applications where processing and memory resources are critical, the solution is a compromise that implements critical sections in assembly language and that uses C language to code the rest. This mixed C-and-assembly programming provides a good balance between ease of coding and efficient implementation. Currently, the efficiency of C compilers is improved significantly, and many optimized assembly-coded DSP libraries allow the user to develop mixed code easily and efficiently. In this book, we use C programs that call assembly-optimized routines for experiments.

Most DSP systems must produce their results in real time. In a real-time DSP application, a sample (or a group of samples known as a frame or block) arrives at the input of the system. The system has to complete the computation of the output associated with a sample (or frame) of an input signal in a period of time that does not exceed the duration of that sample (or frame). For example, in sample processing, each sample is processed before the arrival of the next sample. In block processing, such as FFT, N samples are grouped as one block, and the processing of these N samples can be carried out only after their acquisition. Therefore, real-time block processing incurs longer delays when compared with sample processing.

Most DSP algorithms and applications are developed and verified on general-purpose computers with no constraints on execution time, memory size, and system cost. Preparing the algorithms for practical application requires tradeoffs and integration across the expertise of algorithms, software, and hardware. For example, one may choose whether to perform the algorithm in hardware or software, selecting either speed or flexibility, changing the structure of the algorithm for the specific software or hardware that implements the algorithm, designing custom hardware for the algorithm, or minimizing cost across development and maintenance.

DSP system design involves an understanding of application problems and the design of a system that satisfies all requirements. Implementation of the designed DSP system requires a team of system, hardware, software, and testing engineers. Because DSP systems are computer-based systems, the design can be simulated first in software and then modified, thus greatly reducing product development time. Analyzing and designing an efficient DSP algorithm to meet application requirements is a complex task. A structured approach to DSP system design using computer-aided design (CAD) tools is necessary. In this book, we extensively use MATLAB 6.5 with Simulink, toolboxes, and blocksets from The MathWorks, Inc. [6] for algorithm design and analysis purposes.

1.2.2 The TMS320 Family

In 1982, Texas Instruments [7] introduced the first DSP processor, the TMS32010, in the TMS320 family. Today, it consists of fixed-point and floating-point processors. The 16-bit fixed-point processors include the TMS320C2000 (the C24x and C28x), C5000 (the C54x and C55x), and C6000 (the C62x and C64x) generations. Fixed-point processors are discussed in Chapter 4. The 32-bit floating-point processors consist of C3x, C4x, and C67x generations. Floating-point processors are discussed in Chapter 5. Processors within the same generation of the TMS320 family use the identical core architecture, but have different on-chip memory and peripheral configurations.

The TMS32010 architecture and instruction set was designed to maximize computational power and flexibility from a constrained integrated circuit technology at a

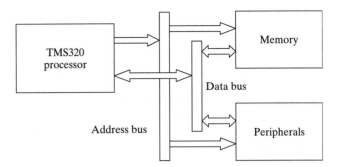

Figure 1.2 External interfaces
for the TMS320 processor

clock rate of 5 MHz in 1982. Today, the fastest C64x increases its execution speed (throughput) up to 1 GHz as a result of architecture enhancements and faster clock cycles. Architectural improvements that are independent of clock speed include using multiple multiply-add computation (MAC) units, performing intensive pipelining and parallel operations, providing more and faster memory with multiple internal buses on chip, and introducing more powerful algorithm-specific instructions and hardware engines for implementing some widely used DSP algorithms.

We have shown in Fig. 1.1 that a DSP system requires an ADC and a DAC as well as the DSP processor. Figure 1.2 illustrates how a DSP system is configured around the TMS320 processor. The major external blocks needed are memory (program and/or data) and peripherals. DSP processors usually provide some on-chip cache, program read-only memory (ROM), data random-access memory (RAM), and peripherals. Peripherals such as the ADC and the DAC can either connect to the data bus by using a dedicated address or to the serial interface if serial ports are available on chip.

In DSP applications, the processor interfaces to the outside world for data transfer, as shown in Fig. 1.1. A wide variety of commercial input/output (I/O) devices are available, and we have to make a selection to satisfy all application requirements from both the hardware and software viewpoints. In most DSP processors, I/O devices are selected and enabled by using control signals decoded from the address bus (lower bits) and dedicated control lines. An extra control signal is needed to differentiate between memory addressing and digital I/O addressing. The task of the software is to set up the correct external device address and activate the I/O control line.

1.2.3 Digital Signal Processing Applications

DSP techniques are replacing analog signal processing methods in many fields. DSP is now a commercial reality and is becoming a standard technique in many areas of telecommunications; biomedical signal analysis and processing; digital control; speech, audio, and video processing; digital instruments; radar and sonar; etc. An understanding of many applications requires knowledge of the field in which they are used. In this book, we introduce simple applications in every chapter. Examples of DSP applications are summarized as follows (adapted from [8]):

Audio

Audio watermarking
Coding and decoding
Effects generators
Surround-sound processing
Three-dimensional audio
  synthesizers

Automotive

Active noise and vibration control
Adaptive ride control
Cellular telephones
Digital radios
Entertainment systems
Navigation and global positioning
Voice commands

Control and Industrial

Disk drive control
Laser printer control
Motor control
Numeric control
Robotics control
Security accesses

Graphics/Imaging

Animation/digital maps
Computer and robot vision
Digital cameras
Image compression/transmission
Image enhancement
Pattern recognition
Three-dimensional rotation

Medical

Diagnostic equipment
Digital hearing aids
Fetal monitoring
Patient monitoring
Ultrasound equipment

Communications

Communication security (scrambling)
Detection
Encoding and decoding
Software radios
Waveform synthesizers

Consumer

Digital answering machines
Digital audio/video players and recorders
Educational tools and toys
Music synthesizers and toners
Pagers
Plug-in for personal digital assistants (PDAs)
Radar detectors

General Purposes

Adaptive filtering
Convolution and correlation
Digital filtering
FFTs and fast cosine transforms
Hilbert transforms
Waveform generations

Instrumentation

Digital filtering
Function generation
Measurement equipment
Pattern matching
Phase-locked loops
Spectrum analysis
Transient analysis

Military

Image processing
Missile guidance
Navigation
Radar and sonar processing
Secure communications

Telecommunications and Voice/Speech

Adaptive equalizers	Speakerphones
Cellular telephones	Speaker verification
Channel multiplexing	Speech enhancement
Data encryption	Speech recognition
Digital public-switching systems	Speech synthesis
Dual-tone multifrequency (DTMF) generation/detection	Text-to-speech
Echo cancellation	Video conferencing
Faxing	Vocoders
Modems	Voicemail
PDAs	Voice morphing
Personal communication systems	Voice-over-Internet protocol

Several practical applications are introduced in following chapters. Additional hands-on experiments and applications are given in Appendix B.

1.3 DIGITAL SIGNAL PROCESSOR ARCHITECTURES

As shown in Fig. 1.1, the DSP processor performs the primary signal-processing functions required by the application. The task of developing an efficient DSP system depends on the DSP hardware and software architectures, including data flow, arithmetic capabilities, memory configurations, I/O structures, programmability, and instruction set of the processor. The processor architecture and the corresponding DSP algorithm must be complementary. For some applications, the algorithm is given, and we have to select a suitable processor. For other applications, the processor is given, and the task is to develop efficient algorithms that satisfy the application requirements. The goal is to develop a DSP algorithm such that the minimum amount of processor resources is used, thus leaving unused resources for improvements to meet future growth.

1.3.1 Introduction

Most DSP processors are designed to perform repetitive MAC operations such as finite-impulse response (FIR) filtering (see Chapters 2 and 6) expressed as

$$y(n) = \sum_{i=0}^{L-1} b_i x(n - i), \qquad (1.3.1)$$

where $\{b_0, b_1, \ldots, b_{L-1}\}$ are filter coefficients, $\{x(n), x(n-1), \ldots, x(n-L+1)\}$ are signal samples, and L is the length of the filter. The computation of output $y(n)$ requires the following steps:

Step 1: Fetch two operands, b_i and $x(n - i)$, from memory.
Step 2: Multiply b_i and $x(n - i)$ to obtain the product.

Step 3: Add the product, $b_i x(n - i)$, to the accumulator.

Step 4: Repeat steps 1, 2, and 3 for $i = 0, 1, 2, \ldots, L - 1$.

Step 5: Store the result, $y(n)$, in the accumulator to memory.

Step 6: Update the pointers for b_i and $x(n - i)$ and repeat steps 1 through 5 for the next input sample.

The generic internal architecture of the DSP processor illustrated in Fig. 1.3 is optimized for the FIR-filtering operations given in Eq. (1.3.1). Most DSP processors employ similar functional blocks, although the detailed interconnection and operations can vary. Compared with general-purpose microprocessors, the most unique feature is the use of parallelism and pipelining for improving processing speed. DSP processors have a number of special processing units supported by multiple dedicated buses, most of which can operate independently and concurrently. As shown in Fig. 1.3, the arithmetic and logic unit (ALU) performs addition, subtraction, and logical operations. The shifter is used for scaling data, and the hardware multiplier and accumulators are used to perform MAC operations. Data address generators (DAGENs) generate the addresses of operands used by instructions. With these available resources, the DSP processor achieves a fast execution speed by performing operations within these units simultaneously. The basic operations of these units are re-examined later.

Most DSP processors use a modified Harvard architecture with two (program and data) or three (program and two data) memory buses, allowing access to filter

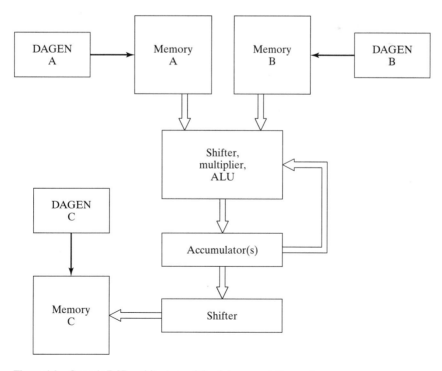

Figure 1.3   Generic DSP architecture of the data computation unit

coefficients and input signals in the same cycle. However, use of the Harvard architecture causes some difficulties in writing C programs that use the full potential of
the MAC structure and multiple memory accesses.

1.3.2 Central Processing Unit Operations

This section introduces the fundamental functions that are performed by the central
processing unit (CPU) of DSP processors. Many functions are common to general-
purpose microprocessors, with the exception of fast MAC unit(s), barrel shifter(s),
multiple memory blocks with supporting buses, and powerful DAGENs.

Multiply-Accumulate

Multiplication operations require several clock cycles on a microprocessor or microcontroller, where they are performed by repetitive shift-and-add operations. To
achieve the speed required by multiplication-intensive DSP algorithms, such as the
FIR filtering given in Eq. (1.3.1), TMS320 processors employ a fully parallel hardware multiplier, which can multiply two items of data within one clock cycle. At the
same time, an adder immediately following the multiplier adds the product from the
previous multiplication operation into a double-precision accumulator. Some
processors such as the C55x provide two MAC units and four 40-bit accumulators. A
number of dedicated instructions that can perform multiply, accumulate, data-move,
and pointer-update operations in a single instruction are built into processors for filtering and correlation algorithms. These instructions take full advantage of the par-
allel-and-pipelining architecture to achieve very fast execution.

Arithmetic and Logic Unit

The basic arithmetic operations performed by DSP processors are addition, subtraction, etc. The logical unit performs Boolean logic such as AND, OR, and NOT operations on individual bits of a data word and executes logical shifts of the entire data
word. The arithmetic and logical units are often combined into a single ALU. Binary
division is usually implemented with a software routine because it involves a repeated
series of shift and conditional subtraction operations.

Shifter

Shifters for scaling operations support shift instructions. The shifter can be used for
prescaling an operand in data memory or the accumulator before an ALU operation or for postscaling the accumulator value before storing it back into data memory.
In addition, it also performs a logical or arithmetic shift of accumulator value. For
example, the TMS320C5000 shifter can produce a left shift of 0 to 31 bits and a right
shift of 0 to 32 bits on the input data in a single cycle.

Pipelining and Parallelism

The instructions that control the operations of the DSP processor require multiple
steps to execute. First, the address of the instruction is generated, and the contents
of program memory at that address are read and decoded. Based on the instruction,

one or more operands are then fetched to provide the required data for executing that instruction. Finally, the results are stored, and the address of the next instruction is computed. Each instruction may take several clock cycles to execute multiple steps of prefetch, decode, operand fetch, execute, and write result. These steps can be cascaded in assembly-line fashion by using pipelining. If each step requires one clock cycle, a sequence of seven-stage instructions (in the TMS320C55x) can be completed at one instruction per clock cycle after the pipeline is full. The pipeline architecture takes advantage of the inherent decomposition of instructions into multiple serial operations. A detailed explanation on the operation of pipelining is presented in Section 3.4.2.

Figure 1.3 shows that cascading the multiplier and ALU allows the simultaneous operation of both. That is, when the multiplier is performing its work at time i to produce $b_i x(n - i)$, the ALU adds the previous product $b_{i-1}x(n - i + 1)$ into the accumulator. The parallel architecture takes advantage of the inherent parallelism of DSP algorithms and applications. As shown in Fig. 1.3, all processing units in the parallel configuration can receive different data streams and execute different operations at each cycle. For example, the multiplier accesses two operands and multiplies them. At the same time, the DAGENs update the address pointers as specified, and the ALU adds the previous product into the accumulator with (or without) rounding.

The technique of overlapping processing steps on adjacent processing units is called pipelining. However, pipelining can present programming difficulties when the execution of an instruction must wait until an earlier result is available. This problem is further discussed in Chapters 3 and 4.

Buses

Buses and other interconnections provide the communication paths among the units that make up a DSP processor. The execution speed of filtering given in Eq. (1.3.1) can be improved further by using separate data buses for each of the two inputs for the multiplier. Instead of requiring an operand fetch for b_i and another on the same data bus for $x(n - i)$, both operands can be fetched simultaneously on two separate data buses, as shown in Fig. 1.3. The two data buses with supporting address buses connect with two separate memories [e.g., memory A for coefficient b_i and memory B for signal $x(n - i)$]. This configuration avoids the conflict of accessing two operands from the same memory at the same time. A third memory with associated address and data buses is used for storing the value in the accumulator back to memory C. In addition to these data memory blocks, there is program memory and its dedicated program and data buses for fetching program instructions to avoid delays in accessing data.

Data Address Generator

As shown in Fig. 1.3, each memory has its own address bus, which originates with a DAGEN. Accessing the sequence of operands, b_i and the corresponding $x(n - i), i = 0, 1, \ldots, L - 1$, is a regular sequential operation where these operands are stored in consecutive locations of memory. Each DAGEN simply increases (or decreases) the address pointer for pointing at the next data item within the same clock cycle when the multiplier and ALU are performing arithmetic operations; as a result, no extra clock cycle is needed to update the address pointer. After accessing the last coefficient, b_{L-1}, the coefficient pointer has to wrap around to b_0 for the next iteration. This operation can be performed by arranging the coefficient buffer in a circular fashion. Therefore, further improvements to the DAGEN include modulo L arithmetic for implementing circular buffers (see Chapters 3 and 4). In addition, the DAGEN supports bit-reversal addressing for computing FFT algorithms. (See Chapter 8.)

1.3.3 Memory Configurations

The two common types of memory available are RAM and ROM. RAM can be written dynamically to as well as read from; thus, it is the most flexible data storage for development purposes. However, unless a battery backup is available, RAM cannot provide permanent memory for program and constant data. ROM, on the other hand, is usually in the form of erasable-programmable ROM (EPROM), which allows the memory contents to be erased and reprogrammed by electrical devices. Modern DSP systems use flash memory instead of EPROM since it can be erased and reprogrammed on board, which is especially useful for product development.

TMS320 memory is organized into three individual spaces: program, data, and I/O. Within any of these spaces, ROM, RAM, or memory-mapped peripherals can reside either on or off chip. Program memory contains the instructions that implement the DSP algorithms. These instructions may include data (constants), memory addresses for program branches, and commands that control the execution of the processor. The data memory stores data (coefficients and signal samples) used by instructions. The I/O memory space interfaces to external memory-mapped peripherals. Most advanced DSP processors provide cache, program, and data memory on chip, as well as support larger memory off chip.

A facility exists on most modern DSP processors for manipulation of data blocks within internal and external memory, which allows large volumes of data or programs to be stored off chip in slower memory and transferred to faster on-chip memory when required. For example, TMS320C2000 processors are built without any dual data memory and buses on chip. A block of internal memory can be software configured to be either program or data memory. The use of this memory for single-cycle execution of a dual-operand fetch and MAC operation is discussed in Section 4.2.

Memory Space

The DSP processor must address and access memory. Many processors use on-chip programmable program memory for volume applications since it is the most efficient means of program memory access. Adding flash memory on chip is a significant step towards promoting low-cost compact applications, especially for development phases and small-volume applications. Some processors also provide a limited amount of on-chip RAM, which can be dynamically configured as program memory. When external program memory is needed, the important issues that must be considered are program memory size and access speed. As the clock rate of DSP processors increases, the access time of external memory must decrease.

An example of the TMS320C5510 memory map is illustrated in Fig. 1.4. As shown in the map, the TMS320C5510 supports a unified memory map such that the program and data accesses occur in the same physical space. The total on-chip memory, including a 32K-byte (Kb) of ROM, is 352Kb (or 176K 16-bit words). Dual-access RAM (DARAM) is composed of 8 blocks of 8Kb each. Each DARAM block can perform two accesses per cycle (two reads, two writes, or one read and one write). The CPU and peripherals can read from and write to a DARAM memory in the same cycle. The DARAM is always mapped in data space and primarily is intended for storing data values. Single-access RAM (SARAM) is composed of 8Kb blocks. Each SARAM block can perform one access per cycle (one read or one write).

When processor speed is not a limiting factor, memory access time can be relaxed by using a slower clock rate. This method slows down the processor cycle time, thus reducing memory access time accordingly. Another effective method is to let the processor run at full speed, but allow a number of wait states for accessing memory. Wait states allow DSP processors to wait for time-consuming external-interface operations. These processors usually have on-chip memory configurable as program memory, thus allowing small programs to be stored and executed at full

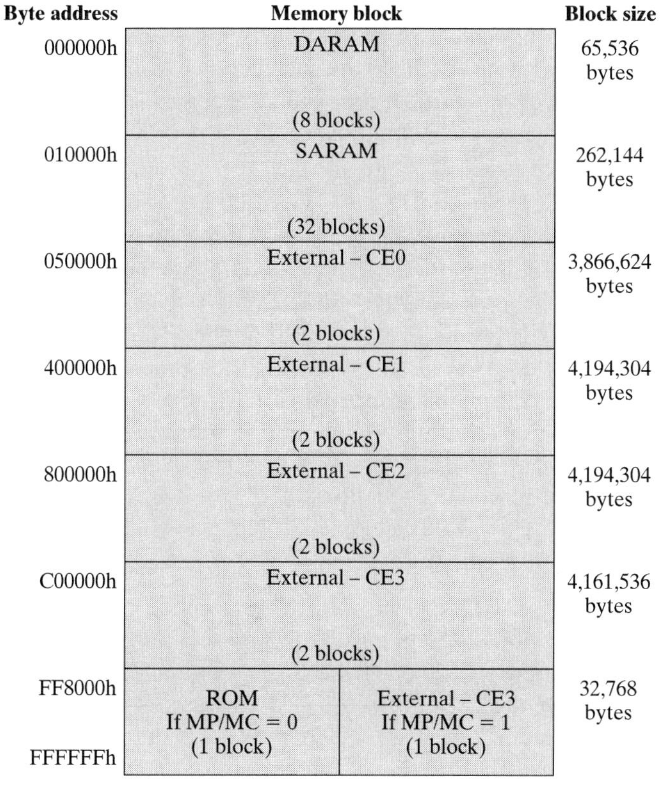

Byte address	Memory block	Block size
000000h	DARAM (8 blocks)	65,536 bytes
010000h	SARAM (32 blocks)	262,144 bytes
050000h	External – CE0 (2 blocks)	3,866,624 bytes
400000h	External – CE1 (2 blocks)	4,194,304 bytes
800000h	External – CE2 (2 blocks)	4,194,304 bytes
C00000h	External – CE3 (2 blocks)	4,161,536 bytes
FF8000h FFFFFFh	ROM If MP/MC = 0 (1 block) — External – CE3 If MP/MC = 1 (1 block)	32,768 bytes

Figure 1.4 Memory map of the TMS320C5510

speed. The initial loading of the program from the external slow program memory can be accomplished using wait states. This method overcomes the need for expensive, fast ROM without sacrificing processing speed.

The amount of on-chip ROM available varies on each processor. Some processors use a small amount of ROM containing a boot loader that boots the faster on-chip or external RAM. Almost all modern DSP processors have on-chip data RAM. For many applications, on-chip memory avoids the need to access external data memory.

TMS320 processors use some data memory locations for internal use, such as interrupt vectors and memory-mapped registers. Memory-mapped registers provide a convenient way for saving and storing registers for context switches and for transferring data between accumulators and other registers.

Cache

Processor memory systems can be optimized with smaller amounts of fast memory placed between the processor and larger amounts of slower memory. This high-speed memory is called cache, which contains the most recently accessed instructions and data from main memory. There are two types of cache: L1 cache, which is small and close to the processor core, and bigger L2 cache. They are commonly used in the microprocessor, and newer generations of DSP processors are also using them to store commonly used program and data sections. For example, the TMS320C55x has 64 bytes of instruction-buffer queue, which can be treated as program cache. Most instructions such as repeated loops can be stored in the cache and are executed repeatedly without fetching the instructions from external memory. The cache speeds up the access of critical code and data, but may suffer from loss of data determinism and increased code complexity.

Memory Move

Digital processing allows the sharing of a given processor among a number of signals by time-sharing, thus reducing the cost of processing per signal. For example, a single DSP processor can process several channels of a T1 line, where 24 channels are combined into one through time-division multiplexing. The multiplexed signal, then, can be fed into a single processor. By switching the processor coefficients corresponding to that channel prior to the arrival of each signal at the input of the processor, the system functions as if each channel has one designated DSP processor. This application requires a very efficient method for moving coefficient and signal buffers between different memory locations.

1.3.4 Peripherals and Input/Output

DSP processors normally provide on-chip peripherals or peripheral interfaces to facilitate the integration of DSP with external devices, such as an ADC, a DAC, or other DSP processors or microprocessors. In addition, some internal peripherals are used to control and manage the clocking of the DSP processors, data transfer mechanism, and power management facilities. Therefore, it is important to receive a quick overview of these peripherals before discussing a more detailed description of the

DSP processor and its programming. Some design examples addressing connections of peripheral devices with DSP processors are given in Appendix C.

Serial Port

The interfacing of a DSP processor to peripheral devices involves the use of data buses, address buses, and control signals. Many processors use a common data bus for external program memory, data memory, digital I/O, and multiprocessor communications. To prevent two devices from driving a common bus at the same time, these devices are equipped with enable/disable control. The time taken by a device to access and use the data bus determines how effectively other peripherals can operate. Most modern DSP processors provide both a serial and parallel I/O capability. The serial port has the advantage of being separate from the data bus. Thus, it is not constrained by stringent access time and possible bus-conflict considerations.

When passing data in and out of the DSP processor, a serial port (typically 16 bits) is used to hold the received and transmitted data so that the DSP processor can operate on them. Modern processors have the following different types of serial ports:

1. A standard serial port uses two registers, a data-transmit register and a data-receive register, to transfer data out of and into the DSP processor, respectively. The processor is interrupted at every data transfer.
2. A buffered serial port (BSP) consists of a full-duplex, doubled-buffered serial-port interface with an autobuffering unit for the fast transfer of data between the serial port and the processor's internal memory. The DSP processor is interrupted if the buffer is either empty or full.
3. The time-division multiplexing serial port enables communication over multiple DSP processors through time-division multiplexing. For example, the TMS320C54x processor can connect with up to seven other DSP devices using the time-division multiplexing serial port. It divides the given time interval into a number of subintervals and allocates a subinterval to each channel.
4. Multichannel BSP (McBSP) provides high-speed, full-duplex, multichannel BSPs. It usually is used to connect a DSP processor with other DSP processors and devices. It is a new feature in the latest generation of TMS320 processors.

A more detailed description of the serial data transfer format and its frame-synchronization methods is given in Section 3.6.3.

Host-Port Interface

The host-port interface (HPI) is used to interface a host processor (such as a microprocessor) with the DSP processor. Both the host and DSP processors exchange information via the DSP processor's on-chip memory, which can be programmed for on-chip data or program memory. Data transfer usually occurs in 8-bit or 16-bit words.

Direct-Memory-Access Controller

The direct-memory-access (DMA) controller is used to control data transfer of the DSP processor-memory space, which includes on- and off-chip memory and peripherals. It operates independently of the processor. The data transfer is done in block format, and the DMA controller sends an interrupt to the DSP processor when the transfer is complete. Typically, the DMA can handle multiple channels (e.g., six channels in the C54x), and the user can assign different or same priority to the channels.

Parallel Port

A parallel port is different from a serial port in that it transmits and receives multiple data bits (typically 8 or 16 bits) all at once. Therefore, it can usually transfer data faster, but it requires more pins for implementation. Another difference between serial and parallel ports is that handshakes or strobe lines are normally used in the parallel communication to indicate proper transfer of data. In the case of serial communication, a synchronization pulse or a clock signal deduced from the data itself is used.

In order to save the number of pins used for parallel communication with external devices and other processors, a DSP processor uses the main data bus as a parallel port. The DSP processor accomplishes this task by reserving a section of addresses for I/O space and using a few general-purpose I/O pins for handshaking functions.

Hardware Timer

Most DSP processors provide programmable hardware timers that are used as sources of periodic interrupts for the processor and that can also be used as software-controlled signal generators. The timer consists of a clock source, a prescalar (to reduce the frequency of the clock), and a down counter. The counter is decremented by one step at every clock edge and interrupts the processor when the counter decrements to zero.

Clock Generator and Phase-Locked Loop

The DSP processor's master clock normally comes from either external-supplied clock sources or from external crystal. When driven by external crystal, the DSP processor must have an internal oscillator that works with the external crystal to generate the required clock. On the other hand, if no internal oscillator is included, an external clock circuit must be implemented.

Some DSP processors include internal-frequency synthesizers or a phase-locked loop (PLL) that can increase the external-supplied clock frequency. Therefore, a PLL allows a higher-frequency internal clock to be generated from a low-frequency external clock, which has the advantages of reducing clock-generated electromagnetic interference and using a cheaper external-clock generator. The PLL comes in hardware PLL or software-programmable PLL. The hardware PLL has limited multiplication factors of one to five, while the software PLL can be changed under software control and provides a wider set of multiplication factors (e.g., ranging from 0.25 to 16 for TMS320C54x processors).

Power Management

With the increasing demand for running DSP-based products with less power and prolonged battery usage time, the DSP processor incorporates power-management features in addition to the conventional low-voltage approach. Several methods are used in power management: (1) clock-frequency control, (2) power-down mode, and (3) disabling of unused peripherals.

Clock-frequency-control mechanisms can be programmed to run a processor at a fraction of its full speed since most applications do not need to operate at full speed all of the time. We can take advantage of this idle time and let the processor run at a reduced speed. Most DSP processors provide ways to put the processor in sleep (or power-down) mode, which consumes less power and still maintains the current CPU contents. Many levels of power-down modes are possible, corresponding to whether or not the processor is halted; the processor clock or peripheral clock is stopped; the external address, data, and control lines are pulled to high impedance state; etc. The DSP processor can be woken either by the peripherals' activity or by external interrupts. Another way to manage power is to turn off unused peripherals.

1.4 SOFTWARE DEVELOPMENTS

This book focuses on the development and implementation of DSP algorithms in order to achieve desired functions for a given application. Once suitable algorithms have been designed, the remaining task is to develop and test the program. A programming language states the algorithm in a manner that precisely defines its operations inside the processor. The algorithm is initially described by means of equations or flowcharts that use symbolic names for the inputs and outputs. This form is suitable for documentation and the generation of test data. In documenting the algorithm, it is sometimes helpful to clarify which inputs and outputs are involved by means of signal-flow diagrams. It is essential to document programs thoroughly with titles and comments because doing so greatly simplifies the task of troubleshooting and also helps with program maintenance. For ease of understanding, it is also important to use meaningful mnemonics for variables, labels, subroutine names, etc.

In many applications, execution time and memory size are decided by the sampling rate and the complexity of the algorithm for the given DSP processor. Figure 1.3 shows the block diagram of the data CPU in the DSP processor. The actual architecture and instruction set of the DSP processor have a major impact on the optimal form of implementation. The programmer's tasks are to move the data between memory and registers and to manage the operation units in such a way that the ALU and MAC operate on the correct data samples to perform the desired processing. The execution time of a particular algorithm depends as much on the efficiency of the code as it does on the instruction cycle time. DSP software development must fully exploit the parallel-processing and pipelining capabilities inherent in the architecture and instruction set of the given processor. Good programming technique plays an essential part in the successful application of DSP. A structured and well-documented approach to programming should be initiated from the beginning. It is important to have an overall plan of the signal-processing tasks that includes memory requirements, processor constraints, program size, execution time, etc. Program and data memory blocks should be arranged to minimize data-access time.

There are two commonly used programming languages for writing DSP code: assembly and C. Because the processor allows many processing units to operate in parallel, high-level languages such as C are inefficient.

1.4.1 Instruction Set

The instruction defines the operations to be executed by the DSP processor at each clock cycle. The instruction set defines the complete set of instructions that the processor performs. The instruction set includes instructions for arithmetic processing, data manipulation, memory movement, logical operations, condition testing, and program-flow control in a manner that optimizes the use of memory and time of execution. Several TMS320 processors support two forms of the instruction set: a mnemonic form and an algebraic form. In this book, we use the traditional mnemonic form.

Arithmetic instructions perform operations using the shifter, multiplier, and ALU. These instructions include add, subtract, absolute value, multiply, multiply-accumulate, multiply-subtract, negate, and left and right shifts. Depending on the processor architecture, there may be a number of operations for specifying the particular registers involved, the role of the carry bit, and whether a sign extension should be invoked. Associated with the multiplier are the instructions of multiply and accumulate with variants to handle intermediate overflow.

Logical operations such as AND, OR, logical shift, test, and exclusive-OR (XOR) are supported by the ALU and are used for masking or setting particular bits of operands. These instructions can also be used for bit manipulations (e.g., test, set, complement, or clear) of any bit (or pair of bits) of a word either in memory or internal registers. Data movement instructions cover the loading and storing of data (or a block of data) from and to data memory, between program and data memory, and between internal registers.

Condition-testing instructions support program-control operations such as IF-ELSE statements in C. Conditions that can be tested vary from processor to processor, but usually include the value of the accumulator (e.g., zero, positive, negative), the state of various input lines, and the state of specific flags such as the carry, overflow, etc. Program-flow control instructions include branches, subroutine calls, returns from subroutines, and interrupts. Repeat, block-repeat, and delayed-branch instructions for minimizing overheads are available on more advanced processors. These instructions usually can be executed conditionally based on the condition-testing results.

1.4.2 Assembly Programs

In order to design the most effective DSP system, code must be carefully optimized to make the most of the available resources from the processor. Assembly language is one step away from the machine code used by the processor; thus, it provides the fastest execution and most compact form of the algorithm. Instructions are specified by mnemonics, internal registers, and predefined symbols. Numbers can be represented in hexadecimal, decimal, or binary forms or by means of user-defined symbols. The assembler in the host computer translates the assembly program into object code that can be downloaded into the processor's program memory for execution. Writing in assembly code allows specification of address generation, usage of the processor's resources (registers and functional units), mode setting, and other housekeeping operations, thus giving the programmer full control of performance.

Assembly instructions typically have fields separated by spaces for a label, a mnemonic, address information, and comments. When parallel operations are supported, additional fields are needed. In order to facilitate cross-referencing between the flowchart and the assembly program, the addresses of all variables and constants should be given symbolic names that correspond to those used in the flowchart.

1.4.3 C Programs

C language is a valuable programming tool for computationally intensive tasks. General-purpose programming languages such as C and the object-oriented C++ are commonly used for compact and portable signal-processing tasks. Object-oriented programming has been receiving strong interest because it encourages the reuse of its software elements (or objects) in future programs. The use of object-oriented C++, which encourages software reuse, can reduce future software development costs.

Most engineers and programmers understand C language, which offers a structured environment for code writing. C has high-level language capabilities such as arrays and functions for implementing DSP algorithms. It also provides low-level language capabilities such as bit-level manipulation, pointers, and direct-hardware I/O. It is often easier to write a program in C so that the algorithm can be tested offline using simulated (or digitized) data. These features make C an ideal language for DSP applications. Most manufacturers of DSP processors provide C compilers with assembly-optimized C-callable functions in libraries to generate very efficient code. Because the majority of the code is in C, the application can be ported to other processors much more easily than an algorithm that is coded completely in assembly language. As DSP processors became faster and more complicated, the justification for full assembly coding decreased. C-compiler efficiency improves continuously, but C programs may never replace the hand-optimized assembly programs completely.

While the subroutine approach is normally favored in DSP programs, it is a question of balancing the overhead in execution time of subroutine calls against readability and against the excess program memory required for repeated sequences when a subroutine approach is not adopted.

1.4.4 Mixing C and Assembly Code

The primary goal of software development is to write a program that is easy to write, understand, debug, maintain, and port on a variety of processors. In addition, we also want to develop a program that requires fewer memory locations and makes the most efficient use of the unique computational resources of a particular processor, thus allowing it to run as fast as possible. These two goals are sometimes contradictory. A high-level C language is easier to write for the given algorithm and to port from one specific processor to another. It also follows the underlying algorithm more directly, thus making it easy to understand and maintain. However, the assembly code generated by the C compiler is generally not as efficient as the assembly code handwritten by the programmer. Handcrafted assembly code is generally faster and smaller, but it takes longer to develop.

In general, most execution bottlenecks occur in a few sections of DSP code, usually in the loops (especially inner loops) of a program. These loops may only occupy 10% of the code, but may take 90% of the time to execute. The best strategy is coding the entire algorithm in C first, identifying the time-critical bottlenecks, and then rewriting only that small percentage of code in assembly language. Because DSP C compilers generate intermediate assembly code for optimization, time-critical portions of code can be identified by using profiling capabilities and can be replaced with handcrafted assembly code. Another method is to use a library of hand-optimized functions coded in assembly language by the engineers or in the run-time library provided by the manufacturers. These assembly routines may be either called as a function or in-line coded into the C program.

Software libraries become important as DSP algorithms become more complicated and computationally demanding. DSP manufacturers usually provide a set of commonly used signal-processing operations in software libraries that are optimally written for a particular processor. Because of the improvement in C-compiler efficiency and the availability of user-friendly integrated software development tools, a mix of C and assembly routines is the most effective way of developing programs for DSP systems. In this book, we use this approach in experiments in later chapters.

1.4.5 Software Development Tools

Software tools for developing new DSP algorithms are slowly improving as the need to design new DSP applications more efficiently becomes important. The manufacturers of DSP processors typically provide a set of software tools for use in developing efficient code for the processors. As illustrated in Fig. 1.5, the basic software tools include a C compiler, an assembler, and a linker. Either a software simulator or hardware emulator can be used to validate the algorithm.

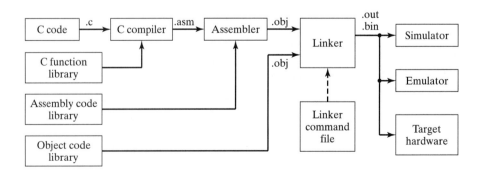

Figure 1.5   Block diagram and relationships of software development tools

In addition, MATLAB is a language for high performance computing that is frequently used in DSP for algorithm design and analysis. MATLAB provides a working environment for managing variables and data I/O and display, as well as for developing applications. It includes basic mathematic and DSP libraries and supports more advanced functions in a variety of toolboxes. A brief introduction to MATLAB and its toolboxes is given in Appendix A.

C Compiler

C compilers of DSP processors usually generate assembly code so that the user can see the assembly instructions generated by the compiler for each line of C source code. The assembly code can then be optimized by the user and fed into the assembler. The TMS320 C compiler conforms to American National Standards Institute (ANSI) C standards, thus providing maximum portability and increased capabilities. The compiler tools come with a complete run-time library that includes functions for standard input and output, string manipulation, dynamic memory allocation, data conversion, and mathematic functions. The compiler provides an optional optimization pass for generating efficient and compact code from the C-source program.

Most DSP C compilers allow the programmer to add in-line assembly instructions and assembly-optimized intrinsics and routines to C programs for generating highly efficient code in time-critical sections of a program. For example, the FIR filtering defined in Eq. (1.3.1) can be calculated on a DSP processor by using repeating multiply-accumulate operations. The amount of effort required by the programmer to write efficient assembly code only for the FIR filter is much less than the effort required to develop the entire program in assembly language. In addition, a library of highly optimized assembly routines for commonly used DSP functions is available in the public domain. For example, a set of assembly-optimized routines such as FIR filtering is available in the C5000 DSP Library (DSPLIB). This code can be used in C programs directly or with minor modifications. We discuss DSPLIB in Section 5.4.5 and use these optimized functions in the library for hands-on experiments given in Chapters 6, 7, 8, and 9.

Assembler

The file containing the assembly program is used as input for the assembler, which typically resides in a host computer. The task of the assembler is to convert the source-code modules, written using assembly-language mnemonics, into executable object-code modules for the DSP processor. The assembler produces two output files. A listing file documents the assembly operations on a particular module or group of modules in terms of data-program-memory locations, assembly errors, etc. The other file is the object file, which can be used as input to a linker. Assembler usually includes macro-library facilities for supporting a modular programming approach and produces relocatable code for a software linker.

Linker

The operation of the linker is to combine separately generated object-code modules and associated files to form a single linked object-code module. The linker performs relocation and resolves external references as it creates the executable module. The advantage of the modular assembler-C compiler-linker process is that only the relevant sections or modules of a program need to be edited and assembled (or compiled) during program development rather than the complete source file. This facility can save considerable time and effort. A number of special commands relating to memory mapping and organization can be implemented in a linker-command file. An example of a linker-command file is given in Section 1.7.

Simulator

The simulator is a program that runs on the host computer. It accepts object-code input and simulates the operations of the DSP processor as it executes its program in nonreal time under the user's control. The user interface shows memory, internal registers, I/O, etc., and the effects there on after each instruction is executed. Break points can be inserted or the program can be single-stepped, and the contents of any of the internal registers, program memory, or data memory can be examined with user-definable display modes. The user can optionally change the contents of any register, flag, or memory location before resuming the simulated execution. Data can be read from and written to designated files for further analysis or processing. Simulators for a particular DSP processor allow the user to determine the performance of an algorithm on a specific target processor before developing physical hardware. It is possible to develop all of the DSP programs for a particular processor before designing or purchasing target hardware.

Real-world signals can only be recorded in disk files first, and then later used by a simulator as test data. Although test data may verify the algorithm performance, the timing of the algorithm under all possible input conditions cannot be tested using a simulator. It is good practice to design a series of test procedures and generate the associated sets of test data during the program-development stage for use on the simulator. Because simulators execute DSP programs on a host computer, the speed is much slower than the program run on the actual DSP processor in real time. It may take a long time for a program to reach a particular breakpoint in a simulator for a complicated algorithm.

Most DSP processors also provide standard debuggers that support standard debugging features. However, with the increasing demand for a more complete development platform with more advanced tools such as block-diagram-based programming, real-time operating systems, graphical tools for memory organization, code debugging and profiling, real-time data logging, heterogeneous and multiprocessor debugging, etc., advanced tools such as Code Composer Studio (CCS) from Texas Instruments are emerging that help the user to develop efficient code faster.

Code Composer Studio

CCS from Texas Instruments is an integrated development environment that incorporates the C compiler, assembler, linker, simulator, debugger, etc., with additional features such as a graphical display for developing DSP software. It is an easy-to-use software tool for supporting software development. It provides an editor for programming and a project manager for building application programs. For software-testing purposes, it provides steps, breakpoints, watch windows, graphic capabilities for data monitoring, and probe points for data-file I/O. CCS also provides profiling capabilities that allow the user to determine the amount of time spent in one portion of program.

In addition, CCS provides the user with a DSP operating system called DSP/BIOS (basic input/output system). DSP/BIOS is a real-time software kernel that allows the user to schedule tasks in the DSP program and monitor the real-time performance of the DSP code without the need to set any breakpoints. Unlike conventional debugging, where the user needs to stop the execution of the DSP with breakpoints, DSP/BIOS provides run-time services and detects run-time glitches that would otherwise not be detected in conventional debugging. In addition, a new tool that links CCS with MATLAB® is briefly introduced in Appendix A.

Before DSP hardware is available, CCS can be used to verify the code. In Section 1.7, we use a simple C5000 program to learn how to use CCS for DSP software testing.

1.5 HARDWARE ISSUES

Besides the software issues described previously, there exist important issues such as selecting necessary hardware components and interfacing external hardware devices to the DSP processor. It is also necessary to configure the hardware with software known as firmware, which is integrated with the DSP software to form a complete DSP system.

1.5.1 Hardware Selection

The same TMS320 family provides different devices to provide the best match for the given application. For example, the devices within the C54x family differ in the number of DSP cores, operating clock frequencies, voltages, on-chip ROM configurations, RAM configurations, type and number of serial ports, and host ports. These variants of DSP devices provide flexibility for users to select the best processor for their needs. Another advantage is that users can program in one device and port code into another device within the same family with minimum effort.

Other than internal peripherals, DSP processors also need to communicate with external devices, such as communication channels, loudspeakers, microphones, and recording and playback devices. Furthermore, DSP processors need to store additional data and programs in external memory and even communicate to another processor via external buses or interconnects. DSP processors also require an external power supply and regulator, as well as a clock generator or oscillating crystal to power and trigger the processor. Therefore, additional hardware circuitries and devices are needed, which are highlighted in Chapter 3.

1.5.2 Hardware Configurations

After a suitable DSP processor has been selected and relevant external hardware has been connected to the processor interfacing pins, a set of programs known as firmware needs to be written to initialize the peripherals of the processor and the external devices. Some important hardware configurations include the following:

1. Setting the clock frequency and programming the internal PLL to drive the processor.
2. Setting the software and hardware wait states of external memory and I/O devices.
3. Setting the sampling frequency, transfer rate, and input and output buffer sizes of the ADC and/or DAC.
4. Setting the communication protocol of the DMA to move data in and out of the processor's internal memory without processor intervention.

1.5.3 Hardware Tools

Once the firmware has been properly programmed, the code is linked and down-loaded into the DSP internal memory via boot-loader software. The firmware is run before the main DSP routine. A set of hardware tools is usually available to test the functionality of the firmware and software. Most DSP manufacturers and vendors have developed different development platforms for meeting different user demands. These development tools have different types, levels of support, and costs, ranging from the most expensive in-circuit emulator to the cheapest DSP starter kit (DSK). Development tools assist the designer in testing DSP algorithms in real time and in prototyping their product. Therefore, a good array of software and hardware tools, peripheral I/O chips, and memories is crucial to the selection of the best DSP processor for a given application.

Emulator

The final check on program validity is to load the program into the DSP processor's program memory and execute it on the target hardware. In order to determine the results from real-time input, emulators allow breakpoints to be set at a particular point in a program to examine registers and memory locations. Before a breakpoint is reached, the DSP program runs at full speed. An in-circuit emulator allows the final hardware to be tested at full speed by connecting to the processor in the user's real-time environment. Cycle counts can be determined between breakpoints, and the hardware and software timing of a system can be examined.

Emulators speed up the development process by allowing the DSP program to run at full speed in a real-time environment. Because emulation is performed in real time using the actual target system, hardware and timing errors can be identified using a logic analyzer. A more detailed explanation of emulators is given in Chapter 3.

Evaluation Module

An evaluation module is a comprehensive development platform that comes with a DSP board that also houses supporting peripheral chips such as coder-decoder (CODEC) for audio interface, joint test action group (JTAG) emulator support, and memories. An evaluation-module board can be plugged into personal computers or workstations. Integrated development software such as CCS is also included, which allows the user to start evaluating code on the target platform and determine if real-time constraints are met.

Digital Signal Processing Starter Kit

The DSK is a scaled-down version of the evaluation module, which is very economical for performing quick code evaluation. It connects easily to a host computer through a parallel port or to the emulator via the JTAG cable. However, the lack of peripherals and memories in the DSK may require the user to build additional I/O or memory daughter cards to compensate for the shortfall.

1.6 SYSTEM CONSIDERATIONS

The traditional distinction between programmable and function-specific DSP design is fading because of custom DSP solutions. DSP system designers can decide which section of a design is best suited for a hardwired approach. Code that must maintain upgradability can be downloaded into the on-chip RAM, and the rest of the program can be masked on on-chip ROM. In addition, algorithm accelerators and custom peripherals can be designed and placed on the same chip.

DSP system design consists of understanding the given applications, developing DSP algorithms, implementing the algorithms in software based on the chosen processor, developing the hardware, integrating the software with the hardware, and analyzing system performance to make sure that it satisfies all requirements. Design of the DSP system using CAD and analysis tools provides an efficient approach to handling complex applications.

The most critical factors that determine program execution time are the sampling rate and the complexity of the algorithm. These factors impose restrictions on the maximum sampling rate, which must be at least twice the maximum frequency of the input signal in order to avoid aliasing. This issue should be considered when assessing the suitability of a DSP system for a given application.

1.6.1 The Choice of Digital Signal Processors

DSP processors are following the path of microprocessors in terms of performance and on-chip integration. At the same time, power consumption becomes an important issue for portable products. A DSP product design is constrained by the following key design goals:

1. Cost of the product
2. Cost of the design
3. Upgradability
4. System integration
5. Power consumption

These design goals play key roles in selecting DSP processors.

The selection of a DSP processor suited to a given application is a complicated task. Some of the factors that might influence choice are cost, performance, future growth, and software and hardware development support. Based on the data format, a DSP device is either a fixed-point or floating-point processor. The data format determines the device's ability to handle signals of different precisions, dynamic ranges, and signal-to-noise ratios. Using floating-point processors can increase the dynamic ranges of signals and coefficients. Floating-point processors are usually more expensive than fixed-point processors, but they are more suitable for high-level C programming. Thus, they are easier to use and allow a quicker time to market. When choosing between fixed-point and floating-point processors, ease of use and software development time are often equally important.

The execution speed of a DSP algorithm is also an important issue when selecting a processor. When performance is the most important factor, the algorithm must be implemented with optimized code written for those processors, and the execution times must be compared. The time to complete a particular algorithm coded in optimized assembly language is called a benchmark. A benchmark can be used to give a general measure of the performance of a specific algorithm for a particular processor. Other related issues include memory size (on chip and externally addressable) and the availability of on-chip peripheral devices such as serial and parallel interfaces, timers, and multiprocessing capabilities. In addition, space, weight, and power requirements must be minimized. A key system constraint is the system cost. Like the case for general-purpose microprocessors, second sourcing, third-party support, and industry standards are other important issues for consideration.

1.6.2 Sampling and Quantization

Many DSP applications deal with analog signals. As illustrated in Fig. 1.6, the analog signal is sampled and quantized (encoded) to a digital signal. Sampling is the process of converting a continuous-time signal into a discrete-time signal, which still has continuous amplitudes at discrete sampling intervals. The amplitudes of discrete-time signals are then quantized into digital values based on the given wordlength N. The ADC performs the processes of sampling and quantization. It is important to note that only digital signals, not discrete-time signals, are available for the DSP system. Discrete-time signals are only used for design and analysis purposes.

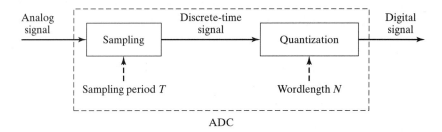

Figure 1.6   The process of analog-to-digital conversion

The sampling period, T, is determined by the frequency contents of the input signal. According to the Shannon sampling theorem, an analog signal with frequency bandlimited to $f \leq f_M$ can be reconstructed accurately from the sampled values if the sampling rate, f_s, is greater than twice the bandwidth, f_M, of the input signal. That is,

$$f_s \geq 2f_M. \tag{1.6.1}$$

If the sampling theorem is violated, the undesired aliasing phenomenon occurs. This issue is discussed further in Chapter 2. The sampling rate (frequency) in Hz is related to the sampling period T in seconds expressed as

$$f_s = \frac{1}{T}. \tag{1.6.2}$$

The sampling rate is usually determined by a given application. For example, a music CD is sampled at 44.1 kHz, and voice in telecommunication is sampled at 8 kHz.

Discrete-time signals assume that an infinite number of bits are used to represent digital values. In practice, only a finite number of bits is used (e.g., 16 bits in most fixed-point processors). Quantization is the process of representing a discrete-time sample by the nearest level that corresponds to an integer scale. This process introduces undesired quantization noise, which is the difference between the discrete-time (true) value and the actual value of a digital signal. The greater the wordlength N used to represent digital samples, the smaller the quantization noise. In general, the signal-to-quantization-noise ratio (SQNR) is given by

$$\text{SQNR} \cong 6N \text{ dB}. \tag{1.6.3}$$

A low value of N also restricts the dynamic range. Any desirable accuracy can be achieved simply by increasing the wordlength, subject to cost constraints. These issues are discussed further in Section 3.3. The theoretical aspects of sampling and quantization are explained in advanced DSP textbooks [9–11].

The speed of processing determines the rate at which the analog signal can be sampled. Bandlimited signals can be sampled if the sampling rate given in Eq. (1.6.1) is satisfied. However, most real-world applications define a required sampling rate of real-life analog signals that may have bandwidths higher than half of the specified sampling frequency. For example, the sampling rate in most telecommunication systems is defined as 8 kHz, but the bandwidth of speech signals may be up to 20 kHz, which is much higher than 4 kHz. To guarantee that the sampling theorem is satisfied, we use an anti-aliasing analog lowpass filter to reduce the bandwidth of the analog signal before the ADC. Even if the input signal is bandlimited below half of the sampling frequency, an anti-aliasing filter is still required to reduce out-of-band noise from aliasing into the desired frequency band.

Ideally, the anti-aliasing filter should have a flat magnitude and a linear-phase response over the bandwidth of the signal and have infinite attenuation above half of the sampling frequency. One of the most commonly used anti-aliasing filters is the switched-capacitor filter, since it can be implemented easily in semiconductors, and its bandwidth can be varied by changing the clock frequency. A technique for relaxing the specification of anti-aliasing filters is oversampling. The sampled signal in a high rate is then filtered by a digital lowpass filter and downsampled to the desired sampling rate. This technique is discussed in Section 6.5.

1.6.3 Digital-to-Analog Conversion

An analog signal can be reconstructed without distortion from its ideally sampled digital signal with DAC and analog lowpass filtering. The ideal reconstruction (or smoothing) filter has a flat-magnitude response and linear-phase characteristics in the passband from DC (direct current) to half of the sampling frequency. Any departure from the ideal filter characteristics introduces spectral distortion. The design of the smoothing filter is easier using the oversampling technique. Switched-capacitor filters are preferred for implementation because of their programmable cutoff frequency and physical size.

1.6.4 Coder-Decoder

The CODEC includes the anti-aliasing filter, ADC, DAC, and reconstruction filter on a single chip. CODECs have been developed primarily for telecommunications; thus, they are serial data converters. They are a very attractive data converter for voice applications when serial ports exist on the processor. A block diagram of the CODEC is illustrated in Fig. 1.7. Both the anti-aliasing filter and the reconstruction filter can be realized using a switched-capacitor filter to save costs and space. As discussed previously, switched-capacitor filters offer programmable cutoff frequencies, making them preferable for applications that need to change sampling rates.

A uniform quantizer provides identical spacing between successive levels throughout the entire dynamic range of the signal. However, many analog signals such as speech have the characteristic that small signal amplitudes occur more frequently than large ones. Thus, a better approach is to use a nonlinear quantizer that compresses the signal using the logarithmic operation, followed by a uniform quantizer. In the reconstruction of the signal, the inverse operation is used to expand the

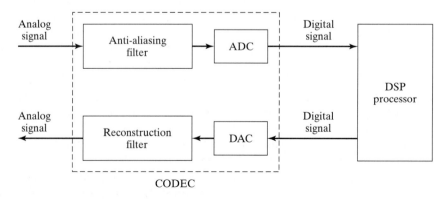

Figure 1.7   Block diagram of the CODEC

signal amplitude. The combined compressing-expanding process is called companding, and the most popular companding standards are μ-law companding used in North America and A-law companding used in Europe. A more detailed discussion of companding techniques is available in digital communication textbooks [12].

If a CODEC with μ-law or A-law companding is used, the binary signal is no longer related linearly to the sample magnitude. In order to perform linear processing of the signal, an expanding process must be performed within the processor to restore a linear relationship. Similarly, the converse process must be performed on the data output to the CODEC to obtain the companded form expected by the output device. Some DSP processors, such as the C5000, support the expansion of input data and the compression of output data in dedicated hardware under software control. When this facility is not available, companding must be implemented in the software based on either the curve fit or table look-up techniques, which involves processing overhead that should be taken into account when choosing data converters.

1.6.5 Speed Optimization

Many DSP applications require real-time processing over long periods of time. For real-time operations, a successive input sample must be processed to produce the corresponding output sample within a sampling interval because data cannot be stored and subsequently processed over a time period longer than the sampling period. The processor must execute the algorithm fast enough to avoid any loss of input data. Execution time can be reduced by precomputing as much information as possible and storing that information in memory. For example, a sine function can be evaluated by a polynomial approximation implemented in a program that uses many clock cycles to execute. However, we can also precompute one period of the sinewave, store those samples into memory, and generate the sine function using a table-lookup technique. In this case, we trade memory space for execution speed.

1.7 EXPERIMENTS AND PROBLEMS

The experiments in this section use a simple TMS320C5000 assembly program for introducing the major functions of CCS for software debugging and testing. Only some of the major features are introduced in this chapter. More advanced capabilities and the debugging of C programs are introduced in Chapter 2 and in the *Code Composer User's Guide* [13]. In this book, we use CCS Version 2.10 for experiments.

If CCS is not installed properly, double-click on the **Setup CCS 2 (C5000)** shortcut on the desktop. When the **Import Configuration** window appears, click on the **Clear** button and click on **Yes** to confirm that the system configuration should be cleared. In the following hands-on experiments, we run similar assembly programs on two different platforms: C55x and C54x simulators. Therefore, under the available configurations, we import either **C55x Simulator** or **C5402 Simulator** and click on the **Close** button at the bottom of the window. We can examine the properties of the configured system by right-clicking on **Simulator** under **My System** and then selecting **Property**. A window appears that highlights the board name, driver location, simulator configuration file, and startup general-extension language file used in the simulator. The preceding settings should not be altered and should be left as the default. Click on **Finish** to close the board-property window. Finally, click on **File** → **Exit** to end the setup process. When prompted to save changes and start CCS on exit, click on **Yes**.

1.7.1 Using Assembly Programs and Linker-Command Files

In this section, we use the TMS320C55x assembly program `exp1c55.asm` (or `exp1c54.asm` for the C54x) to introduce some important features of the CCS used to create, build, debug, and test programs. First, we create a new folder `c:/dsps/chap1` and then copy the assembly-program source file `exp1c55.asm` (or `exp1c54.asm` for the C54x) and the linker-command file `exp1c55.cmd` (or `exp1c54.cmd` for the C54x) into that folder. The data used in the assembly program is generated by the MATLAB program `exp1.m`. These files are available on the companion CD. The C55x assembly code used in this experiment is adapted from the *TMS320C55x DSP Programmer's Guide* [14] and is listed as follows:

```
        .def    x,y,init
x       .usect  "vars",8        ; reserve 8 locations for x
y       .usect  "vars",1        ; reserve 1 location for y

        .sect   "table"
init    .int    0,5792,8191,5792,0,-5792,-8191,-5792
                                ; value of x vector
        .text                   ; create code section
        .def    start           ; label of the beginning of code

* Processor initialization

start
        BCLR    C54CM           ; set C55x native mode
        BCLR    AR0LC           ; set AR0 in linear mode
        BCLR    AR6LC           ; set AR6 in linear mode

* Copy data to vector x using indirect addressing mode

copy
        AMOV    #x,XAR0         ; XAR0 pointing to x0
        AMOV    #init,XAR6      ; XAR6 pointing to table of data
        RPT     #7              ; repeat next instruction 8 times
        MOV     *AR6+,*AR0+     ; copy 8 data to x vector

* Add the first 4 data using direct addressing mode

add
        AMOV    #x,XDP          ; XDP pointing to vector x
        .dp     x               ; notify assembler
        MOV     @x,AC0          ; x0 -> AC0
        ADD     @(x+1),AC0      ; add x1
        ADD     @(x+2),AC0      ; add x2
        ADD     @(x+3),AC0      ; add x3
```

```
* Write the result to memory location y

write
      MOV     ACO,*(#y)    ; ACO -> y
end
      NOP
      B       end          ; stop here
```

The linker assigns the final addresses to the code and data sections. The file that instructs the linker to assign the addresses is called the linker-command file, which is listed as follows:

```
/*  exp1c55.cmd - Linker command file for C55x              */

MEMORY              /* byte address, length in byte */
{
  DARAM : origin=000100h, length=8000h /* dual-access RAM  */
  SARAM : origin=010000h, length=8000h /* single-access RAM*/
}

/* Section allocation                                       */
SECTIONS
{
    vars  :>  DARAM
    table :>  SARAM
    .text :>  SARAM
}
```

The MEMORY directive declares all of the physical memory spaces available in the system. Memory blocks cannot be overlapped. Note that the C55x linker-command file uses byte addresses and byte lengths, while the C54x uses 16-bit word addresses and wordlengths. The SECTIONS directive lists all of the sections contained in the input files and indicates where they should be allocated.

For readers who use the C54x simulator, the preceding C55x assembly code and linker-command file must be modified as follows:

```
    .def   x,y,init
x   .usect "vars",8     ; reserve 8 locations for x
y   .usect "vars",1     ; reserve 1 location for y

    .sect  "table"
init .int   0,5792,8191,5792,0,-5792,-8191,-5792
                        ; value of x vector
    .text               ; create code section
    .def   start        ; label of the beginning of code
start
    NOP
* Copy data to vector x using indirect addressing mode
copy
    STM   #x,AR0        ; AR0 pointing to x0
    RPT   #7            ; repeat next instruction 8 times
    MVPD  #init,*AR0+   ; copy 8 data to x vector
```

```
* Add the first 4 data using direct addressing mode
add
     LD    #x,DP          ; DP pointing to vector x
     LD    @x,A           ; x0 -> A
     ADD   @(x+1),A       ; add x1
     ADD   @(x+2),A       ; add x2
     ADD   @(x+3),A       ; add x3

* Write the result to memory location y
write
     STL   A,*(y)         ; A -> y

end
     NOP
     B     end            ; stop here

/*  exp1c54.cmd - Linker command file for C54x              */

MEMORY
{
  PAGE 0: EPROG:     origin = 0x100,     len = 0x1f80
          EDATA:     origin = 0x2000,    len = 0x1fff
  PAGE 1: IDATA:     origin = 0x80,      len = 0x1f80
}
SECTIONS
{
  vars:    {} > IDATA PAGE 1
  table:   {} > EDATA PAGE 0
  .text:   {} > EPROG PAGE 0
}
```

1.7.2 Creating a Project

Start CCS by double-clicking on the **CCS 2 (C5000)** icon located on the desktop (or from the **Start** menu). The CCS window displays. Select **Project → New** from the menu, and the **Project Creation** dialog box displays. Type exp1c55 (or exp1c54 for the C54x) in the **Project Name** field, type c:/dsps/chap1 in the **Location** field, and then click on the **Finish** button. This action creates a project file called exp1c55.pjt, which stores project settings, references, and files used by the project. Note that the file exp1c55.pjt is now displayed in the **Project View** window.

Select **Project → Add Files to Project** from the menu, and the **Add Files to Project** dialog box displays. From the **Look in** field, browse to the folder c:/dsps/chap1. From the **Files of type** field, select **Asm Source Files**, click on exp1c55, and then click on **Open** (or double-click on exp1c55). Repeat the preceding process, but select **Linker Command File** in the **Files of type** field, select exp1c55, and click on **Open**. In the **Project View** window, click on exp1c55.pjt, and then click on **Source**. The exp1c55.cmd and exp1c55.asm options display. The project manager identifies files by their file extension.

Select **File → Open** from the menu, and the **Open** dialog box displays. From the **Files of type** field, select **Assembly Source Files**. Double-click on the file exp1c55 to open it. Alternatively, double-click on exp1c55.asm in the **Project View**

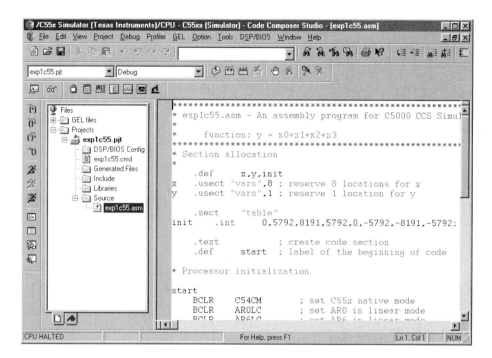

Figure 1.8   CCS window with project information and source code (C55x)

window to open it. The source code appears in the right half of the window, as shown in Fig. 1.8.

1.7.3 Building the Project

Select **Project** → **Build Options** from the menu, and the **Build Options for** exp1c55.pjt dialog box displays. Click on the **Linker** tab. From the **Output Module** field, select **Absolute Executable**. From the **Code Entry Point** field, type in start and click on **OK** to save the new option settings. Note that these steps are only needed for compiling assembly programs.

Select **Project** → **Compile File** from the menu (or click on the **Compile File** button ![btn] on the toolbar) to compile the current source file. A message showing the compiler results displays at the bottom of the CCS window. When there is no compiler error and warning, select **Project** → **Build** from the menu (or click on the **Incremental Build** button ![btn] on the toolbar) to build the current project. This action compiles only the files that have changed since the last build. Alternatively, select **Project** → **Rebuild All** from the menu (or click on the **Rebuild All** button ![btn] on the toolbar) to recompile all files in the current project and relink all of the files in the project to form the executable file. Messages about this process are shown in the bottom window. If a message says the program contains compile errors, scroll up in the **Build** tab area to see a syntax error message, and double-click on the red text that describes the location of the syntax error. The cursor then moves to the line in the source-file window. Correct the errors, select **File** → **Save** from the menu to save the changes, and then choose **Project** → **Build** from the menu. CCS rebuilds files that have been changed.

Note that you may get the following warning message in the build window: "WARNING entry point other than _c_int00 specified". Do not be concerned about this message since the default entry point is usually stated as _c_int00 for C programs, and we are using start in our exercise instead. In addition, instead of using text linker with a .cmd extension, the visual linker (with a .rcp extension) can be used by clicking on **Tools** → **Linker Configuration**. An error occurs if the text-linker file is used and if **Linker Configuration** is set to **Visual Linker**. This error can be solved by double-clicking on the error message. The **New Recipe Wizard** window then appears and guides the user to create the .rcp recipe file that is intended to replace the .cmd file. However, the text linker is the preferred option since it allows the user to enter settings, such as heap size, stack size, directory for the output file, start of the program, etc., in a single file.

Select **File** → **Load Program** from the menu, and the **Load Program** dialog box displays. This dialog box allows the user to select the desired file. Double-click on the **Debug** folder, and the file exp1c55.out appears. Double-click on the file to open it. The Disassembly window displays, as shown in Fig. 1.9. A yellow arrow points at the current instruction (below the label start). Note that the original source code is displayed in the Disassembly window. This source-level debugging capability makes program testing much easier.

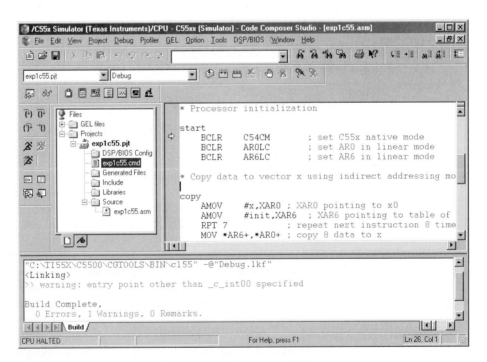

Figure 1.9   Debugger with Disassembly and Build windows (C55x)

1.7.4 Debugging the Program

We can single-step through the code by either clicking on the **Single Step** icon [⊕]
on the debug toolbar or selecting **Debug** → **Step Into** from the menu. The current
instruction (which the yellow arrow points at) executes, and the yellow arrow moves
to the next instruction. As we step through the program using stepping commands,
the related register(s) and/or memory location(s) change according to the instruction. By clicking continuously on the **Single Step** button, we can execute the program step by step. By examining the changes in CPU registers and memory
locations, we can debug the program to make sure each instruction is correct. In
addition, CCS provides a **Step Over** [⊕] command to step through and execute individual instructions in the current function. If the current instruction is inside a subroutine, the **Step Out** [⊕] command completes execution of the subroutine and
returns to the calling function. We can also use the **Run to Cursor** [⁰] command to
run the loaded program until it encounters the cursor position.

Select **Debug** → **Reset CPU** from the menu. This command initializes all registers to their power-up state and halts execution of program. Select **Debug** → **Restart**
and **Debug** → **Run** from the menu (or click on the **Run** button [⊠] on the toolbar) to
execute the program from the current program-counter location. The sign **CPU
RUNNING** displays at the bottom left of the CCS window. Execution continues until
a breakpoint is encountered. We can stop execution of the program by clicking on
the **Halt** button [⊠] on the debug toolbar or by selecting **Debug** → **Halt** from the
menu. The display changes to **CPU HALTED**.

Select **View** → **CPU Registers** → **CPU Registers** from the menu (or click on
Register Window on the toolbar), and the register window displays. To edit the contents of the CPU or peripheral registers, select and **Edit** → **Register** from the menu
(or double-click on a register in the register window). This action opens the **Edit
Registers** dialog box. In the **Register** field, specify the register to be edited by typing its name (or click on the scroll button and choose the appropriate register). The
Value field contains the current value of the register displayed in Hex format. We
can enter another value in this field by typing a new value in decimal, or in Hex
with the prefix 0x, and clicking on the **Done** button. The new value in the modified
register appears in red. Note that in the case of C54x programming, data memory
and program memory share the same DARAM when the OVLY flag is set to 1.
Therefore, if the OVLY flag is not set to 1, we can use the **Edit** → **Register** command
to set it.

1.7.5 Viewing Memory and Graphics

CCS allows the user to view the contents of memory at specific locations. Select
View → **Memory** from the menu (or click on the **View Memory** button on the toolbar), and the **Memory Window Options** dialog box displays. Type in the symbolic
variable name x in the **Address** field, select **16-Bit Signed Int** from the **Format** field,
and then click on **OK**. The memory window appears, as shown in Fig. 1.10. The
addition result 19,775 (0 + 5,792 + 8,191 + 5,792) is stored in memory location y.
The **Memory Window Options** dialog box allows the user to specify various characteristics of the memory window. We can enter the starting address of the memory

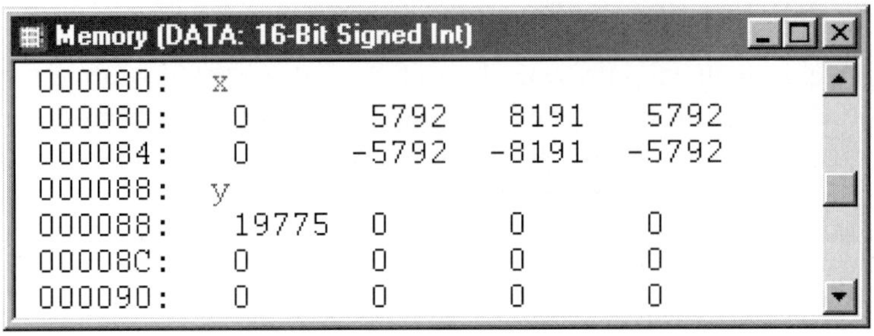

```
Memory (DATA: 16-Bit Signed Int)                    _ □ X
000080:    x
000080:      0        5792      8191      5792
000084:      0       -5792     -8191     -5792
000088:    y
000088:   19775       0          0         0
00008C:      0        0          0         0
000090:      0        0          0         0
```

Figure 1.10   **Memory** window showing data at symbolic addresses x and y

location by symbolic name (e.g., x) or physical address (e.g., 000080h). We can also
select the format of the memory display. To change the starting location or display
format of the memory window, right-click on the active memory window and select
Properties from the pop-up menu. The **Memory Window Options** dialog box opens.
Right-click on the active memory window, and select **Close** to close the memory
window.

Select **View** → **Graph** → **Time/Frequency** from the menu, and the **Graph Property Dialog** box displays. Field names appear in the left column, and we can adjust

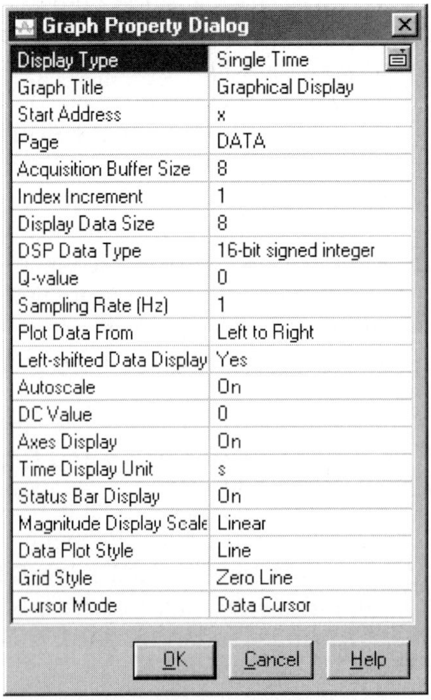

Graph Property Dialog	X
Display Type	Single Time
Graph Title	Graphical Display
Start Address	x
Page	DATA
Acquisition Buffer Size	8
Index Increment	1
Display Data Size	8
DSP Data Type	16-bit signed integer
Q-value	0
Sampling Rate (Hz)	1
Plot Data From	Left to Right
Left-shifted Data Display	Yes
Autoscale	On
DC Value	0
Axes Display	On
Time Display Unit	s
Status Bar Display	On
Magnitude Display Scale	Linear
Data Plot Style	Line
Grid Style	Zero Line
Cursor Mode	Data Cursor

OK   Cancel   Help

Figure 1.11   **Graph Property Dialog** box

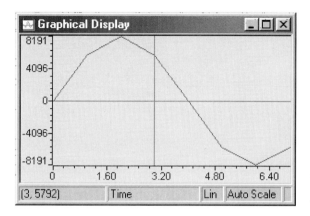

Figure 1.12   **Graphical Display** window showing the time-domain waveform

the values in the right column, as shown in Fig. 1.11. Clicking on the **OK** button opens a graphical display window inside the CCS window, as shown in Fig. 1.12.

1.7.6 Using Breakpoints and the Profiler

When we are developing and testing programs, we often need to check the values of variables and registers during program execution. To set a breakpoint, we place the cursor on the instruction in the Disassembly window, right-click, and select **Toggle breakpoint** from the drop-down menu (or click on the 🖑 button on the toolbar). The red dot at the left of the instruction marks the breakpoint. To clear the breakpoint, place the cursor on the instruction beside the red dot. Right-click and select **Toggle breakpoint**. The red dot disappears, showing that the breakpoint has been cleared.

After verifying the correct operation of exp1c55.asm, we can use CCS to count the number of cycles the code takes to execute. Select **File → Reload Program** from the menu, and the window shown in Fig. 1.9 displays. Select **Profiler → Enable Clock** and **Profiler → View Clock** from the menu, and the clock window with Clock = 0 displays. Set one breakpoint at the beginning of the code (e.g., the instruction after copy) and the other at the end of the code (e.g., the instruction before end). Run to the first breakpoint by selecting **Debug → Run** from the menu. The execution stops at the first breakpoint (the yellow arrow is overlapped with the red dot), and Clock = 17 displays for the C55x (Clock = 6 displays for the C54x). Double-click on the **Clock Window** to clear the cycle count. Run to the second breakpoint by selecting **Debug → Run** from the menu again. As shown in Fig. 1.13 for the C55x, the execution stops at the second breakpoint, and the **Clock Window** displays the number of cycles the code took to execute between the breakpoints (e.g., Clock = 17). Similar results can be observed in the C54x experiment, which shows Clock = 22 cycles.

The profiler allows the user to identify bottlenecks in the code. We thus can optimize the time-critical sections of the code in order to improve its run-time efficiency.

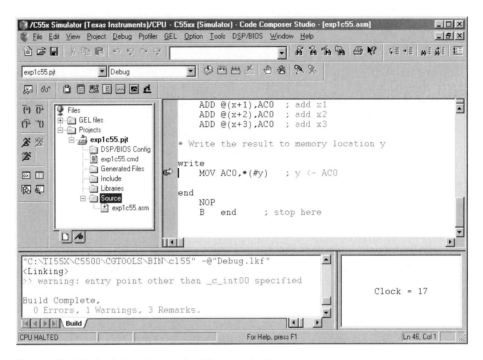

Figure 1.13   Clock window showing the efficiency of code

PROBLEMS

PART A

Problem A–1–1

In addition to viewing the values of variables, it can be very useful to view the contents of CPU registers when debugging assembly programs. Write the contents of PC, AC0, XAR0, and XAR6 in a table where the code between copy and end is executed with a single-step command.

Problem A–1–2

The **Display Type** option in the **Graph Property Dialog** box contains several options in the drop-down menu in the right column. The **FFT Magnitude** option performs an FFT on the data in the display buffer and plots a magnitude-versus-frequency graph. Report the procedure for displaying the magnitude spectrum of input samples starting at x.

Problem A–1–3

Memory values from memory can be stored in a data file using different file formats. Store the eight samples of output starting from x into a data file x.int by selecting **File** → **Data** → **Save** from the menu. Report the procedure for getting the data file x.int, and print it.

Problem A–1–4

Find the program called `try1c55.asm` (for the C55x) or `try1c54.asm` (for the C54x) on the companion CD. Use CCS to find the bug in these programs. Report the steps used for finding and fixing the bug.

Problem A–1–5

As stated in Section 1.3.2, a typical accumulator consists of 40 bits. Modify the assembly programs listed in Section 1.7.1 to save both the most significant and the least significant words in address y using both STH and STL mnemonics (for the C54x) or MOV hi(AC0) and MOV lo(AC0) (for the C55x). Use the **View Memory** option of CCS to examine the values saved in y. Do we need to save the most significant word in this case?

Problem A–1–6

Replace the ADD mnemonic with the MPY mnemonic for the programs listed in Section 1.7.1 to perform multiplication of the second, third, and fourth numbers of the array x. Save both the most significant and the least significant words in address y and view the results. Comment on whether the most significant word is required in this case.

PART B

Problem B–1–7

Name some typical external blocks in DSP systems, and name the processing blocks present inside the DSP processor.

Problem B–1–8

What are the advantages of using programmable DSP processors? How are DSP processors compared with application-specific integrated circuits (ASICs), general-purpose microprocessors, and microcontrollers?

Problem B–1–9

What is the best way to program a real-time DSP application that has limited processing and memory resources? Also, explain the meaning of "real time."

Problem B–1–10

In addition to a faster DSP processor, it is important that data are fed to the processor as fast as the processing rate. Name some of the I/O peripherals present inside the DSP processor that provide fast I/O features.

Problem B–1–11

Explain the functionalities of the following software-development tools: C compiler, assembler, linker, and simulator. Explain the functionalities of the following hardware-development tools: emulator, evaluation module, and DSK.

Problem B–1–12

Explain the functionalities of the following blocks inside the CODEC: anti-aliasing filter, ADC, DAC, companding, and reconstruction filter.

SUGGESTED READINGS

1 Ackenhusen, J. G. *Real-Time Signal Processing: Design and Implementation of Signal Processing Systems*. Upper Saddle River, NJ: Prentice Hall, 1999.

2 Grover, D. and J. R. Deller. *Digital Signal Processing and the Microcontroller*. Upper Saddle River, NJ: Prentice Hall, 1999.

3 Kuo, S. M. and B. H. Lee. *Real-Time Digital Signal Processing: Implementations, Applications, and Experiments with the TMS320C55x*. New York, NY: John Wiley & Sons, 2001.

4 McClellan, J. H., R. W. Schafer, and M. A. Yoder. *DSP First: A Multimedia Approach*. Upper Saddle River, NJ: Prentice Hall, 1998.

5 Steiglitz, K. A. *Digital Signal Processing Primer*. Reading, MA: Addison-Wesley, 1995.

6 The MathWorks, Inc. *http://www.mathworks.com*.

7 Texas Instruments. *http://www.ti.com*.

8 Texas Instruments. *TMS320C54x DSP CPU and Peripherals*. Reference Set, Vol. 1, SPRU131F, 1999.

9 Oppenheim, A. V. and R. W. Schafer. *Discrete-Time Signal Processing*. 2nd Ed. Upper Saddle River, NJ: Prentice Hall, 1999.

10 Proakis, J. G. and D. G. Manolakis. *Digital Signal Processing: Principles, Algorithms, and Applications*, 3rd Ed. Upper Saddle River, NJ: Prentice Hall, 1996.

11 Mitra, S. K. *Digital Signal Processing: A Computer-Based Approach*. 2nd Ed. New York, NY: McGraw-Hill, 2001.

12 Proakis, J. G. *Digital Communications*. 4th Ed. New York, NY: McGraw-Hill, 2001.

13 Texas Instruments. *Code Composer User's Guide*. SPRU296, 1999.

14 Texas Instruments. *TMS320C55x DSP Programmer's Guide*. SPRU376A, 2001.

2

Fundamentals of Digital Signal Processing

This chapter introduces the fundamentals of DSP, including the concepts and principles of signals and systems in time and frequency domains. Rigorous mathematical derivations and descriptions are available in the DSP textbooks listed in the references and are not presented in this chapter. Section 2.1 introduces the notation of digital signals and basic operations. Section 2.2 discusses the concept of the z-transform representation of digital signals and systems. Section 2.3 covers linear, time-invariant systems including FIR and infinite-impulse response (IIR) filters. Section 2.4 introduces frequency analysis, including the discrete Fourier transform (DFT) and its efficient computation algorithm, the FFT. Section 2.5 discusses the basic concepts and processing of random signals. Readers who are already familiar with these topics may skip these sections. Finally, in Section 2.6 we provide a complete DSP software development process involving algorithm selection, MATLAB analysis and realization, C implementation, verification using the Signal-Processing Tool (SPTool), and testing using CCS.

2.1 DIGITAL SIGNALS AND OPERATIONS

As discussed in Chapter 1, most signal-processing tasks are performed in the digital domain due to the increase in speed and flexibility of digital systems. A DSP system processes a digital signal, which appears to the hardware as a sequence of numbers. In order for the DSP system to manipulate an analog signal, the signal must have been sampled and encoded in binary form. Thus, a digital signal (or sequence) is defined as a function of time index n, which corresponds to time nT if the signal is sampled from an analog signal $x(t)$ with a sampling period of T seconds. The sampling period can be expressed as

$$T = \frac{1}{f_s},\tag{2.1.1}$$

where f_s is the sampling rate in Hz (or cycles per second).

It is common practice to normalize the sampling period T to 1 and drop it from the equation. Thus, we can write

$$x(n) = x(nT), \quad -\infty \le n \le \infty,\tag{2.1.2}$$

where n is the time index. Therefore, a digital signal is defined as a function that maps the set of integers into the real line for real-valued signals or into the complex plane for complex-valued signals. In representing realizable sequences, it is assumed that any element of a sequence whose time index is less than zero has a value of zero. This type of signal is called a causal signal and is described as $x(n) = 0$ for $n < 0$.

2.1.1 Basic Signals

There are many commonly used digital signals for DSP systems. We introduce some deterministic signals in this section and study random signals in Section 2.5.

The digital-unit impulse function is defined by

$$\delta(n) = \begin{cases} 1, & n = 0 \\ 0, & n \neq 0 \end{cases}.\tag{2.1.3}$$

Equally spaced (T seconds) train-of-unit impulses constitute an ideal sampling function. Multiplication of an analog signal by the sampling function yields an ideal-sampled discrete-time signal.

The unit-step function is defined by

$$u(n) = \begin{cases} 1, & n \ge 0 \\ 0, & n < 0 \end{cases}.\tag{2.1.4}$$

The unit-step function is very useful in distinguishing causal signals from general signals, $x(n)$, described in Eq. (2.1.2). For example, a causal signal may be described as $x(n)u(n)$.

The rectangular window function is defined as

$$w(n) = u(n) - u(n - N) = \begin{cases} 1, & 0 \le n \le N - 1 \\ 0, & \text{otherwise} \end{cases}.\tag{2.1.5}$$

The rectangular window is useful for expressing digital signals with finite length. For example, a finite-duration signal $x(n), n = 0, 1, \ldots, N - 1$ can be expressed as $x(n)w(n)$.

A digital sinusoidal signal (sinewave) can be expressed as

$$x(n) = A\sin(2\pi f_0 nT) = A\sin\left[\left(\frac{f_0}{f_s/2}\right)\pi n\right]$$

$$= A\sin[F_0\pi n] = A\sin[\omega_0 n], \tag{2.1.6}$$

where A is amplitude, f_0 is the frequency of the sinusoid in Hz, and F_0 is the normalized frequency in cycles per sample defined as

$$F_0 = \frac{f_0}{f_s/2}, \quad -1 \le F_0 \le 1. \tag{2.1.7}$$

The range of normalized frequency is $-1 \le F_0 \le 1$ since $f_0 \le f_s/2$ based on the sampling theorem given in Eq. (1.6.1). The digital frequency, ω_0, in radians per sample is defined as

$$\omega_0 = F_0\pi, \quad -\pi \le \omega_0 \le \pi. \tag{2.1.8}$$

Example 2.1

Generate, plot, and save a sinusoidal signal in the file sn.dat using the MATLAB script exmp2_1.m as follows:

```
A = 2.0;              % amplitude of sinewave
w0 = 0.2*pi;          % digital frequency defined in Eq.(2.1.8)
N = 128;              % length of file
n = [0:N-1];          % time index n
sn=A*sin(w0*n);       % generate sinewave using Eq.(2.1.6)
save sn.dat sn -ascii;% save sn in file sn.dat
plot(n,sn)            % plot the sinewave
axis([0, N-1, -inf, inf])
title('Sinewave');    % title of figure
xlabel('Time index, n'); ylabel('Amplitude');
```

A periodic signal of period N is defined as

$$x(n) = x(n + kN), \tag{2.1.9}$$

where k is an integer.

2.1.2 Basic Operations

An example of a simple DSP system is illustrated in Fig. 2.1, where $x(n)$ is an input sequence, $O[]$ is an operator, $O[x(n)]$ is the operation of input signal $x(n)$, and $y(n)$ is an output sequence. The three basic operations of digital signals are addition, multiplication, and delay. These basic building blocks can be connected to form a signal-flow diagram representation of a DSP system.

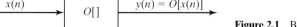

Figure 2.1   Block diagram of a DSP operation

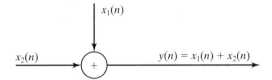

Figure 2.2   Block diagram of an adder

A sample-by-sample addition of two sequences, $x_1(n)$ and $x_2(n)$, is illustrated in Fig. 2.2 and expressed as

$$y(n) = x_1(n) + x_2(n) = x_2(n) + x_1(n), \qquad (2.1.10)$$

where $y(n)$ is the adder output. The addition operation can be easily implemented in MATLAB using the arithmetic operator $+$ if the lengths of $x_1(n)$ and $x_2(n)$ are the same.

A signal can be scaled (amplified or attenuated) by multiplying each sample with a scalar (gain), as illustrated in Fig. 2.3, and can be expressed as

$$y(n) = \alpha x(n), \qquad (2.1.11)$$

where α is the gain factor. When $\alpha > 1$, the input signal is amplified. If $\alpha < 1$, the signal is attenuated. As discussed in Chapter 1, almost all DSP processors provide a hardware multiplier for speeding up the multiplication. If the value of α is a power of 2, Eq. (2.1.11) can be implemented with the shift operation. An arithmetic operator $*$ is used to implement the multiplier in MATLAB.

A signal $x(n)$ can be shifted (delayed) in time by one sample interval (or sampling period T) as illustrated in Fig. 2.4, where the box labeled z^{-1} represents a unit delay. The operation is expressed as

$$y(n) = x(n - 1). \qquad (2.1.12)$$

A delay by M units can be implemented by cascading M delay units in a row, configured as a first-in, first-out signal (or delay) buffer. This topic is discussed further in Section 2.3.

The I/O-difference equation of a DSP system defines the relationship between the input and output signals. The I/O equation consists of mathematical expressions with addition, multiplication, and delay. For example, consider a simple DSP system described by the difference equation

$$y(n) = b_0 x(n) + b_1 x(n - 1), \qquad (2.1.13)$$

Figure 2.3   Block diagram of a multiplier

$$\xrightarrow{\quad x(n) \quad} \quad \alpha \quad \xrightarrow{\quad y(n) = \alpha x(n) \quad}$$

$$\xrightarrow{\quad x(n) \quad} \boxed{z^{-1}} \xrightarrow{\quad y(n) = x(n - 1) \quad}$$

Figure 2.4   Block diagram of a unit delay

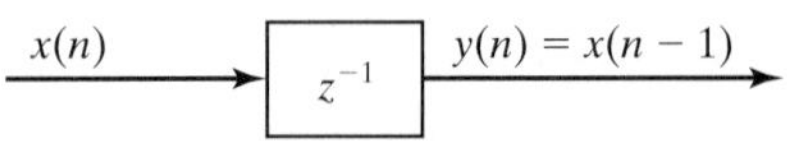

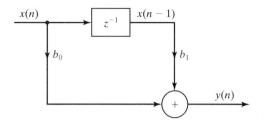

Figure 2.5   Block diagram of the simple DSP system given in Eq. (2.1.13)

where b_0 and b_1 are predetermined constants. The signal-flow diagram of the system, using the three basic building blocks is illustrated in Fig. 2.5. This diagram clearly shows that the output $y(n)$ is computed based on the I/O equation given in Eq. (2.1.13).

The energy of a digital signal is defined as

$$E_x = \sum_{n=-\infty}^{\infty} |x(n)|^2. \tag{2.1.14}$$

A periodic signal defined in Eq. (2.1.9) has infinite energy. For a periodic signal of period N, an appropriate measure of signal is power, which is defined by the average energy expressed as

$$P_x = \frac{1}{N} \sum_{n=0}^{N-1} |x(n)|^2. \tag{2.1.15}$$

An alternate method for describing the power of a signal is decibel (dB). The dB is the unit used to describe the ratio of two powers defined as

$$r = 10 \log_{10} \left[\frac{P_x}{P_y} \right] \text{dB}, \tag{2.1.16}$$

where P_y is the power of $y(n)$. If $x(n)$ is the desired signal corrupted by the noise $y(n)$, r, as defined in Eq. (2.1.16), is called the signal-to-noise ratio.

Example 2.2

Consider the real-valued sinewave of amplitude A given in Eq. (2.1.6). The power of sequence is computed as

$$P_x = \frac{1}{N} \sum_{n=0}^{N-1} x^2(n) = \frac{A^2}{N} \sum_{n=0}^{N-1} \sin^2(\omega n) = \frac{A^2}{N} \sum_{n=0}^{N-1} \frac{1}{2}[1 - \cos(2\omega n)] = \frac{A^2}{2}. \tag{2.1.17}$$

The power of the sinewave sequence generated in Example 2.1 (saved in the file sn.dat) can be calculated using the following MATLAB script (exmp2_2.m):

```
load -ascii sn.dat;    % load saved sinewave in sn.dat
N = length(sn);        % length of sequence sn
Px = sum(abs(sn).^2)/N % estimate power of sequence
```

2.2 THE *z*-TRANSFORM

The *z*-transform is a powerful technique for analyzing digital signals and systems. It is very useful for representing block diagrams and transfer functions of digital systems. The *z*-domain representation is used extensively throughout this book because it provides a notation that relates to the processing of digital signals.

2.2.1 Definitions

The *z*-transform of a digital signal $x(n)$ is defined as

$$X(z) = \sum_{n=-\infty}^{\infty} x(n)z^{-n}, \tag{2.2.1}$$

where z is a complex variable. The set of z values for which $X(z)$ exists is called the region of convergence. In general, the value z can be expressed in polar form as $z = re^{j\theta}$. This representation is used in Section 2.4 to evaluate the frequency response of a digital system.

   If the signal $x(n)$ is causal, the two-sided *z*-transform defined in Eq. (2.2.1) can be modified to a one-sided *z*-transform given as

$$X(z) = \sum_{n=0}^{\infty} x(n)z^{-n}. \tag{2.2.2}$$

Tables of commonly used *z*-transform pairs are available in most DSP textbooks.

Example 2.3

   Consider the causal signal defined as

$$x(n) = a^n u(n).$$

By using Eq. (2.2.2) and using the formulas of the infinite-geometric series

$$\sum_{n=0}^{\infty} x^n = \frac{1}{1-x}, \quad |x| < 1, \tag{2.2.3}$$

the *z*-transform of the signal is calculated as

$$X(z) = \sum_{n=0}^{\infty} a^n z^{-n} = \sum_{n=0}^{\infty} (az^{-1})^n = \frac{1}{1-az^{-1}}, \quad |az^{-1}| < 1.$$

The region of convergence is defined as $|az^{-1}| < 1$ or

$$|z| > |a|,$$

which is the exterior of a circle with radius $|a|$.

2.2.2 Properties of the z-Transform

Some important properties of the z-transform are introduced in this subsection.

If the signal $y(n)$ is a delayed version of $x(n)$ by k samples [i.e., $y(n) = x(n - k)$], the z-transform of $y(n)$ is given as

$$Y(z) = \sum_{n=-\infty}^{\infty} y(n)z^{-n} = \sum_{n=-\infty}^{\infty} x(n - k)z^{-n} = \sum_{n=-\infty}^{\infty} x(n - k)z^{-(n-k)-k}$$

$$= z^{-k} \sum_{m=-\infty}^{\infty} x(m)z^{-m} = z^{-k}X(z). \tag{2.2.4}$$

Thus, the delay of k samples in the time domain corresponds to the multiplication of z^{-k} in the z-domain. A unit delay z^{-1} is shown in Fig. 2.4. This element has the effect of delaying the sampled signal by one sampling period (i.e., T seconds in time).

If $y(n)$ is the linear convolution of two sequences $x(n)$ and $h(n)$ expressed as

$$y(n) = x(n) * h(n) = \sum_{i=-\infty}^{\infty} x(i)h(n - i) = \sum_{i=-\infty}^{\infty} h(i)x(n - i), \tag{2.2.5}$$

the z-transform of $y(n)$ is given as

$$Y(z) = X(z)H(z) = H(z)X(z). \tag{2.2.6}$$

Therefore, convolution in the time domain is equivalent to multiplication in the z-domain.

The linear convolution operation defined by Eq. (2.2.5) can be evaluated by using a graphical approach. The time sequence $h(i)$ is first time-reversed (reflected about the origin) to obtain $h(-i)$. We then shift $h(-i)$ to the right by n samples if $n > 0$ (or to the left by n samples if $n < 0$) to form the sequence $h(n - i)$. Next, we compute the products of $x(i)h(n - i)$ for those i that have nonzero overlapped between $x(i)$ and $h(n - i)$. Finally, summing all of the products $x(i)h(n - i)$ yields $y(n)$.

In general, if $h(n)$ and $x(n)$ are two sequences of length L and N, the resulting sequence $y(n)$ is of length $L + N - 1$. The MATLAB function $y = \text{conv(h, x)}$ implements the linear convolution of two sequences in vectors h and x, with the output sequence stored in vector y.

2.3 DIGITAL SYSTEMS

A digital system can be described by the I/O equation given in Eq. (2.1.13), which expresses the output in terms of input samples and system parameters. Connecting delay units, multipliers, and adders allows us to pictorially represent difference equations as the signal-flow diagram given in Fig. 2.5. An alternate technique for representing digital systems by using the z-transform is introduced in this section.

2.3.1 Linear Time-Invariant Systems

Consider the digital system defined in Eq. (2.1.13) or Fig. 2.5. When the input signal $x(n)$ is the unit-impulse function given in Eq. (2.1.3), the output can be obtained as

$$y(n) = h(n) = \begin{cases} b_0, & n = 0 \\ b_1, & n = 1 \\ 0, & \text{otherwise} \end{cases} , \qquad (2.3.1)$$

where $h(n)$ is the output (response) of a digital system when the input signal is a unit impulse. Therefore, $h(n)$ is called the impulse response of a digital system.

From Eqs. (2.2.5) and (2.3.1), we have

$$y(n) = x(n) * h(n) = \sum_{i=-\infty}^{\infty} h(i)x(n-i) = h(0)x(n) + h(1)x(n-1)$$
$$= b_0 x(n) + b_1 x(n-1), \qquad (2.3.2)$$

which is identical to the I/O equation given in Eq. (2.1.13). Therefore, a digital system can be described by either the I/O equation or the impulse response of the system.

In the mathematical description of the system given in Eq. (2.2.5), it is assumed that the impulse response of the system includes values that occur before any applied input sequences. It is clear that no physical system can produce an output in response to an input that has not yet been applied. A causal system does not provide a response prior to the application of input and is defined as

$$h(n) = 0, \quad n < 0. \qquad (2.3.3)$$

Taking the z-transform of $h(n)$ given in Eq. (2.3.1), we have

$$H(z) = b_0 + b_1 z^{-1}. \qquad (2.3.4)$$

If we take the z-transform of both sides of Eq. (2.1.13), we obtain

$$Y(z) = b_0 X(z) + b_1 z^{-1} X(z) = (b_0 + b_1 z^{-1}) X(z). \qquad (2.3.5)$$

Equations (2.3.4) and (2.3.5) show that

$$H(z) = b_0 + b_1 z^{-1} = \frac{Y(z)}{X(z)}, \qquad (2.3.6)$$

which is called the transfer function of the digital system. The transfer function of a given system is defined as the z-transform of the system's impulse response or as the ratio of the system's output and input in z-domain. The transfer function $H(z)$ describes how the system operates on the input signal $x(n)$ to produce the output signal $y(n)$.

Consider the digital system defined by Eq. (2.1.13) or (2.3.6). If the input signal consists of two components expressed as

$$x(n) = a_1 x_1(n) + a_2 x_2(n), \qquad (2.3.7)$$

the output $y(n)$ can be computed as

$$
\begin{aligned}
y(n) &= b_0 x(n) + b_1 x(n - 1) \\
&= b_0[a_1 x_1(n) + a_2 x_2(n)] + b_1[a_1 x_1(n - 1) + a_2 x_2(n - 1)] \\
&= a_1[b_0 x_1(n) + b_1 x_1(n - 1)] + a_2[b_0 x_2(n) + b_1 x_2(n - 1)] \\
&= a_1 y_1(n) + a_2 y_2(n),
\end{aligned} \qquad (2.3.8)
$$

where $y_1(n) = b_0 x_1(n) + b_1 x_1(n - 1)$ and $y_2(n) = b_0 x_2(n) + b_1 x_2(n - 1)$ are the outputs of the system due to the inputs $x_1(n)$ and $x_2(n)$, respectively. The system that satisfies the superposition property given in Eq. (2.3.8) is called a linear system.

The system is time-invariant if $y(n) = O[x(n)]$, and the shifted version of the input gives the response expressed as

$$
y(n - k) = O[x(n - k)]. \qquad (2.3.9)
$$

A system that is both linear and time invariant is called a linear time-invariant system. This system can be represented in both the time domain and z-domain given in Eqs. (2.2.5) and (2.2.6) as illustrated in Fig. 2.6, where $h(n)$ is the impulse response of the system.

The combination of linear operations can be considered as digital filters. Filtering is the most commonly used signal processing technique. A filter allows certain frequency components of the input signal to pass unchanged to the output while blocking other frequency components. Filters are usually used to remove or attenuate undesired signal components while enhancing the desired portions of the signals. A digital filter is the implementation of I/O equations using software on general-purpose computers, special-purpose hardware such as ASIC chips, or software on DSP processors. Digital filtering can be performed in either the time domain or the frequency domain, which is discussed in Chapter 8. The operation of a digital filter in the time domain is specified completely by the I/O equation, the transfer function, or the signal-flow diagram. There are two classes of digital filters based on the length of the impulse response: FIR and IIR.

2.3.2 Finite-Impulse Response Filters

The system described by Eq. (2.1.13) or (2.3.6) has an FIR of length 2. This system can be generalized to a system with an FIR of length L (i.e., $\{h(i) = b_i,$

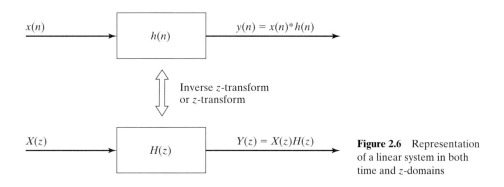

Figure 2.6   Representation of a linear system in both time and z-domains

$i = 0, 1, \ldots, L - 1\}$). Such a filter is called an FIR filter since its response to an impulse input becomes zero after a finite number of L output samples. From Eq. (2.2.5), FIR filters of length L can be described by the I/O equation expressed as

$$y(n) = \sum_{i=0}^{L-1} h(i)x(n - i) = \sum_{i=0}^{L-1} b_i x(n - i)$$

$$= b_0 x(n) + b_1 x(n - 1) + b_2 x(n - 2) + \cdots + b_{L-1} x(n - L + 1), \quad (2.3.10)$$

where the filter coefficients (taps or weights) $\{b_i, i = 0, 1, \ldots, L - 1\}$ are the same as the impulse response of the filter.

By taking the z-transform of both sides of Eq. (2.3.10) and rearranging the terms, we obtain the transfer function

$$H(z) = b_0 + b_1 z^{-1} + b_2 z^{-2} + \cdots + b_{L-1} z^{-(L-1)} = \sum_{i=0}^{L-1} b_i z^{-i}. \quad (2.3.11)$$

By setting $H(z) = 0$ in Eq. (2.3.11), we obtain $(L - 1)$ zeros. Therefore, the FIR filter of length L has order $L - 1$. In practical application, an FIR filter is implemented using the direct form illustrated in Fig. 2.7. The FIR filter requires $2L$ memory locations for storing L input samples and L filter coefficients. The signal buffer $\{x(n), x(n - 1), x(n - 2), \ldots, x(n - L + 1)\}$ is also called a delay buffer or a tapped delay line, which is implemented as a first-in, first-out buffer in memory. The MATLAB function y = filter (b, 1, x) implements the FIR filtering, where vector b contains the filter coefficients $\{b_i\}$, and vectors x and y contain input and output signals.

Consider a simple moving-average filter expressed as

$$y(n) = \frac{1}{L} \sum_{i=0}^{L-1} x(n - i) = \sum_{i=0}^{L-1} \frac{1}{L} x(n - i), \quad (2.3.12)$$

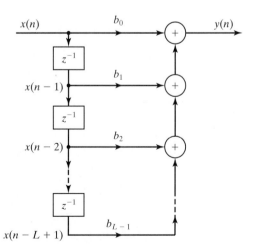

Figure 2.7　Signal-flow diagram of an FIR filter

which is an average of L samples of data in the signal buffer. Compared with Eq. (2.3.10), this is the FIR filter with the coefficients $\{b_i = 1/L, i = 0, 1, \ldots, L - 1\}$.

Example 2.4

Consider the noisy sinewave in file `xn.dat` generated by `xngen.m`, which is listed as follows:

```
N = 64;                         % file length
n = [0:N-1];                    % time index
f = 500;                        % frequency = 500 Hz
fs = 8000;                      % sampling rate = 8000 Hz
omega = 2*pi*f/fs;              % frequency is 0.125 pi
x1n = 1200*sin(omega*n);        % sinewave amplitude A = 1200
x2n = (rand(1,N)-0.5).*400;     % noise amplitude: -200 - 200
xn = round(x1n+x2n);            % rounding to near integer
fid = fopen('xn.dat','w');      % save signal to file xn.dat
fprintf(fid,'%4.0f\n',xn);      % in integer format
fclose(fid);
```

Note that we save the data in integer format, which can be used by CCS, as discussed in Section 2.6.4.

This noisy sinewave may be smoothed (enhanced) by the L-point moving-average filter given in Eq. (2.3.12). The following MATLAB script (`exmp2_4.m`) smoothes the noisy signals and plots both the input and output signals, as shown in Fig. 2.8:

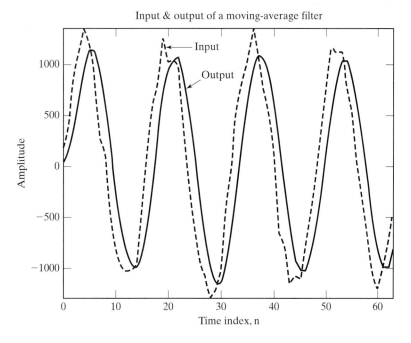

Figure 2.8   Input and output signals of a moving-average filter, where $L = 4$

```
L = 4;                  % moving-average filter of length 4
load -ascii xn.dat; % load noisy sinewave into xn
N = length(xn);     % length of xn vector
n = [0:N-1];        % time index
b = ones(L,1)/L;    % initialize coefficient vector
yn = filter(b,1,xn);% FIR filtering
plot(n,xn,'--b',n,yn,'-r') % plot both input & output
title('Input and output of a moving-average filter');
xlabel('Time index, n'); ylabel('Amplitude');
axis([0 N-1 -inf inf])
```

Example 2.5

The following C code `mov_avg.c` implements the moving-average filter given in Eq. (2.3.12) to enhance the noisy-sinewave data file `xn.dat` used in Example 2.4:

```c
#include <stdio.h>
#include <stdlib.h>
#include <math.h>
const float bnCoef[4] = {0.25, 0.25, 0.25, 0.25};
void main()
{
  void datamov();         // function to update signal vector
  float fir();            // function to perform FIR filtering
/*    Define variables and arrays    */
  int i;                  // array index
  int order = 4;          // length of moving-average filter
  int input;              // input data from integer file
  float xn = 0;           // x(n), input from xn.dat file
  float yn = 0.0;         // y(n), output from FIR filter
  float xnBuf[4];         // signal buffer for FIR filter
/*    Declare file pointers, open files, clear array    */
  FILE *xn_in;                // file pointer of x(n)
  FILE *yn_out;               // file pointer of y(n)
  xn_in = fopen("xn.dat","r"); // open file for input x(n)
  yn_out = fopen("yn.dat","w");// open file for output y(n)
  for (i=0; i<order; i++)
    xnBuf[i] = 0.;          // clear signal buffer
/*    Start of main program    */
  while( (fscanf(xn_in,"%d",&input)) != EOF)
  {                          // read x(n) from data file
    xn = (float)input;       // convert to floating-point
    datamov(xnBuf, order, xn);// refresh signal buffer
    yn = fir(xnBuf, bnCoef, order);// FIR filtering
    fprintf(yn_out,"%d\n",(int)(yn+0.5));
  }        // round to integer and write to output file
  fcloseall();               // close all opened files
}
```

Note that the C functions `datamov.c` and `fir.c` are discussed in Section 6.1.6.

The finite length of the impulse response guarantees that the FIR filters are stable. In addition, a perfect linear-phase response can be easily designed with an FIR filter, allowing a signal to be processed without phase distortion. The disadvantage of

FIR filters is the computational complexity because they may require a higher-order filter. However, as discussed in Chapter 1, the architecture of DSP processors shown in Fig. 1.3 is optimized to implement FIR filters. There are a number of techniques for designing FIR-filter coefficients for given specifications. FIR-filter design methods and tools are discussed in Chapter 6.

2.3.3 Infinite-Impulse Response Filters

If the impulse response of a filter is not a finite-length sequence, the filter is called an IIR filter. Similar to Eq. (2.3.11), the transfer function of the IIR filter is expressed as

$$H(z) = \frac{b_0 + b_1 z^{-1} + b_2 z^{-2} + \cdots + b_{L-1} z^{-(L-1)}}{1 + a_1 z^{-1} + a_2 z^{-2} + \cdots + a_M z^{-M}} = \frac{\sum\limits_{i=0}^{L-1} b_i z^{-i}}{1 + \sum\limits_{m=1}^{M} a_m z^{-m}}, \quad (2.3.13)$$

where the coefficient sets $\{b_i\}$ and $\{a_m\}$ are constants that determine the filter's response. Note that if all of the coefficients a_m are equal to zero, the transfer function $H(z)$ defined in Eq. (2.3.13) is identical to Eq. (2.3.11), an FIR filter.

The time-domain output of $H(z)$ defined in Eq. (2.3.13) can be expressed as

$$y(n) = \sum_{i=0}^{L-1} b_i x(n-i) + \sum_{m=1}^{M} -a_m y(n-m). \quad (2.3.14)$$

The signal-flow diagram given in Fig. 2.9 illustrates this I/O equation. No simple relationship like that in FIR filters exists between the coefficients of the IIR filter and the impulse response $h(n)$. It is important to note that there is a sign change of a_m in the transfer function $H(z)$ given in Eq. (2.3.13) and the I/O difference equation described in Eq. (2.3.14). The signs used in the signal-flow diagram shown in

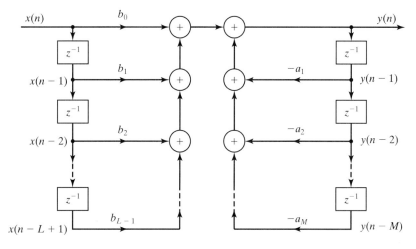

Figure 2.9   Signal-flow diagram of an IIR filter

Fig. 2.9 are identical to Eq. (2.3.14). This direct-form realization of an IIR filter can be treated as two FIR filters and, thus, can be implemented in C using the `fir.c` function used in the previous section, as follows:

```
y1 = fir(xnBuf, biCoef, L);
y2 = fir(ynBuf, amCoef, M);
y  = y1 + y2;
```

The block diagram shown in Fig. 2.9 is called a direct-form I realization. It requires two signal buffers, $\{x(n), x(n-1), x(n-2), \ldots, x(n-L+1)\}$ and $\{y(n), y(n-1), y(n-2), \ldots, y(n-M)\}$. These two buffers can be combined into one using a direct-form II realization. In addition, the simple direct-form implementation of an IIR filter is not used in practical applications due to severe response-sensitivity problems resulting from coefficient quantization, especially as the order of the filter increases. To reduce this effect, a high-order IIR-filter transfer function is factored into second-order sections. These sections are connected in cascade or parallel to form an overall filter.

Factoring the numerator and denominator polynomials of $H(z)$ given in Eq. (2.3.13), we obtain

$$H(z) = \frac{b_0(z - z_1)\ldots(z - z_i)\ldots(z - z_{L-1})}{(z - p_1)\ldots(z - p_m)\ldots(z - p_M)}, \tag{2.3.15}$$

where z_i and p_m denote the zero and pole of $H(z)$, respectively. The system is bounded-input, bounded-output stable if

$$\sum_{n=-\infty}^{\infty} |h(n)| < \infty. \tag{2.3.16}$$

For a causal system, the system is stable if and only if the transfer function has all of its poles inside the unit circle. That is,

$$|p_m| < 1, \quad m = 1, 2, \ldots, M. \tag{2.3.17}$$

In general, IIR filters require fewer coefficients to approximate a desired frequency response than FIR filters. The primary advantage of IIR filters is that sharp frequency cutoff characteristics are achievable with a relatively low-order filter, which saves processing time and/or hardware complexity. However, IIR filters are more difficult to design and implement for practical applications. Stability, finite-precision effects, and nonlinear phases must be considered in IIR-filter designs. The most commonly used IIR filter-design method is the bilinear transform, which converts the existing analog filter from s-domain (Laplace transform) into z-domain. The issues of filter design, IIR filters in a direct-form II realization, filter-design tools, quantization effects, and implementation of IIR filters based on a cascade or parallel connection of second-order sections is discussed further in Chapter 7.

If we take the z-transform of both sides of Eq. (2.3.12) and rearrange the terms, we obtain

$$H(z) = \frac{Y(z)}{X(z)} = \frac{1}{L}\sum_{i=0}^{L-1} z^{-1} = \frac{1}{L}\frac{1 - z^{-L}}{1 - z^{-1}}. \tag{2.3.18}$$

Note that we use the geometric series

$$\sum_{n=0}^{L-1} x^n = \frac{1 - x^L}{1 - x}, \quad x \neq 1 \tag{2.3.19}$$

in the derivation of Eq. (2.3.18). Equation (2.3.18) can be further expressed as

$$Y(z)[1 - z^{-1}] = \frac{1}{L} X(z)[1 - z^{-L}]. \tag{2.3.20}$$

By taking the inverse z-transform of this equation and rearranging the terms, we have

$$y(n) = y(n - 1) + \frac{1}{L}[x(n) - x(n - L)]. \tag{2.3.21}$$

Compared with Eq. (2.3.12), the moving-average filter can be realized by the recursive equation given in Eq. (2.3.21) with the saving of $(L - 1)$ add operations. However, we need $(L + 1)$ memory locations for the signal buffer instead of the L locations required by Eq. (2.3.12).

Consider the moving-average filter given in Eq. (2.3.18). The system transfer function can be rewritten as

$$H(z) = \frac{1}{L} \frac{z^L - 1}{z^{L-1}(z - 1)}. \tag{2.3.22}$$

Therefore, there is a pole at $z = 1, (L - 1)$ poles at $z = 0$, and L zeros at $z_i = e^{j\frac{2\pi}{L}i}, i = 0, 1, \ldots, L - 1$, which are located on the unit circle and separated by the angle $2\pi/L$. Note that the pole at $z = 1$ is canceled by the zero at $z = 1$. Thus, the moving-average filter is still an FIR filter even though its transfer function defined in Eq. (2.3.18) is in rational form and its I/O equation given in Eq. (2.3.21) is in recursive form.

Example 2.6

To calculate zeros and poles of a given transfer function $H(z)$, we can use the MATLAB function `roots` on both the numerator and denominator polynomials. The function `zplane(b, a)` computes and plots both zeros and poles on the z-plane for the given numerator vector b and denominator vector a. The following script (`exmp2_6.m`) shows the poles and zeros of the moving-average filter defined in Eq. (2.3.21), where $L = 8$, as shown in Fig. 2.10:

```
b = [1/8 0 0 0 0 0 0 0 -1/8];   % numerator vector
a = [1 -1];                      % denominator vector
zplane(b, a)                     % shows poles & zeros
```

As shown in Fig. 2.10, the pole at $z = 1$ was canceled by the zero at that same location. Therefore, the transfer function given in Eq. (2.3.18) and the I/O equation given in Eq. (2.3.21) exhibit an IIR form, but the filter is an FIR filter.

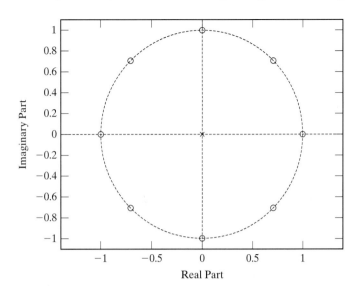

Figure 2.10 Poles and zeros of a moving-average filter, where $L = 8$

Consider the I/O equation of the moving-average filter given in Eq. (2.3.21). The system needs $L + 1$ memory locations for storing $\{x(n), x(n - 1), \ldots, x(n - L)\}$. To reduce memory requirements, we assume $x(n - L) \cong y(n - 1)$, and Eq. (2.3.21) can be simplified to

$$y(n) \cong \left(1 - \frac{1}{L}\right)y(n - 1) + \frac{1}{L}x(n) = (1 - \alpha)y(n - 1) + \alpha x(n), \quad (2.3.23)$$

where $\alpha = 1/L$. The transfer function of this simplified filter is given by

$$H(z) = \frac{\alpha}{1 - (1 - \alpha)z^{-1}}. \quad (2.3.24)$$

This is a first-order IIR filter with a pole at $z = (1 - \alpha)$. Since $1 - \alpha = (L - 1)/L < 1$, the system is guaranteed to be stable. When L is larger, the length of the moving window is longer, and the filter provides better averaging effects. The implementation of this simple IIR filter using MATLAB, C, and TMS320C5000 assembly programs for experiments is given in Section 2.6.

2.4 FREQUENCY ANALYSIS

In Section 2.2, we studied z-domain representations of digital signals and systems. We showed that a complicated convolution in the time domain could be simplified to the multiplication in the frequency domain. Similarly, digital signals and systems can be easily represented and analyzed in the frequency domain. In this section, we use MATLAB extensively to demonstrate various frequency concepts and to analyze digital systems. We also introduce the discrete-time Fourier transform (DTFT) and the DFT of digital signals.

2.4.1 Discrete-Time Fourier Transform

The DTFT of infinite-length signal $x(n)$ is defined as

$$X(\omega) = \sum_{n=-\infty}^{\infty} x(n)e^{-j\omega n}. \tag{2.4.1}$$

The inverse DTFT is defined as

$$x(n) = \frac{1}{2\pi} \int_{-\pi}^{\pi} X(\omega)e^{j\omega n}\, d\omega. \tag{2.4.2}$$

The DTFT $X(\omega)$ is periodic in ω with period 2π. That is,

$$X(\omega) = X(\omega + 2k\pi). \tag{2.4.3}$$

Therefore, we only need one period of $X(\omega), 0 \le \omega \le 2\pi$, or $-\pi \le \omega \le \pi$ for analyzing the digital signals.

Compare Eq. (2.4.1) to the equation for the z-transform given in Eq. (2.2.1). They are equal if z is set as

$$z = e^{j\omega}. \tag{2.4.4}$$

Thus, evaluating the z-transform on the unit circle $|z| = 1$ in the complex z-plane is equivalent to the frequency-domain representation of the sequence.

Similarly, the transfer function $H(z)$ can be evaluated on the unit circle to yield the frequency-domain representation of the system. This equation can be written as

$$H(z)|_{z=e^{j\omega}} = H(\omega) = |H(\omega)|e^{j\phi(\omega)}, \tag{2.4.5}$$

where $H(\omega)$ is called the frequency response of the system $H(z)$. In Eq. (2.4.5), $|H(\omega)|$ is the magnitude (amplitude) response and $\phi(\omega)$ is the phase response. The value $|H(\omega_0)|$ is called the system gain at a given frequency ω_0, which is one of the properties of the z-transform that makes it very useful for analyzing digital signals and systems.

Consider the transfer function of the moving-average filter given in Eq. (2.3.18). From Eq. (2.4.5), the frequency response can be expressed as

$$H(\omega) = \frac{1}{L}\left[\frac{1 - e^{-jL\omega}}{1 - e^{-j\omega}}\right]. \tag{2.4.6}$$

Because

$$e^{j\omega/2}e^{-j\omega/2} = e^{jL\omega/2}e^{-jL\omega/2} = 1$$

and

$$\sin(\omega) = \frac{1}{2j}(e^{j\omega} - e^{-j\omega}),$$

Eq. (2.4.6) becomes

$$H(\omega) = \frac{1}{L}\left[\frac{e^{jL\omega/2} \cdot e^{-jL\omega/2} - e^{-jL\omega}}{e^{j\omega/2} \cdot e^{-j\omega/2} - e^{-j\omega}}\right] = \frac{1}{L}\left[\frac{e^{jL\omega/2} - e^{-jL\omega/2}}{e^{j\omega/2} - e^{-j\omega/2}}\right]\frac{e^{-jL\omega/2}}{e^{-j\omega/2}}$$

$$= \frac{1}{L}\left[\frac{\sin(L\omega/2)}{\sin(\omega/2)}\right]e^{-j(L-1)\omega/2}.$$

Since $\left|e^{-j(L-1)\omega/2}\right| = 1$, the magnitude response is given by

$$|H(\omega)| = \frac{1}{L}\left|\frac{\sin(L\omega/2)}{\sin(\omega/2)}\right|, \qquad (2.4.7)$$

and the phase response is

$$\phi(\omega) = \begin{cases} -\dfrac{(L-1)\omega}{2}, & |H(\omega)| \geq 0 \\[2mm] -\dfrac{(L-1)\omega}{2} \pm \pi, & |H(\omega)| < 0 \end{cases}. \qquad (2.4.8)$$

Example 2.7

The magnitude and phase responses of the system can be computed and displayed by using the MATLAB function `freqz`. The following script (`exmp2_7.m`) plots the responses of the moving-average filter given in Eq. (2.3.18) and is shown in Fig. 2.11:

```
b = [1/8 0 0 0 0 0 0 0 -1/8];% numerator vector
a = [1 -1];  % denominator vector
freqz(b, a)  % plot magnitude and phase responses
```

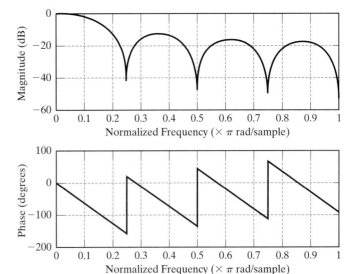

Figure 2.11 Magnitude and phase responses of a moving-average filter, where $L = 8$

The magnitude response displayed in Fig. 2.11 shows that dips occur at radi-an frequencies $\pi/4, \pi/2, 3\pi/4$, and π (or at normalized frequencies 0.25, 0.5, 0.75, and 1), which is the same as the zeros shown in Fig. 2.10. Equation (2.4.8) and the phase response of Fig. 2.11 imply that $\phi(\omega)$ is a piecewise linear function of ω. The time (group) delay function is defined as

$$T_d(\omega) = -\frac{d\phi(\omega)}{d\omega} = \frac{L-1}{2}, \tag{2.4.9}$$

which is independent of ω. The systems (or filters) that have constant $T_d(\omega)$ are called linear-phase systems. The constant time-delay functions cause all sinusoidal components in the input to be delayed by the same amount, thus avoiding phase dis-tortion. The group delay can be calculated and displayed by using the MATLAB function `grpdelay(b,a)`. The result is shown in Fig. 2.12, where a constant group delay of 3.5 samples is observed and can be obtained easily from Eq. (2.4.9) for $L = 8$.

2.4.2 Discrete Fourier Transform

The z-transform and DTFT are defined for infinite-length sequences and are func-tions of the continuous frequency variables z and ω, respectively. These characteris-tics are useful for analyzing digital signals and systems in theory. However, they are difficult to implement from a numerical-computation viewpoint. In this section, we introduce a numerical-computable transform, the DFT. The DFT is a basic opera-tion used in many different DSP applications to transform a sequence of signal sam-ples from the time domain into the frequency domain so that spectral information about the signal can be known explicitly.

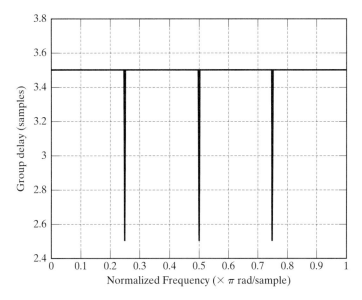

Figure 2.12 Group delay of a moving-average filter, where $L = 8$

If the digital signal $x(n)$ is a finite-duration sequence $\{x(0), x(1), \ldots,$ $x(N-1)\}$ of length N, its DTFT can be modified to DFT, which is expressed as

$$X(k) = X(\omega_k)|_{\omega_k = 2\pi k/N}$$

$$= \sum_{n=0}^{N-1} x(n) e^{-j\left(\frac{2\pi}{N}\right)kn}, \quad k = 0, 1, \ldots, N-1, \tag{2.4.10}$$

where k is the frequency index and ranges from 0 to $N-1$. Usually, the signal $x(n)$ is a real-valued sequence, but the DFT coefficients $X(k)$ are complex values. It is important to note that the DFT defined in Eq. (2.4.10) assumes that the signal is a periodic signal with period N. The DFT is equivalent to evaluating (or sampling) the DTFT, $X(\omega)$ at N equally spaced frequencies $\omega_k = 2\pi k/N$, $k = 0, 1, \ldots, N-1$, thus making it computable using digital computers. The interval between the adjacent frequency samples is called the frequency resolution, which is expressed as

$$\Delta\omega = \frac{2\pi}{N}. \tag{2.4.11}$$

By defining a twiddle factor

$$W_N = e^{-j\left(\frac{2\pi}{N}\right)}, \tag{2.4.12}$$

the DFT given in Eq. (2.4.10) can be modified to

$$X(k) = \sum_{n=0}^{N-1} x(n) W_N^{kn}, \quad k = 0, 1, \ldots, N-1. \tag{2.4.13}$$

The inverse DFT can be expressed as

$$x(n) = \frac{1}{N} \sum_{k=0}^{N-1} X(k) e^{j\left(\frac{2\pi}{N}\right)kn}$$

$$= \frac{1}{N} \sum_{k=0}^{N-1} X(k) W_N^{-kn}, \quad n = 0, 1, \ldots, N-1. \tag{2.4.14}$$

Thus, the inverse DFT is the same as the DFT except for the sign of the exponent and the scaling factor $1/N$.

The DFT is a numerically computable transform that is suitable for computer implementation. However, the complexity is high for a long sequence of large length N since the number of operations is proportional to N^2. Several algorithms have been developed to efficiently compute the DFT. These algorithms, called FFTs, are discussed in Chapter 8.

2.4.3 Properties of the Discrete Fourier Transform

The DFT is often used as an analysis tool for determining the spectra of digital signals. The DFT can be broken into magnitude and phase components as follows:

$$X(k) = \text{Re}[X(k)] + j\,\text{Im}[X(k)] = |X(k)| e^{j\phi(k)}, \tag{2.4.15}$$

where

$$|X(k)| = \sqrt{\{\text{Re}[X(k)]\}^2 + \{\text{Im}[X(k)]\}^2} \qquad (2.4.16)$$

is the magnitude spectrum and

$$\phi(k) = \tan^{-1}\left\{\frac{\text{Im}[X(k)]}{\text{Re}[X(k)]}\right\} \qquad (2.4.17)$$

is the phase spectrum. It is often preferable to measure the magnitude spectrum in dB scale, which is defined as

$$\text{Spectrum in dB} = 20\log_{10}|X(k)|. \qquad (2.4.18)$$

From Eq. (2.4.11), if the sequence $x(n)$ is a sampled signal with sampling rate f_s, the frequency index k corresponds to frequency

$$\omega_k = \frac{2\pi k}{N} = k\Delta\omega \quad \text{(radians per sample)} \qquad (2.4.19)$$

or

$$f_k = \frac{kf_s}{N} = k\Delta f \quad \text{(Hz)} \qquad (2.4.20)$$

for $k = 0, 1, \ldots, N - 1$. Thus, the frequency components can only be discriminated if they are separated by at least $\Delta f = f_s/N$ Hz. This frequency resolution is a common term used in determining the size of the FFT and sampling frequency required to achieve good frequency analysis.

If $x(n)$ is a real-valued sequence and N is an even number, we can show that

$$X(N/2 + k) = X^*(N/2 - k), \quad k = 0, 1, \ldots, N/2, \qquad (2.4.21)$$

where $X^*(k)$ denotes the complex conjugate of $X(k)$. This complex-conjugate property of the DFT demonstrates that only the first $(N/2 + 1)$ DFT coefficients of a real data sequence are independent. Thus, it is common to only plot the magnitude spectrum from $k = 0, 1, \ldots, N/2$, because the rest of the spectrum points from $k = (N/2 + 1)$ to $k = N - 1$ are symmetrical to the points from $k = (N/2 - 1)$ to $k = 1$. From Eq. (2.4.21), the even-symmetrical property applies to the magnitude $|X(k)|$ and $\text{Re}[X(k)]$, while the odd-symmetrical property occurs in phase $\phi(k)$ and $\text{Im}[X(k)]$. It is also shown that $X(0)$ and $X(N/2)$ are real valued.

A limitation of DFT is its inability to handle signals extending over all time. It is also unsuitable for analyzing nonstationary signals (such as speech) that have time-varying spectra. For such a signal, it makes more sense to divide the signal into blocks over which it can be assumed to be stationary and to estimate the spectrum of each block. The topic of spectral analysis is discussed in Chapter 8.

2.4.4 Fast Fourier Transform

The DFT given in Eq. (2.4.13) shows that N complex multiplications are needed to produce one output. In order to compute N outputs, a total of approximately N^2

complex multiplications are required. A 1,024-point DFT requires more than one million complex multiplications and additions.

The FFT is a family of very efficient algorithms for computing the DFT of a sequence. It takes advantage of the fact that many computations are repeated in the DFT due to the periodic nature of the twiddle factor given in Eq. (2.4.12). The FFT is not a new transform that is different from the DFT; instead, it is simply an efficient algorithm for computing the DFT. The ratio of computing cost in terms of number of multiplications is approximately

$$\frac{\text{FFT}}{\text{DFT}} = \frac{\log_2 N}{2N},$$

(2.4.22)

which is 10/2,048 when N is equal to 1,024. On the other hand, the FFT algorithm is more complicated to implement than the DFT because it becomes lengthy when N is not a power of 2. This restriction on N can be overcome by appending zeros at the tail of the sequence to cause N to become a power of 2 without changing the spectrum of the signal.

The FFT algorithm was introduced by Cooley and Tukey in 1965. Since then, many variations of FFT algorithms have been developed. Each FFT has a different strength and makes different tradeoffs between code complexity, memory requirements, and computational speed. These issues are discussed in Chapter 8.

MATLAB provides a function fft for computing the DFT of a vector x that consists of $\{x(0), x(1), \ldots, x(N-1)\}$. The following command

```
Xk = fft(x);
```

performs N-point DFT, where N is the length of vector x and the Xk vector contains N samples of $X(k)$, $k = 0, 1, \ldots, N-1$. If N is a power of 2, an efficient radix-2 FFT algorithm is used; otherwise, a slower mixed-radix FFT algorithm or the direct DFT is used. To avoid slow computation, we can use

```
Xk = fft(x, L);
```

where L is a power of 2. If N is less than L, the vector x is automatically padded with $(L-N)$ zeros at the tail of the sequence to make a new sequence of length L. If N is larger than L, only the first L samples are used for computing DFT.

Example 2.8

Consider the sinusoidal signal generated in Example 2.1. We can use the MATLAB function fft to compute the DFT coefficients $X(k)$. MATLAB provides the functions abs and angle to calculate the magnitude and phase spectra. The following script (exmp2_8.m) can be used to plot the magnitude spectrum of the sinewave, as shown in Fig. 2.13:

```
load -ascii sn.dat;      % load sinewave from file
N = length(sn);          % length of data file
Xk = fft(sn);            % compute DFT coefficients
absXk=20*log10(abs(Xk)); % magnitude spectrum in dB
```

```
phaXk = angle(Xk);          % compute phase spectrum
plot(absXk);                % plot magnitude spectrum
axis([0 N/2 -inf inf])      % from 0 to N/2
title('Magnitude spectrum of the sinewave');
xlabel('Frequency index k');
ylabel('Magnitude in dB');
```

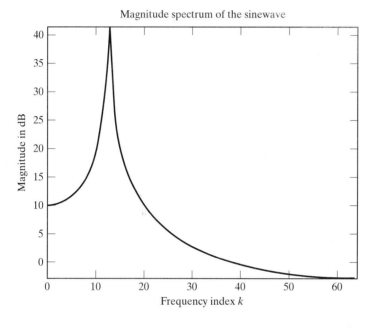

Figure 2.13 Magnitude spectrum of the sinewave generated in Example 2.1

The FFT function in C (`fft.c`) is available on the companion CD and website. The calling sequence is

```
void fft(x,n,SWITCH)
```

The argument `x` is an input complex-data vector, which also holds the output $X(k)$ since the algorithm uses in-place computation, `n` is length of the FFT, and `SWITCH` = `FFT` for computing the FFT (or `SWITCH` = `IFFT` for the inverse FFT).

2.5 RANDOM-SIGNAL PROCESSING

The digital signals introduced in Section 2.1 can be described exactly using mathematical expressions. Since the values of this type of signal are known precisely, they are called deterministic signals. In many practical applications, some signals are difficult to describe using mathematical equations, and their values cannot be predicted exactly. These signals, called random signals, include speech, music, and circuit noise. In addition, the error signal generated by forming the difference between the

ideal-sampled discrete-time signal (infinite precision) and its quantized digital signal (finite precision) generated by practical ADC usually is modeled as a random
signal for analysis purposes. This section briefly introduces some concepts that are
used in the book. Detailed information on random-signal processing can be found in
reference book [5].

2.5.1 Digital Random Signals

A digital random signal can be defined as a sequence of random variables $\{x(n)\}$.
This sequence is called a random (or stochastic) sequence and is characterized by
the associated probability density function or its statistical properties. Random
numbers have some important applications in DSP, such as in simulating noise. Random numbers generated by software or digital hardware can be designed to be near-
ly uncorrelated to each other and have a uniform distribution. These numbers are
called pseudo-random numbers.

MATLAB provides two types of pseudo-random sequences. The function
`rand(1, N)` generates a sequence of length N whose elements are uniformly distributed between [0, 1]. The function `randn(1, N)` generates a Gaussian random
sequence. Most C-compiler libraries have routines such as `int rand()` for generating random numbers.

Example 2.9

A digitized speech signal stored in the file `timit1.asc` can be played and is plotted
(zoomed in 500 samples only) in Fig. 2.14 by the MATLAT script (`exmp2_9.m`) as follows:

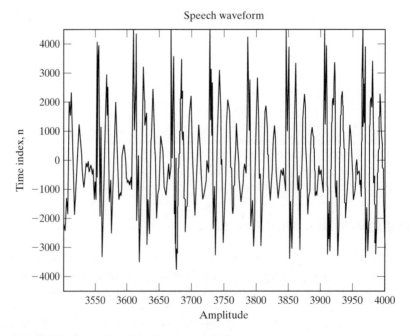

Figure 2.14   Zoomed portion of a speech waveform

```
load -ascii timit1.asc; % load speech data
soundsc(timit1);         % autoscale and play vector
plot(timit1); title('Speech waveform');
axis([3501, 4000, 24500, 4500]);
            % zoom in 500 samples starting from index 3501
xlabel('Amplitude'); ylabel('Time index, n');
```

As noted in the previous section, the DFT is inadequate for analyzing nonstationary signals. The short-time Fourier transform (STFT) breaks the signal sequence into consecutive blocks and performs the FFT of individual blocks over the entire signal. MATLAB provides a function called specgram to compute and plot a time-frequency spectrogram of the input signal. For example, the command

```
specgram(timit1,256);
```

plots the spectrogram of the speech signal timit1.asc using an FFT of length 256, as shown in Fig. 2.15. The spectrogram contains both time and frequency information. Therefore, it can pinpoint the time instance where the signal is active by looking at the color plot (e.g., a darker color indicates higher energy).

2.5.2 Time-Domain Processing

In many practical applications, a complete statistical characterization of a random variable may not be necessary if the average behavior of the random variable is

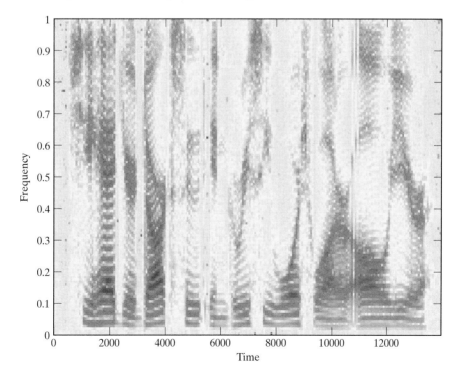

Figure 2.15   Spectrogram of the speech signal timit1.asc

known. Useful measures associated with a random signal are mean, variance, and autocorrelation functions. For stationary signals, the mean (or expected) value is independent of the time and is defined as

$$m_x = E[x(n)] = \lim_{N \to \infty} \frac{1}{2N + 1} \sum_{n=-N}^{N} x(n), \tag{2.5.1}$$

where the expectation operator $E[.]$ extracts an average value. The physical meaning of the mean is the invariant component of the signal that the random variable is supposed to model. For a finite-duration sequence of length N, the mean value may be estimated with the sample mean defined as

$$\hat{m}_x \cong \frac{1}{N} \sum_{n=0}^{N-1} x(n). \tag{2.5.2}$$

The variance of a stationary random signal is the expected value of the square of the random variable after the mean has been removed and is defined as

$$\sigma_x^2 = E[(x(n) - m_x)^2] = E[x^2(n)] - m_x^2. \tag{2.5.3}$$

Note that the expected value of the square of a random variable is equivalent to the average power. The MATLAB function `mean(x)` gives the average of the data in vector x. The function `std(x, 1)` computes the standard derivation, σ_x, based on N samples in vector x.

The statistical properties of random variables depend on their probability distribution (density) functions. The calculation of power, probability of events, and other important characteristics associated with a random variable are closely related to the probability-density function. The two most commonly encountered probability-density functions in DSP applications are the uniform-density function and the Gaussian- (or normal-) density function. The uniform-density function is defined by

$$P_x(X) = \begin{cases} \dfrac{1}{X_2 - X_1}, & X_1 \le X \le X_2 \\ 0, & \text{otherwise} \end{cases}, \tag{2.5.4}$$

where $P_x(X)$ is a probability-density function of random variable x, which takes a value in a specified range from $-\infty$ to X.

The mean of a uniformly distributed random variable x defined by Eq. (2.5.4) is given by

$$m_x = \frac{1}{X_2 - X_1} \int_{X_1}^{X_2} X \, dX = \frac{X_1 + X_2}{2}. \tag{2.5.5}$$

The variance can be computed from Eq. (2.5.3) as

$$\sigma_x^2 = E[x^2] - m_x^2 = \frac{1}{X_2 - X_1} \int_{X_1}^{X_2} X^2 \, dX - m_x^2$$

$$= \frac{X_1^2 + X_1 X_2 + X_2^2}{3} - \frac{(X_1 + X_2)^2}{4} = \frac{(X_2 - X_1)^2}{12}. \tag{2.5.6}$$

The computation of the mean and variance of the Gaussian-density functions is given in Problem B-2-26.

Example 2.10

The pseudo-random numbers generated by the MATLAB function rand are uniformly distributed between [0, 1] (i.e., $X_1 = 0$ and $X_2 = 1$). From Eq. (2.5.5), the mean is 0.5. A zero-mean sequence can be obtained by subtracting 0.5 from the generated numbers as rand (1, N) - 0.5. From Eq. (2.5.6), the variance is 1/12. To obtain a unit-variance random sequence, we should multiply the generated numbers by $\sqrt{12}$.

To generate and check random sequences with a zero mean and a unit variance, we can use the following script (exmp2_10.m):

```
N=1024;                       % length of sequence
vn=(rand(1,N)-0.5)*sqrt(12);  % zero-mean, unit variance
mx=mean(vn)                   % check if mean = 0?
variance=std(vn,1)^2          % check if variance = 1?
```

The correlation between two random variables is defined as the expected value of the product of the two and thus describes how closely they are related. The autocorrelation of a random signal is a function of a time lag k and is defined as

$$r_{xx}(k) = E[x(n + k)x(n)]. \tag{2.5.7}$$

For a finite-duration sequence of length N, autocorrelation values may be estimated by

$$\hat{r}_{xx}(k) = \frac{1}{N - k} \sum_{n=0}^{N-k-1} x(n + k)x(n), \quad 0 \le k < N. \tag{2.5.8}$$

Random signals have infinite energy and finite power expressed as

$$P_x = E[x^2(n)] = \sigma_x^2 + m_x^2 = r_{xx}(0). \tag{2.5.9}$$

In many real-time applications, signal power at time n may be estimated using the moving-average technique expressed as

$$\hat{P}_x(n) = \frac{1}{N}\sum_{i=0}^{N-1}|x(n-i)|^2$$

$$\cong \left(1 - \frac{1}{N}\right)\hat{P}_x(n-1) + \frac{1}{N}|x(n)|^2$$

$$\cong (1-\alpha)\hat{P}_x(n-1) + \alpha|x(n)|^2, \tag{2.5.10}$$

where N is the length of the moving window.

A random signal is called a white random signal if

$$E[x(n)x(k)] = E[x(n)]E[x(k)], n \neq k, \tag{2.5.11}$$

and the autocorrelation function is given by

$$r_{xx}(k) = \sigma_x^2\delta(k) + m_x^2. \tag{2.5.12}$$

A zero-mean $(m_x = 0)$ white random signal is commonly called white noise.

Example 2.11

In this example, we generate a sinusoid embedded in white noise with a signal-to-noise ratio of 10 dB. Using Eq. (2.1.16) and assuming that the variance of white noise is 1 $(P_y = 1)$, we need $P_x = 10$. By substituting this value into Eq. (2.1.17), the amplitude of the sinewave is $A = \sqrt{20}$. Thus, we can generate the noisy sinewave and compute its magnitude spectrum using the following script (exmp2_11.m):

```
N=256; A=sqrt(20); w0=0.2*pi;% define parameters
n = [0:N-1];                  % time index
sn = A*sin(w0*n);             % sine sequence
vn=(rand(1,N)-0.5)*sqrt(12);  % zero-mean, unit-variance white noise
xn = sn+vn;                   % 10 dB signal-to-noise ratio
subplot(2,1,1);               % top figure
plot(n,xn)                    % plot the waveform
title('Waveform of a noisy sinewave');
xlabel('Time index n'); ylabel('Amplitude');
axis([0 N-1 -inf inf])
Xk = fft(xn);                 % compute DFT
absXk = 20*log10(abs(Xk));    % magnitude spectrum in dB
subplot(2,1,2);               % bottom figure
plot(n,absXk)                 % plot the spectrum
title('Spectrum of a noisy sinewave');
xlabel('Frequency index k');
ylabel('Magnitude in dB');
axis([0 N/2 -inf inf])
```

The noisy sinewave and its spectrum are shown in Fig. 2.16.

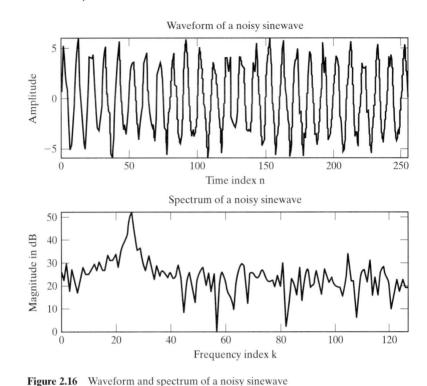

Figure 2.16   Waveform and spectrum of a noisy sinewave

Compare the spectrum of the noisy sinewave with the clean sinewave shown in Fig. 2.13. Notice that the white noise has a flat spectrum over the entire frequency range from 0 to π.

2.6 EXPERIMENTS AND PROBLEMS

In this section, we use the simple IIR filter given in Eq. (2.3.23) to show the process of DSP software development. First, the algorithm is developed and tested by using MATLAB commands and functions. When the algorithm is verified, a C program is written and tested on a computer using the C compiler on that computer. In Section 2.6.3, we use this simple example to introduce an interactive MATLAB tool, SPTool, for filter design and analysis. Finally, the C program is executed using TMS320C5000 C compilers and simulators. This C program is also modified to use probe points in CCS for file-data I/O. Some basic functions of CCS were already introduced in Chapter 1; thus, only special features related to C programs and data I/O using probe points are introduced in this section.

2.6.1 A Simple Infinite-Impulse Response Filter

As illustrated in Fig. 2.6, a digital filter performs a sequence of operations on its input signal $x(n)$ to produce the output $y(n)$. Such a filter might allow certain frequency components to pass while blocking other frequency components. For example, a lowpass filter allows low-frequency components to pass while attenuating high-frequency components.

Assume the signal $x(n)$ consists of a sinewave corrupted by random noise. Our goal is to develop and implement a digital lowpass filter to reduce noise, thus enhancing the sinusoidal component. For simulation purposes, we use the MATLAB script xngen.m used in Example 2.4 to generate $x(n)$, which is a sinewave corrupted by noise. The noisy sinewave is shown as a dotted line in Fig. 2.8. In the MATLAB program xngen.m, we round the signal values from fractional numbers to integer values and save them in a disk file xn.dat, which is used as the input file for the C programs.

The spectrum of the time-domain signal $x(n)$ can be computed using the DFT. The following MATLAB script (spectrum.m) computes the magnitude spectrum of a signal saved in a data file:

```
load -ascii xn.dat;         % load signal from data file xn.dat
N = length(xn);             % size of data in array xn
Xk=fft(xn,N);               % perform N-point DFT of the signal
magX=20*log10(abs(Xk));     % compute magnitude spectrum in dB
plot(magX);                 % plot magnitude spectrum
axis([0 N/2 -inf inf]);     % range from DC to Nyquist
title('Magnitude spectrum');
xlabel('Frequency index'), ylabel('Magnitude, dB');
```

The magnitude spectrum is shown in Fig. 2.17, where the frequency index, 32 ($N/2$), corresponds to π. It clearly shows that the sinusoidal component (a spike) is located at a low frequency and thus can be enhanced by reducing noise components in high-frequency ranges. This objective may be achieved by using a simple lowpass filter.

In this experiment, we use the simple IIR lowpass filter given in Eq. (2.3.23), where $0 < \alpha < 1$. The magnitude response of a filter with $\alpha = 0.25$ can be computed by using the following MATLAB script (mag_res.m):

```
alpha = 0.25; N = 64;       % define parameters
b = [alpha];                % coefficient vector of numerator
a = [1 -(1-alpha)];         % coefficient vector of denominator
H = freqz(b, a, N);         % frequency response of filter
magX=20*log10(abs(H));      % compute magnitude response in dB
plot(magX);                 % plot magnitude response
axis([0 N/2 -inf inf]);     % range from DC to Nyquist frequency
title('Magnitude response');
xlabel('Frequency index'), ylabel('Magnitude, dB');
```

The magnitude response is shown in Fig. 2.18. As compared with Fig. 2.17, we know the sinusoidal component is passed with little attenuation, while noise at a high-frequency

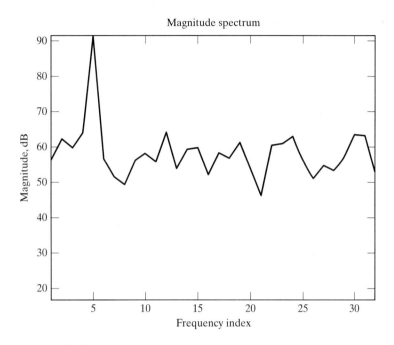

Figure 2.17   Spectrum of a noisy sinewave

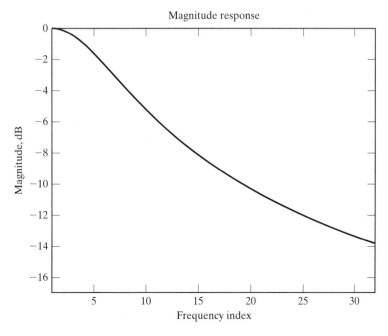

Figure 2.18   Magnitude response of a lowpass filter, where $\alpha = 0.25$

range is further reduced. Thus, this simple filter works for the purpose of reducing the random noise that corrupted the low-frequency sinusoidal signal. The design of higher-performance IIR filters is introduced in Chapter 7.

2.6.2 Software Development

To verify the function of the filter designed in the previous section, we first use the MATLAB function `filter` to implement the lowpass filter given in Eq. (2.3.23). The MATLAB script (`iir1.m`) is listed as follows:

```
alpha = 0.25;                       % define parameter
b = [alpha]; a = [ 1 -(1-alpha)];  % coefficient vectors
load -ascii xn.dat;                 % load xn.dat into vector xn
N = length(xn); n = [0:N-1];        % time index
yn = filter(b, a, xn);              % perform IIR filtering
plot(n,xn,'r-',n,yn,'b--');         % plot both input and output
axis([0 N-1 -inf inf]);
ylabel('Amplitude'); xlabel('Time index');
title('Input and output waveforms');
```

Both the input noisy sinewave and the enhanced output signal are shown in Fig. 2.19. It is clear that this simple lowpass filter is able to reduce noise in the corrupted sinewave. Note that the attenuation of the sinewave amplitude is expected by the magnitude response of the filter shown in Fig. 2.18.

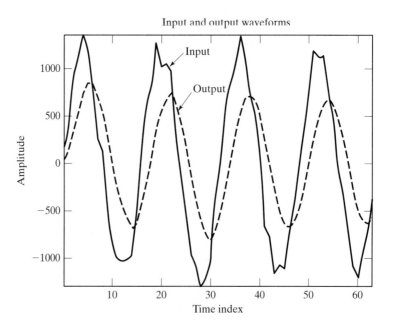

Figure 2.19   Waveforms of noisy input and enhanced output

The simple lowpass filter defined in Eq. (2.3.23) can also be implemented in C
(`iir1.c`) as follows:

```c
#include <stdio.h>
#include <stdlib.h>
#include <math.h>

void main()
{
  /*  Define variable arrays, define and clear variables  */
  int xn = 0;                 // x(n) from xn.dat file
  float yn = 0.0;             // y(n), output from IIR filter
  float alpha = 0.25;         // alpha = 0.25
  float alpha1 = 0.75;        // alpha1 = 1 - alpha = 0.75

  /*    Declare file pointers and open files    */
  FILE *xn_in;                     // file pointer of x(n)
  FILE *yn_out;                    // file pointer of y(n)
  xn_in=fopen("xn.dat","r");   // open file for input x(n)
  yn_out=fopen("yn.dat","w");  // open file for output y(n)

  /*    Start of main program    */
  while( (fscanf(xn_in,"%d",&xn)) != EOF)
  {       // read in x(n) from data file and processing it
    /* IIR filtering: y(n) = (1-alpha)*y(n-1)+alpha*x(n) */
    yn = alpha1*yn + alpha*(float)xn;
    fprintf(yn_out,"%d\n",(int)(yn+0.5));
  }
  printf("Finish");  // mark the completion of execution
  fclose(xn_in);     // close the opened input file
  fclose(yn_out);    // close the opened output file
}
```

The C statement

```c
while( (fscanf(xn_in,"%d",&xn)) != EOF)
```

mimics real-time data I/O by reading one data sample from the file `xn.dat` at each
iteration. After the filter output is generated, we write that output sample to the
data file, which is similar to real-time signal processing.

The C program can be tested with Visual C++ 6.0 on a personal computer
using the following steps:

Step 1: Create a new folder `c:\dsps\chap2` and copy `iir1.c` and `xn.dat` into
the folder.

Step 2: Launch Microsoft Visual C++ 6.0 from the **Start** menu.

Step 3: Select **File → New** from the menu, and the **New** dialog box displays.
Select **Win32 Console Application**, type `iir1` in the **Project name** field,
type `c:\dsps\chap2` in the **Location** field, and then click on **OK**.

Step 4: Select **Project** → **Add To Project** → **Files**, and the **Insert Files into Project** dialog box displays. From the **Look in** field, go to the folder `c:\dsps\chap2`, and double-click on `iir1.c`.

Step 5: Click on **FileView**, **iir1 files**, and **Source Files**, and the file `iir1.c` displays. Double-click on `iir1.c`, and the source code displays in the main window.

Step 6: Click on the **Build** button to compile and link the program. The executable file `iir1.exe` generates in the folder `c:\dsps\chap2\iir1\debug`. Move that executable file to the folder `c:\dsps\chap2`.

Step 7: Click on `iir1.exe` to execute the program. The filter output is written to the new data file `yn.dat`.

Step 8: Use the MATLAB program `plotxy.m` to plot both the input file `xn.dat` and the output file `yn.dat`. The result is identical to that shown in Fig. 2.19.

We have reverified that the C program `iir1.c` is running correctly on a personal computer. In Section 2.6.4, we show the steps in debugging and running the C program under the CCS environment.

2.6.3 Signal Processing Tool

In this section, we use an interactive MATLAB tool, SPTool, in the Signal-Processing Toolbox [10] to perform filtering and analysis. SPTool provides a rich graphic environment for signal viewing, filter design, and spectral analysis. It has a graphical user interface (GUI), and it is an interactive tool that allows the user to analyze signals, design and analyze filters, perform filtering, and analyze signal spectra. A more powerful tool called Filter Design and Analysis Tool (FDATool) provides a more comprehensive collection of features for designing and analyzing filters (especially quantized filters). FDATool is discussed in Chapters 6, 7, and 9, as well as in Appendix A.

To start SPTool, type

```
sptool
```

in the MATLAB command window. A default **SPTool** window appears, as shown in Fig. 2.20. This window is separated into three columns: (1) **Signals**, (2) **Filters**, and (3) **Spectra**. Some default signals, designed filters, and spectrum plots already have been entered into the **SPTool** window.

We can bring in other signals (with their sampling frequencies), filters, or spectra from the MATLAB workspace into the SPTool workspace using the **Import** option under the **File** menu. As shown in Fig. 2.21, the data `xn` has been selected to be imported from the workspace (or disk) into the **Signals** column. The imported data is named `sig1` in this example. If the user clicks on the **OK** button, the original **SPTool** window would be updated with this new signal `sig1`.

The user can view the signal by selecting the name of the signal in the **Signals** list and clicking on the **View** button below the list. To view more than one signal, simply Ctrl-click on the names of the signals and click on **View**.

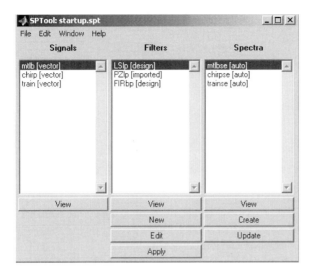

Figure 2.20   SPTool window

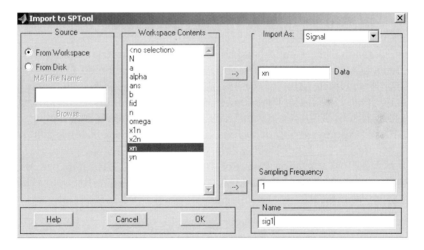

Figure 2.21   Import data into SPTool

For example, Fig. 2.22 shows the corrupted sinewave sig1. The **Signal Browser** window allows the user to display, measure, and analyze certain regions of the time-domain signals. It includes horizontal and vertical markers, markers with tracing and slope, and display peaks and valleys of the signal for measurement and comparison. It also allows the user to play the selected signal by clicking on the icon. A set of zoom buttons allows a closer look at signal features. A **Panner** window, located at the bottom of the **Signal Browser** window, displays the entire signal length and high-lights the portion of the signal that is active in the display window.

The next task is to design an IIR lowpass filter to filter out high-frequency noise components. The simplest way is to import the filter in the MATLAB workspace by

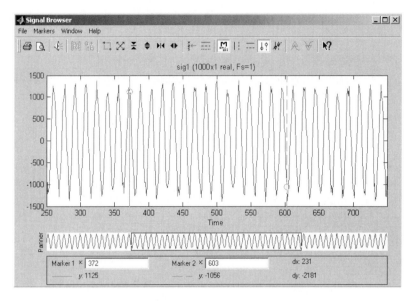

Figure 2.22   Signal Browser window

clicking on **File** → **Import** and importing the numerator (b) and denominator (a) coefficients, as shown in Fig. 2.23. The imported filter is named filt1. The user can examine the characteristics of the imported filter by clicking on the filter, filt1, and then clicking on **View**. The **Filter Viewer** window, shown in Fig. 2.24, allows the user to view the characteristics of a designed or imported filter, including the magnitude response, phase response, group delay, pole-zero plot, impulse response, and step response of the filter.

SPTool also allows the user to design FIR and IIR filters with user-defined specifications, which can be done by clicking on the **New** button below the **Filters** column

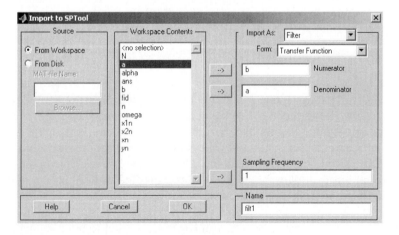

Figure 2.23   Importing the IIR filter's coefficients into SPTool

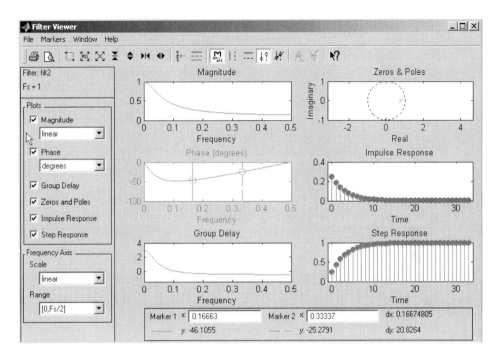

Figure 2.24   Filter Viewer window showing the imported filter, filt1

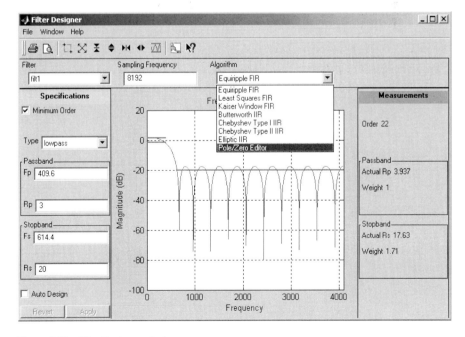

Figure 2.25   Filter Designer window

in the **SPTool** window. The **Filter Designer** window appears as shown in Fig. 2.25, which allows the user to select a filter-design algorithm, filter type, and frequency specification. Once the filter is designed, the coefficients of the filter can be exported to disk or to the workspace by clicking on **File** → **Export** and selecting the filter to be exported.

For example, we can select **Pole/Zero Editor** to specify the poles and zeros of the IIR filter. The window that appears in Fig. 2.26 allows the user to drag and drop poles and zeros in the z-plane. For the first-order IIR filter designed in the previous section, a pole is positioned at 0.75 along the real axis, and a gain of 0.25 is entered in the specification. Once the pole and zero positions are specified, we name the filter `filt2` and view the filter characteristics in the **Filter Viewer** window.

Finally, the designed filter, `filt2`, or the imported filter, `filt1`, can be selected and applied to the input signal, `sig1`, by clicking on the **Apply** button in the **SPTool** window. A new window, shown in Fig. 2.27, is displayed, which allows the user to select the structure of the filter and specify the name of the output signal as

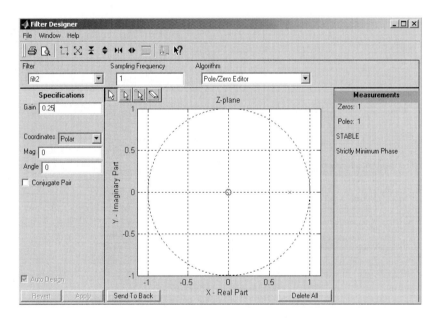

Figure 2.26   Use the **Pole/Zero Editor** to specify the poles and zeros of the filters

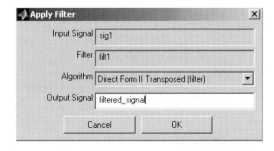

Figure 2.27   **Apply Filter** window

`filtered_signal`. The time-domain plots of the input signal and filtered output signal can be viewed by selecting both `sig1` and `filtered_signal` from the **SPTool** window and clicking on the **View** button below the **Signal** column. The display shown in Fig. 2.28 appears.

The third column in the **SPTool** window is **Spectra**, which is used to analyze and view the frequency content of the signals. The user can use different spectrum-estimation methods, FFT sizes, and windows to create a spectrum plot. (A detailed explanation on spectrum estimation is given in Chapter 8.) Figure 2.29 shows the

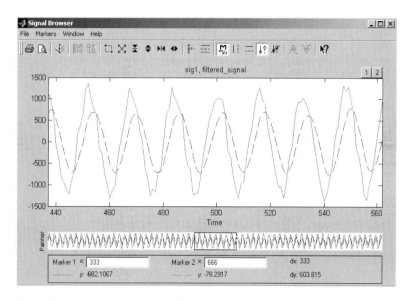

Figure 2.28   The input signal (bold line) and the filtered signal (dashed line)

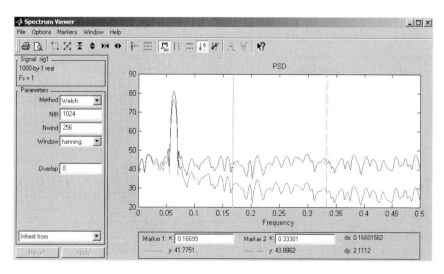

Figure 2.29   Spectra of the input signal (bold line) and the filtered signal (dashed line)

magnitude spectra for the input signal, sig1, and the output signal, filtered_signal. Finally, the SPTool session can be saved in the file chap2.spt for later reference.

2.6.4 Testing a C Program Using Code Composer Studio

In this section, we use the iir1.c program to introduce important features of CCS used to create, build, debug, and test C programs. Some basic settings and functions of CCS already were introduced in Section 1.7. In the following experiments, we run the C program using the C5402 (or C55x) simulator in CCS. The C program iir1.c can be tested without modification. However, there are some minor differences in testing C programs using these simulators.

First, we copy into the working folder the C-program source file iir1.c; the linker-command file c54x.cmd (or c55x.cmd); the assembly file vectorsc54.asm (for the C54x only), which is used to set RESET to branch to the C program's entry point c_int00; and the input-data file xn.dat.

Creating a Project

Start CCS by double-clicking on the **CCS 2 (C5000)** icon located on the desktop. Select **Project** → **New** from the menu, and the **Project Creation** dialog box displays. Type iir1c54 (or iir1c55 for the C55x) in the **Project Name** field, type c:\dsps\chap2 in the **Location** field, and then click on the **Finish** button. This operation creates a project file called iir1c54.pjt (or iir1c55.pjt).

Select **Project** → **Add Files to Project** from the menu, and the **Add Files to Project** dialog box is displayed. From the **Look in** field, browse to the folder c:\dsps\chap2. Select iir1 and then click on **Open**. Repeat this process, but this time select **Linker Command File** from the **Files of type** field, select c54x (or c55x), and click on **Open**. Repeat this process again, but this time choose **Asm Source Files** from the **Files of type** field. Click on the vectorsc54 file to select the required reset file. (Skip the last step for the C55x).

Since this project uses the real-time support library, we have to add it into the project by selecting **Project** → **Add Files to Project**. From the **Look in** field, browse to the compiler library folder c:\ti\c5400\cgtools\lib (or c:\ti\c5500\cgtools\lib for the C55x), choose **Object and Library Files** from the **Files of type** field, and then select rts.lib (or rts55.lib for the C55x). Note that the folder location containing this library may vary at different CCS installations.

Select **File** → **Open** from the menu, and the **Open** dialog box displays. From the **Look in** field, browse to the folder c:\dsps\chap2 and double-click on the file iir1.c to open it. The source code appears in the right half of the window, as shown in Fig. 2.30. Note that it is not required to manually add the include files to the project since CCS finds them automatically when it scans for dependencies as part of the build process.

Building the Project

To ensure that the executable file is located in the folder c:\dsps\chap2, select **Project** → **Build Options**. The **Build Options** dialog box appears. Click on the **Linker** tab,

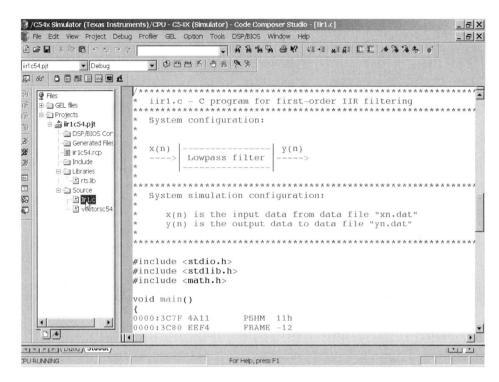

Figure 2.30   CCS window with project information and source code

delete the default directory debug in the **Output Filename** field, and then click on **OK**. Note that we can name the executable file and put it in a directory by changing the **Output Filename** field. (We use iir1c54.out or iir1c55.out for this experiment.)

Click on the **Compile** button 🗔 on the toolbar to compile the current source file iir1.c. A message showing the compiler results displays at the bottom of the CCS window. When there is no compiler error and warning, select **Project → Build** from the menu (or click on the **Rebuild All** button 🗔 on the toolbar) to build the current project. If the project contains errors, correct all of the errors and select **File → Save** from the menu to save the changes. Choose **Project → Build** from the menu again, and CCS rebuilds files that have been changed.

Note that when building the preceding project, two errors may be reported due to the fact that the processor is linked to **Visual Linker** by default. We may solve these errors by double-clicking on the red text in the **Build** window. A **New Recipe Wizard** window appears that guides the user to create the .rcp recipe file. Click on the **Next** button and use the default settings, but change the output file name to iir1c54.out (or iir1c55.out). Once the iir1c54.rcp (or iir1c55.rcp) file has been created, perform **Project → Rebuild All** again.

Debugging the Program

Select **File → Load Program** from the menu, and the **Load Program** dialog box displays. From the **Look in** field, browse to the directory c:\dsps\chap2. When

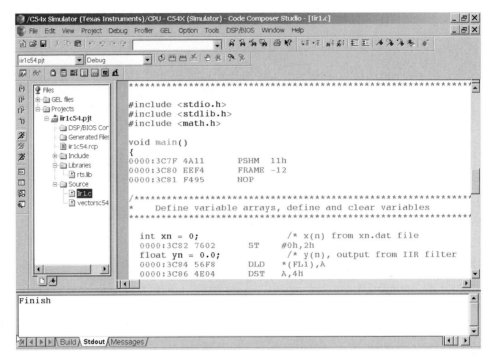

Figure 2.31   Source-level debugger window with mixed source/assembly-language display

the file `iir1c54.out` (or `iir1c55.out` for the C55x) displays, double-click on it. The user is also able to view both C and assembly language in the `Disassembly` window by selecting **View** → **Mixed Source/ASM**. Next, double-click on the C file `iir1.c` in the **Project-View** window to open it from the source folder. The source-level debugger window with mixed C-and-assembly code displays, as shown in Fig. 2.31.

Running the Program

Select **Debug** → **Reset CPU** from the menu to initialize all registers to their power-up state and halt execution of program. Make sure that the input file `xn.dat` is located in the working folder. Click on the **Run** button ⧉ in the toolbar to execute the program from the current program-counter location. Execution continues until a breakpoint is encountered, or execution of the program can be stopped by clicking on the **Halt** button ⧉ in the debug toolbar. Without setting a breakpoint, the program ends, and the word "Finish" is displayed in the **stdout** window at the bottom of window. Note that it may take time to run a floating-point C program with file-data I/O using C5000 simulators.

In order to verify that the C program is correct, we can use the MATLAB script `plotxy.m` again to plot both $x(n)$ and $y(n)$ in the disk files `xn.dat` and `yn.dat`, respectively. The result is identical to that shown in Fig. 2.19.

Preparing to Use Probe Points for Data I/O

The file-data I/O method used in the C program is slow and requires real-time operating system support if the program is running on the target hardware. In the following experiments, we introduce the usage of the file-probing mechanism available in the CCS. This tool is known as the probe point for file I/O. It allows the user to transfer input data from a file located on the host computer to a buffer on the targeted DSP system. It can also be used to transfer output data from a buffer on the targeted DSP system to a file on the host computer.

Before we can proceed with this exercise, we modify the source file iir1.c, and name the new program iir2.c. The main changes are: (1) remove the file pointers, (2) declare data buffers for the input and output, (3) include a new function called dataIO that reads the input signal and writes the output signal, and (4) modify the IIR-filter routine to suit the buffer input. The new C program iir2.c is listed as follows:

```
#include <stdio.h>
#include <stdlib.h>
#include <math.h>

/*  Global declaration  */
int in_buffer[64];
int out_buffer[64];
#define TRUE 1

static void dataIO(void); // function for data I/O

void main()
{ /*  Define variable arrays, define and clear variables  */
  int size = 64;
  float yn = 0.0;        // y(n), output from IIR filter
  float alpha = 0.25; // alpha = 0.25
  float alpha1 = 0.75;// alpha1 = 1 - alpha = 0.75
/*  Declare pointers for data buffers  */
  int *input = &in_buffer[0];
  int *output = &out_buffer[0];
/*  Start of main program  */
  while(TRUE)
  {
    dataIO();          // read in x(n) from data file
    /*  IIR filtering: y(n)=(1-alpha)*y(n-1)+alpha*x(n)  */
    while(size--){
      *output=(int)((alpha1*yn+alpha*((float)*input++))+0.5);
      yn = (float)*output++;
    }
  }
}

/*  Function for dataIO  */
static void dataIO ()
{
  return;
}
```

In addition, the data file needs to include a header of 1651 1 0 0 0 on top of the actual data. This header information is required for the data file xn2.dat to be used by CCS.

Using Probe Points

Start the new exercise by creating a new project file. Rebuild the project and load the new object file into the simulator. Double-click on the iir2.c file in the **Project View** window, and put the cursor on the line dataIO( ) below the main function. Click on the **Toggle Probe Point** icon and the **Toggle breakpoint** icon . A cyan diamond and a red dot mark this line. This point is used for data acquisition. Also, set another breakpoint on the line yn = (float)*output++.

The next step is to connect the file to be used for the file I/O. Choose **File → File I/O**, and the **File I/O** dialog box displays. From the **File Input** tab, click on **Add File**, and the **File Input** dialog box displays. From the **Look in** field, browse to the directory c:\dsps\chap2. Choose the file xn2.dat, and click on **Open**. Change the **Address** to in_buffer (to indicate the starting address for the data), and change the **Length** to 64 (the integer array size). In addition, turn on the **Wrap Around** option that treats the data file as a continuous data stream. Next, click on the **Add Probe Point** tab, and a **Break/Probe Points** dialog box appears. Highlight the line: iir2.c line 67 → No Connection. Click on the down arrow in the **Connect To** field, and select the xn2.dat file from the list. Click on **Replace**, and the proper connection

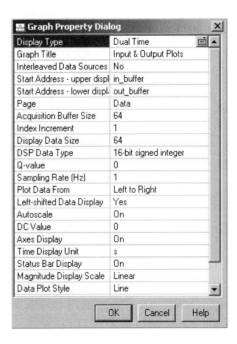

Figure 2.32   Graph Property Dialog options for displaying in_buffer and out_buffer

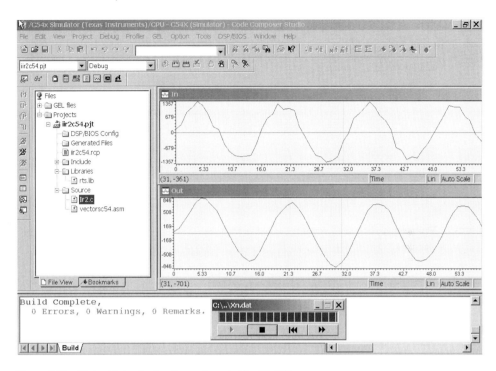

Figure 2.33   Plots obtained using the probe point for file I/O

takes place. A new control window appears to start, stop, rewind, or fast-forward within the data file.

CCS provides a means for viewing data in different graphical display modes. We can view both the input data extracted from the data file xn2.dat and the output data generated by the IIR filter. Select **View → Graph → Time/Frequency**, and the **Graph Property Dialog** box appears. Select **Dual Time** in the **Display Type** field, which plots two graphics. Make the other changes that are shown in Fig. 2.32. Click on **OK**, and empty graph windows for both in_buffer and out_buffer appear. Right-clicking on the graphical windows and unchecking the **Float In Main Window** option can automatically space these windows.

We are now ready to run the program and view the results in the graphical windows. Instead of running the program, we animate the simulation process by clicking on the **Animate** icon ![animate icon]. Instead of halting when the program reaches a breakpoint, as in the case of the **Run** command, the **Animate** command resumes execution after the breakpoint. The simulation can be halted manually by clicking on the **Halt** button ![halt icon]. The final result is shown in Fig. 2.33, where the upper plot displays the data in the in_buffer, and the lower plot displays the processed data obtained in out_buffer.

PROBLEMS

PART A

Problem A–2–1

CCS also supports the display of a magnitude spectrum. From the **Graph Property Dialog** box, select **FFT Magnitude** in the **Display Type** field. Show the spectra of input and output signals shown in Fig. 2.33.

Problem A–2–2

Compare the efficiency of the C programs `iir1.c` and `iir2.c` by using the Profiler in CCS.

Problem A–2–3

The digital signal $x(n)$ consists of two real-valued sinusoidal components expressed as

$$x(n) = A_1 \sin(\omega_1 n) + A_2 \sin(\omega_2 n) + v(n),$$

where $v(n)$ is zero-mean and unit-variance white noise. Complete the steps that follow:

(a) Compute A_1 and A_2 such that $A_1 = 2A_2$ and the signal-to-noise ratio is 10 dB.
(b) Let $\omega_1 = 0.1\pi$ and $\omega_2 = 0.2\pi$. Generate 512 points $x(n)$, using MATLAB.
(c) MATLAB provides the function `y = filter(b, a, x)` to filter the data in the input vector x with the IIR filter specified by numerator coefficient vector b and denominator coefficient vector a and then stores the filtered output in vector y. Design a moving-average filter of order $L = 10$ using Eq. (2.3.21). Filter the signal $x(n)$ generated in step (b) and plot both $x(n)$ and $y(n)$, using MATLAB.
(d) Use the MATLAB functions `fft`, `abs`, and `log10` to compute and plot the magnitude spectra of $x(n)$ and $y(n)$, using a dB scale. Plot the spectra only from frequencies 0 to π.
(e) Explain why the sinusoidal component at frequency $\omega_2 = 0.2\pi$ has been attenuated. (Hint: Use `freqz` to show the magnitude response of the filter.)
(f) Repeat steps (c) and (d), using the IIR filter given in Eq. (2.3.23) for $L = 10$. Compare the results with (c) and (d) and explain the differences.

Problem A–2–4

Redo Problem 3 using SPTool.

Problem A–2–5

Redo step (c) of Problem 3 by using a C/C++ program to implement the moving-average filter. Test the C program by using a C compiler such as Visual C/C++6.0 on the host computer.

Problem A–2–6

Redo Problem 5 by using the C55x (or C54x) C compiler in CCS.

Problem A–2–7

Given the speech file `timit1.asc` used in Example 2.9, Complete the steps that follow:

(a) Use the power estimator given in Eq. (2.5.10) to estimate the power of speech with different window lengths of 4 ms, 32 ms, and 512 ms. Plot the original waveform along with these three power estimates on the same figure.

(b) Add white noise with mean 0 and variance 20,000 into the original speech. Replay the original and corrupted speech using the MATLAB function `soundsc`.

(c) Repeat step (a) using the noisy speech.

(d) Use SPTool to design a digital filter to reduce the noise.

Problem A–2–8

Refer to Example 2.11. Replace the uniformly distributed noise with Gaussian distributed noise with a variance of 1. Plot the magnitude spectrum and compare it to the results shown in Fig. 2.16. What happens when the variance is reduced to 0.1 while the signal-to-noise ratio still is maintained at 10 dB?

Problem A–2–9

The convolution function `conv` in MATLAB can be used to perform FIR filtering. Use both the `filter` and `conv` functions to filter the input data vector $x = [2 -3\ 6\ 8 -1\ 2\ 3 -4]$ with the coefficient vector $h = [1\ 2\ 3\ 2\ 1]$. Is the output derived from the `conv` identical to that from `filter`? If not, how can the two outputs be the same?

Problem A–2–10

Deconvolution is the opposite of the process of convolution. It is used to recover the system response, $h(n)$, given the input $x(n)$ and the output $y(n)$. Use the `deconv(y,x)` function to compute the system response, given $y = [6\ 31\ 47\ 6 -47\ 1\ 41\ 8 -18 -1\ 8\ 2]$ and $x = [3\ 1\ 1\ 7\ 0\ 1\ 4\ 2]$. Using the computed system response, perform the `conv` function to verify that the preceding output y can be derived. If the preceding vectors can be considered as polynomial coefficients, what are the equivalents of the `deconv` and `conv` functions?

Problem A–2–11

Besides being used as a white-noise detector, the autocorrelation function can also be used to detect periodic signals embedded in noise. Use the generated noisy sinewave in Example 2.11 and the MATLAB autocorrelation function `xcorr` to determine the period of the sinewave, and resynthesize the sinewave using the derived period.

Problem A–2–12

Generate a 3-second sinewave of 2,000 Hz in MATLAB using a sampling frequency of 16,000 Hz. Complete the steps that follow:

(a) Use the MATLAB `soundsc` function to play the generated sinewave.

(b) Save the sinewave as a wave file with extension `.wav` by using the `wavwrite` function. The wave file can be saved as 16 bits per sample.

(c) Record the file size of the saved wave file. Examine the size of the file saved if an 8-bit per sample is used.

(d) Read the wave file back into the MATLAB workspace by using the wavread function.

(e) Plot the wave file and compare it with the original data file generated by MATLAB.

PART B

Problem B–2–13

Find the z-transform of

$$x(n) = -a^n u(-n - 1)$$

and plot the region of convergence for $0 < a < 1$.

Problem B–2–14

Find the z-transform of

$$x(n) = \begin{cases} a^n, & 0 \leq n \leq N - 1 \\ 0, & \text{otherwise} \end{cases}$$

and plot the poles and zeros of $X(z)$ for $a > 0$ and $N = 8$.

Problem B–2–15

Use the time-shifting property given in Eq. (2.2.4) to find the z-transform of the following two sequences: $x(n) = \delta(n - k)$ and $x(n) = u(n - k)$.

Problem B–2–16

Use the z-transform to compute the linear convolution of the following two sequences: $x(n) = u(n)$ and $h(n) = a^n u(n)$.

Problem B–2–17

Use the z-transform to compute the linear convolution of the following two sequences:

(a) $x(n) = \{1, 1, 1, 1\}$

(b) $h(n) = \{1, 1, 1\}$.

Problem B–2–18

Find the DTFT of the rectangular window function given in Eq. (2.1.5).

Problem B–2–19

A digital system is defined by the I/O equation $y(n) = x(n) + x(n-1)$. Complete the steps that follow:

(a) Sketch the signal-flow diagram of the system.
(b) Find the frequency response of the system.
(c) Compute the magnitude and phase responses of the system.

Problem B–2–20

Find the transfer function of the systems described by the following difference equations:

(a) $2y(n) + y(n-1) + 0.9y(n-2) = x(n-1) + x(n-2)$.
(b) $y(n) - 0.5y(n-1) + 0.8y(n-2) = x(n-1) + 0.75x(n-2)$.

Problem B–2–21

Find the difference equations that describe the systems with the following transfer functions:

(a) $H(z) = \dfrac{1 + 0.25z^{-1} - 0.5z^{-2}}{1 + 0.5z^{-2}}$.

(b) $H(z) = \dfrac{z}{(2z-1)(4z-1)}$.

Problem B–2–22

A digital system is defined by the I/O equation $y(n) = x(n) + ay(n-1)$. Complete the steps that follow:

(a) Sketch the signal-flow diagram of the system.
(b) Find the frequency response of the system.
(c) Compute the magnitude response of the system and sketch it for $a = 0.9$ and $a = 0.5$.

Problem B–2–23

A three-point moving-average filter is defined by the I/O equation

$$y(n) = \tfrac{1}{3}[x(n) + x(n-1) + x(n-2)].$$

Complete the steps that follow:

(a) Find the frequency response of the system.
(b) Compute the magnitude and phase responses of the system.

Problem B–2–24

The transfer function of an all-pass filter is given by

$$H(z) = \frac{b + z^{-1}}{1 - az^{-1}}.$$

Find the value of b such that the magnitude response $H(\omega) = 1$ for all ω.

Problem B–2–25

The output of a linear, time-invariant system is expressed as

$$y(n) = 3(-1)^n u(n) + (1/3)^n u(n),$$

when the input is expressed as

$$x(n) = (-1/3)^n u(n).$$

Find the transfer function and impulse response of the system.

Problem B–2–26

Find the mean value and the variance of the Gaussian- (or normal-) distributed random variable expressed as

$$f_x(X) = \frac{1}{\sigma_x \sqrt{2\pi}} e^{-(X - m_x)^2/2\sigma_x^2}.$$

Problem B–2–27

Find the mean value and the variance of the exponentially distributed random variable expressed as

$$f_x(X) = \begin{cases} \dfrac{1}{b} e^{-(X-a)/b} & X > a \\ 0 & X < a \end{cases}.$$

Problem B–2–28

Let x be a weighted sum of N uncorrelated random variables x_i defined as

$$x = \sum_{i=1}^{N} a_i x_i.$$

Find the mean value and variance of x.

Problem B–2–29

An analog sinewave with frequency $f_0 = 100$ Hz is sampled at sampling rate of 1,000 Hz. What is the normalized frequency F_0 in cycles per sample and the digital frequency ω_0 in radians per sample?

Problem B–2–30

The digital sinewave defined in Problem 29 is taken for the DFT with $N = 100$. What is the frequency resolution of the DFT spectrum? If the magnitude spectrum defined in Eq. (2.4.16) is computed, what is the k that corresponds to the maximum value of $|X(k)|$?

SUGGESTED READINGS

1 Oppenheim, A. V., R. W. Schafer, and J. R. Buck. *Discrete-Time Signal Processing*. 2nd Ed. Upper Saddle River, NJ: Prentice Hall, 1999.

2 Mitra, S. K. *Digital Signal Processing: A Computer-Based Approach*. 2nd Ed. New York, NY: McGraw-Hill, 2001.

3 Proakis, J. G. and D. G. Manolakis. *Digital Signal Processing: Principles, Algorithms, and Applications*. 3rd Ed. Upper Saddle River, NJ: Prentice Hall, 1996.

4 Orfanidis, S. J. *Introduction to Signal Processing*. Upper Saddle River, NJ: Prentice Hall, 1996.

5 Hayes, M. H. *Statistical Digital Signal Processing and Modeling*. New York, NY: Wiley, 1996.

6 Bateman, A. and W. Yates. *Digital Signal Processing Design*. Rockville, MD: Computer Science Press, 1989.

7 Kuo, S. M. and D. R. Morgan. *Active Noise Control Systems: Algorithms and DSP Implementations*. New York, NY: Wiley, 1996.

8 Grover, D. and J. R. Deller. *Digital Signal Processing and the Microcontroller*. Upper Saddle River, NJ: Prentice Hall, 1999.

9 McClellan, J. H., R. W. Schafer, and M. A. Yoder. *DSP First: A Multimedia Approach*. Upper Saddle River, NJ: Prentice Hall, 1998.

10 Kuo, S. M. and B. H. Lee. *Real-Time Digital Signal Processing: Implementations, Applications and Experiments with the TMS320C55x*. New York, NY: Wiley, 2001.

11 Shenoi, K. *Digital Signal Processing in Telecommunications*. Upper Saddle River, NJ: Prentice Hall, 1995.

12 The MathWorks. *Signal Processing Toolbox for Use with MATLAB*. Version 5, 2000.

3

Implementation Considerations

In Chapter 2, the discrete-time signal values and system parameters are represented using infinite-precision numbers. In practical DSP systems, we have to use digital hardware with finite wordlength to represent and process digital signals. In this chapter, we discuss practical considerations for implementing DSP algorithms on digital systems for real-time applications and provide various examples, experiments, and exercise problems for understanding the issues that are critical to DSP implementation.

3.1 INTRODUCTION

Digital systems, filters, and algorithms are implemented on digital hardware with finite wordlength. Practical DSP implementation requires special attention because of potential quantization and arithmetic errors, as well as the possibility of overflow. These effects always must be taken into consideration in DSP system design and implementation for practical applications.

A DSP processor's data format determines its ability to handle signals of different precisions, dynamic ranges, and SQNRs. Section 3.2 reviews the data representations in fixed-point and floating-point formats and compares the major differences between the fixed-point and floating-point processors, such as precision and dynamic range. Section 3.3 introduces important finite-wordlength effects, which are especially crucial for using fixed-point DSP processors.

In order to write efficient programs for DSP applications, we must understand the internal operations of DSP processors and the manner in which processors manipulate data. Programming issues are highlighted in Section 3.4, and real-time implementation considerations are presented in Section 3.5. A brief introduction of the interfacing hardware connected to the DSP processor is described in Section 3.6, and some design examples of connecting peripheral devices with DSP processors are given in Appendix C. In Section 3.7, we use Simulink with Fixed-Point Blockset and CCS to study finite-wordlength effects.

3.2 DATA REPRESENTATIONS AND ARITHMETIC

In this section, we define how numbers are represented in digital systems with different formats and how different arithmetic operations are performed in DSP processors. The purpose of this section is to provide the necessary background for programming both fixed-point and floating-point processors in later chapters.

3.2.1 Fixed-Point Numbers and Arithmetic

In fixed-point DSP processors, a number is represented with a series of binary digits (1s and 0s). There are many different binary number systems such as (1) sign magnitude, (2) one's complement, and (3) two's complement. Table 3.1 shows examples representing 4-bit signed numbers in three different formats. The leftmost bit is called the most significant bit (MSB), which represents the sign of the

TABLE 3.1   Binary Representations of 4-Bit Signed Numbers

Decimal value	Sign magnitude	One's complement	Two's complement
+7	0111	0111	0111
+6	0110	0110	0110
+5	0101	0101	0101
+4	0100	0100	0100
+3	0011	0011	0011
+2	0010	0010	0010
+1	0001	0001	0001
+0	0000	0000	0000
−0	1000	1111	—
−1	1001	1110	1111
−2	1010	1101	1110
−3	1011	1100	1101
−4	1100	1011	1100
−5	1101	1010	1011
−6	1110	1001	1010
−7	1111	1000	1001
−8	—	—	1000

number. A sign bit of 0 indicates a positive number, while a sign bit of 1 indicates a negative number.

Table 3.1 shows that these three binary formats have the same representation for positive numbers (from +0 to +8). The difference is in representing the negative numbers. Note that there are +0 and −0 in sign-magnitude and one's complement representations. However, in the commonly used two's complement, there is only one zero. Thus, it frees up one combination for representing −8, which cannot be represented in the other two formats. Therefore, signed numbers are represented using two's complement format in most DSP systems.

All binary representations have different weights associated with each bit. The weights depend on the position of the binary point. Different binary-point positions form different fractional number representations. For the integer values given in Table 3.1, the binary point is assumed to be at the right of the rightmost bit, which is called the least significant bit (LSB).

Integers versus Fractional Numbers

Different notations are used to represent different binary formats. The $Qm.n$ convention uses m bits to represent the integer portion of the number and n bits to represent the fractional portion. By defining N as the total number of bits, we have $N = m + n + 1$. For an N-bit signed number in $Qm.n$ format, the MSB (b_{N-1}) is the sign bit. For example, a 16-bit number that uses 1 sign bit and 15 bits for the fractional value is called the Q0.15 (or Q.15 or simply Q15) format. A 16-bit integer uses 15 bits for integer portion and is called the Q15.0 format.

The binary values of N-bit in the $Qm.n$ representation shown in Fig. 3.1 can be computed using the equation

$$x = (-b_{N-1}2^{N-1} + b_{N-2}2^{N-2} + b_{N-3}2^{N-3} + \ldots + b_0)2^{-n}$$
$$= -b_{N-1}2^m + \sum_{k=0}^{N-2} b_k 2^{k-n}. \tag{3.2.1}$$

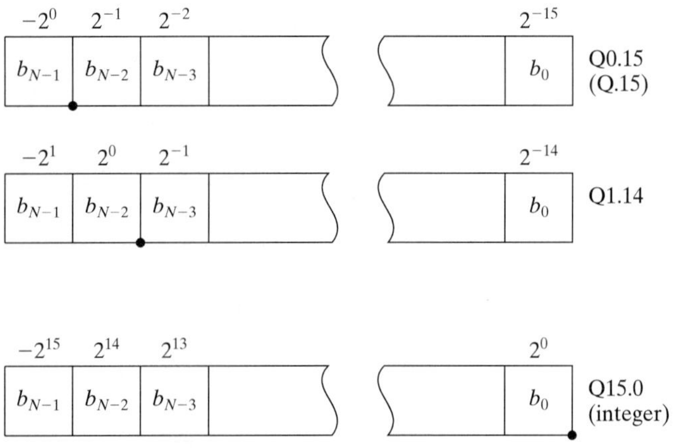

Figure 3.1   Different $Qm.n$ fixed-point formats

TABLE 3.2   Dynamic Ranges of Integer and Fractional 4-Bit Numbers

Unsigned integer	Signed integer
Smallest value: 0000 = (0) Largest value: 1111 = (15)	Most positive value: 0111 = (+7) Least negative value: 1000 = (−8)
Unsigned fractional	Signed fractional
Smallest value: .0000 = (0) Largest value: .1111 = (0.9375)	Most positive value: 0.111 = (+0.875) Least negative value: 1.000 = (−1)

For an integer, $n = 0$ and $m = N - 1$. Thus, Eq. (3.2.1) can be modified as

$$x = -b_{N-1}2^{N-1} + b_{N-2}2^{N-2} + \ldots + b_1 2^1 + b_0 2^0$$

$$= -b_{N-1}2^{N-1} + \sum_{k=0}^{N-2} b_k 2^k, \tag{3.2.2}$$

where b_{N-1} is the sign bit.

Table 3.2 shows the dynamic ranges of 4-bit-integer and fractional representations for both signed and unsigned numbers. The number inside the parentheses indicates its decimal value. It is important to note the difference in terms of the most positive and the least negative signed numbers that can be represented using different formats.

Dynamic range in dB is defined as the ratio between the largest number and the smallest number (excluding zero) that can be represented. It is expressed as

$$\text{Dynamic range in dB} = 20 \log_{10}\left(\frac{Max}{Min}\right). \tag{3.2.3}$$

In fixed-point-integer representation, *Max* is 2^N for unsigned numbers and 2^{N-1} for signed numbers, while *Min* is 1 for both numbers. For the fractional Q.15 format, *Max* is $1 - 2^{-N}$ for unsigned numbers, and $1 - 2^{-N+1}$ for signed numbers, while *Min* is 2^{-N} and 2^{-N+1} for unsigned and signed numbers, respectively. Therefore, a 16-bit fixed-point unsigned number always has 96 dB of dynamic range (or 90 dB for signed numbers) regardless of the Q format used for number representation.

Precision is defined as the smallest step (or difference) between two consecutive numbers that can be obtained for a given number of bits. A summary of the dynamic range and precision for a 16-bit number using different formats is shown in Table 3.3. In addition, Table 3.4 compiles a list of dynamic range and precision using

TABLE 3.3   Dynamic Range and Precision of a 16-Bit Number Using Different Formats

	Dynamic range	Dynamic range in dB	Precision
Unsigned integer	0 to 65,536	$20 \log_{10}(2^{16}) = 96$ dB	1
Signed integer	−32,768 to 32,767	$20 \log_{10}(2^{15}) = 90$ dB	1
Unsigned fractional	0 to 0.99998474	96 dB	2^{-16}
Signed fractional	−1 to 0.99996948	90 dB	2^{-15}

TABLE 3.4 Dynamic Range and Precision of 16-Bit Numbers Using Different Q Formats

Format	Largest positive value	Least negative value	Precision
Q0.15	0.999969482421875	−1	0.00003051757813
Q1.14	1.99993896484375	−2	0.00006103515625
Q2.13	3.9998779296875	−4	0.00012207031250
Q3.12	7.999755859375	−8	0.00024414062500
Q4.11	15.99951171875	−16	0.00048828125000
Q5.10	31.9990234375	−32	0.00097656250000
Q6.9	63.998046875	−64	0.00195312500000
Q7.8	127.99609375	−128	0.00390625000000
Q8.7	255.9921875	−256	0.00781250000000
Q9.6	511.984375	−512	0.01562500000000
Q10.5	1023.96875	−1,024	0.03125000000000
Q11.4	2047.9375	−2,048	0.06250000000000
Q12.3	4095.875	−4,096	0.12500000000000
Q13.2	8191.75	−8,192	0.25000000000000
Q14.1	16383.5	−16,384	0.50000000000000
Q15.0	32,767	−32,768	1.00000000000000

different Q formats. Note that the Q.15 format is commonly used in DSP systems, and data must be properly scaled so that their value lies between −1 and 0.999969482421875. We examine the advantage of using Q.15 format in performing arithmetic operations in DSP systems later.

Most fixed-point DSP processors use two's complement fractional numbers in different Q formats. However, assemblers only recognize integer values. Thus, the programmer must keep track of the position of the binary point when manipulating fractional numbers in assembly programs. The following steps convert a fractional number in Q format into an integer value that can be recognized by the assembler:

Step 1. Normalize the fractional number to the range determined by the desired Q format.

Step 2. Multiply the normalized fractional number by 2^n, where n is the total number of fractional bits.

Step 3. Round the product to the nearest integer.

Example 3.1

Assume that a coefficient used by an assembly program is 1.18. Use the preceding procedure to convert the coefficient into an integer value that can be recognized by a DSP assembler using Q.15 format with $n = 15$ ($2^n = 32,768$): Follow these steps:

Step 1. Normalize the number to the range between ±1; thus, $1.18/2 = 0.59$.

Step 2. Multiply 0.59 by 2^{15}; thus, $0.59 \times 32,768 = 19,333.12$.

Step 3. Round the decimal value 19,333.12 to obtain $19,333 = $ 4B85h (or 0x4B85).

TABLE 3.5 Scaling Factors and Dynamic Ranges for 16-Bit Numbers Using Different Q Formats

Format	Scaling factor (2^n)	Range in Hex (Decimal value)
Q0.15	$2^{15} = 32,768$	7FFFh (0.99) $\rightarrow$ 8000h (-1)
Q1.14	$2^{14} = 16,384$	7FFFh (1.99) $\rightarrow$ 8000h (-2)
Q2.13	$2^{13} = 8,192$	7FFFh (3.99) $\rightarrow$ 8000h (-4)
Q3.12	$2^{12} = 4,096$	7FFFh (7.99) $\rightarrow$ 8000h (-8)
Q4.11	$2^{11} = 2,048$	7FFFh (15.99) $\rightarrow$ 8000h (-16)
Q5.10	$2^{10} = 1,024$	7FFFh (31.99) $\rightarrow$ 8000h (-32)
Q6.9	$2^9 = 512$	7FFFh (63.99) $\rightarrow$ 8000h (-64)
Q7.8	$2^8 = 256$	7FFFh (127.99) $\rightarrow$ 8000h (-128)
Q8.7	$2^7 = 128$	7FFFh (511.99) $\rightarrow$ 8000h (-512)
Q9.6	$2^6 = 64$	7FFFh (1023.99) $\rightarrow$ 8000h ($-1,024$)
Q10.5	$2^5 = 32$	7FFFh (2047.99) $\rightarrow$ 8000h ($-2,048$)
Q11.4	$2^4 = 16$	7FFFh (4095.99) $\rightarrow$ 8000h ($-4,096$)
Q12.3	$2^3 = 8$	7FFFh (4095.99) $\rightarrow$ 8000h ($-4,096$)
Q13.2	$2^2 = 4$	7FFFh (8191.99) $\rightarrow$ 8000h ($-8,192$)
Q14.1	$2^1 = 2$	7FFFh (16383.99) $\rightarrow$ 8000h ($-16,384$)
Q15.0	$2^0 = 1$(Integer)	7FFFh (32,767) $\rightarrow$ 8000h ($-32,768$)

The arithmetic result obtained by a DSP processor is in the integer form. It can be interpreted as a fractional value by dividing by 2^n. For example, $19,333/32,768 = 0.589996$. Note that there is a difference (error) from the actual value, 0.59, due to the rounding in step 3. In DSP implementation, it is not always necessary to use Q.15 format throughout the DSP algorithm; instead, we can use different Q formats for different dynamic range requirements. Table 3.5 shows different scaling factors that are required when operating in different Q formats. It also summarizes the dynamic range and its corresponding hexadecimal (Hex) values used in the DSP processor.

There is a tradeoff between the dynamic range and the precision of the number when using different Q formats. Q.15 has the best precision (2^{-15}), but the narrowest dynamic range (±1). The integer Q15.0 has the worst precision (1), but the widest dynamic range from $-32,768$ to $+32,767$.

Binary Addition and Multiplication

In this section, we explore the addition and multiplication performed by a fixed-point DSP processor. All of the numbers are represented using the two's complement format. First, we use the following two examples to demonstrate binary addition in both fractional and integer forms.

Example 3.2

The following examples show the addition of two 4-bit numbers represented in Q.3 format (the numbers inside parentheses are decimal values):

1. $0.100 \ (0.5) + 0.011 \ (0.375) = 0.111 \ (0.875)$, no overflow
2. $0.101 \ (0.625) + 0.011 \ (0.375) = 1.000 \ (-1)$, overflow
3. $1.100 \ (-0.5) + 0.111 \ (0.875) = 0.011 \ (0.375)$, no overflow
4. $1.100 \ (-0.5) + 1.011 \ (-0.625) = 0.111 \ (0.875)$, overflow

Example 3.3

The following examples show the addition of two 4-bit numbers in integer representation (or Q3.0 format):

1. $0100 \ (4) + 0011 \ (3) = 0111 \ (7)$, no overflow
2. $0101 \ (5) + 0011 \ (3) = 1000 \ (-8)$, overflow
3. $1100 \ (-4) + 0111 \ (7) = 0011 \ (3)$, no overflow
4. $1100 \ (-4) + 1011 \ (-5) = 0111 \ (7)$, overflow

These examples show that overflow may occur when adding two fractional and integer numbers. In case 2 of Examples 3.2 and 3.3, adding two positive numbers results in a negative number. In case 4 of both examples, adding two negative numbers results in a positive number. We call both cases overflow. We introduce several methods for avoiding overflow in Section 3.3.3.

When multiplying two 4-bit numbers in Q.3 format requires a 7-bit word in Q.6 format to store the product, and there is no overflow. Finite-precision errors occur when limited numbers of bits are used to represent the result, which is less than that required for exact representation. In the case of fractional multiplication, the MSBs contain the more important results: thus, the LSBs can be truncated (or rounded off) to fit the size of memory.

Example 3.4

The following simple examples show multiplication of two 4-bit numbers in integer and fractional representations:

1. *Integer multiplication:* $0111 \ (7) \times 0110 \ (6) = 0101010 \ (42)$. When we store only the most significant 4-bit number, $0101 \ (5)$, in memory, the error is $42 - 5 = 37$. Thus, we must store the full 7-bit number to obtain the exact answer.
2. *Fractional multiplication:* $0111 \ (0.875) \times 0110 \ (0.75) = 0101010 \ (0.65625)$. When we store the upper 4-bit number, $0101 \ (0.625)$, in memory, the error is $0.65625 - 0.625 = 0.03125$.

Example 3.4 shows that fractional multiplication provides a more accurate result (less error) if we have to truncate (or round) the double-precision product into a single-precision product for output or storage. This is the main reason we use fractional numbers for fixed-point DSP implementations.

The above multiplication can be extended to the 16-bit precision commonly used in fixed-point DSP processors. The result of multiplying two Q.15 numbers always is contained within ±1, and the significant result can be stored within the most significant 16 bits. Unlike integer (or Q15.0) multiplication, care must be taken to store all of the product bits.

Note that when multiplying two Q.15 numbers, the product is a 32-bit word with an extra sign bit. The binary point is between b_{30} and b_{29}. Therefore, a left shift of 1 bit is required to get rid of the extra sign bit in the product. After that, we can save the upper 16-bit number into a 16-bit memory location. This idea can also be extended to a mixed-Q format multiplication. In this case, we must understand where to locate the binary point and perform the necessary left shift to get rid of the extended sign bits.

Binary Division

DSP processors are seldom needed to perform division, and the hardware implementation of division is expensive. Therefore, most processors do not provide a single-cycle divide instruction supported by the hardware. The TMS320 family provides a single-cycle 1-bit divide operation using the conditional subtraction instruction. For an N-bit fractional number, fractional division can be realized by repeating the conditional subtraction instruction $(N - 1)$ times. In the case of integer division, conditional subtraction must be conducted N times.

3.2.2 Floating-Point Arithmetic

In this section, we introduce floating-point number formats used by floating-point DSP processors. Examples of floating-point processors (TMS320C3x and TMS320C67x) are introduced in Chapter 5. Floating-point formats allow numbers to be represented with a large dynamic range. Therefore, floating-point arithmetic can reduce the problem of overflow that occurs in fixed-point arithmetic.

Floating-Point Formats

There are many ways to represent floating-point numbers. Generally, the number is represented in three fields: Sign bit(s), exponential bits, and mantissa bits. The Institute of Electrical and Electronics Engineers (IEEE) introduces a standard for representing floating-point numbers in 32-bit (single precision), ≥43-bit (extended single precision), 64-bit (double precision), and ≥79-bit (extended double precision) formats. These formats are stated in the IEEE-754 standard [1].

As shown in Fig. 3.2(a), the IEEE single-precision floating-point format is expressed as

$$x = -1^s \times 2^{(\text{exp}-127)} \times 1.\text{man}, \tag{3.2.4}$$

where the sign bit is b_{31}, exponent bits (exp) are 8 bits from bits b_{30} to b_{23}, and mantissa bits (man) are 23 bits from b_{22} to b_0. The double-precision format shown in Fig. 3.2(b) can be expressed as

$$x = -1^s \times 2^{(\text{exp}-1023)} \times 1.\text{man}, \tag{3.2.5}$$

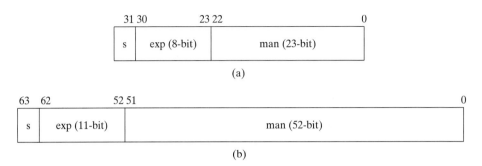

Figure 3.2   (a) IEEE-754 single-precision and (b) double-precision floating-point formats

where the sign bit is b_{63}, exponent bits are 11 bits from bits b_{62} to b_{52}, and mantissa bits are 52 bits from b_{51} to b_0.

The exponential value is used to offset the location of the binary point left or right by a certain amount. In the IEEE standard, the exponent value is also being biased by a value of 127 (single precision) or 1,023 (double precision) to obtain positive and negative offsets. There is a set of rules for representing special floating-point data types [1].

Floating-point DSP processors automatically scale the number to obtain the full range representation of the mantissa, which is done by increasing or decreasing the exponent value for small or large numbers, respectively. In other words, the floating-point processor tracks the number and adjusts the value of the exponent.

Table 3.6 shows the dynamic range and precision of IEEE floating-point numbers. In general, the dynamic range of the floating-point number is determined by the size of the exponent. It provides 6 dB for each exponent level. For example, the single-precision format has 8 exponent bits with 255 ($2^8 - 1$) levels and a dynamic range of 1,530 (6×255) dB. It is also noted that overflow (or underflow) is still possible in floating-point operations if the number exceeds the dynamic range summarized in Table 3.6.

For a given wordlength, a floating-point number has a larger dynamic range compared to the fixed-point number with the same wordlength because the exponent field is able to scale the floating-point numbers to very large or very small values. However, since the exponent bits consume part of the wordlength, fewer bits are left for the mantissa (which determines precision) compared to the number of bits used in the fixed-point representation.

TABLE 3.6   Summary of Dynamic Ranges and Precision

	Dynamic range	Dynamic range in dB	Precision
Single precision	1.18×10^{-38} to 3.4×10^{38}	~1,530 dB (6 dB $\times$ 255)	2^{-23}
Extended single precision	1.18×10^{-38} to 3.4×10^{38}	~1,530 dB	2^{-31}
Double precision	2^{-1022} to 2^{1024}	$6 \times (2^{11} - 1) \sim 12,282$ dB	2^{-52}

Modern floating-point processors such as TMS320C67x and ADSP2106x support IEEE floating-point data formats. This support guarantees that algorithms developed on IEEE-compatible computers are portable across DSP processors without concern for the possible instability introduced by bias rounding or inconsistent error handling. However, some DSP processors such as the TMS320C3x use a different type of floating-point representation, which is introduced in Chapter 5.

Floating-Point Addition and Multiplication

In order to perform floating-point addition, we have to adjust the exponent of the smaller number to match that of the bigger number. We define the two numbers as

$$x = -1^{sx} \times 2^{(\exp_x - 127)} \times 1.\mathrm{man}_x \tag{3.2.6}$$

and

$$y = -1^{sy} \times 2^{(\exp_y - 127)} \times 1.\mathrm{man}_y, \tag{3.2.7}$$

where sx and sy represent the sign bit, $\exp_x$ and $\exp_y$ represent the exponent, and man_x and man_y represent the mantissa of x and y, respectively. The addition of these two numbers can be expressed as

$$z = x + y$$
$$= \begin{cases} [-1^{sx} \times 1.\mathrm{man}_x + (-1^{sy} \times 1.\mathrm{man}_y \times 2^{-(\exp_x - \exp_y)})] \\ \quad \times 2^{\exp_x - 127}, & \textit{if } |x| \geq |y| \\ [-1^{sy} \times 1.\mathrm{man}_y + (-1^{sx} \times 1.\mathrm{man}_x \times 2^{-(\exp_y - \exp_x)})] \\ \quad \times 2^{\exp_y - 127}, & \textit{if } |y| > |x|. \end{cases} \tag{3.2.8}$$

Example 3.5

We are given two floating-point numbers

$$x = 2.44 = -1^0 \times 2^{(128-127)} \times 1.22$$

and

$$y = -12.16 = -1^1 \times 2^{(130-127)} \times 1.52.$$

Since $|y| > |x|$,

$$x + y = [-1^1 \times 1.52 + (-1^0 \times 1.22 \times 2^{-(130-128)})] \times 2^{130-127}$$
$$= -1 \times 1.215 \times 2^3 = -9.72.$$

Therefore, we obtain the floating-point sum of $s = 1$, $\mathrm{man} = 0.215$, and $\exp = 3 + 127 = 130$.

Most floating-point processors perform automatic normalization so that numbers are properly shifted and aligned. The programmer just needs to take care of the overflow problem. However, due to the enormous dynamic range, scaling is rarely needed. Therefore, floating-point processors are easier to use than their fixed-point counterparts, but are more expensive.

Floating-point multiplication can be carried out in a more straightforward manner. The mantissas of the two numbers are multiplied, while the exponent terms are added without the need to align them. Given the same two numbers shown in Eqs. (3.2.6) and (3.2.7), the multiplication of these two numbers can be expressed as

$$z = x \times y$$
$$= [(-1^{sx} \times 1.\text{man}_x) \times (-1^{sy} \times 1.\text{man}_y)] \times 2^{(\exp_x + \exp_y - 254)}. \tag{3.2.9}$$

Example 3.6

We are given two floating-point numbers

$$x = 2.44 = -1^0 \times 2^{(128-127)} \times 1.22$$

and

$$y = -12.16 = -1^1 \times 2^{(130-127)} \times 1.52.$$

The multiplication of these two numbers is expressed as

$$x \times y = [(-1^1 \times 1.52) \times (-1^0 \times 1.22)] \times 2^{(128+130-254)}$$
$$= -1 \times 1.8544 \times 2^4 = -29.6704.$$

Therefore, we obtain the floating-point product of $s = 1$, man = 0.8544, and exp = $4 + 127 = 131$.

The experiments given in Chapter 2 show that we can run a floating-point C program on fixed-point processors. However, emulating floating-point arithmetic in fixed-point DSP processors tends to be very cycle intensive because the emulator routine needs to take care of the exponent and mantissa. It is only worthwhile to perform emulation on small sections of data that require high dynamic range and precision. A better approach is to use block floating-point format, which is described next.

Block Floating-Point Format

An increase in precision and dynamic range can also be obtained by grouping data with a similar range of values with a single exponent value. As shown in Table 3.7, a group of data can share a common exponent, and significant memory is saved when storing only one exponent as compared with formal floating-point numbers. The exponent can be determined by the data elements in the block with the largest magnitude. Therefore, a common exponent is used for a block of numbers with different mantissas.

The block floating-point format is a compromise between the fixed-point and floating-point formats and is used for manipulating a large array of data such as FFTs. Some DSP processors support this format with special instructions such as exponent detection and normalization.

TABLE 3.7 Example of a Group of Data that has a Similar Range of Values and that Shares a Single Exponent for that Block of Data

Data	Data group
0.123	0.123 → Common exponent gain = 1
0.230	0.230
−0.122	−0.122
0.225	0.225
1.34	0.134 → Common exponent gain = 10
−2.65	−0.265
−1.034	−0.1034
0.998	0.0998
1.564	0.1564
0.001	0.01 → Common exponent gain = 0.1
−0.002	−0.02
−0.023	−0.23
0.01	0.1

3.2.3 Fixed-Point versus Floating-Point Format

We have discussed the differences between fixed-point and floating-point representations, their precisions, and their dynamics ranges in previous sections. In real-world applications, the resolution of ADCs has continued to increase (e.g., from 16 bits to 24 bits for audio applications), which requires DSP processors to provide higher precision to handle signals with larger wordlength. Table 3.8 summarizes the key differences between fixed-point and floating-point DSP processors. It highlights the differences between key characteristics, features, development processes, and programming efforts for these processors.

TABLE 3.8 Guidelines in Selecting Fixed-Point or Floating-Point DSP Processors

Fixed-point processors	Floating-point processors
16- or 24-bit devices	32-bit devices
Limited dynamic range	Large dynamic range
Overflow and quantization errors must be resolved	Easier to program since no scaling is required
Poorer C-compiler efficiency; normally programmed in assembly	Better C-compiler efficiency; can be developed in C
Long product development time	Quick time to market
Faster clock rate	Slower clock rate
Less silicon area is required; functional units are simpler	More silicon area is required; functional units are complex
Cheaper	More expensive
Lower power consumption	Higher power consumption

TABLE 3.9   Applications of Fixed-Point and Floating-Point Processors

Fixed-point processors	Floating-point processors
Disk drive and motor control	Image processing in radar, sonar, and seismic applications
Consumer audio applications such as MP3 players, multimedia gaming, and digital cameras	High-end audio applications such as ambient acoustics simulators, professional audio encoding/decoding, and audio mixing
Speech coding/decoding and channel coding	Sound synthesis in professional audio and video coding/decoding
Communication devices such as modems and cellular phones	Prototyping

In general, fixed-point processors are used for large-volume products such as modems, wireless phones, and hard-disk drives. Floating-point processors are preferred for limited-volume and high-performance products such as professional audio equipment. Table 3.9 highlights some common applications for fixed-point and floating-point processors. For example, radar, sonar, and some commercial applications such as speech recognition require a larger dynamic range in order to discern selected signals from noisy environments.

3.3 FINITE-WORDLENGTH EFFECTS

Finite-wordlength effects occur when the wordlength of the memory (or register) is less than the required precision needed to store the actual values. These effects introduce noise into DSP systems and create non-ideal system responses. In this section, we highlight the following finite-wordlength effects:

1. Limited precision when the analog signal is converted to the digital form. The input quantization error is discussed in Section 3.3.1.
2. Limited precision in representing system coefficients and parameters. The coefficient quantization problem is discussed in Section 3.3.2.
3. Limited dynamic range used in arithmetic operations that result in overflow (or underflow). The overflow problem and its solutions are discussed in Section 3.3.3.
4. Rounding (or truncation) of double-precision data to single precision data for storage in a register or memory. We discuss roundoff and truncation errors in Section 3.3.4.

3.3.1 Input Quantization

As introduced in Section 1.6.2 and shown in the top diagram of Fig. 3.3, an ADC produces a digital signal by quantizing a continuous-valued discrete-time signal to form a discrete-valued digital signal. For an N-bit ADC with full-scale voltage of V_{FS}

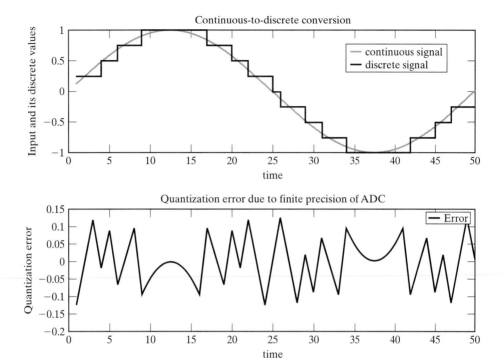

Figure 3.3 The top diagram depicts the continuous-valued signal and its digital representation after ADC; the bottom plot shows that quantization errors are distributed between $\pm\Delta/2$ (0.125)

(peak to peak), the physical quantization interval (step) is defined as

$$\Delta = \frac{V_{FS}}{2^N}. \tag{3.3.1}$$

For example, in the case of $V_{FS} = 2\text{V}$ and $N = 3$, the quantization step $\Delta = 0.25\text{V}$.

The difference between the analog-valued discrete-time signal and the corresponding discrete-valued digital signal forms the quantization error (or noise), $e(n)$. For a linear quantizer, quantization errors are uniformly distributed between $\pm\Delta/2$ (in the preceding example, it is ±0.125) and are shown in the bottom diagram of Fig. 3.3. The diagram also shows that the mean of $e(n)$ is zero.

From Eq. (2.5.6), the variance of the quantization noise is given by

$$\sigma_e^2 = \Delta^2/12. \tag{3.3.2}$$

In the case of sinewave input, the input signal power can be expressed as

$$P_x = (V_{FS}/2)^2/2 = V_{FS}^2/8. \tag{3.3.3}$$

By replacing V_{FS} with Δ given in Eq. (3.3.1), we obtain

$$P_x = \Delta^2 2^{2N}/8. \tag{3.3.4}$$

Therefore, the SQNR is given as

$$\text{SQNR} = 10 \log_{10}\left(\frac{P_x}{\sigma_e^2}\right) = 10 \log_{10}\left(\frac{\Delta^2 2^{2N}/8}{\Delta^2/12}\right) = 10 \log_{10}\left(2^{2N} \times \frac{3}{2}\right)$$

$$= (6.02N + 1.76) \text{ dB.} \tag{3.3.5}$$

This equation implies that every bit increase results in a 6 dB gain of SQNR. However, this amount is the theoretical maximum because the ADC is assumed to be perfect in the above derivation. In practice, we seldom achieve this amount due to the noise commonly associated with ADC. Furthermore, the signal amplitude seldom stays at the same maximum level (such as sinewave) all of the time. Increasing the number of bits used in the ADC can reduce the quantization error introduced by the ADC.

An effective approach in improving the theoretical SQNR of Eq. (3.3.5) uses the oversampling technique, which spreads out the power of quantization noise across a much wider frequency band. If the desired sampling frequency is f_s, we oversample the analog signal by a factor of M using the higher sampling rate Mf_s. The quantization noise power is reduced by a factor of M after the decimation with digital lowpass filtering. Therefore, the improved SQNR using oversampling method is

$$\text{SQNR} = (6.02N + 1.76 + 10 \log_{10} M) \text{ dB.} \tag{3.3.6}$$

This equation shows that for every four times of oversampling, a 6 dB improvement of SQNR results. According to Eq. (3.3.5), this result can also be interpreted as an increase of 1-bit resolution with every four times of oversampling.

The effective number of bits required to attain the required signal-to-noise-plus-distortion ratio (SNDR) is

$$N = \frac{\text{SNDR} - 1.76}{6.02}, \tag{3.3.7}$$

where the SNDR is derived from the measurement expressed as

$$\text{SNDR} = 10 \log_{10}\left[\frac{\text{Power of fundamental}}{\text{Power of } 1^{\text{st}} \text{ 10 harmonics} + \text{noise}}\right] \text{(dB).} \tag{3.3.8}$$

Equation (3.3.7) is derived based on the assumption that the amplitude of the actual input signal swings in full scale and SNDR $\approx$ SQNR. If the signal amplitude is less than full scale, more bits will be needed to achieve the same level of performance. Thus, Eq. (3.3.7) is modified as

$$N = \frac{\text{SNDR} - 1.76}{6.02} + 3.32 \log_{10}\left(\frac{1}{\text{Fraction of full scale}}\right). \tag{3.3.9}$$

It is noted that when an N-bit ADC is used, a $6N$ dB SQNR must be maintained throughout the DSP arithmetic operations. Therefore, it is important that all

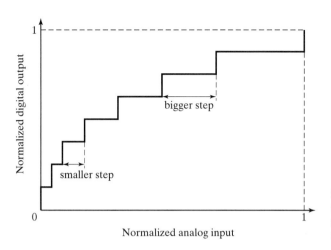

Figure 3.4 Nonlinear companding characteristics between the analog input and the digitized output

intermediate calculations be performed at a higher precision than N-bit. For example, when a 14-bit ADC is used, a 16-bit processor can be used to maintain the 14-bit SQNR of the ADC.

Another group of nonlinear quantizers such as μ-law and A-law CODEC commonly is used in digital telephony systems. As shown in Fig. 3.4, the nonlinear quantizer uses fine (small) quantization steps for a low-level signal and coarser quantization steps for larger signals. The nonlinear quantizer is used to balance the low SQNR of a small signal with that of a large-level signal.

In telecommunication systems, the voice signal is first quantized using 8-bit CODEC by the companding law (μ-law or A-law) to reduce the transmission bandwidth. The receiver then expands this nonlinear, compressed signal back to the linear signal (13 bits for A-law or 14 bits for μ-law CODEC) for further processing. The expansion characteristic is the inverse of the companding curve shown in Fig. 3.4.

3.3.2 Coefficient Quantization

As discussed in Chapter 2, DSP systems are designed based on the assumption that system parameters have infinite precision. However, these parameters are represented with a finite number of bits in DSP systems. For example, quantization of filter coefficients can affect pole/zero locations, thus altering the frequency response of digital filters. This effect is especially critical for IIR filters, where the movement of the pole locations may lead to instability. A possible solution is to use more bits to represent the coefficients. For example, we may use double precision (32 bits) to represent filter coefficients in a 16-bit processor that represents signal samples in 16 bits.

We can use the MATLAB FDATool under the Signal Processing Toolbox to examine the coefficient's quantization effects. FDATool is an easy-to-use and versatile tool that allows quick examination of various quantization errors. It allows the user to design a filter with coefficients in different [N, n] formats and examine fixed-point or

TABLE 3.10   Representation of Ideal Coefficients in Different Wordlengths

Index	Ideal coefficients	[16, 15] or Q.15	[10, 9] or Q.9	[8, 7] or Q.7
(0,39)	−0.00430702004568	−0.00430297851563	−0.003906250	−0.007812500
(1,38)	−0.01308561664228	−0.01309204101563	−0.013671875	−0.015625000
(2,37)	−0.01651509138705	−0.01651000976563	−0.015625000	−0.015625000
(3,36)	−0.00643681346024	−0.00643920898438	−0.005859375	−0.007812500
(4,35)	0.00981299580448	0.00982666015625	0.009765625	0.007812500
(5,34)	0.01080509015771	0.01080322265625	0.011718750	0.007812500
(6,33)	−0.00655909098803	−0.00656127929688	−0.005859375	−0.007812500
(7,32)	−0.01680168434011	−0.01681518554688	−0.017578125	−0.015625000
(8,31)	0.00064690809975	0.00064086914063	0	0
(9,30)	0.02246173646836	0.02246093750000	0.023437500	0.023437500
(10,29)	0.01013863182521	0.01013183593750	0.009765625	0.007812500
(11,28)	−0.02566675360315	−0.02566528320313	−0.025390625	−0.023437500
(12,27)	−0.02656981085405	−0.02658081054688	−0.027343750	−0.023437500
(13,26)	0.02304448932000	0.02304077148438	0.023437500	0.023437500
(14,25)	0.05039554410987	0.05038452148438	0.050781250	0.046875000
(15,24)	−0.00927120611197	−0.00927734375000	−0.009765625	−0.007812500
(16,23)	−0.08790401571487	−0.08789062500000	−0.087890625	−0.085937500
(17,22)	−0.03376590386367	−0.03375244140625	−0.033203125	−0.031250000
(18,21)	0.18733401904142	0.18734741210938	0.187500000	0.187500000
(19,20)	0.40150715664058	0.40151977539063	0.402343750	0.398437500
Sum of squared error		**3.09×10^{-9}**	**1.46729×10^{-5}**	**1.864×10^{-4}**

floating-point representations with rounding and overflow modes. The results of changing the coefficient formats and modes can be shown in the plots of magnitude response, pole-zero, impulse response, and step response. This capability provides a quick analysis in determining the finite-precision errors of designed filters, redesigns the filter to reduce quantization errors, and ensures the stability of IIR filters. We provide further information about FDATool in Appendix A and use it for designing and analyzing filters in Chapters 6 and 7.

An example that illustrates coefficient quantization is shown in Table 3.10. The quantized coefficients for a symmetrical FIR filter of length $L = 40$ are obtained by using different $[N, n]$ formats, where N is the total number of bits and n is the number of fractional bits. The sums of squared errors for the cases of representing 40 coefficients in 16-bit, 10-bit, and 8-bit wordlengths are listed in the bottom row of table. It is obvious that when the number of bits used to represent the coefficients is reduced, larger quantization errors occur.

As mentioned before, the total number of bits used to represent coefficients affects the frequency response of the filter. Figure 3.5 shows the magnitude responses of the coefficients given in Table 3.10 using different Q formats and compares them to the ideal coefficients (reference filter) represented by 64-bit double precision. Figure 3.5 shows that performance is degraded when a smaller wordlength is used to represent the coefficients. By representing the FIR filter coefficients using finite wordlength, errors occur at the zero positions, thus changing the magnitude

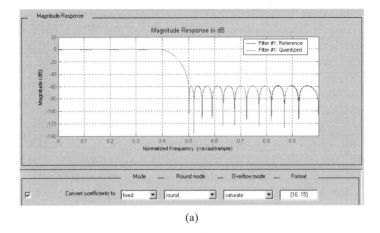

(a)

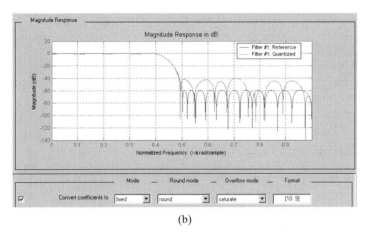

(b)

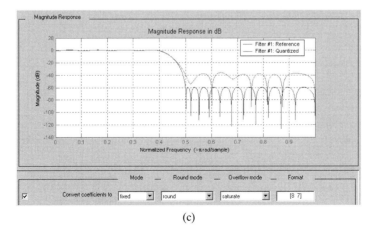

(c)

Figure 3.5   Magnitude responses of a filter using (a) [16, 15], (b) [10, 9], and (c) [8, 7] formats versus the reference filter using 64-bit precision

response of the reference filter (64-bit). The coefficient quantization errors were also affected by using different filter structures for the realization of the filter. Coefficient quantization is particularly critical for IIR filters, where the position of the poles determines the stability of the filter.

As discussed in Section 2.3.3, it is rare that direct forms are used to implement IIR filters. Usually, IIR filters are implemented as a cascade of second-order sections, also commonly known as biquad filters. Therefore, we have to group the poles and zeros of the high-order IIR filter given in Eq. (2.3.15) in pairs to form second-order sections expressed as

$$H_l(z) = \frac{b_{0,l} + b_{1,l}z^{-1} + b_{2,l}z^{-2}}{1 + a_{1,l}z^{-1} + a_{2,l}z^{-2}}, \quad l = 1, 2, \ldots, L/2. \tag{3.3.10}$$

This equation assumes L is an even number. An additional first-order section must be added when L is an odd number. Some common rules for grouping and pairing these poles/zeros into the second-order sections and determining the order of these sections are summarized as follows:

1. Complex-conjugate poles and zeros are grouped in the same section to form real-valued coefficients for each second-order filter.
2. Poles that are close to the unit circle on the z-plane are paired with the zeros that are close to them.
3. After the above grouping and pairing in cases 1 and 2, we can cascade the second-order sections in order of increasing gain.

Finally, the designed-filter coefficients may be greater than $+1$ or less than -1. When representing coefficients in Q.15 format, these coefficients can be halved (or scaled down) to values less than 1 or split into multiple multiply-add operations. For example, if the coefficient value is 1.6, we can break the value down into two coefficients of value 0.8 and multiply by the same input.

3.3.3 Overflow and Solutions

Examples of showing overflow were given in Examples 3.2 and 3.3. In general, adding two N-bit numbers results in a sum of $N + 1$ bits. For adding M numbers, $(N + \log_2 M)$ bits are required to represent the sum in order to prevent overflow. For example, a 40-bit accumulator with 8 guard bits supports up to 256 (2^8) additions of 32-bit numbers without overflow. Underflow occurs when the result of the arithmetic operation is less than the smallest number that can be represented in the given dynamic range. For example, if the result is 2^{-16}, it is too small to be stored using the Q.15 format, since the smallest number that can be represented is 2^{-15}.

Multiplication of two N-bit numbers in Q.15 format does not result in overflow for fractional representation, even when the product is represented in N-bit. However, the one exception is multiplying -1 by -1 in a fractional format. For other Q formats, overflow occurs if an insufficient number of bits is used to represent the final result. For example, when using Q15.0 (or integer) format, $2N$ bits are required to store the result; otherwise, overflow occurs. When dividing an N-bit number by an M-bit number, the result of the division requires as many bits as the numerator.

Several techniques can be used to prevent overflow in DSP operations, including saturation mode, scaling of the input signal, and guard bits in accumulators.

Using Saturation Mode

A DSP processor can operate with either saturation mode turned on or off. Setting a particular bit in the status register of the processor (e.g., overflow mode bit OVM = 1 for the TMS320C2000 and C5000) turns on the saturation mode. Figure 3.6 shows the input and output waveform passing through the nonlinear operation of the saturation arithmetic. When the arithmetic results exceed the maximum positive value that can be represented (e.g., 7FFFh for the 16-bit processor), the processor sets (or clips) the result to that maximum positive value. When the result of adding two negative numbers is smaller than -1 (e.g., 8000h), the hardware limits (or clips) the result to -1.

For some applications, saturation mode may be turned off when adding an array of two's-complement numbers because overflow may occur during the intermediate stages. However, if the final result is within the given dynamic range, the result is still correct. Intermediate overflows on different addition nodes can be allowed as long as the final result is within $[-1, (1 - 2^{-15})]$ for Q.15 representation.

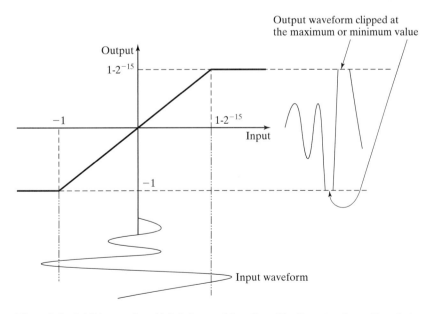

Figure 3.6   Addition results with infinite precision clipped by the saturation arithmetic to prevent overflow

Scaling the Input Signal

In the case of digital filtering, there are many multiply-accumulate operations. Therefore, there is a possibility of overflow when operating in Q.15 format. An effective approach in preventing overflow is scaling down the signal at certain nodes in the filter and then scaling the final result back to the original level.

We examine the scaling factor required to scale down the input signal or coefficients of the FIR filter. The maximum gain of the FIR filter with coefficient b_i is $\sum_{i=0}^{L-1} |b_i|$ for broadband signals, $(\sum_{i=0}^{L-1} b_i^2)^{1/2}$ for a less stringent gain; and $\max|H(\omega)|$ for narrowband signals. Therefore, a scaling (or attenuation) factor A must be introduced to either the input signal or the filter coefficients to ensure that $|y(n)| < 1$. Scaling down the coefficients degrades the performance of the filter since smaller coefficients introduce severe coefficient quantization errors, as discussed in Section 3.3.2. Therefore, the attenuation factor A is usually applied to the input signal of the filter. The scaling factor is given as

$$A < \frac{1}{x_{max} \sum_{i=0}^{L-1} |b_i|} \tag{3.3.11}$$

or

$$A < \frac{1}{x_{max}\left(\sum_{i=0}^{L-1} b_i^2\right)^{1/2}} \tag{3.3.12}$$

for broadband signals, where x_{max} is the maximum value of input samples. For narrowband signals, the scaling factor is limited by

$$A < \frac{1}{\max[|X(\omega)\|H(\omega)|]}. \tag{3.3.13}$$

For broadband signals, there are two possible scaling factors. The scaling factor given in Eq. (3.3.11) is the most conservative one and guarantees no overflow. The scaling factor given in Eq. (3.3.12) is less stringent and allows a higher dynamic range for the input signal. For narrowband signals, the magnitude response scaling given in Eq. (3.3.13) can be carried out very effectively. Besides scaling the input, we can also apply different Q formats for representing filter coefficients.

A potential drawback with scaling is the further reduction of the SQNR given in Eq. (3.3.5). If the input signal is scaled by the factor A, SQNR is further reduced by $20 \log_{10} A$ dB. For example, SQNR is reduced by 6 dB if $A = 0.5$, which is the equivalent of losing 1-bit resolution. Therefore, a tradeoff between preventing overflow and reducing SQNR must be considered.

Using Guard Bits in Accumulators

Modern 16-bit, fixed-point DSP processors, such as the C5000, generally have 40-bit accumulators to support a sequence of additions (such as in FIR filtering) without overflow. For the 40-bit accumulator, the additional 8 guard bits allow up to 256 (2^8) additions before overflow can occur.

In general, programming in fixed-point processors requires extra effort to prevent overflow. The methods of using saturation arithmetic, scaling, mixed-Q arithmetic, and guard bits of accumulators can be applied in different combinations depending on the nature of the problem.

3.3.4 Rounding and Truncation

Addition and multiplication are two commonly used operations in DSP systems. Multiplying two 16-bit numbers requires a 32-bit register (or memory) to store the result (product). If we continue to multiply this 32-bit result with another 16-bit

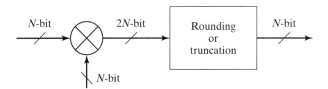

Figure 3.7   Illustration of the need for rounding (or truncation) to prevent wordlength growth.

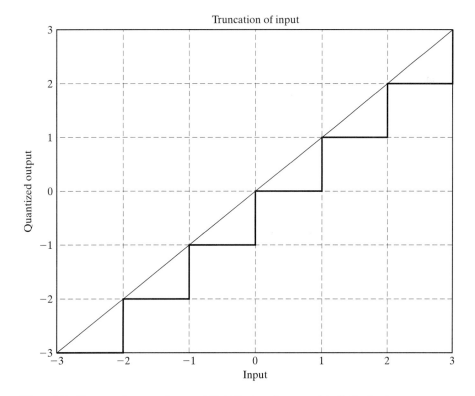

Figure 3.8   The error between the actual (light line) and truncated (solid line) values

number, we require 48 (32 + 16) bits to store the result. Repeat the multiplications and the subsequent intermediate results continue to require more bits. In order to prevent growing wordlength for subsequent multiplications, rounding (or trunca-tion) must be carried out, as shown in Fig. 3.7. Unfortunately, rounding limits the wordlength growth at the expense of increased roundoff errors.

Truncation can be done by assigning the number to the largest quantization level that is less than or equal to the actual data. For example, the actual values 2.9 and 2.1 are truncated to 2 since the quantization level is 1 for the integer format. Figure 3.8 highlights the difference between the actual values (light line) and the truncated values (bold line). The truncation error $e(n)$ lies in the range of

$$0 \leq e(n) < \Delta, \qquad (3.3.14)$$

and the mean of the truncation error is equal to $\Delta/2$.

Rounding is an alternative to truncation. Numbers can be rounded up or down, as shown in Fig. 3.9. For example, the value 2.9 is rounded to 3, whereas the value 2.1 is rounded to 2. The rounding error $e(n)$ lies in the range of

$$-\Delta/2 \leq e(n) < \Delta/2, \qquad (3.3.15)$$

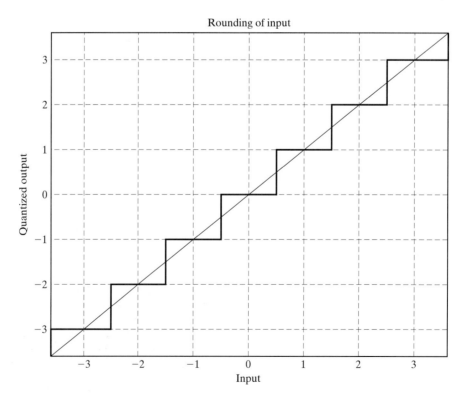

Figure 3.9   Roundoff error between the actual (light line) and rounded (solid line) values

and the mean of the rounding error is zero. Since truncation introduces an undesired bias effect, as shown in Fig. 3.8, rounding is generally used in practice. In most DSP processors, instructions are available to perform rounding after multiplications.

3.4 PROGRAMMING ISSUES

DSP programming is becoming an important skill, and DSP software engineers are in high demand. To keep up with the pace of fast-changing DSP technologies, we must be familiar with programming trends and issues in today's DSP processors. In this section, we explore some general addressing modes and programming concepts used in the TMS320 family. Detailed programming considerations for specific DSP processors are further discussed in Chapters 4 and 5.

3.4.1 Addressing Modes

Instructions use different addressing modes for accessing operands from memory and registers. In this section, we highlight some commonly used addressing modes in TMS320 processors. In particular, we use the immediate, direct, and indirect addressing modes of the TMS320C54x as examples.

The general syntax of an assembly-language source statement contains the following four ordered fields:

```
[label][:]    mnemonic    [operand list]    [; comment]
```

The elements inside the brackets are optional. The optional label is commonly used in assembly programming to refer to a memory location, value of data, address of a subroutine, etc. It always starts in the first column. The required mnemonic field specifies the actual operation performed by the processor. The operand list field is a list of operands, which can be a constant, a symbol, or a combination of constants and symbols in an expression. The comment field is optional, but is very important for documenting the program.

Immediate Addressing

Immediate addressing allows the programmer to operate on an actual value. An example of loading an immediate value (30h) into the accumulator (ACC) A is

```
LD    #30h,A
```

The prompt sign, #, denotes an immediate value, and the load instruction, LD, copies a 16-bit constant (or word from memory) to a CPU register.

Direct Addressing

Direct addressing uses the data-page pointer (DP) to select a main data page in memory and uses the 7 bits of the data memory address (dma) from the instruction as an offset. As shown in Fig. 3.10, the most significant 9 bits from the DP specify 1 of the 512 pages, and the remaining 7-bit DMA addresses 1 of the 128 locations at

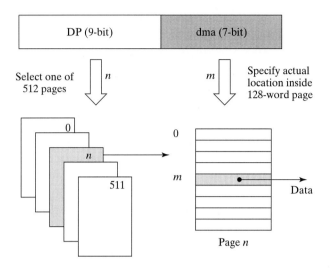

Figure 3.10   Concept of direct addressing used in C2000 and C54x processors

that page. This forms a 16-bit memory address to access data in memory. A simple example of using direct-addressing mode is given as follows:

```
LD     #x,DP   ;  set DP to the page that contains "x"
STL    A,@x+10 ;  store ACC A low to the location at "x+10"
```

The symbol @ denotes the direct address of a variable, the instruction ST (store) is used for transferring a word from a CPU register to a memory location, and L indicates the lower 16 bits of the ACC.

Indirect Addressing

Indirect addressing uses the auxiliary registers (ARn, where n is the number from 0 to 7 for C2000, C3x, and C5000 processors) as address pointers. A symbol * is used in front of the auxiliary register to indicate indirect-addressing mode. Note that in indirect addressing, the auxiliary register points at the data at the memory location.

Circular Addressing

As discussed in Section 1.3.2, a circular buffer is useful when performing linear convolution, FIR filtering, and correlation operations. As shown on the left-hand side of Fig. 3.11, a linear buffer requires all data to be shifted by one position down in the tapped-delay line at each sampling period, the oldest data is discarded, and the new data arrives at $x(n)$. A circular buffer only updates the pointer in modulo fashion without shifting the physical data in memory. As illustrated on the right-hand side of Fig. 3.11, the circular pointer points at the oldest data (❶) and replaces this data with the new data (❷). Circular addressing resets (wraps around) the addressing pointer to the top of the buffer (❸) once the pointer reaches the end of the buffer.

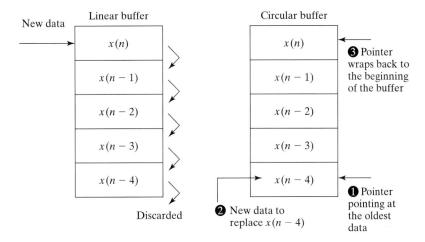

Figure 3.11   Difference between the linear buffer and the circular buffer

Zero-Overhead Looping

Looping is a common operation in most DSP algorithms. Software looping requires the program to keep track of the loop index and loop counter and to check for the end of the loop. Most DSP processors provide zero-overhead hardware loop-control mechanisms for handling single-and block-repeat operations. It allows a single instruction and a group of instructions to be repeated $N + 1$ times (a maximum of 65,536 times for a 16-bit processor), where N is the operand stated in the repeat instruction that will be loaded into the repeat counter. Since all loop operations are done using hardware, programmers only need to specify the loop-counter value and the starting and ending addresses of the loop. An example of a single-repeat operation is given as follows:

```
RPT      #15
MPY      *AR1+,*AR2+,A   ; repeat MPY instruction 16 times
```

An example of a block-repeat operation is given as follows:

```
    STM      #9,BRC           ; repeat block for 10 times
    RPTB     end-1
    LD       #0,A             ; start of block
    MPY      *AR1+,*AR2+,A
    STH      *AR3+            ; end of block
end:
```

Bit-Reversal Addressing

Bit-reversal addressing is used to scramble or unscramble the order of the input or output for decimation-in-time (DIT) or decimation-in-frequency (DIF) radix-2 FFT algorithms. For example, in the DIF radix-2 FFT shown in Fig. 3.12, the output data must be rearranged from their scrambled order to their natural order. In the case of

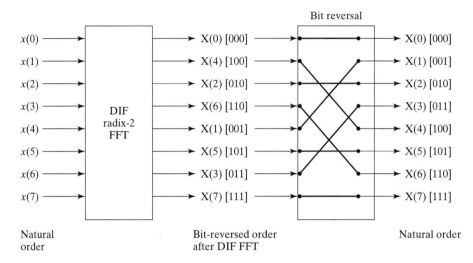

Figure 3.12   Block diagram of the DIF radix-2 FFT (numbers in square brackets are unsigned binary integers)

the DIT radix-2 FFT, the input samples must be rearranged from their natural order to their bit-reversal order before being applied to the FFT algorithm. Bit-reversal addressing is carried out in TMS320 processors using reverse-carry arithmetic. That is, when an increment is added to the address register, the carry bit is propagated to the right. Alternatively, some DSP processors use a simple approach for writing the bit pattern of the address in reversed order (e.g., 100b becomes 001b) to scramble and unscramble the data sequence, or vice versa.

Program-Memory Addressing

The program counter (PC) is the register that points to the program memory to fetch the next instruction. Under normal program execution, the PC is automatically incremented when fetching the next instruction. Several operations alter the PC in different fashion, including (1) branches, (2) function calls, (3) returns, (4) interrupts, (5) repeats, and (6) conditional operations. The program address generator uses a set of registers to modify the PC. These registers contain (1) a repeat counter for single-repeat operations, (2) a repeat counter for block-repeat operations, (3) a block-repeat starting address, and (4) a block-repeat ending address. We discuss these operations for specific processors in Chapters 4 and 5.

3.4.2 Concepts of Pipelining

Pipelining is used intensively to increase the performance of DSP processors. By breaking an instruction into different phases of operation and executing several phases from different instructions in parallel, the time needed to execute each instruction can be reduced. Pipelining maximizes the usage of the independent resources in the processor at any given time. Table 3.11 shows the level of the

TABLE 3.11   Pipeline in Different TMS320 Processors

DSP processor	Pipeline phases
TMS320C2000	F – D – R – X (4 levels)
TMS320C3x	F – D – R – X (4 levels)
TMS320C54x	PF – F – D – A – R – X (6 levels)
TMS320C55x	PF1 – PF2 – F – PD (4-level fetch) D – AD – AC1 – AC2 – R – X – W (7-level execution)
TMS320C6000	PG-PS-PW-PR-DP-DC-E1-E2-E3-E4-E5 (11 levels)

pipeline for different TMS320 processors. The pre-fetch (PF) phase stores the address of the instruction to be fetched, while the fetch (F) phase loads the operation code of the instruction. The decode (D) phase decodes the fetched instruction to determine the type of memory access and control sequence for the processor. The tasks in the access (A) phase include reading the address of the operand and modifying the auxiliary registers and stack pointer if required. The read (R) phase reads the data from the data buses and also writes data to the data buses, if required. The final execution (X) phase executes the instruction and also completes the write (W) process, if required. Other abbreviations used in the table include address (AD); access (AC); pre-decode (PD); program address generate (PG); program address send (PS); program address ready wait (PW); program-fetch packet-receive (PR); instruction dispatch (DP); instruction decode (DC); execute stages 1–5 (E1 to E5).

Pipelining in DSP processors allows different functional units to work simultaneously in the same clock cycle by overlapping different phases from different instructions. Figure 3.13 shows the execution of six C54x instructions and the pipeline associated with these instructions.

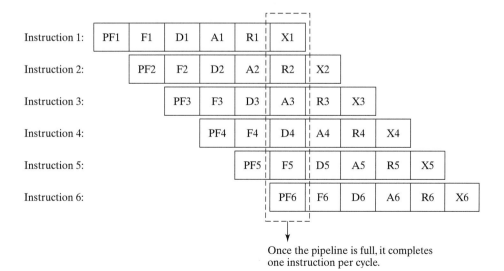

Once the pipeline is full, it completes one instruction per cycle.

Figure 3.13   Six pipeline stages of the TMS320C54x

Pipelining provides very fast throughput, but requires some attention in programming to avoid pipeline conflict. The problem arises when different instructions share resources within the same cycle. In addition, the program flow is affected by a set of instructions that branch or call the subroutine. These operations affect the pipeline because the processor needs to flush out the current pipeline in order to load the new set of instructions. The no-operation instruction is used to fill up the empty slot while the processor determines the new destination. A new set of instructions can be used to avoid the flushing of the pipeline. These instructions are known as delayed branch instructions, which allow the instructions following the delayed branch instruction to be executed normally without needing to flush the pipeline.

3.4.3 Instruction Cache

Instruction cache is becoming a popular feature for modern DSP processors. It is a small and fast on-chip memory that is used to store frequently used code, such as the instructions inside a repeated loop. The instruction cache eliminates the time spent in fetching the instructions from the slow off-chip program memory, thus freeing memory access for a data read (or write). Table 3.12 summarizes the availability of cache memory and their sizes for different TMS320 processors.

3.4.4 Hardware and Software Interrupts

Interrupts are signals that cause the processor to suspend its current program and to execute a special subroutine called the interrupt service routine (ISR). DSP processors can respond to the following sources of interrupt:

1. On-chip peripherals such as serial ports, timer interrupts, etc., that acquire and transmit data samples from CODEC
2. External interrupt lines such as resets, nonmaskable interrupts, and other peripheral interrupts
3. Software interrupts that are generated when trap instructions are executed or when arithmetic exceptions such as divided by zero and overflow have occurred

The sequence of operations following the detection of an interrupt is illustrated in Fig. 3.14.

TABLE 3.12   Cache Size for TMS320 Processors

TMS320 processor	Cache size
TMS320C3x	2 × 32 words
TMS320C55x	64 bytes
TMS320C6000	On-chip program memory acts as instruction cache; C6211 has 8K bytes of L1 program and data cache and has 64K bytes of L2 cache

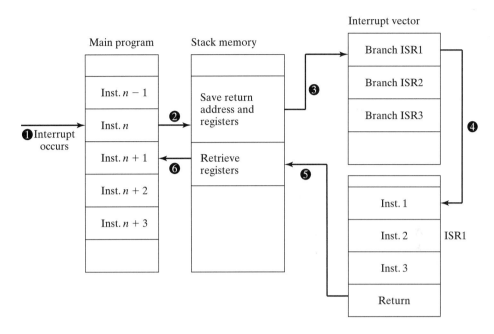

Figure 3.14 Block diagram describing the flow of the interrupt process: ❶ When an interrupt is detected, the processor completes its current instruction n. ❷ The CPU pushes the contents of the PC (which points to the next instruction) and status registers into the stack. The stack is the reserved memory for storing the values of return PC and registers. ❸ Depending on the type of interrupt, the interrupt vector specifies the address of the corresponding ISR. The interrupt vector is a set of memory that contains addresses associated with the specific ISR. It typically occupies two to four words for each interrupt, and contains a branch instruction to the corresponding ISR. ❹ A new set of instructions in the ISR is executed. ❺ The last instruction in the ISR is returned from the interrupt. Upon completion of the ISR, the processor retrieves the values of the PC and the registers in the stack. The CPU then fetches and executes the next instruction $(n + 1)$, as shown in ❻. It continues to execute the main program until it encounters another interrupt, after which time steps ❶ through ❻ are repeated.

The definitions of timing when the processor enters the ISR are illustrated in Fig 3.15. Interrupt latency is the time from when an interrupt event occurs to the time the processor starts saving its return PC and other registers. Interrupt latency is not always predictable, since the current interrupt may be blocked by a higher-priority interrupt. Interrupt response extends beyond the interrupt latency by requiring extra time for saving the processor contents. Interrupt recovery is the time required by the CPU to restore its PC and other registers.

Figure 3.16 shows a detailed example of an interrupt for an I/O operation when an ADC acquires a digital sample, which triggers an interrupt request to the processor. The processor needs some time to respond to this interrupt and then obtains the sample value from the input port before processing the digitized input signal. The processed signal is stored to the output port before returning to the main program. All these operations must be completed within the specified

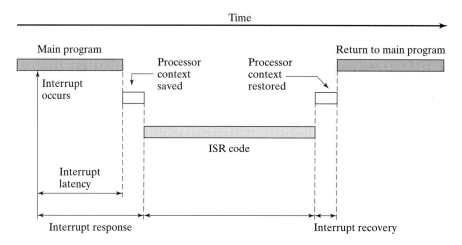

Figure 3.15   Definitions of timing during an interrupt

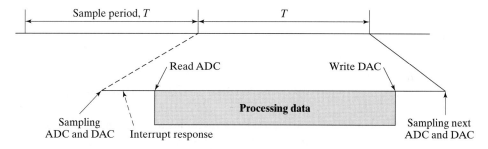

Figure 3.16   Detailed interrupt timing at each sampling interval for an I/O operation

sampling period T. Note that the sampling operations of the ADC and DAC usually are carried out at the same time since they are synchronized and since they use the same ISR.

3.5 REAL-TIME IMPLEMENTATION CONSIDERATIONS

Consider a DSP system that produces an output signal $y(n)$ based on an input signal $x(n)$ expressed as

$$y(n) = T[x(n)]. \tag{3.5.1}$$

Upon arrival of the input sample $x(n)$, the DSP processor performs the computation described by the algorithm, which requires a certain computational time. In addition to this sample-by-sample processing, some DSP systems also perform block processing. In this case, N samples of input signals are collected over a period of time. This set of input samples is referred to as a frame (or block) of length N. The whole signal block is processed to produce a set of output samples. For example, the FFT shown in Fig. 3.12 shows a block processing that processes a block of eight input samples and produces a block of eight DFT coefficients.

3.5.1 Signal Converters

The ADC is a device that converts an analog signal $x(t)$ to a digital signal $x(n)$. A DAC performs the reverse process, converting a digital signal $y(n)$ to an analog signal $y(t)$. Some DSP processors have integrated these converters on chip, while others use off-chip ADC and DAC. As introduced in Section 1.6.4, a CODEC is a device that combines the ADC and DAC with the associated anti-aliasing and smoothing filters on a single chip. Many parameters must be set in the ADC, DAC, and CODEC in order to digitize the right samples at the right time. These parameters are summarized as follows:

1. *Bit resolution*: Most ADC, DAC, and communication CODECs are 8-, 10-, 14-, 16-, or 18-bit devices, while audio CODEC has a higher resolution of 16-, 18-, 20-, or 24-bit. The gain of dynamic range is about 6 dB per bit. Thus, a 16-bit converter has a dynamic range of 96 dB.
2. *Sampling rate*: Most ADCs and DACs can be operated at a higher sampling rate (more than 10 MHz) compared to the CODEC, which usually operates in the audio range of up to 48 kHz. The sampling rate must be more than twice the bandwidth of the signal. A rule of thumb is to add at least 20% more of its highest frequency component of interest.
3. *SNDR*: SNDR is the ratio of the signal power to the power of the converter-induced noise and harmonic distortion. This ratio is used to measure the quality of the converter and is defined in Eq. (3.3.8).

3.5.2 Stream Processing

Stream (or sample-by-sample) processing starts upon the arrival of the input sample $x(n)$, performs an identical set of operations at each sampling interval, and completes the operations before the next sample arrives. All operations must be completed within one sampling period. This is called the real-time constraint of stream processing. In order to meet this real-time constraint, the computational time T_s must satisfy

$$T_s \leq T - T_H, \tag{3.5.2}$$

where T_H is the hardware overhead time that includes both ADC and DAC and the data transfer between the DSP processor and the I/O devices and where T is the sampling interval.

The advantages of stream processing include the following:

1. All results are kept current within the sampling period.
2. Delay between the input and the output is kept to the theoretical minimum.
3. Storage of the input and output samples is also kept to the theoretical minimum.

The disadvantage of stream processing is the overhead of reading and writing each data sample. Therefore, processors must be fast enough to complete all operations before the next input sample arrives.

3.5.3 Block Processing

In block processing, incoming input samples $x(n)$ are first stored in a memory buffer. After N samples have arrived, the entire block of data samples is processed at once to produce the output signal $y(n)$. Therefore, block processing is conducted for every NT seconds. As illustrated in Fig. 3.17, the processing of data begins after the last sample of the input block has arrived. The following activities (shown in the

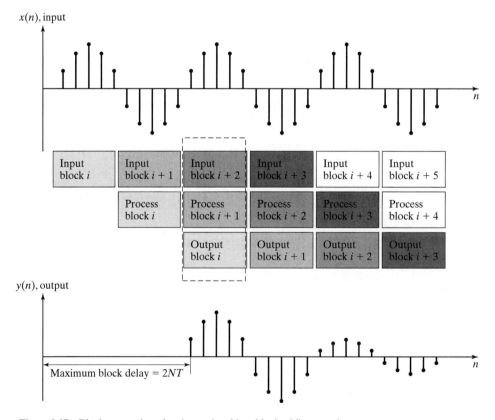

Figure 3.17   Block processing of an input signal in a block of five samples

dotted box) are performed simultaneously:

1. The processing of N samples in the current block, $i + 1$.
2. Collection and storage of the input samples for the next block, $i + 2$.
3. Output of the data samples in the previous block, i.

In order to meet the real-time constraint for the block processing shown in Fig. 3.17, the computation time for block processing T_b has to satisfy

$$T_b \leq NT - T_o, \tag{3.5.3}$$

where T_o is the overhead for block processing. The overhead of I/O operations is reduced in modern processors that have powerful DMA controllers, which we discuss later.

The advantages of block processing include the following:

1. It allows a slower processor to keep up with input samples. These samples may be buffered and used in computation after all of the input has ceased.
2. It reduces the overhead of read/write operations to memory. It performs the I/O operations in every NT seconds instead of T seconds for stream processing.

However, the disadvantages of block processing are the $2NT$ samples delay, as illustrated in Fig. 3.17, and the need for additional memory storage. For example, $4N$ memory locations are required for input and output samples using the double-buffering method. Furthermore, the processor must compute the current block of data while receiving data samples from the I/O devices, store them at the right locations for the next block processing, and output the previous block of data samples in a sequential fashion, which requires switching between data processing and I/O processing for the DSP processor.

3.5.4 Vector Processing

Vector (or multichannel) processing takes in multiple inputs at any given time. A simple example is stereo input, where the audio CODEC takes in the left- and right-channel inputs at the same time. These inputs can undergo the same or different processing operations. This example can be extended to surround-sound systems such as Dolby Digital, where 5.1 channels (front left, front right, center, rear left, rear right, and a limited-frequency channel of bass signal) must be sampled at the same time. The vector samples can be captured either in sample or block manner, as described in previous sections.

3.5.5 Benchmarks

The performance of some TMS320 processors in terms of speed, memory, and voltage is summarized in Table 3.13. In this subsection, we show several examples that illustrate the selection of the DSP processor to match the performance requirements for different applications.

TABLE 3.13   TMS320 Rating in Terms of Speed, Memory Size, and Voltage/Power Consumption

DSP processors	Speed (instruction cycle)	On-chip memory	Voltage (core) / Power consumption
C2000	25–50 nsec	544 × 16 DARAM 2–4K × 16 SARAM 4K–32K × 16 PROM	3.3V–5V ~0.144 W
C3x	13.3–40 nsec	512–34K × 32 RAM 4K × 32 ROM (C30)	3.2V–5V 1.2W–0.2W
C54x	6.25–33nsec	4K–128K × 16 DARAM 0K–256K × 16 SARAM 4K–16K × 16 ROM	1.8V–5V 0.012W–0.2W
C55x	5–6.25 nsec	32K × 16 DARAM 128K × 16 SARAM 16K × 16 ROM	1.6V ~0.1W
C62x	1.67–3.3 nsec	16K–96K × 32 PRAM 32K–256K × 32 DRAM	1.5V/3.3V 0.8W–2.1W
C64x	~0.4 nsec	512 × 256 L1 P cache 4K × 32 L1 D cache 4K × 256 L2 cache	1.2V/3.3V
C67x	3–5 nsec	(C6701) 16K × 32 PRAM 16K × 32 DRAM (C6711, C6712) 128 × 256 L1 P cache 512 × 64 L1 D cache 8K × 64 L2 cache	1.8V/3.3V 0.7W–1.4W

Example 3.7

In a CD player, the sampling rate is 44.1 kHz, and each audio sample is quantized to 16 bits. (1) If the CD player requires a 300-tap FIR filter, what kind of DSP processor has an adequate million-instructions-per-second (MIPS) rating to perform this task? (2) What is the size of memory required to store up to 30 minutes of digitized stereo signal?

Solution

1. To perform FIR filtering of length $L = 300$, the processor needs 300 MACs per input sample. Most processors can execute one MAC in a single instruction cycle. Therefore, the processor requires 300×44.1 K $= 13.23$ MIPS. A DSP processor with 20 MIPS is required when additional overheads are included.
2. Total numbers of bits per second is 16 bit/sample $\times$ 44.1 K sample/second $=$ 705.6 kilobits per second (Kbps). The RAM size required to store 30 minutes of data samples per channel is 705.6 kbps $\times$ 30 $\times$ 60, which is about 1.27 giga bits (or 80 mega words). Therefore, about 160 M words is required for storing a dual-channel stereo signal.

Example 3.8

A bandlimited analog signal is sampled at 16 kHz over a time interval of 10 seconds. The power spectrum of the signal will be estimated using the FFT-based approach. (1) What is the length of the data record in terms of number of samples N? What length of the radix-2 FFT is required? (2) An alternative approach is to divide the data record into many blocks of length M. What is the minimum M required to achieve the desired frequency resolution of 15 Hz?

Solution

1. The total number of data records is $16\,K \times 10 = 160,000$ samples. Therefore, the FFT length required is $2^{18} = 262,144$. This large-size FFT takes a long time to compute, and it is meaningless if the signal changes with time.
2. As discussed in Chapter 2, the resolution of the spectrum is $\Delta = f_s/N$. Therefore, $N = f_s/\Delta = 16,000/15 = 1,067$. The nearest radix-2 FFT order is $M = 2,048$.

Example 3.9

The typical fixed-point C54x described in Table 3.13 can execute a single instruction in 10 nsec for a 100-MIPS processor. One of the routines given in the DSP library can execute an L-tap FIR filter in $\sim 38 + L$ instruction cycles per input sample. The same routine executes a block of N input samples in about $38 + N(4 + L)$ instruction cycles. (1) What is the maximum bandwidth of the signal that can be filtered with an FIR filter of order $L = 255$ in real time for stream processing and block processing with $N = 16$? (2) If a speech signal is sampled at 8 kHz, what is the highest FIR-filter order that may be used in real time for both stream processing and block processing?

Solution

1. Stream processing of a 255-tap FIR filter requires 293 $(38 + 255)$ instruction cycles to compute one output sample. The maximum sampling rate is $f_s < 1/(293 \times 10\,\text{nsec}) \sim 341$ kHz. Thus, the maximum bandwidth of the signal is about 170 kHz $(f_s/2)$. For block processing with size $N = 16, 38 + 16 \times (4 + 255)$ instruction cycles are required to complete a block of 16 samples. Therefore, $f_s(\text{max}) = 16/(4,182 \times 10\,\text{nsec}) \sim 382$ kHz. The maximum bandwidth of the signal is about 191 kHz.
2. The sampling frequency, f_s, is 8 kHz, and the sampling interval is $1/8,000 = 0.125$ msec. Since $(38 + L) \times 10\,\text{nsec} < 0.125$ msec for real-time stream processing, $L < 12,462$. Similarly, for real-time block processing of 16 samples, $(38 + 16 \times (4 + L)) \times 10\,\text{nsec} < 16 \times 0.125\,\text{msec}, L < 12,493$.

3.6 HARDWARE INTERFACING

The core of the DSP processor is surrounded by a set of peripherals that link the processor to other devices. Figure 3.18 shows some of the commonly used hardware interfacing blocks that help the processor control and communicate with external devices. This section introduces the functionalities of these blocks. Some practical examples of hardware interfacing are given in Appendix C.

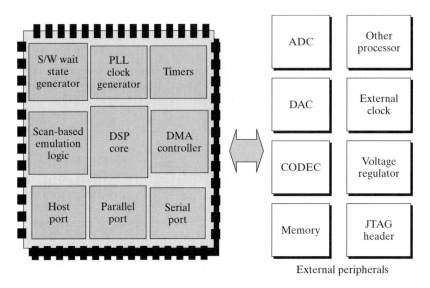

Figure 3.18   Block diagram of internal and external peripherals

3.6.1 External-Memory Interfacing

The DSP processor usually provides limited on-chip memory for program and data storage. External memory provides storage for program code, data, boot-tables, etc., that are too large to fit into internal memory. However, the penalty for storing programs and data off chip is slower access time and limited bus availability for external data transfer. Figure 3.19 shows a generic diagram of the DSP processor interfacing to external memory.

The latest generation of the TMS320 processor offers an advanced external-memory interface that provides a glueless interface to a variety of external memory such as static RAM, dynamic RAM, first-in-first-out memory, EPROM, flash, field programmable gate array, and ASICs. An external memory interface is able to interface with 8-, 16-, or 32-bit wide memories and aligns the bytes used in the system. The external memory interfaces with the DSP processor via the enhanced DMA controller, and program-memory and data-memory controllers.

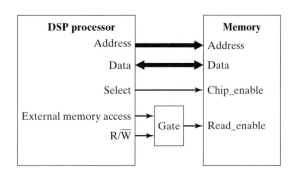

Figure 3.19   Connecting the DSP processor to external memory

As discussed in Chapter 1, wait states are required when the processor is waiting for memory access or other slower devices. The reasons for requiring wait states are to (1) interface with slower off-chip memory, (2) perform multiple accesses of single-access memory, (3) share the common bus among other processors, and (4) provide a smooth transition between two external memories. Wait states can be inserted via a wait-state register. In TMS320 processors, the wait-state generator can extend external bus cycles from 1 to 14 clock cycles. More than 14 clock cycles can be implemented in hardware via the ready line in TMS320 processors.

3.6.2 Timers and a Master Clock

Most DSP processors have programmable internal timers, which are used as clocks that generate periodic interrupts, benchmark the elapse time run by code, and generate signals with software control. A block diagram of a typical timer is illustrated in Fig. 3.20, which shows a clock source, a prescalar, and a down counter. The master clock can be derived from (1) an externally generated clock signal, (2) an on-chip oscillator with an externally connected crystal, or (3) an on-chip PLL or frequency synthesizer that produces a master clock from the external clock signal.

The master clock in the DSP processor typically generates a clock at a rate of 100 MHz or higher. The clock rate may be equal to the MIPS rating of some DSP processors. In other processors, multiple clocks are required for each instruction. For example, a processor with a clock rate of 100 MHz that needs two clock cycles for executing each instruction only operates at 50 MIPS.

The prescalar divides the source-clock frequency by some integer values. The down counter counts down on every rising clock edge from the prescalar's output until its preloaded value reaches zero. At this point, the counter interrupts the processor. A 4-bit prescalar commonly is used to divide the master clock frequency by a factor of 1 to 16. The typical size of the down counter is 16 bits, which provides up to $2^{16} = 65,536$ steps in counting down. Once zero is reached, the timer interrupts the processor and reloads the down counter with a user-specified value. Table 3.14 shows some timers and their associated registers that are available on TMS320 processors.

3.6.3 Serial-Port Interfacing

In a typical DSP application, the processor needs to handle multiple sources of data from other devices. The serial port operates 1 bit at a time and commonly is used in

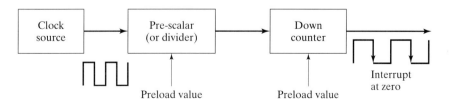

Figure 3.20   Typical timer circuits

TABLE 3.14   Typical Timers and Their Registers

DSP processor	The number of timers and the size of their registers
TMS320C2000	One 16-bit timer and down counter, a prescalar (4 bits), and a watchdog timer
TMS320C3x	Two 32-bit timers (divide the internal clock by 2)
TMS320C5000	One or two 16-bit timers and down counters and a prescalar (4 bits)
TMS320C6000	Two 32-bit timers (divide the internal clock by 4)

sending and receiving data between the processor and the ADC, DAC, and CODEC; communicating with other processors; and interfacing with other external peripherals. Therefore, DSP processors must have the ability to receive and transmit serial data in real time without interrupting their internal operations with memory accesses. A DMA controller is usually present in the latest DSP processors to transfer data between memory and the I/O ports without interrupting the processors. A high-speed serial port is usually available in most DSP processors to handle audio and telecommunication signals at a rate of more than 10 Mbit/sec. As shown in Fig. 3.21, four lines are required in a synchronous serial port: clock, data (transmit and receive), and frame synchronization.

Connecting two devices via the synchronous serial link requires a serial clock that synchronizes both devices and a frame synchronization signal to deliver the data at the required sampling frequency. The serial clock and frame synchronization signal can be generated from the CODEC, which acts as a serial bus master. Table 3.15 shows some common characteristics and features of serial ports in TMS320

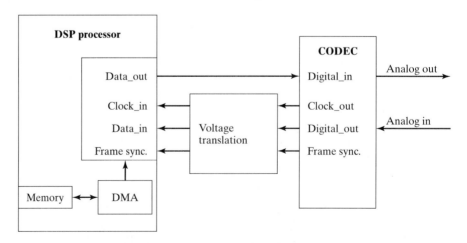

Figure 3.21   Interfacing the DSP processor with the CODEC, where the voltage translation is required only for different operating voltages

TABLE 3.15 Serial Ports Supported by TMS320 Processors, where CLKOUT is the Master Clock Output Signal

DSP processor	Features and characteristics
TMS320C2000	Synchronous (8- or 16-bit) and asynchronous (8-bit) serial ports Synchronous: Maximum bit rate up to CLKOUT/2 Asynchronous: Maximum bit rate up to CLKOUT/16
TMS320C3x	Synchronous (8-, 16-, 24-, or 32-bit) serial port Maximum bit rate up to CLKOUT/2
TMS320C5000	Synchronous (8- or 16-bit) serial port Maximum bit rate up to CLKOUT/4
TMS320C6000	Two to three synchronous (8-, 12-, 16-, 20-, 24-, or 32-bit) serial ports Maximum bit rate up to 100 Mbit/second

processors. An asynchronous serial port does not need a synchronous clock and is rarely used in DSP processors.

When interfacing between a DSP processor and a CODEC that operate at different voltages, voltage translation is needed if those signal lines originated from the CODEC. In most cases, the signals originating from the DSP processor, such as Data_out, do not need to be translated since the high and low voltages from the DSP processor are compatible with the CODEC.

Some TMS320 processors provide a BSP, which is a full-duplex, double-buffered serial port. The BSP supports direct communication with the ADC, DAC, and other serial devices with little glue logic. It provides an autobuffering mechanism that transfers data between the serial port and the processor's internal memory. Furthermore, the BSP provides the flexibility to define the frame synchronization signal, clock rate, polarity of the clock, and data-frame transmission.

A further extension to the BSP is the McBSP, which allows the DSP to handle multiple channels (e.g., the C5000 supports up to 128 channels). In the multi-channel mode, the McBSP provides one or multiple devices per time slot to transfer data within a frame-synchronizing period, which is done by time-division multiplexing the data stream. In addition to the features found on the standard BSP, the McBSP interfaces directly to popular digital communication channels such as T1/E1 framers, and other high-speed serial devices. Most DSP processors also support A-law (or μ-law) companding in serial-port hardware without the conversion overhead in software. A detailed description of the McBSP is given in Appendix C.

3.6.4 Direct Memory Access Controller

The DMA controller transfers data between internal and external memories, peripherals, and external devices without the intervention of the processor. Most TMS320 processors have DMA channels that move data from source to destination locations via dedicated I/O ports. Each DMA channel has the following set of registers that must be initialized prior to data transfer:

1. The source-address register contains the address of the source memory element that is transferred.
2. The destination-address register contains the address of the destination memory element that is transferred.
3. The element count register contains the number of elements (8, 16, or 32 bits) to be transferred within a frame.
4. The synchronization select-and-frame-count register contains synchronization events that trigger the DMA transfer and the number of frames to be transferred. A frame can contain 1 to 64K elements. A block is a collection of frames, and up to 64K frames can be supported within a block. For example, a block can contain three frames, and each frame can consist six elements, as illustrated in Fig. 3.22.
5. The transfer-mode control register determines the transfer modes of the DMA channel, which include channel priorities, generation of an interrupt to the processor based on the transfer events, and initialization of a new transfer.

There are many important uses of DMA channels. One example is in program paging, which is used when program code is too large to fit into on-chip memory and needs to be run on internal memory to meet real-time requirements. In this application, the DMA channel brings a block of code from external memory to internal memory while the DSP processor is executing another block of code. We can perform this action by partitioning internal memory into at least two sections and performing the necessary switching by the processor without disruption. Another application of the DMA channel is to transfer and execute data in a ping-pong (or double-buffering) fashion. This scheme is described for block processing

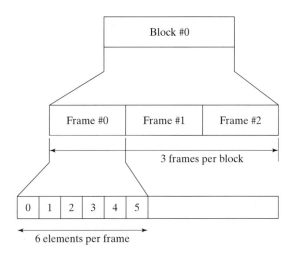

Figure 3.22 Examples of using six elements per frame and three frames per block for the DMA transfer

in Section 3.5.3. In this application, the DMA channel is used to deliver N samples of data in and out of the buffers, while the DSP processor operates on data in the current buffer. The roles of these buffers are changed for the next block of processing. A detailed description of DMA initialization is given in Appendix C.

3.6.5 Parallel-Port Interfacing

A parallel port is used to transmit and receive multiple data bits (typically 8 or 16 bits) at a time. Therefore, the parallel port can transfer more data at a faster rate compared to the serial port. However, it requires more pins for transferring multiple bits and handshake lines for synchronization. In TMS320 processors, parallel ports use part of the data bus by assigning an address space for I/O communication over the external bus. This approach saves the pin count, but complicates the interfacing of external devices. An alternate approach is to have a separate set of pins for the parallel port to simplify the interfacing, but at the expense of requiring more pins.

3.6.6 Host-Port Interfacing

Host communication is important when a general-purpose microprocessor, a micro-controller, or another DSP processor is used to supervise the DSP processor. Host ports are typically 8- or 16-bit bidirectional ports that communicate with other processors that have different (or the same) bus standards. An HPI is a dedicated parallel interface that allows the host to communicate with the DSP processor. Figure 3.23 shows the common signals associated with the HPI. Data is transferred via the 8- or 16-bit data bus, while the read/write line specifies the direction of the data transfer. The ready line generated by the DSP processor informs the host that the HPI is ready for a new transfer, while the interrupt line interrupts the host. The host device communicates with the TMS320 processor's memory through three dedicated HPI registers (address, data, and control). The host device has full control of the three registers, but the DSP processor can only access the HPI control register. The HPI usually has its own dedicated DMA channel and port within the DMA controller. More information on interfacing signals and the HPI's registers can be found in Appendix C.

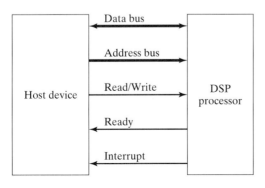

Figure 3.23   Block diagram of the interfacing signals between the DSP processor and the host device

3.6.7 Multiprocessing Techniques

Multiprocessing is necessary when a single processor is not able to support the high processing demand or handle multiple data streams in real-time applications. Many real-time signal-processing algorithms can be decomposed temporally or spatially and mapped into multiple processors to achieve higher real-time performance in some demanding applications. Some DSP processors provide special communication ports to connect several processors. For example, the TMS320C40 has six link (or communication) ports. Figure 3.24 shows the block diagram of different configurations for connecting one processor with another.

The communication port generally provides the bi-directional data bus, address bus, and bus-request lines. Due to the massive pins used in transferring 32-bit data at one transfer, multiple transfers of 8-bit packets are used to move one 32-bit word in the C40. The communication port is usually operated under the DMA, which allows data at the port to be written into the processor memory automatically without interrupting the processor. The maximum transfer rate is 20 Mbyte/sec for the communication port in the C40, and the maximum combined transfer rate of $20 \times 6 = 120$ Mbyte/sec can be achieved theoretically.

TMS320 processors also support interconnection between processors via the McBSP, which serves as a high-speed data communication port. For example, as shown in Fig. 3.25, processor #1 can be configured as a master processor when transmitting data to the other processors. It generates the necessary clock and frame-synchronization signals during the transmission frame, while processor #2 acts as a slave processor that receives data. Similarly, the McBSP of processor #2 is configured as the master during its transmission.

3.6.8 Power-Supply Regulator

Power-supply requirements are normally different for different parts of the DSP processor. Therefore, a power supply controller is needed to deliver the correct

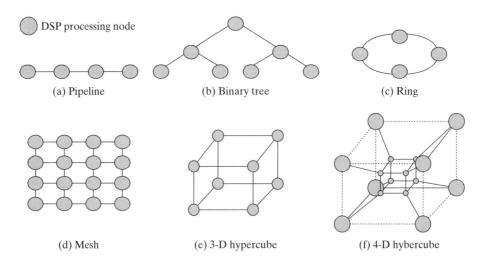

(a) Pipeline	(b) Binary tree	(c) Ring
(d) Mesh	(e) 3-D hypercube	(f) 4-D hybercube

Figure 3.24   Different parallel configurations for a DSP processor with six communication ports

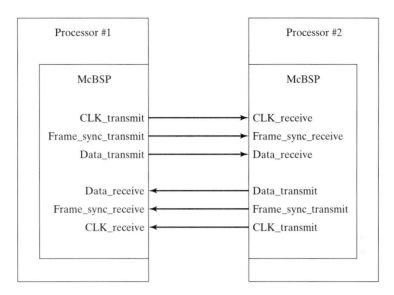

Figure 3.25   McBSP connection between two DSP processors

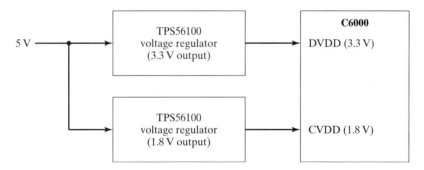

Figure 3.26   Dual power supplies from a single 5V supply

voltages to different parts of the processor. For example, we can use the TPS56100 power-supply controller to deliver 3.3V to the external-interfacing peripherals (such as serial port) and 1.8V to the C6000 DSP core and internal circuitry (such as the PLL clock-generation circuitry), as shown in Fig. 3.26. The advantages of using a dual power supply are the ability to connect the external-interfacing pins of the DSP processor to external logic circuitry without requiring additional level-shifting circuits and the ability to operate the DSP core at a lower voltage. However, an issue arises in using a dual power supply when considering the sequence of turning on individual power supplies. The ideal case is to turn on the two supplies simultaneously, but this may not be possible. For the C6000, the higher supply voltages delivered to the DSP core must be turned on before delivering power to the external-interfacing pins of other peripherals. The sequence is reversed for the C5000 DSP processor.

3.6.9 Emulator-Interconnect Standard

The emulator accesses the DSP processor and its peripherals without requiring additional hardware and software to be installed in the DSP system, which allows debugging of the DSP system while the processor is running at full speed. The hardware consists of an emulator card that is inserted into the host-computer's bus. The other end of the emulator card is connected to an emulator-cable pod that contains a test-bus controller, which controls the scan information of the target processor. This is known as scan-based in-circuit emulation. The debugging software installed in the host computer allows the user to download code to the targeted DSP system and perform limited debugging tasks.

TMS320 processors use the IEEE standard 1149.1 scan-based emulator known as JTAG. It consists of pins for (1) receiving and sending serial data in and out of the DSP processor, (2) clocking the internal emulation logic, (3) setting the processor mode, and (4) enhancing debugging capabilities such as benchmarking, breakpoints, and profiling. The JTAG standard also supports multiprocessor debugging.

3.7 EXPERIMENTS AND PROBLEMS

This section uses several experiments to demonstrate some important concepts in performing fixed-point arithmetic. The main objective of these experiments is to enhance the understanding of several fixed-point considerations introduced in Sections 3.2 and 3.3.

3.7.1 Experiments Using MATLAB

In this section, we use MATLAB tools Simulink and Fixed-Point Blockset to study finite-wordlength effects. Simulink is a powerful software package for modeling, simulating, and analyzing dynamic systems. For modeling purposes, Simulink provides a GUI for building simulation models as interconnections of block diagrams. Some basic information on using Simulink is included in Appendix A.

Overflow Analysis and Solutions

In this experiment, we use a simple example to show the usage of saturation arithmetic, scaling, and guard bits. Figure 3.27 shows the block diagrams for multiplying and adding two sinewaves using fixed-point arithmetic. We can open this Simulink example by typing `fxpmult_add` in the MATLAB command window. Two fixed-point conversion blocks convert the double-precision inputs to the Q.15 fractional format using `sfrac(16)` as the **output data type**. We also check **saturate to max and min when overflows occur** on these boxes. The data types of the output for FixPt Sum and FixPt Product are set to `sfrac(32)` to specify that the arithmetic results are stored in Q.31 format in a 32-bit accumulator. Note that the saturation mode is unchecked in order to show a wrapping effect on overflow. Finally, click on the play button ▶ to run the simulation.

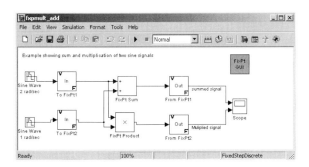

Figure 3.27 Multiplying and adding two sinewaves using Simulink with Fixed-Point Blockset

The results are displayed in the scope as shown in Fig. 3.28(a), and the statistics of the arithmetic can be viewed in Fig. 3.28(b) by double-clicking on the **FixPt GUI** block. The summed signal is distorted because the Q.15 fractional addition results in overflow. In the case of Q.15 fractional multiplication, no overflow is observed due to the contained dynamic range of fractional multiplication, as explained in Section 3.2.1. The Fixed-Point Blockset interface tool displays the statistics of the arithmetic. The arithmetic properties of the associated block appear after the simulation. These properties indicate the number range, number type, and scaling applied to the signal. When the signal overflows or saturations occur, a message is displayed to indicate the number of times overflows or saturations have occurred.

Figure 3.28(b) shows that more than 20 saturations are occurring in the To FixPt1 and To FixPt2 blocks. Saturations occur when +1 is entered into the blocks that only take in Q.15 numbers. An effective method to prevent these saturations is to limit (or scale) the sinewave values to less than 1. Furthermore, more than 3,000 overflows have occurred in the Fixpt Sum block in the simulation.

The overflow errors can be replaced with less severe saturation errors by double-clicking on the Fixed-Point Blockset interface window and checking on the saturation mode for addition. As introduced earlier, a better technique to prevent overflow and saturation in Q.15 addition is to use guard bits in the accumulator. For example, by using sfrac(32, 1) in the Fixpt Sum block, 1 guard bit is introduced and no overflow is observed.

Rounding and Truncation

Truncation can be performed by assigning a number to the largest quantization level that is less than or equal to its actual value. In the fractional-number format, simply ignore the bits to the right of the binary point. For example, truncate a Q2.5 number 3.5 (011.10000) to 3 (011) by ignoring the bits after the binary point. However, when truncating a two's complement negative number, we cannot consider only the integer portion of the number. For example, truncating the Q2.5 number -0.5 (111.10000) results in -1(111) instead of 0. Figure 3.8 highlights the difference between the actual values (light line) and the truncated values (bold line). As discussed in Section 3.3.4, truncation is seldom used in DSP arithmetic due to the larger quantization error and the undesired biased effects. A better method for limiting

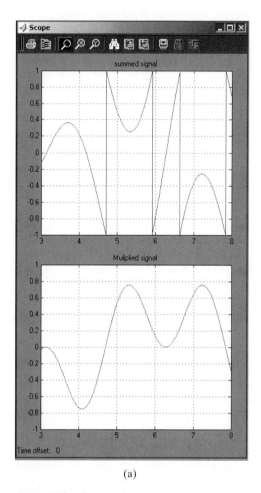

(a)

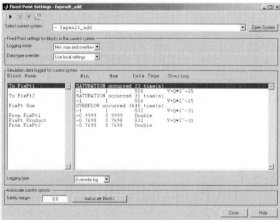

Figure 3.28 Simulink simulation results, where (a) shows the sum and product of two sinewaves, and (b) shows the statistics of the arithmetic

(b)

wordlength is to round the number to the nearest value that can be represented. For example, the value 2.6 is rounded up to 3, whereas the number 2.4 is rounded down to 2.

There are four commonly used rounding methods: (1) round toward nearest, (2) round toward zero, (3) round toward ceiling, and (4) round toward floor. These rounding schemes can be simulated easily in Fixed-Point Blockset or by using MATLAB functions. For example, MATLAB provides four different functions to round a number to its nearest integer in accordance with the aforementioned rounding methods: (1) `round`, (2) `fix`, (3) `ceil`, and (4) `floor`. Fixed-Point Blockset also provides an option to select one of the four rounding modes as (1) nearest, (2) zero, (3) ceiling, and (4) floor.

An example of quantizing sinewave values using the Q5.2 format illustrates the behavior of different rounding schemes. In Fig. 3.29(a), the numbers are rounded to the nearest level. This scheme has the smallest error of all rounding modes, and roundoff errors are symmetric about the positive and negative numbers. Rounding

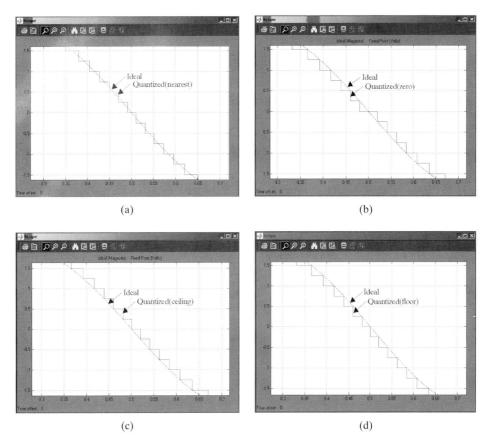

(a)   (b)

(c)   (d)

Figure 3.29 The performance of different rounding modes: Round toward (a) nearest, (b) zero, (c) ceiling, and (d) floor

towards zero is shown in Fig. 3.29(b), where numbers are rounded to smaller numbers. Therefore, a negative cumulative-bias error is introduced in positive numbers, and a positive cumulative-bias error is introduced in negative numbers. In Fig. 3.29(c), rounding towards the ceiling results in a positive cumulative-bias error for all numbers, while rounding towards the floor in Fig. 3.29(d) results in a negative cumulative-bias error.

As previously described, rounding towards the floor is similar to truncation. Among the four rounding schemes, only rounding towards the nearest has a zero-mean error. In TMS320 DSP processors such as the C2000, adding 8000h to the accumulator performs the rounding operation.

Limit Cycle in Infinite-Impulse Response Filters

The output of an IIR filter may oscillate indefinitely when a small constant value is input to the filter. This phenomenon, known as the limit cycle, occurs because of the quantization errors introduced by rounding or truncation. In this section, we use the first-order IIR filter described in Section 2.6.1 to examine the causes and remedies of the limit cycle.

The IIR filter is implemented using Simulink with Fixed-Point Blockset (fxlimitcycle.mdl), as shown in Fig. 3.30. An input value of 0.5 is fed into the IIR filter. The summing and gain blocks of the IIR filter are set to sfrac(8) with different rounding schemes. The effect of different rounding schemes is examined one by one. Furthermore, the limit-cycle effect is examined for both lowpass (using Gain = 0.75) and highpass (using Gain = −0.75) filters.

When Gain is set to −0.75, a highpass filter is implemented. The outputs of the IIR filter are plotted in Fig. 3.31(a) to 3.31(d) for rounding to (a) floor, (b) ceiling, (c) zero, and (d) nearest, respectively. Double-click on the respective block, and choose the desired rounding modes for Add, Gain, Gain1, and FixPt to Dbl blocks. In order to observe the initial response of the output, set the simulation stop time to 0.05 second. The output oscillates within a range of constant levels known as a deadband. The deadband has the largest value when round to floor is used and the

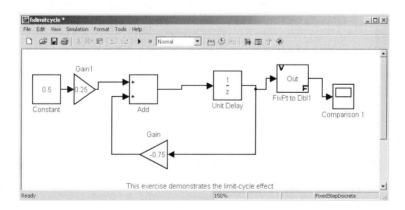

Figure 3.30   Simulink implementation to investigate the limit cycle

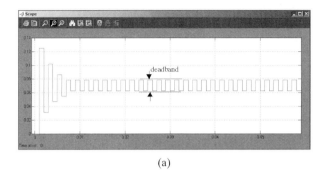

(a)

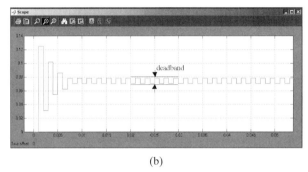

(b)

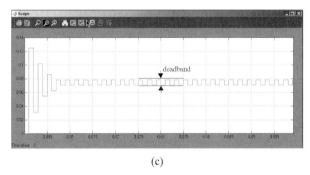

(c)

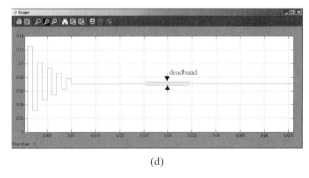

(d)

Figure 3.31 Limit cycle of an IIR filter due to different rounding methods: Round to (a) floor, (b) ceiling, (c) zero, and (d) nearest

smallest value for round to nearest. When the wordlength is increased to sfrac(16), the deadband reduces to insignificant values for all rounding schemes.

We can also examine the limit-cycle effect for the lowpass filter by simply changing Gain to 0.75. No significant deadband is observed for a lowpass filter with a small wordlength due to the fact that the lowpass filter is filtering out undesired high-frequency oscillations.

3.7.2 Experiments Using Fixed-Point C

As presented in Section 2.6.2, an IIR filter is implemented in C (iir1.c) using the floating-point data type, float, which is based on the single-precision, 32-bit IEEE-754 floating-point format. In this section, we investigate techniques for converting a floating-point C program to fixed-point C code, which results in a more efficient implementation on fixed-point DSP processors such as the C5000. Since there is no data type in C to represent the fixed-point fractional format exactly (such as the Q.15 format), we have to represent variables and constants using 16-bit-integer (int) data types. In addition, multiplying two 16-bit numbers results in a 32-bit product, which can be represented using the long data type. The fixed-point C program iir1fx.c is listed as follows:

```
#include <stdio.h>
#include <stdlib.h>
#include <math.h>

void main()
{
  /* Define variable arrays, define and clear variables */
  int xn = 0;              // x(n) from xn.dat file
  long yn = 0;             // float yn = 0.0
  int alpha = 8192;        // alpha = 0.25*32768
  int alpha1 = 24576;      // alpha1 = 1 - alpha = 0.75*32768

  /* Declare file pointers */
  FILE *xn_in;                     // file pointer of x(n)
  FILE *yn_out;                    // file pointer of y(n)
  xn_in = fopen("xn.dat","r"); // open file for input x(n)
  yn_out = fopen("yn.dat","w");// open file for output y(n)

  /* Start of main program */
  while( (fscanf(xn_in,"%d",&xn)) != EOF) // read in x(n)
  { /* IIR filtering: y(n) = (1-alpha)*y(n-1)+alpha*x(n) */
    /*yn = alpha1*yn+ alpha*(float)xn; */
    yn = (((long)(int)alpha1*(long)(int)yn)>>15)
       +(((long)(int)alpha*(long)(int)xn)>>15);
    fprintf(yn_out,"%d\n",(int)(yn)); // round y(n) to integer
  }
  printf("Finish");
  fclose(xn_in);
  fclose(yn_out);   // close all opened files
}
```

The main changes in fixed-point C (`iir1fx.c`) code over floating-point C code (`iir1.c`) are as follows:

1. The `float` data types are changed to `long` for `yn` and to `int` for `alpha` and `alpha1`. As discussed in Section 3.2.1, floating-point numbers can be converted to a 16-bit Q.15 integer by multiplying the `float` numbers by 32,768.

2. Floating-point arithmetic can be converted to fixed-point arithmetic with Q.15 data. Q.15 multiplication (e.g., Q.15 × Q.15 → Q.30) can be implemented in fixed-point C as `((long)(int)alpha1*(long)(int)yn)>>15`. In this example, both `alpha1` and `yn` are represented using Q.15 format, and the product is represented using Q.30 format with 2 sign bits. The result can be left-shifted by 15 bits (to get rid of an extra sign bit) before being added to another multiplication result.

3. The output (result) is converted into a 16-bit integer and stored sequentially in the file `yn.dat`. We can compare the output file with the one generated using the floating-point C code given in Chapter 2.

4. The C5000 intrinsics (discussed in the next section) provided by the TMS320C54x (or C55x) C compiler can be used to improve the fixed-point C code, compare the result obtained, and profile the time it takes to complete the same tasks using different programming approaches.

3.7.3 Experiments Using the C5000 CCS

In this section, we need to modify the given C-source files for different experiments and observe the changes under the CCS environment. The experimental steps (create a project, compile, load, and run the program) used in debugging the code under CCS are introduced in the experiments shown in Chapters 1 and 2. These C programs can be used for both C54x and C55x experiments using different linker-command files and libraries. The linker-command file `exp3c54.cmd` (or `c55x.cmd` for the C55x) is used for all of the experiments in this section.

Different Q Formats

In this experiment, we represent a series of numbers using Q.15, Q1.14, and Q2.13 formats, as shown in the program listing of `exp3c5x_1.c`. This source file and linker-command file are added into the project, compiled, and run under the CCS environment using the steps described in previous chapters. The `exp3c5x_1.c` is listed as follows:

```
#include <stdio.h>
#include <stdlib.h>
#include <math.h>
#include <intrindefs.h>
/* Define global arrays and variables */
int num_q15[4] = {16384, 8192, 4096, 2048};// Q15 numbers
int num_q14[4] = {8192, 4096, 2048, 1024}; // Q14 numbers
int num_q13[4] = {4096, 2048, 1024, 512};  // Q13 numbers
int sum_q15, sum_q1_14, sum_q2_13;         // different format
int i;
```

```
void main()
{ /* Using _sadd intrinsic with OVM turn on */
  sum_q15 = num_q15[0];     // Q15 format
  for (i=0; i<3 ; i++){
    sum_q15 = _sadd(num_q15[i+1],sum_q15);
  }
  sum_q1_14 = num_q14[0];   // Q14 format
  for (i=0; i<3 ; i++){
    sum_q1_14 = _sadd(num_q14[i+1],sum_q1_14);
  }
  sum_q2_13 = num_q13[0];   // Q13 format
  for (i=0; i<3 ; i++){
    sum_q2_13 = _sadd(num_q13[i+1],sum_q2_13);
  }
}
```

Note that intrinsics (which produce assembly-language instructions) are specified with a leading underscore and can be called just like any C functions. When using intrinsic functions, the header file `intrindefs.h` needs to be included in the C program. The intrinsic `_sadd(int src1, int src2)` adds two 16-bit integers and produces a 16-bit result. More information on C-compiler intrinsics can be found in the *TMS320C55x Optimizing C/C++ Compiler User's Guide* [10].

Once the program `exp3c5x_1.c` is compiled and loaded, the user can run the program, open the watch window in the CCS by clicking on **View → Watch Window**, and enter the variables to observe, as shown in Fig. 3.32. Note that the results of adding the same four numbers using different Q formats are represented differently due to the fact that when converting into integers, the same number under different Q formats has different values. It is also important to understand the differences in representing an integer in different Q formats. For example, the number 16,384 (#4000h) equals 0.5 in Q.15 format, but represents 1 in Q1.14 and 2 in Q2.13. In the case of Q.15, Q1.14, and Q2.13 formats, numbers are confined between [0.99 to −1],

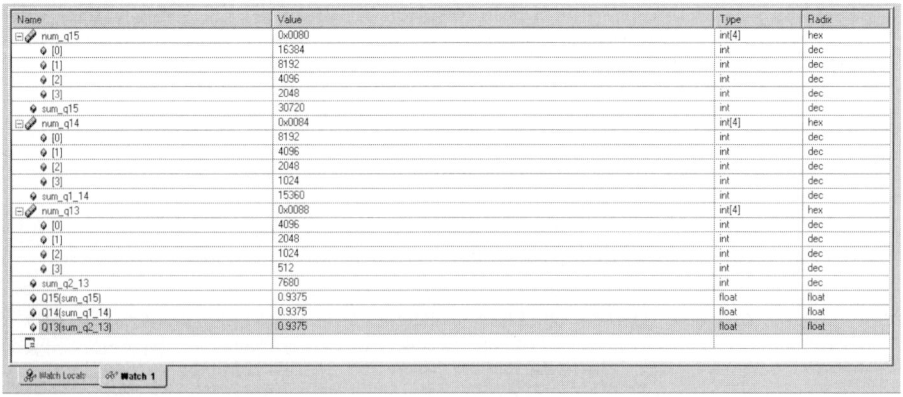

Name	Value	Type	Radix
num_q15	0x0080	int[4]	hex
[0]	16384	int	dec
[1]	8192	int	dec
[2]	4096	int	dec
[3]	2048	int	dec
sum_q15	30720	int	dec
num_q14	0x0084	int[4]	hex
[0]	8192	int	dec
[1]	4096	int	dec
[2]	2048	int	dec
[3]	1024	int	dec
sum_q1_14	15360	int	dec
num_q13	0x0088	int[4]	hex
[0]	4096	int	dec
[1]	2048	int	dec
[2]	1024	int	dec
[3]	512	int	dec
sum_q2_13	7680	int	dec
Q15(sum_q15)	0.9375	float	float
Q14(sum_q1_14)	0.9375	float	float
Q13(sum_q2_13)	0.9375	float	float

Figure 3.32 Result of adding four numbers represented in different Q formats

[1.99 to -2], and [3.99 to -4], respectively. Therefore, we can obtain the actual value 0.9375 for all cases by dividing the obtained integer values 30,720 (Q.15) by 32,768, 15,360 (Q1.14) by 16,384, and 7,680 (Q2.13) by 8,192. In this example, the results are the same. However, there may be cases where lower Q formats are not able to represent the result accurately, since additional bits have been used to represent the integer portion. As discussed earlier, there is always a tradeoff between dynamic range and precision in fixed-point representations.

The watch window does not display variables in Q formats. We can solve this problem by loading a GEL file `qvalue.gel` [9] available on the companion CD and website. Click on **File → Load GEL**, and select `qvalue.gel`. The `qvalue.gel` is a handy tool to display variables with a specific Q format in the watch window. Once the watch window is opened, we can display the Q.15 value of the variable `y` by entering `Q15(y)` in the watch window, as shown in Fig. 3.32.

Coefficient Quantization

As stated in Section 3.3.2, we can represent filter coefficients in fixed-point DSP processors by multiplying the floating-point coefficients designed by MATLAB with the correct scaling factor. In this experiment, we multiply the filter coefficients listed in Table 3.10 with the scaling factor 32,768 and store the coefficients in the `q15_coeff` array. The program then performs multiply-accumulate operations on these coefficients with a constant value `#7FFFh` (0.999969482 in Q.15). The final sum of the products is stored in the variable `out`. The C program `exp3c5x_2.c` is listed as follows:

```
#include <stdio.h>
#include <stdlib.h>
#include <math.h>
#include <intrindefs.h>
/*  Define global arrays and variables  */

int q15_coeff[40] = {-141, -429, -541, -211, 322, 354, -215, -551,
21, 736, 332, -841, -871, 755, 1651, -304, -2880, -1106, 6139,
13157, 13157, 6139, -1106, -2880, -304, 1651, 755, -871, -841, 332,
736, 21, -551, -215, 354, 322, -211, -541, -429, -141};
int input = 32767;
int i;
long out=0;

void main()
{   /*  Use _smac intrinsic with OVM, FRCT set   */
  for (i=0; i<40; i++){
    out = _smac(out,q15_coeff[i],input);
  }
}
```

Figure 3.33 shows the watch window that displays the contents of the variables after executing MAC operations 40 times. Note that the intrinsic function `_smac(long src, int op1, int op2)` multiplies op1 and op2, shifts the product left

Figure 3.33 Watch window after executing 40 consecutive MAC operations

by 1 bit, and then adds the result to the src. This result is presented in 32 bits with both the OVM and FRCT modes being set. It is sufficient to save the most significant 16 bits of the final value of src, which is 70B1h (or 0.88 in Q.15).

It is a good exercise to enter the coefficient values stated in Table 3.10 using Q.9 and Q.7 formats. Examine and compare the results of the multiplication using different Q formats.

Overflow

This experiment illustrates a case where intermediate overflow is allowed during a series of additions. In the C program exp3c5x_3.c, a sequence of numbers is declared as integers. In the program, the C intrinsic function _sadd(int src1, int src2) is used to add eight 16-bit integers with overflow mode (OVM) set and produces a saturated 16-bit result. A watch window can be set up, as shown in Fig. 3.34. With the overflow mode being set, we can single-step through the C code to observe the process of addition. Note that the result obtained after seven additions is −16,384, which is incorrect. Why is this the case? The C program exp3c5x_3.c is listed as follows:

```
#include <stdio.h>
#include <stdlib.h>
#include <math.h>
#include <intrindefs.h>

/*  Define global arrays and variables  */
int num[8] = {19660, 19660, 6554, -9830, -32768, -29491, 9830, 6554};
int i,y;
void main()
{  /*  Understand the difference between _sadd and _lsadd  */
   /*  _lsadd can be treated as OVM mode been turned off    */
   y = num[0];
   for (i=0; i<7; i++){
     y = _sadd(num[i+1],y);
   }
}
```

We can also check whether the arithmetic results in overflow by observing the overflow flag, which is displayed in the CPU register window. If the overflow flag is 0, the final result is correct; otherwise, the result has overflowed. The user is encouraged to modify the above C code using the _lsadd(long src1, long src2) intrinsic to add eight 16-bit numbers and observe the result again.

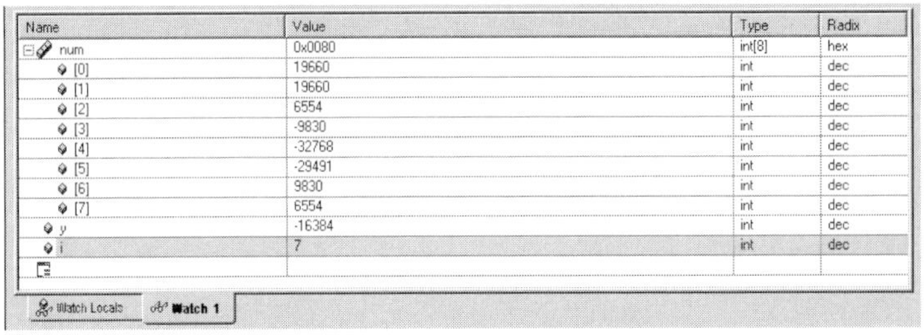

Name	Value	Type	Radix
num	0x0080	int[8]	hex
[0]	19660	int	dec
[1]	19660	int	dec
[2]	6554	int	dec
[3]	-9830	int	dec
[4]	-32768	int	dec
[5]	-29491	int	dec
[6]	9830	int	dec
[7]	6554	int	dec
y	-16384	int	dec
i	7	int	dec

Watch Locals   **Watch 1**

Figure 3.34   Watch window that shows the numbers in an array and the result

Scaling and Saturation Mode

This experiment shows the effect of scaling sinusoidal input values that are larger than the dynamic range of the Q.15 format. The original data samples stored in the array `sine` are scaled by a factor of 1/8, which is implemented by shifting left by 3 bits. The shifted data samples are stored back into the same memory locations. This operation is carried out using the C intrinsic function `_sshl(int src1, int src2)`, which shifts `src1` left by `src2` bits and produces a 16-bit result. The OVM flag is set to allow saturation. The C program `exp3c5x_4.c` is listed as follows:

```
#include <stdio.h>
#include <stdlib.h>
#include <math.h>
#include <intrindefs.h>

/*  Define global arrays and variables  */
int sine[8] = {0, 5792, 8191, 5792,0,-5792,-8191,-5792};
int scale = 3;
int i;

void main()
{  /*  Understand the difference between _sshl and _lsshl  */
   /*  _lsshl is treated as OVM OFF if long src1 is used  */
  for (i=0; i<8; i++){
    sine[i] = _sshl(sine[i],scale);      // OVM mode turned ON
    // sine[i] = _lsshl(sine[i],scale);  // OVM mode turned OFF
  }
}
```

The intrinsic function in the code can be changed to `_lsshl(long src1, int src2)` to simulate the effect of turning off OVM mode when `src1` is declared as `long` instead of `int`. Figure 3.35 shows the scaled-data samples when OVM mode is (a) not set and (b) set. When OVM mode is cleared to 0 (using intrinsic `_lsshl`), the overflowed sample changes the sign, thus producing a totally different signal, which is seen clearly in Fig. 3.35(a). However, when OVM mode is set to 1 (using intrinsic `_sshl`), data samples that exceed these limits are clipped to

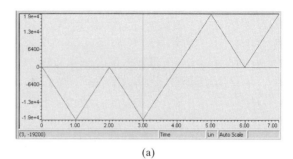

(a)

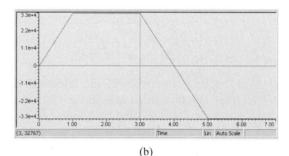

Figure 3.35   Effects of turning OVM
mode (a) off and (b) on

(b)

the maximum positive (#7FFFh) or maximum negative (#8000h) values. Therefore, the polarity of the original signal is maintained with the upper and lower
parts of the waveform being clipped to +32,767 (#7FFFh) and −32,768 (#8000h),
as shown in Fig. 3.35(b).

Rounding

In this experiment, we explore rounding operations in C5000 programming. Rounding is carried out in the processor by adding a constant number of #8000h to the
lower word of the accumulator. As explained in Section 3.3.4, rounding does not
result in undesired bias as compared with truncation, which simply ignores the contribution from the lower word. We can use the C program exp3c5x_5.c to observe
the difference between the results of using multiply without the rounding intrinsic
function _smpy(int src1, int src2) and multiply with the rounding intrinsic function _smpyr(int src1, int src2). When performing MAC operations in sequence,
only the final result needs to be rounded. The C program exp3c5x_5.c is listed as
follows:

```
#include <stdio.h>
#include <stdlib.h>
#include <math.h>
#include <intrindefs.h>

int num[2] = {8520, 26214}; // 8520 is 0.26, 26214 is 0.79
int y1, y2;
```

```
void main()
{ /* Understand the difference between _smpy and _smpyr */
  /* OVM and FRCT are both set */
  y1 = _smpyr(num[0],num[1]);    //with rounding
  y2 = _smpy(num[0],num[1]);     //without rounding
}
```

In the code, the numbers 0.26 and 0.79 are multiplied with or without rounding. Build, load, and run the code. Observe the different values obtained in integers y1 and y2.

PROBLEMS

PART A

Problem A–3–1

Show the steps involved in multiplying −1 by −1 in Q.3 and Q1.3 formats. Show why Q.3 produces an incorrect result, and show how Q1.3 overcomes this problem. Also explain how the DSP processor handles this problem. In this experiment, we consider the case of multiplying −1 by −1 in Q.15 format, which is highlighted in Section 3.3.3. It is obvious that −1 multiplied by −1 results in +1; however, since +1 cannot be represented in Q.15 format, we need to consider this special case in fixed-point DSP programming. We can prevent the incorrect result by setting the OVM flag, which is automatically carried out using the C-compiler intrinsic function _smpy(int src1, int src2). This operation multiplies two numbers, src1 and src2, shifts the product left by 1 bit, and produces a saturated 16-bit result. The user can refer to the C code in exp3c5x_6.c for more information. Build, load, and run the code in the CCS, and then observe the result of the multiplication. Is the result accurate? Modify the code using the intrinsic function _lsmpy(int src1, int src2). What is the result now?

Problem A–3–2

IEEE-754 double-precision (64-bit) format is used in the MATLAB computing environment. In order to see the binary-bit pattern for the floating-point number representation, we can use the MATLAB command format hex, which displays the hexadecimal value of the number. Interpret the hexadecimal value in terms of its sign, exponent, and fractional values for the numbers 0.25, −10.45, and 1.76×10^3.

Problem A–3–3

Perform double-precision 32 × 32-bit multiplication in the DSP processor using Q.31 format. What precision of the register is required to store the result? Use Fixed-Point Blockset to simulate the 32-bit multiplication.

Problem A–3–4

A sixth-order IIR filter has the following transfer function:

$$H(z) =$$

$$\frac{0.0153 + 0.0320z^{-1} + 0.0573z^{-2} + 0.0640z^{-3} + 0.0573z^{-4} + 0.0320z^{-5} + 0.0153z^{-6}}{1.0000 - 2.6383z^{-1} + 4.3444z^{-2} - 4.4716z^{-3} + 3.1345z^{-4} - 1.3781z^{-5} + 0.3157z^{-6}}.$$

Use the filter design and analysis toolbox to represent the preceding direct form to a cascade of second-order filters, and represent these coefficients in Q.15 format in a 16-bit fixed-point DSP processor. Examine the different ways of grouping, pairing, and ordering the biquad sections. Implement the preceding IIR filter using the C5000 simulator.

Problem A–3–5

A DSP processor has a 40-bit accumulator, which consists of 8 guard bits. Determine the maximum length L for FIR filters that can be used in order to avoid overflow in the accumulator. How can we extend the length L besides using guard bits? Use Fixed-Point Blockset to simulate the preceding exercise.

Problem A–3–6

Use MATLAB functions to implement coefficient quantization for the set of coefficients shown in Table 3.10 using the following modes:

(a) Fixed-point representations: $[16, 15]$, $[10, 9]$, and $[8, 7]$
(b) Roundup for rounding mode
(c) Saturation for overflow mode

Problem A–3–7

Implement the multiply-accumulate functional unit using Fixed-Point Blockset, assuming Q.15 format is used, and use different wordlengths for the multiply-add unit, with and without guard bits.

Problem A–3–8

What happens if the code `iir1fx.c` listed in Section 3.7.2 is changed to the following?

```
yn = (((long)(int)alpha1*(long)(int)yn) +
      ((long)(int)alpha*(long)(int)xn))>>15;
```

PART B

Problem B–3–9

Calculate 6-bit representations of the numbers 0.4375 and -0.4375 in (a) sign-magnitude, (b) one's complement, and (c) two's complement forms.

Problem B–3–10

Perform the following arithmetic using 4-bit two's complement format:

(a) subtract 0.625 from 0.25
(b) subtract 0.25 from 0.625

Problem B–3–11

Find the Q.7 representation for the following decimal numbers: $0.2, 0.435, -0.72$, and 0.81640625. Repeat this exercise for a Q.15 representation.

Problem B–3–12

Truncate and round the Q.7 numbers obtained in Problem 11 to Q.5 format. Also, compute the corresponding truncation and roundoff errors.

Problem B–3–13

Represent the following integer values in Q.15 and Q8.7 formats:

(a) 0x1234
(b) 0x0000
(c) 0xFFFF
(d) 0x9123
(e) 0xABCD

Problem B–3–14

Perform binary multiplication of the following Q.3 numbers:

(a) -0.5×0.875
(b) -0.375×-0.625

Determine the quantization error in storing the results using a Q.6 format.

Problem B–3–15

Compare fractional (Q.3) multiplication of 0.625×0.375 with the integer multiplication of 7×6. Examine the errors if only 4 bits are available to store the result.

Problem B–3–16

A Q1.6 number is multiplied by Q.7 number. Determine a suitable wordlength to store the multiplication result and determine the Q format used. How many extra sign bits are generated in the product?

Problem B–3–17

Find the smallest and largest positive number that can be represented in the IEEE-754 single-precision floating-point format. Repeat the same problem using double-precision floating-point format.

Problem B–3–18

Compute the dynamic range and precision of Q.15 and Q7.8 fixed-point formats. What is the arithmetic accuracy when using these two formats? Compare the dynamic range of a 32-bit IEEE-754 floating-point number and a 32-bit fixed-point number. Comment on their respective precisions (assuming a suitable Q format is used). Finally, comment on the dynamic range and precision of a block floating-point format using an 8-bit scaling factor and 16-bit mantissa.

Problem B–3–19

A multichannel surround-sound system consists of five channels: front left (FL), front right (FR), front center (FC), surround left (SL), and surround right (SR).

These channels must be downmixed to two channels using the following downmix equations:

$$\text{Downmixed left channel: } L = a \times FL + b \times FC + c \times SL$$
$$\text{Downmixed right channel: } R = a \times FR + b \times FC + c \times SR$$

where the downmix coefficients are a $= 1.0$, b $= 0.707$, and c $= 0.707$.

In order to prevent arithmetic overflow, the downmix coefficients must be scaled down under the worst-case consideration when all channels are delivering signal at full scale. Perform the following tasks:

(a) Multiply the scaling factor by all downmix coefficients.
(b) Compute the scaled coefficients using 5-bit precision. (Note that all coefficients are positive.)
(c) Compute the coefficient-quantization error.
(d) Compute the gain reduction in dB for each channel.

Problem B–3–20

The image resolution for a digital TV is 484×427 pixels, and the frame rate is closed to 30 Hz. Calculate the pixel (8-bit) rate and the bit rate in processing one frame of a black-and-white picture. What is the bit rate for a color picture using 24-bit resolution? What is the size of the video RAM required to store 1 minute of video clips? What is the MAC time for a processor when performing 30 MACs per frame operations? Can a single processor with 200 MIPS handle this operation?

Problem B–3–21

A 16-bit fixed-point processor has an instruction cycle time of 10 nsec. A set of DSP libraries has been written for some commonly used DSP algorithms. One of these functions is the FIR-filtering routine, which can perform sample-by-sample or block processing for an L-tap FIR filter. The number of instruction cycles required to perform this function is given as $38 + N \times (4 + L)$, where $N = 1$ for sample-by-sample processing and where N is the number of samples used per block for block processing. In addition, the scalar number denotes the overhead required to set up the memory pointers, zero the accumulator, etc. The code size for this function is given as 42 words. Determine the maximum length of filter that can be implemented for stream- and block-processing methods.

Problem B–3–22

Determine the maximum sampling frequency and the bandwidth of signal that can be filtered with a 127-order FIR filter in real time using (1) sample-by-sample and (2) block ($N = 32$) processing on the above processor. If an audio-signal waveform is sampled at 44.1 kHz, what is the maximum length of a FIR filter using the sample-by-sample and block ($N = 32$) modes? (Hint: Use the instruction cycles given in Problem 21.)

Problem B–3–23

A DSP processor is operating at 100 MHz and is connected to a 20-bit CODEC, which generates a serial clock of 256 times the sampling frequency. Determine the number of audio channels that can be connected to the DSP within a sampling frequency of 48 kHz.

Note that the particular frame format uses the first 16 bits of the bit stream as a tag and the remaining bits as a data stream.

Problem B–3–24

Determine the oversampling factor required to implement a 1-bit converter that has the same SQNR as a 16-bit converter.

SUGGESTED READINGS

1 IEEE. *IEEE Standard for Binary Floating-Point Arithmetic*. IEEE Standard 754-1985, pp 1-17, 1985.

2 The MathWorks. *MATLAB User's Guide*. Version 6, 2000.

3 Texas Instruments. *TMS320C54x DSP: CPU and Peripherals, Reference Set*. Vol. 1, SPRU131F, 1999.

4 Texas Instruments. *TMS320C54x DSP: Enhanced Peripherals Reference Set*. Vol. 5, SPRU302, 1999.

5 Lapsley, P., J. Bier, and A. Shoham. *DSP Processor Fundamentals: Architectures and Features*. New York, NY: IEEE Press, 1999.

6 Ackenhusen, J. G. *Real-Time Signal Processing: Design and Implementation of Signal Processing Systems*. Upper Saddle River, NJ: Prentice Hall, 1999.

7 Kuo, S. M. and B. H. Lee. *Real-Time Digital Signal Processing: Implementations, Applications, and Experiments with the TMS320C55x*. New York, NY: Wiley, 2001.

8 Texas Instruments. *TMS320C54x Optimizing* C/C++ *Compiler User's Guide*. SPRU103F, 2001.

9 Texas Instruments. *Q-Values in the Watch Window*. Application Report, *SPRA109*, 2002.

10 Texas Instruments. *TMS320C55x Optimizing* C/C++ *Compiler User's Guide*. SPRU281C, 2001.

11 The MathWorks. *Using Simulink*. Version 4, 2000.

4

Fixed-Point Digital Signal Processors

This chapter introduces the architecture, the assembly programming, and applications for fixed-point DSP processors from Texas Instruments. The TMS320 fixed-point family consists of three generations: the C2000 (C24x and C28x), C5000 (C54x and C55x), and C6000 (C62x and C64x). Each generation of the TMS320 family has a similar CPU with a variety of memory and peripheral configurations. We introduce important concepts of DSP programming from the application viewpoint. Detailed architectures and instruction sets of these DSP processors can be found in user's guides that are available in the **Help** menu of CCS and on the Texas Instruments website, which is listed in Appendix E.

4.1 INTRODUCTION

Many companies have developed digital signal processors with architectures and instruction sets specifically tailored for DSP applications. In 2000, DSP processors became the largest processor segment, passing the traditionally strong 8-bit microcontroller market. The rapid growth in DSP processors is not a surprise considering the commercial advantages in terms of the potentially fast, flexible, and low-cost design capabilities offered by these processors. Major advanced applications for these processors include networks and communications. Networks have become the

154

enabling techniques for wired and wireless infrastructure systems and devices, as well as for voice (or video) over networks that support streaming multimedia over the Internet.

DSP applications have become very diversified, from large-scale communications infrastructures to handheld, portable devices and low-cost consumer products. The TMS320C2000 generation, which provides on-chip integration of powerful computational abilities with microcontroller-like peripherals, is optimized for control applications such as motor control in order to drive system costs down. The C5000 generation is optimized for consumer products with ultrapower efficiency and allows new features to be added to miniaturized devices from mobile Internet to other portable consumer electronics. The C6000 generation is optimized for the highest performance and ease of use in high-level language programming. C6000 processors support computation-intensive applications such as image processing, multichannel broadband infrastructures for 3G wireless communications, digital-subscriber loops, and cable.

DSP processors have become integral to many high-volume designs in which time to market is critical. However, efficient DSP software is difficult to write. Developing optimum software for the designed algorithm on a given DSP processor requires an understanding of the processor architecture, instruction set, addressing modes, and software development techniques. We briefly introduce these topics in this chapter and use many assembly instructions and programs as examples to help explain the use of these processors. Therefore, we introduce the basic concepts of assembly programming first in this section. A detailed introduction of TMS320 assembly language tools is available in the *TMS320C1x/C2x/C2xx/C5x Assembly Language Tool User's Guide* [1].

4.1.1 Source-Statement Format

TMS320 assembly-language tools use common-object file format (COFF) to support modular programming. COFF files contain separate sections of code and data that can be loaded into TMS320 memory spaces. TMS320 assembly programs consist of source statements that contain four ordered fields. The general syntax for the source statement is

```
[label][:]   mnemonic   [operand list]   [; comment]
```

where [x] indicates that x is optional. The mnemonic field cannot start in column 1. Statements must begin with a label, a blank, an asterisk, or a semicolon. One or more blanks must be used to separate each field. For ease of understanding and code maintenance, use meaningful symbolic names for variables, labels, subroutines, etc. The fields are described as follows:

Label field: A label associates a symbolic name (address) with a location in the program. A line that is labeled in the assembly program then can be referenced with that symbolic name, which is useful for modular programming and branch instructions. Labels are optional, but if used they must begin in the first column.

Mnemonic field: The mnemonic field can contain an assembler directive, a macro directive, a macro call, or an instruction. The TMS320 instruction set supports both DSP-specific and general-purpose applications. TMS320 user's (and reference) guides provide detailed individual instructions for the specific TMS320 processor.

Operand field: The operand list field is a list of operands. An operand can be a constant, a symbol, or a combination of constants and symbols in an expression. Constants can be represented in binary, decimal, or hexadecimal format. For example, a binary constant is a string of binary digits (0s and 1s) followed by the suffix b, and a hexadecimal constant is a string of hexadecimal digits (0-9 and A-F) followed by the suffix h (or the prefix 0x). Symbols include labels, register names, and the symbols defined in the assembly code using assembler directives.

Comment field: Comments are notes about the logic and function of the program. A comment can begin with either a semicolon or an asterisk if it starts in column 1. Comments that begin in any other column must begin with a semicolon.

4.1.2 Assembler Directives

The smallest module of an object file is a section, which is a block of code or data (constants or variables). Once the DSP algorithm has been programmed in terms of sections of code and data, it is necessary to add some assembly directives to the source code in order to manipulate these sections, such as specifying the start location of the program in program memory and defining constants, etc. Assembler directives can be used to control the assembly process and to enter data into the program.

A COFF file always contains three default sections: a .text section that contains executable code, a .data section that contains initialized data, and a .bss section that reserves space for uninitialized data. The .text directive tells the assembler to begin assembling source code into the .text section, which normally contains executable code. The .data directive usually contains data, such as a sinewave table, pre-initialized variables, etc. In addition, the .sect directive defines an initialized section and tells the assembler to begin assembling source code into that named section. It can be used to partition a large program into logical sections. For example, it can separate the subroutines from the main program or separate constants that belong to different tasks.

Uninitialized sections such as signal and I/O buffers reserve space in the TMS320 memory, which usually is allocated into RAM. A program can use this space at run time for storing variables. Uninitialized data areas are reserved by using the .bss and .usect directives. For example,

```
        .bss      symbol, size
```

or

```
symbol    .usect    "section_name", size
```

The `symbol` (symbolic name) points to the first location of the reserved memory space, `size` specifies the number of words to be reserved, and `section_name` tells the assembler the named section in which to reserve space.

The `.word` (or `.int`) directive places one or more 16-bit integer values into consecutive words in the current section, which allows us to initialize memory with pointers to variables or labels. Similarly, the `.byte` directive places 8-bit values into the current section.

The `.def` directive makes a symbol global (known to external files) and indicates that the symbol is defined in the current file. External files can access the symbol by using the `.def` directive. A symbol can be a label or a variable.

Assembly-time directives equal meaningful symbolic names to constant values or strings. The `.set` (or `.equ`) directive assigns a value to a symbol; thus, that symbol in the assembly code becomes a constant in the machine code. The `.set` directive has the form

```
symbol     .set   value
```

where `symbol` must appear in the label field. This type of symbol is known as an assembly-time constant. It can be used in source statements in the same manner as a numeric constant.

4.1.3 Software-Development Processes

As introduced in the experiments in Chapter 2, we first developed DSP applications using high-level languages, MATLAB, and floating-point C programs. We then used data files stored in the host computer for testing the performance of algorithms for a given application. The testing data may be generated with a software signal generator or digitized from a setup that mimics real-world applications. At this stage, we can develop C code using a C compiler such as Microsoft Visual C/C++ without any knowledge of the target DSP processor that executes the C code in real applications.

There is a growing need for using fixed-point DSP systems that meet power, cost, and size restrictions. Therefore, after the performance of the floating-point C programs is satisfied, we can develop a fixed-point C program that mimics the operations of the fixed-point DSP processor that will be used for implementation. When we convert a C program from floating point to fixed point, we use quantization to perform the conversion. Some basic principles of writing fixed-point C programs for fixed-point processors is introduced in Section 4.4. After making the fixed-point C code functional, we can improve its performance by selecting high-level-optimization compiler options available for TMS320 C compilers.

In the third stage, we focus on optimizing C code to achieve better performance. We can use CCS profiling tools to identify any inefficient sections in the fixed-point C code. Some techniques, such as user-specific data types, modify the C code to better suit the given processor architecture, use already optimized functions available in DSPLIBs, and use special functions such as intrinsics provided by C compilers. After modifying the code, we use the profiling tools again. If the code is still not as efficient as we want, we have to use optimized assembly code.

In the final stage, we identify the time-critical sections of the C code and rewrite them as C-callable assembly-language functions. After making the mixed C and assembly code functional, we still can optimize the assembly-language functions by using techniques such as placing instructions in parallel, rewriting or reorganizing code to avoid pipeline conflicts and delays, and minimizing stalls in instruction fetching.

We follow the software development processes introduced in this section for experiments and applications given in Chapters 6, 7, 8, and 9 and in Appendix B.

4.2 TMS320C2000

C2000 fixed-point processors include the TMS320F24x, C24x, LF24x, LC24x, and C28x, where F and C denote the flash and the ROM device option, respectively. Flash-based processors offer a reprogrammable capability useful for applications requiring field upgrades and for prototyping of applications that migrate to ROM-based processors in production. The CPU of these processors is based on the TMS320C2xx, with additional architectural enhancements and on-chip peripherals to greatly improve overall performance. These processors offer an array of on-chip memories (summarized in Table 4.1) and different peripherals tailored to meet specific price/performance points required by various applications. The on-chip peripherals include fast analog-to-digital converters, event managers for pulse-width modulation and I/O features, and timers. By integrating memory and peripherals onto a single chip, C2000 processors greatly reduce system costs and save circuit-board space.

In this section, we briefly introduce architecture, programming, and system issues. Detailed descriptions of processor architecture and instruction set can be found in the *TMS320F/C24x DSP Controllers CPU and Instruction Set Reference Guide* [2], and information on the peripheral devices is available in the *TMS320F243/F241/F242 DSP Controllers, Systems and Peripherals Reference Guide* [3].

TABLE 4.1   On-Chip Memories of C2000 Processors

Processor	RAM (16-bit)	ROM (16-bit)	Flash (16-bit)
F240	544	0	16K
C240	544	16K	0
F241	544	0	8K
C242	544	4K	0
F243	544	0	8K
LF2407	2.5K	0	32K
LF2406	2.5K	0	32K
LF2402	544	0	8K
LC2406	2.5K	32K	0
LC2404	1.5K	16K	0
LC2402	544	4K	0

4.2.1 Architecture Overview

The TMS320 family has an architecture designed specifically for real-time DSP applications. The C2000 generation combines this low-cost, high-performance processing capability with microcontroller peripherals optimized for digital-control applications. The C2000 architecture is built around two major buses: the program bus and the data bus. The program bus carries the instruction code and immediate operands from program memory. The data bus interconnects various elements such as the CPU to data memory. This Harvard architecture maximizes processing power by allowing simultaneous access to instructions and data for full-speed execution. The C2000 also modified the Harvard architecture by supporting data transfer between program memory and data memory. This architecture permits coefficients to be stored in program memory, therefore eliminating the need for a separate coefficient ROM.

The C2000 processor consists of three execution units that operate concurrently: the program controller, the CPU, and the DAGEN. The C2000 processor executes some critical DSP functions that other general-purpose microprocessors typically implement in software or microcode. For example, the device contains hardware for single-cycle 16 × 16-bit multiplication, data shifting, and address manipulation. The high-speed CPU allows the system designer to execute algorithms in real time rather than approximate results with look-up tables. In addition, several advanced peripherals optimized for digital control applications have been integrated to provide a powerful single-chip microcontroller. These execution units, along with their associated on-chip memory, are described in the following sections.

For maximum throughput, C2000 processors implement a four-stage pipeline. The internal data- and program-bus structure is divided into six 16-bit buses: the program-address bus, the data-read-address bus, the data-write-address bus, the program-read bus, the data-read bus, and the data-write bus. Because the same program and data buses are used to communicate with external programs, data, or I/O space, the number of cycles required for executing an instruction depends on whether the operand is from internal or external memory. Maintaining data memory on chip and using fast external memory achieves the highest throughput.

To explain the architecture of C2000, we have to use assembly instructions. To be familiar with the instruction format, we use the ADD instruction with the following different syntax as examples:

```
[label]    ADD    dma[,shift1]           ; direct addressing
[label]    ADD    ind[,shift1[,ARn]]     ; indirect addressing
[label]    ADD    #k                     ; short immediate addressing
[label]    ADD    #lk[,shift2]           ; long immediate addressing
```

where

$0 \leq$ dma ≤ 127 is the data-memory address (or a symbol defined by a .equ or .set directive) using direct-addressing mode,

$0 \leq$ shift1 ≤ 16 (default is 0) is the left-shift count,

$0 \leq$ ARn ≤ 7 specifies the address register for the next instruction,

$0 \leq k \leq 255$ is the 8-bit short immediate data,

$-32768 \leq lk \leq 32767$ is the 16-bit long immediate data,

$0 \leq shift2 \leq 15$ is the left-shift count, and `ind` indicates indirect-addressing mode, which is discussed later.

These operands are the most frequently used operands in C2000 assembly instructions. Operands may be constants or assembly-time expressions referring to memory, I/O ports, register addresses, pointers, shift counts, or a variety of other constants.

4.2.2 Central Processing Unit

As illustrated in Fig. 4.1, the major components of the CPU include an input (scaling) shifter, a hardware parallel multiplier, a product shifter at the output of the multiplier, an ALU, an accumulator (ACC), an output shifter at the output of the ACC, and an auxiliary-register arithmetic unit (ARAU) with eight auxiliary registers AR0–AR7. The 32-bit ACC can be split into two 16-bit registers, upper ACC, ACCH (b31-b16), and lower ACC, ACCL (b15-b0), for storage in 16-bit data memory.

Input Shifter

The 32-bit input shifter produces a left shift of 0 to 16 bits on the 16-bit input data from either data or program memory. The shift count can be specified by either a constant embedded in the instruction word or by the four LSBs of the temporary register (TREG). The LSBs of the 32-bit output are filled with zeros, while the MSBs either may be filled with 0s or sign extended, depending on the value of the sign-extension-mode (SXM) bit of the status register ST1. The SXM bit can be set to 1 with the instruction SETC SXM or cleared to 0 with the instruction CLRC SXM.

The input shifter can be invoked with the instructions that load into, add to, or subtract from the ACC. For example, in the instruction

```
LACC    dma,shift1
```

the 16-bit data from the specified data memory at address `dma` is left-shifted by `shift1` bits and then loaded into the ACC. Similarly, in the instruction

```
ADD     dma,shift1
```

the data is left-shifted by `shift1` bits and then added to the ACC. In both cases, the shift operation is included in the instructions, thus requiring no cycle overhead.

In the previous examples, the size of the shift is specified with a constant `shift1` embedded in the instruction. The four LSBs of the TREG can also specify the shift count. For example,

```
LACT    dma
```

results in a left-shift of 0 to 15 bits, depending on the value of the four LSBs in TREG, which can be updated using the program. Using TREG contents as a shift count allows the scaling factor to be determined dynamically based on the system's performance.

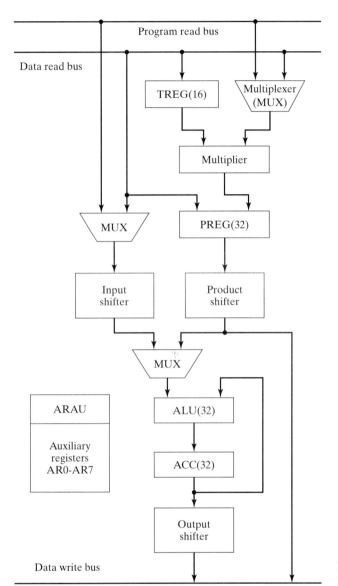

Figure 4.1 Block diagram of the central processing unit

Example 4.1

The input shifter aligned the 16-bit data from memory to the 32-bit ALU. The shifter can be used for multiplying signals by an integer that is a power of two. For example, the equation $x3 = x2 + 16 \times x1$ can be implemented as follows:

```
LACC    x2          ; ACC <- x2
ADD     x1,4        ; ACC + 16*x1
SACL    x3          ; x3 <- ACCL = 16*x1+x2
```

Example 4.2

The input shifter can also be used for dividing a signal by an integer that is a power of two. For example, the equation $x2 = x1/8$ can be implemented as

```
LACC   x1,13      ; load ACC with x1 left-shift 13 bits
SACH   x2         ; x2 <- ACCH = x1/8
```

which is equivalent to right-shifting 3 bits of $x1$. In general, dividing by 2^M is equivalent to right-shifting M bits of data, which can be implemented by left-shifting $(16-M)$ bits of data and then saving the upper 16 bits of the ACC (ACCH).

Multiplication Unit

As shown in Fig. 4.1, the multiplication unit consists of four elements: a 16×16 hardware multiplier to compute a 32-bit product in a single clock cycle, a 16-bit TREG that holds one of the operands for the multiplier, a 32-bit product register (PREG) that holds the product from the multiplier, and a product shifter that shifts the product from the 32-bit PREG to the ALU. Executing the instruction MPY #0 can clear the PREG. The product from the PREG can be transferred to the ACC with the instruction PAC (load ACC with PREG) or to 16-bit data memory with the SPH (store PREG high) and the SPL (store PREG low) instructions.

As shown in Fig. 4.1, the multiplier accepts two inputs. One input is always from the TREG, while the other is from either data or program memory. If both operands are located in data memory, one operand must be loaded into the TREG first. To load one operand first, use the instruction

```
LT     dma
```

which loads the data of the specified data memory at address dma into the TREG.

The more powerful instructions that load the TREG in preparation for multiplication are LTA (load TREG and accumulate previous product), LTP (load TREG and store PREG in ACC), and LTS (load TREG and subtract previous product). For example, the instruction

```
LTA    dma
```

loads the data from the specified data memory at address dma into the TREG, and the contents of the PREG are shifted as defined by the product-shift mode (PM) and then added to the ACC. When the code executes a sequence of multiply-add operations, the CPU supports the pipelining of the TREG load operations with multiplier and ALU operations using the previous product.

In addition, the instruction

```
LTD    dma
```

performs the LTA operation plus a data-move (DMOV) operation. This instruction fetches the data for multiplication from the specified data memory at address dma and also writes that data to the next higher address. This is an efficient method for refreshing the signal buffer for FIR filtering, as discussed in Chapter 2. In order to

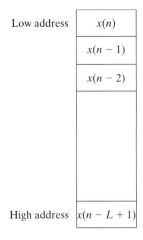

Low address $x(n)$

$x(n-1)$

$x(n-2)$

High address $x(n-L+1)$

Figure 4.2   Memory configuration for data-move instructions

move data using the DMOV, LTD, and MACD (MAC plus a data-move operation) instructions, the data samples in the signal buffer should be arranged as illustrated in Fig. 4.2, where the first (or the most recent) sample $x(n)$ is stored at the lowest memory address in the buffer, the previous sample $x(n-1)$ is located at the memory of next higher address, and so on.

The MPY (multiply) instruction provides the second operand for the multiplier. For example, the instruction

```
MPY     dma
```

multiplies the contents of the TREG register with the contents of the data memory at location dma. As discussed in Chapter 3, multiplication of two Q.15 numbers results in Q.30 format, which has 30 fractional bits and 2 sign bits. To store the result as a Q.15 number into memory, a left shift of 1 bit is needed to eliminate the extra sign bit and then save the leftmost 16 bits.

The product shifter left shifts the data in the PREG. The PM bits of the status register ST1 specify the function of the product shifter, as summarized in Table 4.2. The PM bits can be set with the instruction

```
SPM     constant     ; set product shift mode
```

where $0 \leq$ constant ≤ 3. The output of the PREG can also be right-shifted 6 bits to enable the execution of up to at least 128 consecutive multiply-accumulate operations without the possibility of overflow.

TABLE 4.2   Product-Shift Modes

PM	Resulting Shift
00	No shift
01	Left shift 1 bit
10	Left shift 4 bits
11	Right shift 6 bits

The more powerful instructions that multiply and accumulate (or subtract) the previous product are MPYA (multiply and accumulate previous product) and MPYS (multiply and subtract previous product). For example, in the instruction

```
MPYA    dma
```

the data in the TREG is multiplied by the data in the data memory, and then the product is stored in the PREG. At the same time, the previous product in the PREG is shifted as defined by the PM bits and then added to the ACC.

Arithmetic and Logic Unit Section

The ALU is a general-purpose arithmetic and logic unit that performs addition, subtraction, Boolean logic, and bit manipulation (testing, shifting, and rotating) operations. As shown in Fig. 4.3, the ALU section consists of the ALU, an ACC, and an output shifter. The result of the ALU operation is stored in the ACC. For example,

```
ADD     dma[,shift1]    ; add data in memory to ACC with shift
APAC                    ; add PREG to ACC
```

In the first example, the contents of the data memory at address dma are left-shifted shift1 bits by the input shifter and then added to the ACC. In the second example, the contents of the PREG are shifted by the product shifter as specified by the PM bits and then added to the ACC.

As shown in Fig. 4.3, the 32-bit ACC can be partitioned into two 16-bit segments: the ACCH (b31-b16) and ACCL (b15-b0). The ACC is capable of shifting or rotating its contents. The carry bit (C) is affected by (1) additions to and subtractions from the ACC and (2) single-bit shifts and rotations of the ACC value. The carry bit allows more efficient computation of extended-precision arithmetic.

The output shifter at the output of the ACC provides a left shift of 0 to 7 bits. This shift is performed while the data is transferred from the ACC to data memory

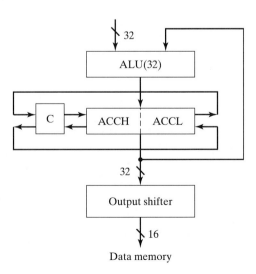

Figure 4.3 Block diagram of the ALU section

for storage, as shown in Fig. 4.3. When the output shifter is used on the ACCH (bits 16-31), the MSBs are lost and the LSBs are filled with bits shifted from the MSBs of the ACCL. When the output shifter is used on the ACCL, the MSBs are discarded and the LSBs are filled with zeros.

Example 4.3

If the value of the ACC is 12345678h. After the execution of the instructions

```
SACL    data1,4    ; store ACCL with 4-bit shift
SACH    data2,4    ; store ACCH with 4-bit shift
```

the value of the data memory at address `data1` is 6780h. The value of the data memory at address `data2` is 2345h, where 5h comes from the 4 MSBs of ACCL.

As mentioned in Chapter 3, the multiplication of two numbers in Q.15 format results in a number of Q.30 format. Setting the PM bits to 1 enables the product shifter to eliminate the extra sign bit. An alternative is to clear the PM bits to 0 and add the product in Q.30 format into the ACC. Thus, the data in the ACC has 2 sign bits. We can get rid of the extra sign bit when the data is saved into data memory using the instruction

```
SACH    yn,1    ; save ACCH to yn and left shift 1 bit
```

which has the advantage of having 1 guard bit in the ACC if consecutive multiply-add operations are performed.

Auxiliary Register Arithmetic Unit

As illustrated in Fig. 4.4, the C2000 processor has eight auxiliary registers (AR0-AR7) for addressing data memory. In the indirect addressing mode, the 16-bit auxiliary register (AR) specified by the 3-bit auxiliary register pointer (ARP) is called the current AR and is used as the address pointer to data memory. For example, when the value of ARP is 3, AR3 is the current AR register used as the address pointer. In indirect addressing, any location in the 64K data-memory space can be accessed via a 16-bit address contained in the current AR.

The main function of the ARAU is to perform address arithmetic operations for updating auxiliary registers in parallel with data operations performed by the ALU. The eight LSBs of the instruction register (IR), the contents of AR0, or the data memory can be used as one of the inputs to the ARAU. The other input is the current AR pointed at by the ARP. As a result, accessing a sequence of data does not require the ALU to generate data addresses, thus allowing the CPU to perform data operations in parallel.

The auxiliary registers and ARP can be loaded from data memory, the ACC, the PREG, or an immediate operand defined in the instruction. A specific ARn can be selected as a current AR for the address pointer by loading the 3-bit ARP with a value from 0 through 7. For example, the instruction MAR *,AR1 loads 1 into ARP such that AR1 will be used as a current AR after this instruction.

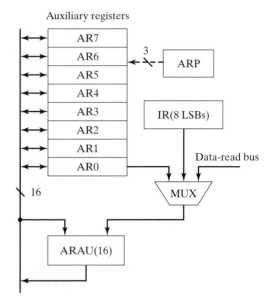

Figure 4.4   Indirect addressing with ARAU

The ARP can also be updated by many instructions using the indirect-addressing mode with the following general syntax:

```
mnemonic    ind[,shift[,ARn]]
```

where the last optional ARn loads the value of n into the ARP register for specifying the next AR. For example, assuming the current ARP is 0, the instruction ADD *,1,AR2 reads the data in the data memory pointed at by AR0, shifts it left 1 bit, and then adds the result to the ACC. At the same time, the ARP is loaded with the value 2. Thus, AR2 is selected as the next AR without an extra cycle.

The AR can be initialized with the instruction LAR (load auxiliary register). For example, the instruction

```
LAR     AR1,#k (or #lk)
```

loads AR1 with the 8-bit value k (or the 16-bit value lk). In addition, the contents of the specified data memory can be loaded into the designated AR using the direct-or indirect-addressing mode.

The ARAU performs the seven indirect addressing operations listed in Table 4.3. For example, if ARP = 1, the instruction

```
ADD     *+
```

adds the data in the data memory pointed at by AR1 to the ACC, and the value of AR1 is then incremented by 1. If the operand is * −, the contents of AR1 are decremented by 1 after the access. In addition, the instruction

```
ADD     *0+
```

TABLE 4.3 C2000 ARAU Operations

Operand	Operation
*	None
*+	ARn ← ARn + 1
*−	ARn ← ARn − 1
*0+	ARn ← ARn + AR0
*0−	ARn ← ARn − AR0
*BR0+	ARn ← ARn + AR0 with bit reversed carry
*BR0−	ARn ← ARn − AR0 with bit-reversed carry

adds the contents of the current AR to an unsigned 16-bit integer contained in AR0. If the operand is *0−, the contents of AR0 are subtracted from the current AR after the access. The operands *BR0+ and *BR0− generate a bit-reversed address used for FFT algorithms.

In the direct-addressing mode, the instruction contains the lower 7 bits of the dma. As illustrated in Fig. 4.5, this field is concatenated with the 9 bits of the DP register to form the full 16-bit data-memory address. In this scheme, the DP points to 1 of 512 pages (each page has 128 words) of data memory, and the dma points to the desired word within that page, see Fig. 3.10 as an example.

The DP can be initialized by using the LDP (load data memory page pointer) instruction. For example, the instruction

```
LDP     #k
```

loads the constant k into the DP, where $0 \leq k \leq 511$. The DP can also be loaded using the LST (load status register) instruction. To develop efficient code using direct addressing, keep frequently used data on the same page to avoid having to update the DP register, which requires at least two clock cycles.

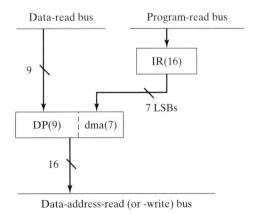

Data-address-read (or -write) bus

Figure 4.5 Data-address generation using direct-addressing mode

Example 4.4

The data xn is located at data-memory address 0302h, which can be expressed in binary as 0000 0011 0000 0010b. In this case, the 7 LSBs represents 2 and the 9 MSBs represents 6, which indicates that the data is located at address 2 on page 6. To access xn in the program, we can use the assembly directive

```
xn    .set    2
```

to set the symbol xn equal to 2. In the program, the following instructions

```
LDP     #6        ; DP <- 6
LACC    xn        ; ACC <- xn
```

load the value of xn from the data-memory address 0302h into the ACC.

Status Registers

Two status registers for the C2000 processors, ST0 and ST1, contain the status and control bits. These registers can be stored to and loaded from data memory, thus allowing the status of the processor to be saved and restored for executing subroutines.

The LST instruction writes to ST0 and ST1, and the SST (store status register) instruction reads from them. The individual bits of these registers can be set or cleared with the SETC and CLRC instructions. For example, the interrupt mode bit (INTM) is set by using

```
SETC    INTM        ; set interrupt mode bit
```

Note that when the INTM bit is set to 1, all nonmaskable interrupts are disabled.

The OVM controls saturated arithmetic as discussed in Chapter 3. When overflow occurs in overflow-saturation mode (i.e., the OVM bit is set to 1), the overflow flag is set and the ACC is set to either its most positive value (07FFFFFFFh) or its most negative value (080000000h), depending on the direction of the overflow. When the OVM bit is cleared to 0, overflow occurs if the value in the ACC is larger than or equal to 1 or is less than −1. The saturation mode can be set by using

```
SETC    OVM         ; set overflow saturation mode
```

4.2.3 Program Control

The program controller is an independent execution unit that controls program flow, which involves determining the order of instructions to be executed. Hardware resources include the PC, the stack, and the status registers. Software-program control elements are branches, calls, returns, conditional instructions, repeats, and interrupts.

Program Counter and Stack

The C2000 processor has a 16-bit PC that points to the next instruction to be executed in program memory. When the instruction is fetched from that address and loaded into the IR, the PC is incremented automatically. The PC can be loaded in different ways depending on program flow. When the code is executed sequentially, the PC is incremented by 1 or 2 if the current instruction uses one or two words, respectively. When the branch is executed, the PC is loaded with the long immediate address included in the branch instruction. In the case of a subroutine call, the address of the next instruction (PC + 2) is pushed onto the stack, and then the PC is loaded with the long immediate value in the call instruction (CALL, CALA, or CC). The return instructions (RET and RETC) pop the return address stored in the stack back into the PC; thus, the program returns to the calling (or interrupting) sequence of code.

The stack is used during interrupts and subroutine calls. The C2000 hardware stack is 16 bits wide and eight levels deep. When the CPU executes a call instruction or an interrupt forces the CPU into an ISR, the return address is automatically stored at the top of the stack. When the subroutine or ISR is completed, a return instruction transfers the return address from the top of the stack into the PC. The stack is accessible through the use of PUSH and POP instructions. Two additional instructions, PSHD and POPD, push a data memory value onto the stack and pop a value from the stack to data memory, respectively. These instructions build a software stack in data memory for the nesting of subroutines or interrupts beyond eight levels.

The contents of the ACCL may be loaded into the PC to implement computed branch operations and subroutine calls. This task can be accomplished with BACC (branch to address specified by ACC) or CALA (call subroutine at location specified by ACC) instructions. These instructions provide the ability to branch to a computed address based on the performance of the system.

Branch Operations

The C2000 processor provides several branch instructions for implementing program-control structures used in C programs. In general, there are two types of branches, unconditional and conditional, as illustrated in Fig. 4.6. Unconditional

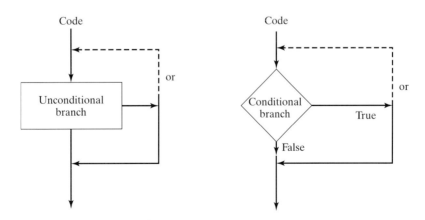

Figure 4.6   Program flow of unconditional (left) and conditional (right) branches

branch instructions, such as B (branch unconditionally), load the PC with the specified program-memory address, thus forcing execution to jump to the instruction at that address. For example,

```
B   dst_label      ; branch unconditionally to dst_label
```

where dst_label is a label used to identify the first instruction of the destination module. In this example, the address loaded into the PC comes from the second word (16-bit immediate address dst_label) of the branch instruction.

Similar to branches, calls and returns also break sequential program flow. A branch instruction only changes the program flow to another location, while a call instruction saves the return address to the hardware stack. Every subroutine including ISR uses a return instruction at the end, which pops the return address from the stack back to the PC. Thus, the program branches back to the instruction following the call instruction.

In the case of conditional branches, the conditions listed as operands of the conditional-branch instruction are evaluated first. If any one of the conditions is false, the PC is incremented such that the instructions below the branch instruction are executed. Otherwise, a branch is taken. For example,

```
BCND    pma,[cond1][,cond2][,…]      ; branch conditionally
```

branches to program memory at address pma (16-bit address) if all of the specified conditions are met.

When multiple conditions are defined in the operands of the conditional instructions, all conditions must be met for the branch to be taken. The conditional-branch, call, and return instructions test the conditions listed in Table 4.4.

TABLE 4.4 Conditions Tested in Conditional Instructions

Operand	Condition	Description
EQ	$ACC = 0$	ACC equal to 0
NEQ	$ACC \neq 0$	ACC not equal to 0
LT	$ACC < 0$	ACC less than 0
LEQ	$ACC \leq 0$	ACC less than or equal to 0
GT	$ACC > 0$	ACC greater than 0
GEQ	$ACC \geq 0$	ACC greater than or equal to 0
C	$C = 1$	Carry bit set to 1
NC	$C = 0$	Carry bit cleared to 0
OV	$OV = 1$	ACC overflow detected
NOV	$OV = 0$	No overflow detected
BIO	BIO is low	BIO pin is low
TC	$TC = 1$	Test/control flag set to 1
NTC	$TC = 0$	Test/control flag cleared to 0

For example, the following instruction

```
BCND     pma,TC
```

performs the condition branch to the program memory at address pma if TC = 1. Note that only certain combinations of conditions can be used.

Example 4.5

In writing assembly code, if the values of variables $x1 > x2$, execute the code module labeled with label1. If $x1 = x2$, execute the module called label2. Otherwise, execute the code module that starts at label3. The program-flow control can be implemented using C2000 instructions as follows:

```
         LACC    x1            ; ACCL <- x1
         SUB     x2            ; ACCL = x1-x2
         BCND    label1,GT     ; branch to label_1 if x1>x2
         BCND    label2,EQ     ; branch to label_2 if x1=x2
label3:  ...                   ; execute code here if x1<x2
```

The subroutine call can also be executed conditionally. The CC (call conditionally) instruction operates like the BCND (branch conditionally) except that the return address is pushed onto the stack, which allows the return instruction RET at the end of the subroutine to pop the stack in order to return to the main program. The conditional return instruction RETC allows a subroutine to have multiple return paths based upon the data being processed.

The conditional branch is a two-word instruction. If all of the conditions defined in the operands of the instruction are met, the PC is loaded with the second word, and the CPU starts refilling the pipeline with instructions at the branch address. Because the pipeline has been flushed, the branch instruction has an effective execution time of four cycles if the branch is taken. If any of the conditions are not met, the pipeline controller allows the next instruction (already fetched) to be decoded, so the effective execution time of the branch is two cycles.

Looping

The simplest method to implement loops is to use the BANZ (branch on auxiliary register not zero) instruction. The syntax of BANZ is

```
[label]   BANZ     pma[,ind[,ARn]]
```

where the 16-bit program-memory address pma can be either a label used in the assembly program or a numeric address (between 0 and 65,535). The program branches to the designated program memory if the value of the current AR is not zero. Otherwise, the processor executes the next instruction following the BANZ instruction. The processor also modifies the current AR as specified, and the default modification is decremented by one. To use the loop with N iterations implemented by BANZ, initialize an AR as a loop counter with the value $N - 1$ prior to loop entry.

Example 4.6

To add four consecutive data entries, start at address 60h in data memory, to the ACC. We can use the following code:

```
        MAR    *,AR0         ; select AR0 as current AR
        LAR    AR1,#3        ; AR1=3, use it as loop counter
        LAR    AR0,#60h      ; AR0=60h is address pointer
SUM     ADD    *+,AR1        ; ACC<-ACC+(AR0), AR0+1, ARP=1
        BANZ   SUM,*-,AR0    ; branch to SUM if AR1≠0,
        ...                  ; AR1 <- AR1-1, ARP=0
```

In this example, the loop counter AR1 is initialized to the value 3, so the loop is executed 4 times. The address pointer AR0 is initialized to 60h and is incremented by 1 each time the loop is executed. Therefore, the data sample in data-memory locations 60h-63h are added to the ACC. Note that the loop counter is decremented by the BANZ instruction, while the address pointer is incremented with the ADD instruction.

4.2.4 Programming Issues

As discussed in previous chapters, DSP involves some special functions such as repeated multiply-add operations for FIR filtering. The architecture and instructions optimized for this application are introduced in this subsection.

Repeat Operations

The repeat instruction (RPT) executes the next instruction $N + 1$ times, where N is an operand of the repeat instruction. For example, the following instructions

```
    RPT    #k      ; repeat next instruction k+1 time
    ADD    *+      ; add k+1 samples to the ACC
```

load the 8-bit immediate value k into the 16-bit repeat counter (RPTC). When the RPTC is loaded with a number k, the following instruction is executed k + 1 times. The RPTC can also be loaded with a 16-bit value in data memory using either direct- or indirect-addressing mode, which allows the instruction following the RPT to be repeated a maximum of 65,536 times. The next instruction is pipelined when the repeat feature is used, and it effectively becomes a single-cycle instruction, although multiple cycles normally are needed. For example, the TBLR (table read) instruction may take three or more cycles to execute, but when the instruction is repeated, one datum can be read at every cycle. Note that not all instructions can be repeated.

Example 4.7

The sum of squared N consecutive samples expressed as

$$E_x = \sum_{i=0}^{N-1} x^2(i) \tag{4.2.1}$$

can be implemented using the following code:

```
        LACC    #0          ; clear ACC
        MPY     #0          ; clear PREG
        MAR     *,AR1       ; ARP <- 1
        LAR     AR1,#xn     ; AR1 points to the first sample
        RPT     #N-1        ; repeat next instruction N times
        SQRA    *+          ; sum of squared values
        APAC                ; accumulate the last product
```

Since the SQRA instruction adds the contents of the PREG to the ACC before squaring the addressed data-memory value and storing the product into the PREG, we have to clear the PREG and ACC first. At the end of the repeat operations, we have to add the last product in the PREG to the ACC with the APAC instruction.

Multiply-Add Operations

The multiply-add instructions (MAC and MACD) fully use the computational bandwidth of the multiplier, allowing both operands to be processed simultaneously. As shown in Fig. 4.1, these two operands can be transferred to the multiplier during each cycle via the program-read and data-read buses. The MAC instruction (1) adds the previous product, shifted as defined by the PM, to the ACC, (2) loads the TREG with the contents of the specified data memory, (3) multiplies the value in the TREG by the contents of the specified program memory, and (4) stores the product in PREG. When the MAC instruction is repeated, the program-memory address contained in the PC is incremented by 1 during each repetition, which facilitates the single-cycle multiply-add operation when used with the RPT instruction. In this case, the PC generates the coefficient addresses at program memory, while the data-memory addresses are generated by the ARAU, which allows the repeated instruction to sequentially access the values from the coefficient table and the data buffer.

Example 4.8

The output of an FIR filter of length L is computed as

$$y(n) = \sum_{i=0}^{L-1} b_i x(n - i). \tag{4.2.2}$$

The signal buffer $\{x(n), x(n - 1), \ldots, x(n - L + 1)\}$ is located in data memory space and is arranged as shown in Fig. 4.2. Assume that COEF is the symbolic address of the coefficient buffer located in the program memory and that the current AR1 points to the beginning of the signal buffer as shown in Fig. 4.7. We can use the following instructions to implement FIR filtering:

```
        LACC    #0          ; clear ACC
        RPT     #L-1        ; repeat the next instruction L times
        MAC     COEF,*+     ; multiply and accumulate previous product
        APAC                ; accumulate the last product
```

The MAC instruction multiplies the signal $x(n - i)$ in data memory pointed at by AR1 with the coefficient b_i in program memory pointed at by PC, which is initialized to be pointing at the first coefficient at program-memory address COEF.

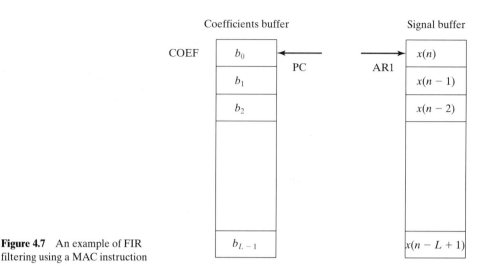

Figure 4.7 An example of FIR filtering using a MAC instruction

The MACD instruction performs all of the tasks of MAC plus a data-move operation, which copies the contents of the specified data-memory address, to the next higher data-memory address, such as the DMOV instruction, to support FIR filtering operations further. Note that the DMOV portion of the MACD instruction does not function with external data memory.

Example 4.9

To efficiently use the MACD instruction for FIR filtering, the memory configuration shown in Fig. 4.7 must be modified to that shown in Fig. 4.8. The FIR filtering can be implemented as:

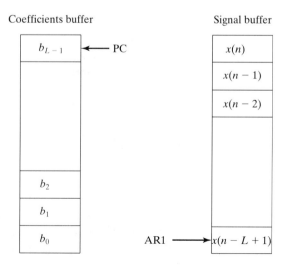

Figure 4.8 Memory layout for FIR filtering using the MACD instruction

```
PRT    #L-1       ; repeat the next instruction L times
MACD   COEF,*-    ; multiply and accumulate with data move
APAC              ; add last product in PREG to ACC
```

In this example, AR1 points to the oldest sample $x(n - L + 1)$ located at the highest address and PC points to the last coefficient b_{L-1} at the lowest address. The PC is used to step through the coefficients in the coefficient table. The address pointer AR1 with a decrement option steps through the signal buffer, starting with the oldest one. As the addressed data is fetched, it is also written to the next higher location in data memory.

4.2.5 System Issues

C2000 processors are designed to meet a wide range of digital motor and other embedded-control applications. This family is complimented with a wide range of on-chip peripherals and ROM (or flash memory), plus on-chip DARAM and optional SARAM. The peripherals include timers, serial peripheral interfaces, and safety features such as watchdog timers.

Memory Configurations

The total memory address range of C2000 processors is 192K words, which is organized into three individually selectable spaces: (1) 64K-word program memory contains instructions and data such as filter coefficients, (2) 64K-word data memory holds data used by the instructions, and (3) 64K-word I/O space interfaces to external peripherals. The 16-bit address and data buses, along with $\overline{PS}$, $\overline{DS}$, and $\overline{IS}$ space-select signals, support addressing of 64K words in the program, data, and I/O space, respectively.

As shown in Table 4.1, C2000 processors include different amounts of on-chip memory. On-chip memories can be configured by software in or out of the data-memory map. When these memories are mapped into data space, the processor automatically accesses them when addressing within their bounds. When an address is outside these bounds, the processor generates an external access. The on-chip memory has higher performance because no wait states are required and because there is better flow within the pipeline of the CPU. Using on-chip memory, the CPU with six independent buses can execute a program fetch, data read, and data write on the same machine cycle. In addition, on-chip memory has lower power consumption.

Internal memory is configured with the following memory modules: DARAM, SARAM, flash (or ROM), and boot ROM. An example of the LF2407 memory map is illustrated in Fig. 4.9. The 544 words of DARAM are partitioned into three blocks: block 0 (B0, 256 words), block 1 (B1, 256 words), and block 2 (B2, 32 words). Both B1 and B2 blocks are located only in data-memory space. The block B0 has 256 words from 200h to 2FFh if B0 is configured as data memory (CNF = 1) or from 0FF00h to 0FFFFh in program space if B0 is configured as program memory (CNF = 0).

At reset, the DARAM block B0 (256 words) is mapped into local data space. Block B0 can be reconfigured into program space by setting the CNF (DARAM

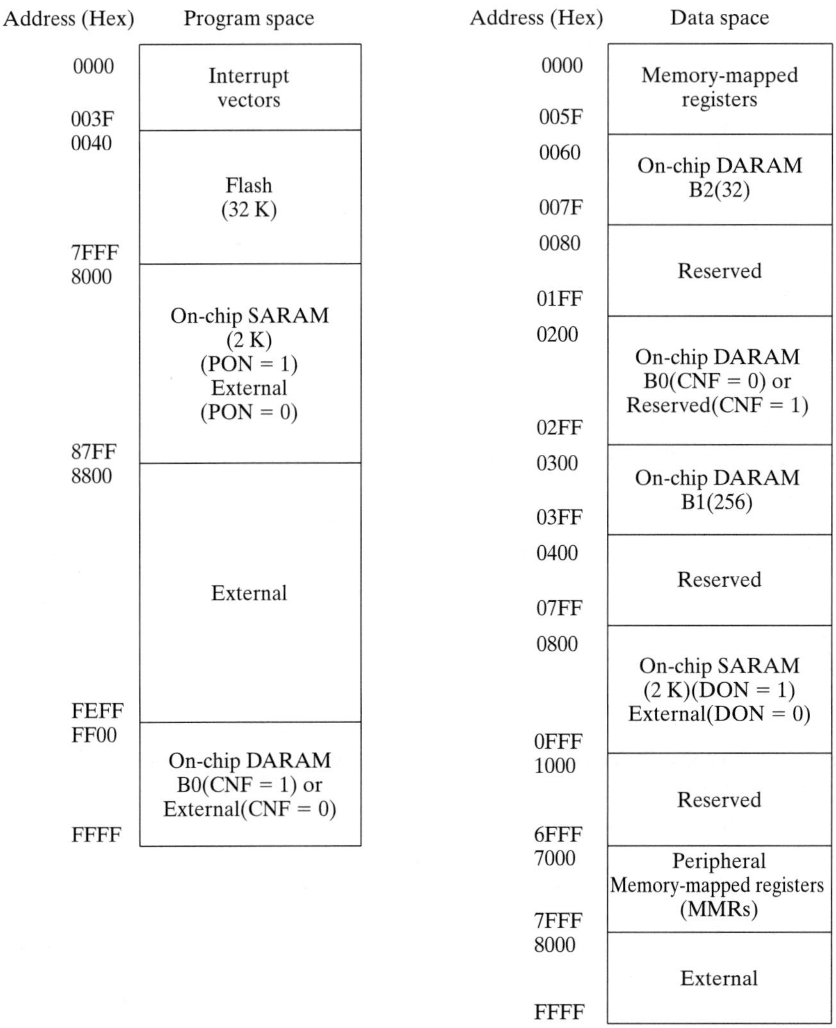

Address (Hex)	Program space
0000	Interrupt vectors
003F	
0040	Flash (32 K)
007F	
7FFF	
8000	On-chip SARAM (2 K) (PON = 1) External (PON = 0)
87FF	
8800	External
FEFF	
FF00	On-chip DARAM B0(CNF = 1) or External(CNF = 0)
FFFF	

Address (Hex)	Data space
0000	Memory-mapped registers
005F	
0060	On-chip DARAM B2(32)
007F	
0080	Reserved
01FF	
0200	On-chip DARAM B0(CNF = 0) or Reserved(CNF = 1)
02FF	
0300	On-chip DARAM B1(256)
03FF	
0400	Reserved
07FF	
0800	On-chip SARAM (2 K)(DON = 1) External(DON = 0)
0FFF	
1000	Reserved
6FFF	
7000	Peripheral Memory-mapped registers (MMRs)
7FFF	
8000	External
FFFF	

Figure 4.9   TMS320LF2407 memory map in microprocessor mode

configuration bit) in the ST1 register to 1. For example, the instruction

```
SETC    CNF    ; CNF <- 1
```

configures B0 as program memory, and the instruction

```
CLRC    CNF    ; CNF <- 0
```

configures B0 as data memory, which allows dynamic configuration of B0 using software. It is important to note that when the MAC and the MACD instructions use coefficients stored in block B0, the CNF bit must be set to 1.

Flash memory (32K in program space) provides an attractive alternative to masked-program ROM. Like ROM, flash is nonvolatile. However, it has the advantage of having a field-reprogramming ability. Thus, this type of memory is useful for prototyping, early field testing, and single-chip applications.

Memory Management

C2000 processors can address a large memory space (192K words), but only have a limited amount of on-chip memory. Several instructions are available for moving blocks of data or programs from slower off-chip memory to on-chip memory for faster program execution. In addition, data can be transferred from on-chip memory to off-chip memory for storage. For example, the instruction

```
BLDD    source,   destination
```

transfers data from data memory (internal or external) to data memory. The word in data memory at address source is copied to data memory at address destination. The source and destination blocks do not have to be entirely on chip or off chip. The repeat instruction RPT can be used with the BLDD instruction to move consecutive words in data memory. When used with RPT, BLDD becomes a single-cycle instruction once the RPT pipeline is started. Similar to BLDD, the block move instruction BLPD moves data from program memory to data memory.

Example 4.10

The following code moves 256 coefficients from external memory starting at address 8800h to internal-data memory B1 starting at address 300h:

```
LAR    AR7,#300h   ; AR7 <- 300h
MAR    *,AR7       ; ARP <- 7
RPT    #255        ; repeat next instruction 256 times
BLDD   #8800h,*+   ; block move of 256 data
```

In the example code, the source address specified by the long immediate constant (e.g., 8800h) is stored in the PC. Because the PC is incremented by 1 automatically during each repetition, it is able to access a series of data in memory.

Another method of transferring a word from program memory to data memory is to use the TBLR instruction. In addition, the instruction TBLW transfers a word from data memory to program memory. In these two instructions, the 16 bits of the ACCL define the program-memory address. Since the address comes from the ACC, these instructions can specify a calculated (rather than a predetermined) location of data in program memory for transfer. When repeated with the RPT instruction, the TBLR and TBLW effectively become single-cycle instructions.

Interrupts

The C2000 processor supports one nonmaskable interrupt and six prioritized interrupt requests. The software-programmable interrupt structure supports flexible on-chip and external-interrupt configurations to meet real-time interrupt-driven applications. There are three interrupt sources: reset (hardware or software initiated), hardware-generated interrupts (requested by external pins or by on-chip peripherals), and software-generated interrupts. Because the CPU does not have sufficient capacity to handle all peripheral-interrupt requests, six core interrupts (INT1-INT6) are expanded using the peripheral-interrupt expansion controller to manage peripheral-interrupt requests from different sources.

There are two CPU registers for controlling interrupts: (1) the interrupt-flag register, which contains flag bits that indicate when maskable-interrupt requests have reached the CPU, and (2) the interrupt-mask register, which contains mask bits that enable or disable the corresponding interrupts. In addition, the INTM in the status register (ST0) enables (INTM = 0) or disables (INTM = 1) all of the maskable interrupts.

When the CPU receives an interrupt request, it does not know which peripheral caused the request. To enable the CPU to recognize these events, a unique peripheral-interrupt vector is generated in response to an active interrupt request. This vector is loaded into the peripheral-interrupt register in the peripheral-interrupt expansion controller. It can then be read by the CPU and used to generate a vector to branch to the ISR, which corresponds to the event being acknowledged.

4.2.6 An Application: Phase-Locked Loop

In this section, we introduce an application of the TMS320F2407 for implementing a first-order IIR filter and a nonlinear median filter used for all-digital PLL.

Introduction to Phase-Locked Loops

A PLL is a circuit that synchronizes an output signal generated by an oscillator with a reference (or input) signal in frequency as well as in phase. In the synchronized state, the phase error between the oscillator's output signal and the reference signal is zero or very small. If the phase error increases, the circuit acts on the oscillator in such a way that the phase error is again reduced to a minimum. Thus, the phase of the output signal is actually locked to the phase of the reference signal.

The PLL consists of three basic functional blocks, as illustrated in Fig. 4.10. The phase detector compares the phase of the voltage-controlled oscillator output $y(t)$ with the phase of the reference signal $x(t)$ and develops a difference signal $e(t)$, which is approximately proportional to the phase error. The signal $e(t)$ consists of

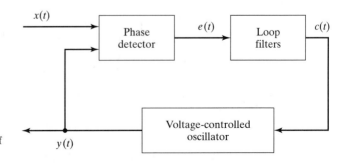

Figure 4.10   Block diagram of the PLL

the desired DC (direct current, a constant signal) component and undesired noise, which must be reduced by the loop filters. The voltage output $c(t)$ is used to control the frequency and phase of the voltage-controlled oscillator.

The first PLL chips that appeared around 1965 were purely analog devices. An analog multiplier was used as the phase detector, and the loop filter was built from a passive or active RC filter. This type of PLL is called linear PLL. The first digital PLL, which appeared around 1970, used digital circuits such as XOR gates or JK flip-flops for the phase detector. The all-digital PLL is built exclusively from digital function blocks. In this application, we introduce a DSP-based all-digital PLL that uses the TMS320F2407 for implementing loop filters. Detailed information on PLL is available in the reference book [4].

Loop Filters

The interference presented in $e(n)$ consists of two components: impulse noise and broadband noise. A lowpass filter is very efficient for attenuating noise at frequency ranges higher than the desired DC component, and a nonlinear median filter is very effective for reducing impulse noise. For the purpose of testing the loop filters, we first use the MATLAB script engen.m to generate the testing signal $e(n)$ and save it to the data file en.dat. The waveform is shown in Fig. 4.11, where the desired DC value (2,047) is corrupted by a 60 Hz hum, broadband white noise, and impulse noise. The moving-average filter implemented as the first-order IIR filter introduced in the experiments in Chapter 2 can attenuate the broadband noise and 60 Hz

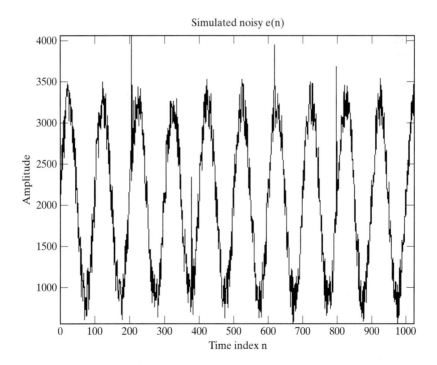

Figure 4.11   Interference (impulse and broadband noise) at the output of a phase detector

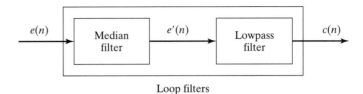

Figure 4.12 Block diagram of loop filters

Loop filters

hum effectively. However, the energy presented in the impulse noise still influences the linear filter output. The most efficient technique for reducing impulse noise is a nonlinear median filter.

As shown in Fig. 4.12, the loop filters for PLL consists of two filters: a nonlinear median followed by a linear lowpass filter. In this application, the 5-point median filter is used. In the filter, a first-in first-out buffer of size 5 is used to store signals $e(n)$, $e(n-1)$, ..., $e(n-4)$. These samples are moved to a new sorting buffer, where the elements are ordered by magnitude. The output of the median filter $e'(n)$ is simply the middle of the five numbers in the sorting buffer. Although median filters remove impulse noise in the signal, they often fail to reduce the undesired broadband noise and 60 Hz hum. A good compromise is to use a linear filter with a narrow bandwidth such as the moving-average filter given in Eq. (2.3.23).

The loop filters shown in Fig. 4.12 are implemented in dpll.c. Similar to the iir1.c given in Section 2.6.2, the C program opens and reads the disk file en.dat that contains $e(n)$ generated by engen.m. It then processes the signal using the 5-point median filter and the moving-average filter and saves the output $c(n)$ to the file cn.dat. The MATLAB program plotec.m is used to plot both $e(n)$ and $c(n)$, and the figure shows that the algorithm is effective for preserving the desired DC component.

Implementation Considerations

We can implement the loop filters on the TMS320F2407 using the assembly-language program dpll.asm. Note that the IIR filter given in Eq. (2.3.23) is simplified to

$$c(n) = (1 - \alpha)c(n - 1) + \alpha e'(n)$$
$$= c(n - 1) - \alpha c(n - 1) + \alpha e'(n). \qquad (4.2.3)$$

When the coefficient α is a negative power of 2, that is if

$$\alpha = 2^{-m}, \qquad (4.2.4)$$

$\alpha e'(n)$ can be implemented by shifting $e'(n)$ m bits to the right. Since the accumulator of the TMS320C2000 processor is 32 bits, shifting m bits to the right and loading the result into the high word of the accumulator can be done by shifting $(16 - m)$ bits to the left and loading the result into the low word of the accumulator. The portion of C2000 assembly code used to implement this algorithm is listed as follows:

```
LPF:    LACC    cn_high,16      ; ACCH <= c(n-1), high word
        ADDS    cn_low          ; ACC = c(n-1)
```

```
SUB    cn_high,m8      ; -a*c(n-1)
ADD    sort1_buf2,m8   ; + a*e'(n)
SACH   cn_high         ; save c(n), high word
SACL   cn_low          ; save c(n), low word
```

To test the TMS320C2000 assembly program dpll.asm using CCS 2000, we create a new folder and copy the assembly-program source file dpll.asm, the linker command file dpll.cmd, the register header file f240regs.h, and the input data file xn.int into that folder. Note that the C program asc2int.c is used to convert the ASCII data file en.dat into an integer format file en.int that can be included in the TMS320 assembly program.

4.3 TMS320C54x

The TMS320C54x generation consists of power-efficient, fixed-point DSP processors that are targeted for portable devices such as cellular phones, MP3 players, digital cameras, and computer-telephony applications. In this section, we briefly introduce architecture, programming, and system issues. Detailed descriptions of these topics are available in TMS320C54x reference sets [5, 6].

4.3.1 Architecture Overview

Like the C2000 generation, the C54x processor is based on the modified Harvard architecture with separate program and data spaces. Table 4.5 shows the program and data memories available for C54x processors. The DARAM and SARAM can be configured as either data memory or data/program memory. The DARAM is usually used for storing frequently used data and program code because it can be

TABLE 4.5   Memories Available for TMS320C54x Processors

Processors	ROM (16-bit)	DARAM (16-bit)	SARAM (16-bit)	Data/program memory (16-bit)
C5401	4K	16K	0	64K/1M
C5402	4K	16K	0	64K/1M
C5404	16K	8K	56K	64K/8M
C5407	128K	40K	0	64K/8M
C5409	16K	32K	0	64K/8M
C5410	16K	8K	56K	64K/8M
C5416	16K	64K	64K	64K/8M
C5420 (Dual core)	0	16K × 2	84K × 2	256K/256K
C5421 (Dual core)	4K	32K × 2 + 120K (share)	32K × 2	64K/256K
C5441 (Four core)	0	32K × 4	64K × 4 + (128K share) × 2	4 × 96K/2 × 128K

accessed twice per instruction cycle. The on-chip ROM contains a boot-loader program for initializing the processor to a known state and starting up the application code. Compared with Table 4.1, we find that a significant amount of memory, especially DARAM, is available on C54x processors. However, as DSP applications become more complex, limited on-chip memory may not be sufficient for storing code and data. Expanding memory space by using slower memory off chip solves this problem. The last column of Table 4.5 shows the maximum addressable memory space that can be accessed by C54x processors.

These memories are connected to four sets of internal address and data buses: one set of program-read buses, two sets of data-read buses C and D, and a data-write bus E, as shown in Fig. 4.13. Therefore, a maximum of one instruction fetch, two data reads, and one data write can be performed within one cycle. The C54x processor supports sustainable memory transfers of 200 million words for reads and 100 million words for write per second in a 100 MHz processor. Compared with the C2000, the C54x provides the second data bus D to support dual-data accesses.

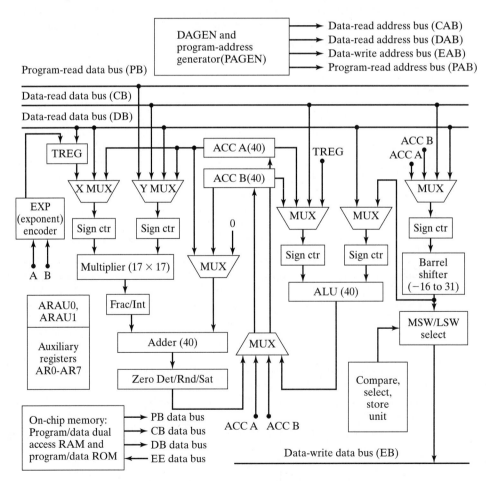

Figure 4.13   Block diagram of the TMS320C54x internal hardware

Figure 4.13 shows the block diagram of the TMS320C54x processor architecture. The CPU contains the following functional units: (1) a 40-bit ALU; (2) two 40-bit accumulators (ACC A and ACC B); (3) a barrel shifter; (4) a 17 × 17-bit multiplier and a 40-bit adder; (5) a compare, select, and store unit; (6) an exponent encoder; and (7) DAGEN and PAGEN units.

Arithmetic and Logic Unit

The 40-bit ALU performs two's-complement arithmetic and logical operations. Data from memory, TREG, or ACCs, as well as output from the barrel shifter can be fed into the ALU. The output from the ALU is stored in ACC A or B. The ALU supports dual 16-bit operations. The lower 32 bits of the ALU can be treated as two parallel 16-bit ALUs, thus making it able to perform two 16-bit additions (or subtractions) in one cycle. This task is achieved by setting the C16 field of the ST1 register.

The ALU supports both saturation and sign extension, which can be activated by setting the OVM bit and SXM bit in the ST1 register, respectively. Saturation limits the value being loaded from the ALU to the ACC. The most positive and negative 32-bit numbers to be loaded into the ACC are 7FFFFFFFh and 80000000h, respectively. Two flags, OVA and OVB, in the ST0 register are used to indicate whether ACC A and B, respectively, have overflowed. In sign-extension mode, the upper bits are sign extended. For example, a negative 16-bit data, F000h, is sign-extended (SXM = 1) to 32-bit data as FFFFF000h or is zero-filled (SXM = 0) to 0000F000h. ALU also supports add-with-carry and subtract-with-borrow operations.

Accumulators

Compared with the C2000, C54x processors not only extend the accumulator from 32 bits to 40 bits, but also increase the number of ACCs from one to two. The 40-bit ACCs A and B are used to store the output from the ALU or the multiply-add unit. Both ACC outputs can be fed into the adder and ALU inputs, but only ACC A output can be fed into the 17 × 17 multiplier for multiply-add operations. The 40-bit ACCs are partitioned into three parts: (1) lower 16 bits (AL and BL), (2) upper 16 bits (AH and BH), and (3) 8 guard bits (AG and BG). Guard bits create additional headroom for computation overflow, as explained in Chapter 3.

The contents of the ACC can be stored in data memory by using a set of store instructions. AL (or BL) can be stored using the STL (store ACC low into memory) instruction, while AH (or BH) can be stored using the STH (store ACC high into memory) instruction. The contents of the ACC can be shifted (between left 15 bits and right 16 bits) before storing in data memory. Similarly, the ACCs can be loaded using the LD instruction.

Barrel Shifter

The input to the barrel shifter can come from the ACCs or memory. The output of the barrel shifter is either to the ALU or memory. The barrel shifter can perform a left shift of 0 to 31 bits and a right shift of 0 to 16 bits on the input data. The main functions of the barrel shifter are scaling of signals, logical or arithmetic shifting of accumulator values, bit extraction, extended arithmetic, and prevention of overflow. Programming examples are given in a later section to illustrate the effects of shifting.

Multiple-Add Computation Unit

The C54x processor has a 17-bit $\times$ 17-bit hardware multiplier coupled with a 40-bit adder to form a multiply-add unit. The multiply-add unit is further linked to the ACC to form a MAC unit. Similar to the C2000 processors, special instructions (MAC, MACA, MACD, MACP, and MACSU) are used to perform the multiply-add operation in one cycle. The output of the adder is passed through a unit that detects a zero or an overflow in the result and performs saturation or rounding according to the programming mode.

Fractional multiplication is usually performed in fixed-point DSP processors. In C54x processors, setting the FRCT bit in the ST1 register to 1 supports Q.15 fractional multiplication. The multiplier output is shifted left by 1 bit to eliminate the extra sign bit. The processor also supports double-precision multiplication (e.g., 32-bit $\times$ 32-bit) by providing instructions for mixing signed-unsigned multiplications. Several MAC instructions are coupled with the repeat instructions for performing multiply-add operations that access data and coefficients in one cycle. More examples of these powerful instructions are given in Section 4.3.3.

Compare-Select-Store Unit

A new compare-select-store unit is included in C54x processors to implement the Viterbi algorithm efficiently. It compares the high and low words of the ACC, selects the larger word, and stores it into data memory. A 16-bit transition register is used to track the comparison history for determining the optimum path in the Viterbi's state trellis and the decoded code.

Address-Generation Unit

The PAGEN unit provides the address of the current instruction. This information is stored in the PC and is increased automatically when the subsequent instruction is fetched. However, instructions such as branch, call, return, etc., redirect the PC to another location. The last column of Table 4.5 shows that the newer generation of the C54x processors can address up to 8M words. The addressing of large memory space is made possible by using a 7-bit page register, XPC, which selects up to 128 pages. The PC points at the instruction inside the selected 64K-word page. Together with the PC, 23-bit addresses are generated for some load and store instructions.

The DAGEN unit calculates the addresses of operands using the following 16-bit registers for data-memory reads and writes: auxiliary registers (AR0-AR7), the block-size register (BK), and the stack pointer (SP). The BK register is used for defining the size of the circular buffer. The auxiliary register can also be used as a circular pointer for modulo addressing, and only one circular buffer can be defined at any time in C54x processors. A detailed description of the circular buffer is given in Section 4.3.4. The SP is used to access the stack memory, which is memory space reserved for storing the returned address, registers, etc., during subroutine calls and interrupts. Stack operations are explained further in Section 4.3.2.

4.3.2 Addressing Modes

In this section, we highlight some commonly used addressing modes include the indirect, stack, and I/O addressing modes. The immediate- and direct-addressing modes are discussed in Section 3.4.1.

Indirect Addressing

Indirect addressing uses the auxiliary registers (ARn, $n = 0, 1, \ldots, 7$) as address pointers. The symbol * in front of the auxiliary register indicates the current AR, and the indirect-addressing mode is used. Address pointers can be modified using one of the options summarized in Table 4.6. Compared with the C2000 ARAU operations listed in Table 4.3, C54x processors provide additional pre-increment, pre-decrement, and circular-addressing operations. We can directly specify the current AR used for the address pointer without updating the ARP, which is required for C2000 processors.

An example of using indirect addressing is illustrated as follows:

```
STM   #2,AR0     ; store the value 2 into AR0
STM   #x,AR1     ; store the address of x into AR1
STL   A,*AR1+0   ; stored AL to memory pointed at by AR1
                 ; AR1 is then increased by 2 since AR0=2
```

Note that in indirect addressing, memory locations are pointed at by auxiliary registers, which contain the addresses of memory operands. The value of the auxiliary register can be altered with the pointer modifications listed in Table 4.6. However, AR0 is the only register that can be used for pre- and post-incrementing when the offset is bigger than 1.

Absolute Addressing

Absolute addressing allows the user to specify the 16-bit address of the data space. Therefore, instructions based on absolute addressing must encode the 16-bit

TABLE 4.6 Summary of Indirect-Addressing Operations

Operation	Operand	Description
Post-increment (read & write)	*ARn+ *ARn+0	Pointer increases by 1 after access Pointer increases by AR0 after access
Post-decrement (read & write)	*ARn− *ARn−0	Pointer decreases by 1 after access Pointer decreases by AR0 after access
Pre-increment (write only)	*+ARn *+ARn0	Pointer increases by 1 before access Pointer increases by AR0 before access
Pre-decrement (write only)	*−ARn *−ARn0	Pointer decreases by 1 before access Pointer decreases by AR0 before access
Circular buffer	*ARn+%	Address pointer will wrap around once it reaches the end address of the buffer
Bit reversal	*ARn+0B *ARn−0B	Use for FFT by adding or subtracting AR0 to ARn with reverse carry

address, which requires at least two words for the instruction. There are four different methods for implementing absolute-addressing mode:

1. Using dma
2. Using a program-memory address (pma)
3. Using a port address (pa)
4. Specifying the exact 16-bit address *(lk)

To be familiar with absolute addressing, we use the following sets of examples:

Using data memory addressing

```
MVDK    *AR1+,1000h     ; move data from memory pointed by AR1 to
                        ; data memory at 1000h (destination)
MVKD    1000h,*AR1+     ; move data from memory 1000h (source) to
                        ; data memory pointed by AR1
MVDM    1000h,AR2       ; move data from memory 1000h to AR2
MVMD    AR2,1000h       ; move data from AR2 to data memory
                        ; at 1000h
```

Note that the data address, 1000h, can be replaced with a symbol to specify an address in data space. In the MVDK and MVKD instructions, either an auxiliary register or a data-page pointer must be used to specify the other address in the data space.

Using program memory addressing

```
LD      #2,DP           ; set data page to 2
MVPD    1000h,2h        ; move word from program memory at address
                        ; 1000h to data memory at address 102h
MVDP    3h,1000h        ; move word from data memory at address
                        ; 103h to program memory at address 1000h
MACP    *AR2+,coeff,A
```

In the last example, the processor multiplies the data pointed at by AR2 by the value at program-memory address coeff and adds the product to the ACC A. The data pointer AR2 is then incremented by one, while the program pointer is incremented automatically. The pma can be specified with either a fixed program address or a symbol, while the data address comes from either an auxiliary register or a data-page pointer. The advantage of storing data in program memory is an increase in data space, which allows programmers to access data simultaneously from both program and data memories.

Port addressing

```
PORTR   100h,input
PORTW   output,101h
```

The first instruction reads a 16-bit value from an external I/O port at address 100h to the data-memory location at symbolic address input. The second instruction writes a 16-bit value from the data memory at symbolic address output to the external I/O port at address 101h.

Specifying the exact 16-bit address $*(lk)$

```
LD  *(TEMP),A   ; Load data in address TEMP into ACC A
```

Unlike direct and indirect addressing, the user can use the absolute-addressing mode to access any location in the data space without needing to initialize the DP or ARn. However, the tradeoff is that absolute addressing increases the instruction length for storing a 16-bit address. Therefore, absolute addressing is not suitable for addressing within a loop.

Memory-Mapped Register Addressing

In C54x processors, registers are located in memory at data page zero (DP = 0). This set of registers, known as memory-mapped registers (MMRs), can be accessed without wait states. MMR-addressing mode is used to modify the MMR without affecting the current DP value, thus avoiding the overhead of writing to the DP register. An example of accessing an MMR is given as follows:

```
.mmregs         ; assembler directive to include MMR
STM   #2,AR2    ; store a value 2 into AR2 register
```

Note that the assembler directive .mmregs must be included to access the MMR. Assembler directives, which are preceded by a period, are used to initialize variables and constants, define the starting and ending addresses of program, declare a certain mode, and reserve memory for variables.

Stack Addressing

A stack is a predefined memory space that is used to store the PC during interrupts and subroutine calls. It is also used to store the contents of registers and important data values during interruption to the main program. A stack usually is defined in on-chip memory for quick access, and it is filled from the highest to the lowest memory address. In the C54x, the processor uses the SP to point at the last used location in the stack. When programming in assembly language, the stack can be defined in the main program as follows:

```
size  .set    200h            ;stack size is set to 200h
stk   .usect  "STACK",size
      .sect   "code"
      STM     #stk+size,SP
```

SP is initialized to point at the top of the stack. When programming in C, the boot.asm file in the runtime-support library contains a routine to set up the stack.

Five instructions in the C54x allow the user to access the stack and manage the stack pointer:

- PSHD or POPD pushes or pops a data memory value onto or from the stack

- PSHM or POPM pushes or pops an MMR onto or from the stack
- FRAME offset is used to increase (or decrease) the stack pointer without affecting the contents of the stack. The 7-bit offset constant contains a number in the range of −128 (decrement) to +127 (increment)

The push operation performs a pre-decrement of the SP before storing the data while the pop operation performs a data extract before a post-increment of the SP. Therefore, data stored in the stack is operated in a last-in, first-out manner.

Example 4.11

In this example, we show the changes of the stack after applying the following instructions (the four digits in front of the instruction represent the pma):

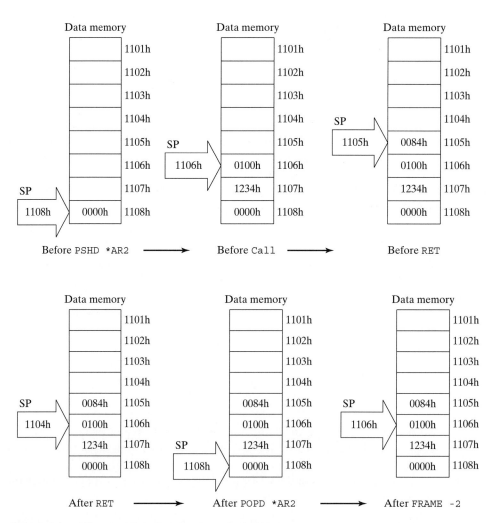

Figure 4.14   Movement of the stack pointer and data in the stack

```
0080        PSHD  *AR2      ; push the data pointed by AR2 onto stack
0081        PSHM  AR0       ; push the contents of AR0 onto stack
0082        CALL  subr      ; push return address 0084 onto stack
0084        POPM  AR0       ; pop the content back to AR0
0085        POPD  *AR2      ; pop the content to memory pointed by AR2
0086        FRAME -2        ; decrease the stack pointer by 2
                           ; without affecting the stack contents
0087  subr: LD #2,DP        ; start of the subroutine
0088        ST #1234h,100h  ; store 1234h to memory 100h
0090        RET             ; pop the return address and return to
                           ; the main routine
```

Figure 4.14 shows the data movement and changes in the stack pointer while the preceding code is run. The sequence starts from the upper-left corner of the diagram. Note that AR2 points to the memory location that contains 1234h. This data is then stored in the stack. The instruction PSHM AR0 pushes the contents of AR0 (0100h) onto the stack. Data is popped out of the stack in a last-in, first-out fashion. Finally, the FRAME instruction alters the stack pointer without affecting the contents of the stack.

Accessing 32-Bit Data

C54x processors also provide instructions with 32-bit word operands. This 32-bit data is stored in contiguous-memory locations, and special instructions (e.g., double load, add, subtract) are used to access 32-bit data in a single cycle. These instructions, with the prefix D, are listed in Table 4.7. Dual 16-bit operations can be carried out in one cycle by setting the C16 field of the ST1 register.

In 32-bit data accesses, the first word is treated as the most-significant word, and the second word is treated as the least-significant word. It is important to note

TABLE 4.7   Instructions with 32-Bit Word Operands

Instruction	Description
DADD	Double-precision add or dual 16-bit adds to accumulator
DADST	Double-precision load with T add or dual 16-bit load with T add/subtract
DLD	Long-word load to accumulator
DRSUB	Double-precision subtract or dual 16-bit subtract from long word
DSADT	Long load with T subtract or dual 16-bit load with T subtract/add
DST	Store accumulator in long word
DSUB	Double-precision subtract or dual 16-bit subtract from accumulator
DSUBT	Long load with T subtract or dual 16-bit load with T subtract

that if the most significant word is at an even address (e.g., 200h), the least significant word is at the next-higher address (e.g., 201h). However, if the first-accessed word is at an odd address (e.g., the most-significant word is at address 201h), then the least-significant word must be extracted from the previous lower address (e.g., 200h).

4.3.3 Instruction Set

This section introduces some commonly used C54x mnemonic instructions. A detailed description of the individual instructions is available in the *TMS320C54x DSP Mnemonic Instruction Set Reference Set* [6]. The C54x instruction set can be divided into five types of operations:

1. Load and store operations
2. Arithmetic operations
3. Logical operations
4. Program-control operations
5. Special instructions

In the following subsections, we use several simple examples to illustrate the usage of the preceding operations.

Load/Store Operations

The basic instructions for moving data are LD (e.g., load data into ACC), LDM (load MMR into ACC), ST (store T or immediate value into memory), STM (store immediate value into MMR), STL (store ACC low into memory), and STH (store ACC high into memory). These instructions can perform the necessary data movement in the processor using any of the addressing modes described in the previous section.

Example 4.12

The following instructions show the usage of the load and store operations:

(A)	LD	#2,DP
(B)	LD	@20h,B
(C)	LDM	AR1,A
(D)	ST	#1000h,@20h
(E)	STL	A,@21h
(F)	STM	#10h,AR1
(G)	LD	*AR1,B

In instruction (A), the DP is loaded with the constant 2 to point at page 2. After initializing the DP, instruction (B) loads ACC B with the contents of the data memory at 120h (100h + 20h, where 100h comes from the DP). In instruction (C), the contents of AR1 are loaded into ACC A. In instruction (D), the immediate value 1000h is stored in data memory at address 120h. In instruction (E), the contents of the AL are stored in data memory at address 121h. Note that an optional shift can be inserted between the two operands, such as

```
STL   A,-8,@21h
```

which performs an 8-bit right shift followed by storage of the AL into data memory. Positive- and negative-shift values represent left and right shifts, respectively. In instruction (F), a constant 10h is stored in the AR1 register to initialize the subsequent instruction (G), which loads the data at memory pointed at by AR1 into ACC B.

Arithmetic Operations

Arithmetic operations consist of add, subtract, multiply, multiply-accumulate, multiply-subtract, arithmetic shift, double-precision arithmetic, etc. Add instructions include ADD (add to ACC), ADDC (add to ACC with carry), ADDM (add long immediate 16-bit value to memory), and ADDS (add to ACC with sign-extension suppressed). Similar sets of subtract instructions are SUB (subtract from ACC), SUBB (subtract from ACC with borrow), and SUBC [conditional subtract (commonly used in divide operations)].

Example 4.13

In this example, some important add operations are given as follows:

```
(A)   RSBX   SXM
(B)   SSBX   OVM
(C)   LD     #7FFFh,16,A
(D)   ADD    #0FFFFh,A
(E)   ADD    #1,A
(F)   ADDM   #1234h,*AR1
```

Instruction (A) turns off sign-extension mode, while instruction (B) turns on overflow mode, which limits the result to a maximum value of 7FFFFFFF and a minimum value of 80000000 in the ACC. In instruction (C), the constant 7FFFh is left-shifted by 16 bits before being loaded into ACC A. Therefore, ACC A contains 7FFF0000h. Instruction (D) adds 0FFFFh to ACC A, resulting in 7FFFFFFFh in ACC A. Instruction (E) adds a constant 1 to ACC A, resulting in 80000000h, which is −1. However, since the OVM is turned on, the ACC is limited (or clipped) to its maximum value of 7FFFFFFFh. Finally, instruction (F) adds a constant 1234h to the data memory pointed at by AR1. As a side note, the C54x does not provide any instruction that allows subtraction from data memory. Therefore, we can use ADDM with a negative number to implement subtraction from memory operations.

The instruction MPY[R] performs multiplication with rounding if the optional suffix R is used for rounding, as in MPYR. The instructions MPYA, MPYU, and SQUR perform multiply by ACC A, unsigned multiply, and square operation, respectively. A further extension to the multiply operation is the set of multiply-add and multiply-subtract instructions, which include MAC[R] (multiply and accumulate with or without rounding), MACA[R] (multiply by ACC A and accumulate with or without rounding),

MACD (multiply by program memory and accumulate with delay), MACP (multiply by program memory and ACC), MACSU (multiply signed by unsigned and accumulate), MAS[R] (multiply and subtract with or without rounding), MASA[R] (multiply by ACC A and subtract with or without rounding), SQURA (square and accumulate), and SQURS (square and subtract).

Example 4.14

Multiplication is an important operation in DSP. In this example, we examine the usage of the following multiply operations:

```
(A)    STM    #100h,T
(B)    MPY    *AR1,A
(C)    MPYU   *AR1,B
(D)    MPY    *AR2,#1000h,A
(E)    LD     #1000h,16,A
(F)    MPYA   B
```

In instruction (A), the T register is loaded with an immediate value of 100h. The next instruction (B) multiplies the contents of the T register by the contents in the data memory pointed at by AR1 and puts the product into ACC A. In instruction (C), unsigned multiplication is carried out using the MPYU instruction. Only positive numbers (ranging from 0 to 0FFFFh) are used in this multiplication. Instruction (D) combines the loading of the T register with the multiplication in a single instruction. It loads the T register with the value at data memory pointed at by AR2, multiplies the contents of the T register by a constant 1000h, and then stores the product in ACC A. Instruction (E) loads the high word of ACC A with 1000h, and instruction (F) multiplies the value 1000h (in AH) with the contents of the T register and stores the result in ACC B.

Example 4.15

When implementing FIR and IIR filters, MAC operations are carried out as follows:

```
(A)    LD     *AR1,T
(B)    MAC    *AR2+,A
(C)    MAC    *AR2+,*AR3+,B
(D)    MACA   *AR3,A
```

Instruction (A) loads the T register with the data pointed at by AR1. Instead of using multiply followed by add, instruction (B) combines these two operations into one. It multiplies the contents of the T register by the data memory pointed at by AR2, adds the product to the value in ACC A, and stores the result back in ACC A. Further, we can combine the loading of the T register in a single instruction as in (C), which loads the T register with the data pointed at by AR2, multiplies the contents of the T register by the data pointed at by AR3, and adds the product to ACC B. In fact, we can repeat instruction (C) N times by using the instruction

```
RPT   #(N-1)
```

before instruction (C). In instruction (D), the data pointed at by AR3 is multiplied by the high word of AH and then accumulated in ACC A.

Logical Operations

Logical operations allow the user to test, alter, and manipulate data bits. They commonly are used to enable or disable a flag, mask out certain bits, and check status flags and conditions. The instructions include AND (logical AND data or a constant with the ACC), ANDM (logical AND data with the contents of data memory), OR (logical OR data or a constant with the ACC), ORM (logical OR data with the contents of data memory), XOR (logical XOR data or a constant with the ACC), XORM (logical XOR data with the contents of data memory), and CMPL (complement the ACC). Logical AND uses zeros to clear the specific bits of the ACC or data to zero, while logical OR uses ones to set specific bits to 1. XOR with ones complements the specific bits, while XOR with zeros remains unchanged.

Example 4.16

In this example, we examine the effects of performing logical operations as follows:

(A)	LD	#33AAh,A
(B)	CMPL	A
(C)	AND	#FFh,A
(D)	OR	#F0F0h,A
(E)	XOR	#FF00h,A

In instruction (A), ACC A is loaded with the value 33AAh. Instruction (B) complements the data in ACC A to FFFFCC55h. Instruction (C) performs an AND operation of value FFh with the value FFFFCC55h, which results in 00000055h in ACC A. In instruction (D), logical OR between F0F0h and 00000055h results in 0000F0F5h. In instruction (E), logical XOR between 0000F0F5h and FF00h results in 00000FF5h. Note that only logical AND operates on the whole 32 bits of the ACC, while logical OR and XOR only operate on the lower 16 bits of the ACC.

C54x processors provide an additional set of logical operations that allows the user to test individual bits of the word stored in data memory directly without requiring the ACC. These instructions include BIT (copy the bit under test to the bit TC in the register ST0), BITT (copy the bit specified by the T register to the bit TC), and BITF (bit-test field specified by an immediate value). It is important to note that in bit operations, bit 0 represents the MSB and bit 15 represents the LSB. For example,

```
BIT    *AR1,0    ; copy bit 0 from the data memory pointed at by AR1
                 ; to TC
BITT   *AR1      ; copy a bit specify by TREG of data pointed at by
                 ; AR1 to TC
BITF   *AR1,#3h  ; test bits 0 and 1 (#3h) of data memory pointed at
                 ; by AR1
```

Program-Control Operations

Program-control operations allow the user to control the program flow by altering the values of the PC. These operations include branches, calls, returns, single and block repeats, interrupts, and RESET.

C54x processors support two types of branch operations: conditional and unconditional. Unconditional branch instruction B redirects the PC to another part of the program. For example,

```
        ADD    #10h,A      ; add 10h to the ACC A
        B      next        ; branch unconditionally to next
        LD     #10h,A      ; following instructions are skipped
  ...
next:   LD     #80h,B      ; program continues from here
```

The conditional branch is executed only when all of the listed conditions are met. The syntax for the conditional branch is

```
BC     destination,condition1[,condition2,condition3]
```

The processor branches to destination if all of the listed conditions are true; otherwise, the CPU executes the next instruction. The conditions or the types of tests available in the C54x are listed in Table 4.8. Up to three conditions can be tested simultaneously in a single BC instruction. However, the three-condition test is limited to the conditions listed in groups C, D, and E (shown in the third column of Table 4.8), and conditions in the same group cannot be tested simultaneously. Two condition tests are available for the conditions labeled as A and B. That is, we can test condition A and condition B, but we cannot test the two conditions in the same group.

TABLE 4.8   Conditions for Conditional Instructions

Condition	Description	Grouping for conditions
BIO	$\overline{\text{BIO}}$ is low	C
NBIO	$\overline{\text{BIO}}$ is high	C
C	C = 1	D
NC	C = 0	D
TC	TC = 1	E
NTC	TC = 0	E
AGT, BGT	(A) > 0, (B) > 0	A
AGEQ, BGEQ	(A) ≥ 0, (B) ≥ 0	A
ALT, BLT	(A) < 0, (B) < 0	A
ALEQ, BLEQ	(A) ≤ 0, (B) ≤ 0	A
AEQ, BEQ	(A) = 0, (B) = 0	A
ANEQ, BNEQ	(A) ≠ 0, (B) ≠ 0	A
AOV, BOV	A overflow, B overflow	B
ANOV, BNOV	A no overflow, B no overflow	B

A simple example to test whether a data value is greater than 100h follows:

```
         LD      *AR1,A
         SUB     #100h,A
         BC      greater,AGT   ; if result >0, branch to greater
less:    end                   ; no processing
greater: LD      *AR1,A        ; branch here and continue
         ADD     #10h,A
```

Other commonly used program-control operations are subroutine calls and returns. It is always a good practice to divide a large program into several subroutines for better management and debugging of the program. Subroutines can be programmed in C54x assembly language as follows:

```
start:    LD      2,DP
          ...
          CALL    add_sub    ; go to subroutine add_sub
          LD      #100h,A    ; continue here at the end of subroutine
          ...                ; some instructions
add_sub:  ADD     #10h,A     ; first instruction in subroutine
          ...                ; some instructions
          RET                ; return to the main program
```

When the instruction CALL is encountered, the CPU first pushes the return address into the stack and then redirects the PC to the new address (in the above example, add_sub). The subroutine is executed until it encounters the RET instruction, which returns control to the main program that called the subroutine. This process is carried out automatically by popping the return address out from the stack. The CALL instruction is similar to the unconditional branch, except that the PC is temporarily transferred to the subroutine and returns to the main routine after executing the subroutine.

The instruction CALA calls the subroutine at the location specified by ACC A. This instruction is especially useful in calling a specific subroutine based on the dynamic computation result in ACC A. The C54x processor also provides a conditional-call instruction, CC, which uses the same set of conditions listed in Table 4.8. For example,

```
ADD     *AR1,A
CC      subroutine,ANOV
```

In the preceding code, if no overflow occurs in ACC A, the subroutine is called. Otherwise, the PC continues executing the instructions after the CC instruction.

Repeating a single instruction or a block of instructions is often necessary for implementing DSP kernel algorithms such as FIR filtering, FFT, matrix-vector multiplication, adaptive filtering, etc. These algorithms usually consist of multiply-add operations, which must be executed repeatedly. The instructions

```
RPT     #N
RPTZ    #N
```

allow the instruction that follows to be executed (N + 1) times. The only difference between these two instructions is that RPTZ also clears the accumulator to zero before the start of the repeat operations. When the repeat instruction is performed, the PC repeats the loop until the counter runs down to zero. An example of a single repeat is given as follows:

```
RPT    #15
MPY    *AR2+,*AR3+,A      ; repeat this instruction 16 times
```

It is important to note that when using the instruction with dual data-memory operands, such as the instruction MPY *AR2+,*AR3+,A, the auxiliary registers used are restricted to AR2, AR3, AR4 and AR5, and the modifiers can only be none, +, −, and +0%.

The repeat of a single instruction in C2000 processors is extended to the repeat of a block of instructions in C54x processors. The RPTB instruction is used for block-repeat operations. The user needs to define the block-repeat counter (BRC) and specify the end address of the repeat block as follows:

```
     STM    #9,BRC         ; repeat block 10 times
     RPTB   end-1          ; block repeat up to the label end
     LD     #0,A
     MPY    *AR2+,*AR4+,A
     STH    *AR3+
end:
```

The final class of instructions that affect the PC is interrupt and RESET. Interrupts can be triggered by hardware signals or software. They are an asynchronous process that can occur anytime. When interrupts occur (such as signaling the receipt of a new sample from the ADC), the processor needs to finish its current instruction before releasing its control to an ISR for I/O operations. The C54x processor supports both software and hardware interrupts. A software interrupt is requested by program instructions such as INTR, TRAP, and RESET. A hardware interrupt is requested by a signal from external hardware or from internal hardware such as on-chip peripherals.

There are two types of C54x interrupts: maskable and nonmaskable. A maskable interrupt can be switched on or off using software settings. A nonmaskable interrupt cannot be turned off by software. For example, the C5402 processor has 28 maskable interrupts that include 14 software interrupts, 4 external interrupts, and 10 on-chip peripherals' interrupts. It also has two nonmaskable interrupts: NMI and RESET. Reset operations can be implemented in hardware or software using the instruction RESET. Reset is commonly used to initialize the processor to a known state, such as initiating the PC to 0FF80h. A detailed description of all of the status bits after RESET is given in the reference set [5].

An interrupt vector (table) is a memory space in ROM used to store the addresses of ISRs that service specific interrupts. Each interrupt entry in the vector occupies four words of program memory and contains the branch instruction that branches to the ISR. The ISR is different from the subroutine in that the return instruction is RETE instead of RET. Furthermore, interrupts can occur anytime, while the CALL instruction occurs at a fixed position in the main program. Interrupts are also necessary for saving the status registers (ST0 and ST1) of the processor as part of the ISR because the ISR may alter the current status of the main program. The instructions PSHM ST0 and PSHM ST1 are usually used at the beginning of the ISR to save these registers in the stack, and the instructions POPM ST1 and POPM ST0 are executed before the RETE instruction of the ISR. Note that popping is carried out in reverse order.

Special Instructions

C54x application-specific instructions include FIRS (symmetric FIR filtering), LMS (adaptive filtering), and SQDST (square distance). These special instructions combine many simpler instructions into one, thus speeding up the operations.

The FIRS instruction can be applied as

```
FIRS    *AR3,*AR4,#coeff
```

This instruction is equivalent to performing the following two instructions in parallel:

```
    MACA    #coeff,B,B      ; B = B +(AH*coeff)
||  ADD     *AR3+,*AR4+,A   ; A = (*AR3 + *AR4)<<16
```

The value in AH is multiplied by the value in program memory, and the product is added into ACC B. At the same time, the instruction adds the data-memory operands pointed at by AR3 and AR4, shifts the sum left by 16 bits, and stores the final result in ACC A. A detailed explanation of FIR filtering is given in Chapter 6.

The LMS (least mean square) instruction can be used as follows:

```
LMS    *AR3+,*AR4+
```

This instruction performs the following operations:

```
     MAC    *AR3+,*AR4+,B    ; B = B+(*AR3+ * *AR4+)
||{ADD     *AR3,16,A         ; A = (*AR3+A) << 16
    RND    A}                ; A = round(A)
```

The first instruction performs the multiply-add operation of FIR filtering. To avoid the conflict of accessing the same data and coefficient, the multiply-add operation operates on the first coefficient and data and places the result in ACC B. At the same time, the second coefficient element is added to ACC A. The result is left-shifted by 16 bits and rounded before being placed into ACC A. In this way, the filtering and the adaptation of coefficients are carried out in ACC B and A, respectively. A detailed explanation of adaptive filtering is given in Chapter 9.

An example of computing the square of the distance between two vectors pointed at by AR1 and AR2 is

```
SQDST   *AR3+,*AR4+
```

This instruction can be interpreted as

```
SQUR    A,A              ; A = AH*AH
ADD     A,B              ; B = A+B
||SUB   *AR3,*AR4,A      ; A = ((*AR3)-(*AR4))<<16
```

Note that the square and subtract operations do not operate on the same result in ACC A. The square operation operates on the previous pair of data, while the subtraction operation operates on the current pair of data pointed at by AR3 and AR4. There is no conflict since these operations access different data.

4.3.4 Programming Considerations

This section introduces some commonly used addressing modes for optimizing C54x assembly code. These special addressing features include circular addressing and bit-reversal addressing.

Circular Addressing

The C54x architecture only supports a single circular buffer by providing one circular-buffer-size register (BK) for specifying the size of the circular buffer. The size of the circular buffer is only limited by the size of on-chip memory. The circular buffer must be aligned to a 2^N boundary, where N is the smallest integer that satisfies $2^N > $ BK.

For example, to define a circular buffer of 32 words, BK must be loaded with 32 and aligned to a 6-bit boundary since $2^6 > 32$. The generic code for defining the circular buffer is listed as follows:

```
STM    #32,BK              ; initialize circular buffer size
...
RPT
MAC    *AR2+0%,*AR3+,A
```

As shown in Fig. 4.15, the circular pointer is AR2, with the increment-index register AR0. In this example, AR0 and AR2 are initialized to 1 and 1000h, respectively. The symbol % indicates that AR2 is addressing a circular buffer, which wraps around to the beginning of the memory when the end of the buffer (1020h) is reached.

Bit-Reversal Addressing

As introduced in Section 3.4.1, bit-reversal addressing is an important feature of the DSP processor that reorders the sequence of data samples in the buffer for FFT algorithms. In the C54x, a bit-reversal operation can be performed by setting AR0 to half of the FFT size and adding it to the index register, with the carry bit propagating to the right. For example, to perform bit-reversed addressing on an 8-point FFT, AR0 is set to 4, and AR2 is used to point at the data address for the 8-point FFT with operand *AR2+0B, where B specifies the bit-reversal addressing. The process is explained in Table 4.9. Starting from the first index 0, subsequent bit patterns are added with the contents of AR0. Note that reverse-carry addition ($+'$) is different from normal binary addition ($+$) because the carry bit is propagating from the left

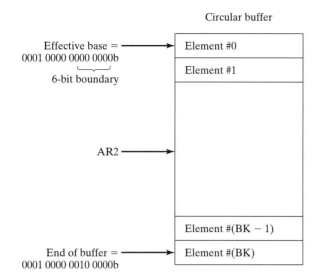

Circular buffer

Effective base = ⎯⎯⎯⎯⎯→ Element #0
0001 0000 0000 0000b

6-bit boundary

Element #1

AR2 ⎯⎯⎯⎯⎯→

Element #(BK − 1)

End of buffer = ⎯⎯⎯⎯⎯→ Element #(BK)
0001 0000 0010 0000b

Figure 4.15   Circular-buffer implementation in the C54x processor

TABLE 4.9   Bit-Reversal Addressing in the C54x Processor, where the
FFT Size is 8

Bit pattern (AR2)	Bit-reversal adding	Bit-reversed pattern	Bit-reversed index
0000	0000	0000	0
0001	0000+' 0100	0100	4
0010	0100+' 0100	0010	2
0011	0010+' 0100	0110	6
0100	0110+' 0100	0001	1
0101	0001+' 0100	0101	5
0110	0101+' 0100	0011	3
0111	0011+' 0100	0111	7

to the right, as opposed to normal addition, where the carry bit propagates from the right to the left. The last column of the table contains the bit-reversed index of samples in the data buffer.

4.3.5 System Issues

C54x processors are used in many low-power and low-cost applications. These applications also demand a high processing speed with a wide range of on-chip memory and peripherals. These hardware functioning units are introduced in this section.

Memory Configurations

Similar to C2000 processors, C54x processors have a total memory space of 192K words. This space is organized into three specific memory segments: (1) 64K words of program that store instructions and constants, (2) 64K words of data that store data used by instructions, and (3) 64K words of I/O that interface external memory-mapped peripherals. Different C54x processors have different on-chip memory types, as summarized in Table 4.5. RAM usually is mapped into the data space and has the option (OVLY = 1) to map into the program space. ROM is mapped into the program space using the microcomputer mode (MP/$\overline{\text{MC}}$ = 0). It can also map a portion of the program space into the data space using DROM = 1, which allows tables that are stored in ROM to be accessed as data. Figure 4.16 shows the memory map of the TMS320C5402 processor, where MP indicates microprocessor mode and MC indicates microcomputer mode.

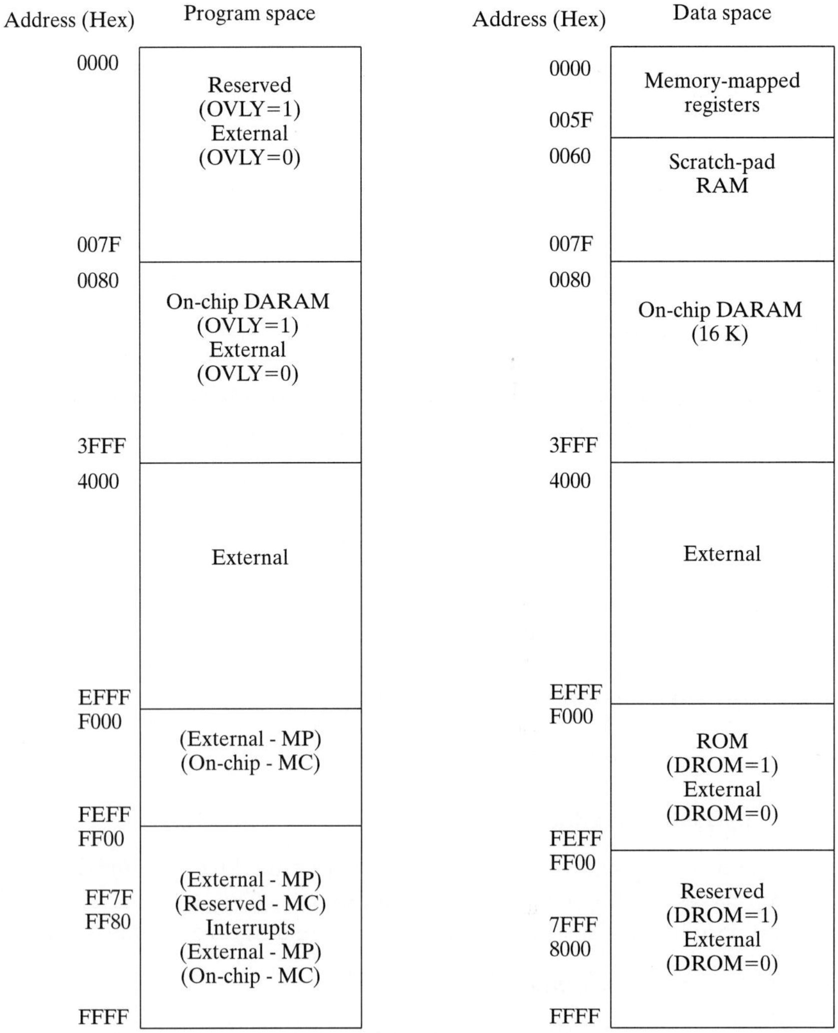

Address (Hex)	Program space	Address (Hex)	Data space
0000	Reserved (OVLY=1) External (OVLY=0)	0000 005F	Memory-mapped registers
		0060	Scratch-pad RAM
007F		007F	
0080	On-chip DARAM (OVLY=1) External (OVLY=0)	0080	On-chip DARAM (16 K)
3FFF		3FFF	
4000	External	4000	External
EFFF F000	(External - MP) (On-chip - MC)	EFFF F000	ROM (DROM=1) External (DROM=0)
FEFF FF00	(External - MP) (Reserved - MC) Interrupts (External - MP) (On-chip - MC)	FEFF FF00 7FFF 8000	Reserved (DROM=1) External (DROM=0)
FF7F FF80			
FFFF		FFFF	

Figure 4.16   Memory map for the TMS320C5402 processor

Some C54x processors use a paged-memory technique to extend program-memory space. For example, the C5402 DSP processor allows extensions to 1M words using 20-bit address lines. A single page of the program space contains 64K words, and 16 pages are required to address the 1M words, as shown in Fig. 4.16. In order to access the entire 1M words of program memory, the OVLY bit must be set to 0. Therefore, all of the program code is stored in external memory. If we need to access some commonly used routine, the OVLY = 1 option maps to the lower 32K words of the program-space pages.

External Memory and I/O Interface

As shown in Fig. 4.17, C54x processors have four sets of internal address/data buses, but only one external address/data bus. Therefore, four pairs of internal buses are multiplexed onto a single external-bus pair. The external data bus is 16 bits wide, but the external address bus ranges from 16 bits to 23 bits wide. Therefore, up to 8M words of paged-program memory can be addressed in some processors. The external memory interface can support only one access per cycle (one external read or two external writes). However, accessing slow external devices requires wait states. Wait states of up to 14 machine cycles can be programmed into the processor to extend external-bus cycles.

On-Chip Peripherals

All C54x processors have general-purpose I/O pins, a timer, a clock generator, a software-programmable wait-state generator, and a programmable bank-switch module. The device-specific peripherals are (1) a HPI, which comes in 8-bit standard, 8-bit enhanced, or 16-bit enhanced, and (2) a serial port, which can be a combination of synchronous serial ports, BSPs, McBSPs, or time-division multiplexed serial ports.

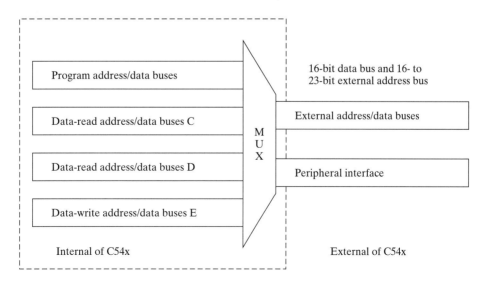

Figure 4.17   Internal and external buses of the C54x processor

The HPI allows the host processor to read and write 1K words of on-chip memory in the C54x processor. The newest generation of C54x processors has a 16-bit enhanced HPI and also supports the lower 8-bit HPI version. There are two operation modes: shared-access mode and host-only mode. In shared-access mode, both the host and DSP processor can access HPI memory. In host-access mode, only the host can access HPI memory, while the C54x processor remains in idling mode, thus saving power consumption. A typical transfer rate of the HPI is around 64 Mbit/sec using a C54x processor running at 40 MHz in shared-access mode. The transfer rate is even higher (160 Mbit/sec) when the processor is running in host-only mode.

Four different serial-port interfaces are available on C54x processors. The basic serial port is a synchronous-serial interface that supports 8-, 10-, 12-, or 16-bit data transfer at a rate equal to the processor's master-clock rate. The BSP operates in either a synchronous-serial-port interface or an autobuffering unit. The synchronous-serial port is the same as the basic serial port, while the auto-buffering unit reserves 2K words of on-chip DARAM. In the C54x processor, the autobuffering unit transfers 2K words of data between the serial port and memory automatically and only interrupts the processor when the buffer is either half full or empty.

The McBSP is similar to the BSP, except that it has the capability to support up to 128 channels per port. The McBSP is designed for connecting many high-speed devices. The McBSP of the C54x processor contains many advanced features such as full-duplex communication, multibuffered data registers, independent framing and clocking for receive and transmit, a flexible clock generator, internal or external shift clocking, and the capability to support A-law and μ-law compression and decompression hardware. A more detailed description of the initialization and operations of the McBSP, DMA, and HPI can be found in Appendix C.

There are three low-power modes (IDLE1, IDLE2, and IDLE3) in the C54x processor. The clock is turned off when any of the low-power modes are activated. IDLE3 mode switches off most on-chip peripherals, including the oscillator and the PLL, and it takes a longer time to wake up. IDLE2 mode turns off most on-chip peripherals except the oscillator and PLL, and it can be awakened by an external interrupt. In IDLE1 mode, on-chip peripherals are on, and any interrupt can wake up the processor. Another way to reduce power is to put the address, data, and control lines in a high-impedance state.

4.4 TMS320C55x

The TMS320C55x is the most power-efficient DSP processor. It is software compatible with existing C54x processors, but it offers faster speed and improved code density at a low system cost. The C55x processor's low power consumption and high performance support feature-rich, miniaturized personal and portable applications, such as cell phones, programmable digital-audio devices, digital cameras, and digital hearing aids. Detailed information on the C55x architecture and on programming and system issues can be found in [7–10].

4.4.1 Architecture Overview

The C55x architecture achieves high performance through increased parallelism by providing additional hardware, more built-in instructions, user-programmed parallel functions, and additional instructions that take advantage of the expanded hardware to accomplish more operations in fewer clock cycles. Proper instructions are scheduled to run automatically in parallel by the C55x assembly optimizer and C compiler. The processor provides advanced automatic power management for all peripherals, memory arrays, and individual CPUs. In effect, the C55x increases idle domains and continually monitors the parts of the chip that are being used, turning parts off when they are not needed.

The C55x CPU provides dual MAC units (instead of the one unit in C54x processors), each capable of 17-bit $\times$ 17-bit multiplication and 40-bit addition (or subtraction) with optional saturation in a single cycle. The C55x also provides four general-purpose 40-bit accumulators instead of the two accumulators in the C54x. A 40-bit ALU is supported by an additional 16-bit ALU for optimizing parallel operations. In addition, four new 16-bit data registers are used for simple computations. The CPU uses software stacks that support 16-bit and 32-bit push and pop operations for data storage, retrieval, automatic context saving (in response to function calls or interrupts), and restoration. To make sure that throughput can attain the theoretical maximum made possible by the new computational hardware, the CPU supports 12 internal buses instead of the 8 buses in the C54x. These buses allow the CPU to perform up to three data reads and two data writes in a single cycle.

The C55x processor supports variable-length instructions for improved code density. Instruction length may be 8, 16, 24, 32, 40, or 48 bits. The instruction-buffer unit (IU) performs 32-bit program fetches (instead of 16-bit) from memory and stores instructions for the program-flow unit (PU). The PU decodes the instructions, directs tasks to the address-data-flow unit (AU) and data-computation unit (DU), and manages the fully protected pipeline. All data-space addresses are generated in the AU. As illustrated in Fig. 4.18, these functional units exchange programs and data with each other and memory through 12 internal buses. The memory-interface unit mediates all data transfers between the CPU and data or I/O space.

Program fetches are performed using the 24-bit program-read address bus (PAB) to carry the address of the instruction and the 32-bit program-read data bus (PB) to deliver 4 bytes of code to the IU. The functional units read data from the memory space (or I/O space) via three 16-bit data-read data buses, BB, CB, and DB, with the associated 24-bit data-read address buses, BAB, CAB, and DAB, respectively. Single-operand reads are performed on DB. Instructions that read two operands at once use both DB and CB. Bus BB provides a third data read from internal memory to the DU, which provides coefficients for dual MAC operations. Two 16-bit data-write data buses, EB and FB, transfer operation results from the functional units of the CPU to data memory or I/O space. The data-write buses also have the associated 24-bit data-write address buses EAB and FAB.

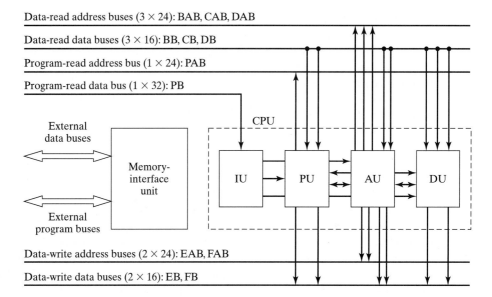

Data-read address buses (3 × 24): BAB, CAB, DAB

Data-read data buses (3 × 16): BB, CB, DB

Program-read address bus (1 × 24): PAB

Program-read data bus (1 × 32): PB

Data-write address buses (2 × 24): EAB, FAB

Data-write data buses (2 × 16): EB, FB

Figure 4.18   TMS320C55x functional-block diagram

4.4.2 Central Processing Unit

As shown in Fig. 4.18, the CPU consists of four functional blocks: IU, PU, AU, and DU. These functional units exchange programs and data with each other and exchange programs and data with memory through one 32-bit program bus (PB), five 16-bit data buses (BB, CB, DB, EB, FB), and six 24-bit address buses (PAB, BAB, CAB, DAB, EAB, FAB). This parallel bus structure enables one 32-bit program read, three 16-bit data reads, and two 16-bit data writes per clock cycle.

Instruction-Buffer Unit

As illustrated in Fig. 4.19, the IU receives 4 bytes of code from the 32-bit PB during each CPU cycle, puts them into the instruction-buffer queue of size 64 bytes, decodes 1 to 6 bytes of code using the instruction decoder, and passes data to the PU, AU, and DU for the execution of instructions. Note that the instruction-buffer queue provides 6 bytes at a time to the instruction decoder, which decodes 1 to 6

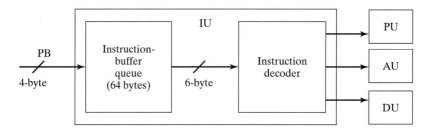

Figure 4.19   Block diagram of IU

bytes of code, and dispatches operations to the other functional units in the CPU. In addition to facilitating instruction pipelining, the instruction-buffer queue supports a local block-repeat instruction that executes a block of code within the queue. It also supports speculative fetching of instructions while a condition is tested for conditional-branch, call, and return instructions. The instruction decoder identifies instruction boundaries so that it can decode 1 to 6 bytes of instructions and determines whether the CPU has to execute two instructions in parallel.

Program-Flow Unit

The PU receives instructions from the IU, generates all program-space addresses, and also controls the sequence of instructions. The PU contains program-flow registers, block-repeat registers, single-repeat registers, interrupt registers, and four status registers (ST0_55–ST3_55). The CPU usually generates sequential addresses, but it is able to generate nonsequential addresses based on immediate data from the IU and register values from the DU. It performs nonsequential program-flow actions, including (1) interpreting conditions for conditional instructions, (2) determining branch addresses, (3) initiating interrupt servicing when an interrupt is requested, (4) managing single- and block-repeat operations, and (5) managing execution of parallel instructions.

Address-Data-Flow Unit

The AU contains the DAGEN and all of the registers necessary to generate addresses for reads from or writes to data space. DAGEN supports both linear and circular addressing for instructions that use indirect-addressing modes. The AU also contains a 16-bit ALU that is capable of performing addition, subtraction, comparison, saturation operations, Boolean-logic operations, arithmetic and logical shifts, and absolute-value calculations. It also tests, sets, clears, and complements AU registers and memory bits, as well as modifies, moves, and rotates register values.

Eight extended auxiliary registers (XAR0-XAR7), an extended coefficient data pointer (XCDP), an extended stack pointer (XSP), and an extended system-stack pointer (XSSP) are used as 23-bit address pointers. As illustrated in Table 4.10,

TABLE 4.10 Extended Auxiliary Registers, a Coefficient Data Pointer, and Stack Pointers

	b22-b16	b15-b0
XAR0	AR0H	AR0
XAR1	AR1H	AR1
XAR2	AR2H	AR2
XAR3	AR3H	AR3
XAR4	AR4H	AR4
XAR5	AR5H	AR5
XAR6	AR6H	AR6
XAR7	AR7H	AR7
XCDP	CDPH	CDP
XSP	SPH	SP
XSSP	SPH	SSP

each 7-bit high part (ARnH, CDPH, or SPH) is used to specify the main data page (0–127) for accesses to data space. Each low part can be used as a 16-bit offset to the specified main data page to form a 23-bit address or a bit address used by instructions that access individual bits or bit pairs in the specific register. For example, the CPU can concatenate ARnH and ARn to form the extended XARn, denoted as XARn (ARnH:ARn). The auxiliary registers AR0-AR7 are used in AR indirect-addressing mode. They can also be used as general-purpose registers or counters. Basic arithmetic, logic, and shift operations can be performed on AR0-AR7 by the 16-bit ALU in AU.

The coefficient data pointer (CDP) is used in the CDP indirect-addressing mode and in the coefficient indirect-addressing mode that provides a third operand for the dual-MAC operations in DU. The AU also contains a data-page register XDP(DPH:DP) to support direct-addressing modes and a peripheral data-page register (PDP) for accessing I/O space. Details of the addressing modes are discussed in Section 4.4.3.

The AU includes four 16-bit general-purpose temporary registers (T0-T3). Temporary registers can be used to (1) hold one of the memory multiplicands for multiply, multiply-accumulate, and multiply-subtract instructions; (2) hold the shift count used in addition, subtraction, and load instructions performed in the DU; and (3) hold the transition metric of Viterbi butterfly operations. In addition, they can track more pointer values by swapping the contents of the auxiliary registers and the temporary registers using a swap instruction. The four temporary registers together with the 16-bit ALU make it possible to perform simple operations in parallel with the main computational units in the DU.

Data-Computation Unit

The DU contains the primary computational units of the CPU: a 40-bit barrel shifter, a 40-bit ALU, two MAC units, four 40-bit accumulators, and two transition registers.

The DU shifter accepts immediate values from the IU, connects with memory and I/O space, and connects with the AU, PU, and DU registers. The shifter performs the following tasks:

- Shifts 40-bit accumulator, 16-bit register, memory, or I/O-space values up to 31 bits to the left or 32 bits to the right. The shift count can be supplied as a constant in the instruction or can be read from one of the temporary registers.
- Shifts 16-bit immediate values up to 15 bits to the left with the shift count as a constant in the instruction.
- Rotates register values.
- Rounds and/or saturates accumulator values for storing to data memory.

In addition, the shifter normalizes accumulator values, extracts and expands bit fields, and performs bit counting.

The 40-bit ALU performs addition, subtraction, comparison, rounding, saturation, Boolean logic operations, and absolute-value calculations. It can perform two arithmetic operations simultaneously when a dual 16-bit arithmetic instruction is

TABLE 4.11 Partitions of Accumulators

	ACxG b39-b32	ACxH b31-b16	ACxL b15-b0
AC0	AC0G	AC0H	AC0L
AC1	AC1G	AC1H	AC1L
AC2	AC2G	AC2H	AC2L
AC3	AC3G	AC3H	AC3L

executed. In addition, it moves register values, as well as tests, sets, clears, and complements DU register bits.

The DU contains two MAC units that support multiplication and addition/subtraction. In a single cycle, each MAC can perform 17-bit $\times$ 17-bit multiplication (fractional or integer) and 40-bit addition or subtraction with optional 32-/40-bit saturation. The three dedicated read buses and two dedicated write buses support single-cycle dual-MAC operations. Details on dual-MAC operations are discussed in Section 4.4.5.

The DU contains four 40-bit accumulators (ACx): AC0, AC1, AC2, and AC3. Any instruction that uses an accumulator can be programmed to use any one of the four. As summarized in Table 4.11, each ACx is partitioned into a low word (ACxL), a high word (ACxH), and 8 guard bits (ACxG). We can access each of these portions individually by using addressing modes that access memory-mapped registers.

Instruction Pipelines

The C55x instruction pipeline is a protected pipeline that has two independent segments: the fetch pipeline and the execution pipeline. The fetch pipeline consists of four stages:

- *PF1 (prefetch 1)*   Present the program address to memory.
- *PF2 (prefetch 2)*   Wait for memory to respond.
- *F (fetch)*   Fetch an instruction packet and place it in the instruction-buffer queue.
- *PD (predecode)*   Predecode instructions in the queue to identify the beginning and the end of instructions and then check if the instructions are parallel.

The execution pipeline performs instruction decoding, data accessing, execution, and data storage in the following seven stages:

- *D (decode)*   Decode a single instruction (or a pair of instructions) from the instruction-buffer queue and dispatch instructions to the appropriate CPU functional units.
- *AD (address)*   Read/modify registers involved in data-address generation and perform operations that use the 16-bit ALU in the AU.
- *AC1 (access 1)*   Send addresses on the appropriate address buses for memory-read operations.

- *AC2 (access 2)*   Allow one cycle for memory to respond to read requests.
- *R (read)*   Read data from memory and evaluate the conditions of conditional instructions.
- *X (execute)*   Execute operations in the AU and DU.
- *W (write)*   Write data to the memory (or I/O) space.

Multiple instructions are executed simultaneously in the pipeline, and different instructions may perform modifications to memory, I/O space, and register values at different phases. In an unprotected pipeline (such as the C54x), simultaneous execution of instructions could lead to pipeline conflicts such as reads and writes at the same location out of the intended order. The C55x pipeline is protected because it automatically insert's inactive cycles between instructions that would cause conflicts. For example, if an instruction is supposed to write to a location, but a previous instruction has not yet read from that location, extra cycles are inserted so that the read occurs first. This protection mechanism effectively prevents pipeline conflicts, but at the cost of wasting CPU cycles. The *TMS320C55x DSP Programmer's Guide* [11] suggests several methods for minimizing the number of cycles that get inserted for pipeline protection.

4.4.3 Addressing Modes

The TMS320C55x supports three basic addressing modes (absolute, direct, and indirect) that access data memory, MMRs, register bits, and I/O space. Each addressing mode uses one or more operands with the general syntax elements summarized in Table 4.12.

Memory Space

The TMS320C55x uses unified program/data space and separated I/O space. The CPU reads instructions from memory that uses program-space addresses. Data space consists of general-purpose memory and MMRs. I/O space is separated from program/data space and is only available for duplex communication with peripherals.

TABLE 4.12   Summary of Syntax of Operands

Syntax elements	Instruction accesses data as
Smem	A single word (16-bit) from memory, I/O space, or MMRs
Lmem	A long word (32-bit) from memory or MMRs
Xmem and Ymem	Two simultaneous words from data memory
Cmem	A single word from data memory
Baddr	One or two bits in AC0-AC3, AR0-AR7, and T0-T3

TABLE 4.13   Generic Memory Map of the TMS320C55x

Main data-page number	Data-space addresses	Data/Program memory	Program-space addresses
0	00 0000h-00 005Fh	MMRs	00 0000h-00 00BFh
	00 0060h-00 FFFFh		00 00C0h-01FFFFh
1	01 0000h-01 FFFFh		02 0000h-03FFFFh
2	02 0000h-02FFFFh		04 0000h-05FFFFh
: :	: :		: :
127	7F 0000h-7F FFFFh		FE 0000h-FF FFFFh

A generic memory map for the TMS320C55x is illustrated in Table 4.13, which shows that all 16M bytes of memory are addressable as program space or data space. When the CPU reads the instruction code from program memory, it uses 24-bit addresses that are assigned to individual bytes. However, when the processor accesses data space, it uses 23-bit word addresses so that the LSB on the address bus is forced to 0. Thus, it always accesses 16-bit words starting at even addresses. Data space is divided into 128 main data pages (0–127), and each main data page has 64K words. On data page 0, the first 96 words (00 0000h to 00 005Fh) are reserved for MMRs. For the detailed memory map, including internal memory and external memory configurations, refer to the data sheet for the specific processor.

The CPU uses word addresses to read (or write) 8-bit (byte), 16-bit (word), or 32-bit (long-word) values in data space. Some dedicated instruction operands select high or low bytes of particular words. For example,

```
MOV     high-byte(Smem),dst      ; load high byte
MOV     src,low_byte(Smem)       ; store low byte
```

It is important to note that the C55x provides a new general-purpose instruction MOV for move (copy) operations instead of using load and store instructions in the C54x. Word addresses are also used to access bytes in data space. When the CPU accesses long words, the address used for the access is the address of the most significant word of the 32-bit value.

Absolute-Addressing Modes

Absolute-addressing mode allows programmers to reference a memory location by supplying a complete (or part of an) address as a constant in an instruction. Three absolute-addressing modes are available: k16, k23, and I/O. It is important to note that an instruction using absolute-addressing mode cannot be executed in parallel with another instruction because of a multibyte extension for encoding an address in an instruction.

The k16 absolute addressing uses the operand *abs(#k16), where k16 is a 16-bit unsigned constant. As illustrated in Table 4.14, the DPH (data-page extension

TABLE 4.14 Summary of k16 Absolute-Addressing Mode

DPH (b22-b16)	k16 (b15-b0)	Data space
0	0 - FFFFh	Main page 0: 00 0000h–00 FFFFh
1	0 - FFFFh	Main page 1: 01 0000h–01 FFFFh
2	0 - FFFFh	Main page 2: 02 0000h–02 FFFFh
. . .	. . .	. . .
127	0 - FFFFh	Main page 127: 01 0000h–01 FFFFh

register) points to 1 of 128 main pages and k16 provides the address inside that main page. Thus, the DPH and k16 are concatenated to form a 23-bit data-space address.

The k23 absolute-addressing mode uses the operand *(#k23), where k23 is a 23-bit unsigned constant, which is encoded as a 3-byte extension to the instruction. The k16 operand *abs16(#k16), or the k23 operand *(#k23), can be used to access data memory in any instruction with the syntax elements Smem and Lmem.

Example 4.17

Assuming the value of DPH = 5h, we find that the syntax

```
MOV   Smem,dst
```

can be used for the following instructions:

```
(A)   MOV     *abs16(#2004h),T1
(B)   MOV     *abs16(#AR2),T2
```

In instruction (A), the 23-bit address is DPH:k16 = 052004h. Thus, the CPU loads the value at memory address 052004h into the T1 register. In instruction (B), the CPU loads the contents of AR2 into T2. Since the address of AR2 is 000012h, the address generated for accessing this MMR is 000012h, regardless of the current DPH value.

In addition, an example syntax

```
MOV   dbl(Lmem),pair(TAx)
```

can be used for the following instruction:

```
(C)   MOV     dbl(*abs(#2004h)),pair(T1)
```

In this example, the CPU loads the values at addresses 052004h and 052005h into T1 and T2, respectively.

Instructions (A) and (C) may be replaced by the following instructions using the k23 operands:

```
MOV   *(#052004h),T1
MOV   dbl(*(#052004h)),pair(T1)
```

In these two instructions, the DPH is not used for supplying the most significant 7 bits of the addresses. However, a 3-byte (instead of a 2-byte) extension is encoded to the instruction.

I/O absolute-addressing mode uses the `port()` operand qualifier to enclose a 16-bit unsigned constant in parentheses, such as `port(#k16)`. For example, the following instruction

```
MOV     port(#4),AR1
```

loads the value from the I/O space at address 0002h into AR1.

Direct-Addressing Modes

Direct-addressing mode allows programmers to reference a location using an address offset. Four direct-addressing modes are available: (1) DP direct for accessing a memory location or an MMR, (2) SP direct for accessing stack values in data memory, (3) register-bit direct for accessing 1 register bit or 2 adjacent register bits, and (4) PDP direct for accessing a location in the I/O space. When accessing an MMR rather than a data-memory location, use the `mmap()` qualifier to enclose the operand.

DP Direct-Addressing Mode   The DP direct mode uses the data-page register, DP, and the associated extension register, DPH. The CPU can concatenate the two to form an extended DP that is called XDP. We can load DPH and DP individually, or we can use an instruction that loads XDP. A process similar to that shown in Table 4.14 generates the 23-bit address in DP direct-addressing mode, where the 7 MSBs are taken from the DPH to select 1 of the 128 main data pages from 0 to 127. However, the 16 LSBs are the sum of the value in the DP and a 7-bit offset (`Doffset`) calculated by the assembler. DP identifies the starting address of a 128-word local data page within the main data page, and `Doffset` points to the location inside the local data page. The calculation of `Doffset` is different for accessing data memory or an MMR. For accessing data memory, we have

```
Doffset = (Daddr - .dp)&7Fh
```

where `Daddr` is the 16-bit local address, `.dp` is a value we assign using the `.dp` assembler directive, and the symbol `&` indicates a bitwise AND operation.

Example 4.18

The following instructions load the 16-bit value at address 03FFF4h into register T2 using DP direct-addressing mode:

```
AMOV    #03FFF0h,XDP    ; DPH <- 03h, DP <- FFF0h
.dp     #0FFF0h         ; assign .dp to FFF0h
MOV     @0FFF4h,T2      ; load the value at local address
                        ; FFF4h into T2
```

In this example, `Doffset` = (FFF4h − FFF0h)&7Fh = 04h. Thus, at run time, the 23-bit data-space address is DPH:(DP + Doffset) = 03:(FFF0h + 0004h) = 03FFF4h.
   Note that `.dp` generally matches DP. When DP = `.dp`, `Doffset` is equal to `Daddr`. The last instruction in the example may be replaced by the following instruction to load two words into T2 and T3:

```
MOV     dbl(@0FFF4h),pair(T2)
```

The DP direct mode can assess an MMR by using the qualifier `mmap()` to indicate access to MMRs. For example,

```
MOV    mmap(@AR0),T2      ; loads the value of AR0 into T2
MOV    mmap(@AC0L),AR2    ; loads the contents of AC0L into AR2
```

The `mmap()` qualifier forces the DAGEN to access the main data 0, regardless of the value in the DPH.

SP Direct-Addressing Mode   In SP direct-addressing mode, the 7 MSBs are supplied by SPH and the 16 LSBs are the sum of the SP value and a 7-bit offset that is specified in the instruction. For example, when SP = FF00h and SPH = 0h, the following instruction

```
MOV    *SP(5),T1
```

loads the value of data memory at address 00 FF05h into T1. Note that in the first main data page, the locations at addresses 000000h − 00005Fh are reserved for MMRs and thus cannot be used as data stacks.

Register-Bit Direct-Addressing Mode   In register-bit direct-addressing mode, the offset given in the operand `@bitoffset` is an offset from the LSB of the register. This addressing mode can be used only by the register-bit test, set, clear, and complement instructions. In addition, only the registers AC0-AC3, AR0-AR7, and T0-T3 can be accessed by these instructions. We use the register-bit direct operand `@bitoffset` to access a register bit if an instruction has the following syntax element:

```
BSET   Baddr,src
```

For example,

```
BSET    @0,AC1      ; sets bit 0 of AC1
```

PDP Direct-Addressing Mode   In PDP direct-addressing mode, the 9-bit PDP selects one of the 512 peripheral data pages (0 to 511). Each page has 128 words (0 to 127), and a particular word can be selected by specifying a 7-bit offset in the instruction. We use the PDP direct operand `@Poffset` to access I/O space by using an instruction with the `port()` qualifier. For example, if PDP = 1, the following instruction

```
MOV    port(@0),T1
```

loads the value at address 0080h (PDP:Poffset) of I/O space into T1.

Indirect-Addressing Modes

Similar to the C54x, indirect-addressing mode allows programmers to reference a memory location using a pointer for linear or circular addressing. The C55x supports four indirect addressing modes: (1) AR indirect mode, which uses one of eight auxiliary registers (XAR0-XAR7), (2) dual AR indirect mode, which accesses two data-memory locations, (3) CDP indirect mode, which uses the CDP, and (4) coefficient indirect mode, which supports instructions that access a coefficient in data memory and two other data-memory values at the same time. It is important to know which operands can be used for a given instruction in indirect-addressing mode.

AR Indirect-Addressing Mode   AR indirect-addressing mode uses an auxiliary register XARn to point at data. In order to generate 23-bit addresses, ARn provides the 16 LSBs, and the associated register ARnH provides the 7 MSBs. For example, in the following instruction

```
MOV     *AR4,T2
```

the value at the data-memory location pointed at by XAR4 is loaded into T2. For accessing the desired memory value in data space, we first use an instruction that loads the address of that value into XARn. When accessing a register bit, the selected ARn contains a bit number. When accessing I/O space, the selected ARn contains the complete I/O address since the I/O space uses 16-bit addresses.

The operands available for AR indirect-addressing mode depend on the ARMS bit of the status register ST2_55. When the ARMS bit is 0, C55x is in DSP mode, which allows efficient execution of DSP-intensive applications. If ARMS is 1, the CPU is in control mode for control-system applications to optimize code size. DSP-mode operands are listed in Table 4.15, where all addition and subtraction is done in modulo 64K since we cannot address data across main data pages without changing the value in ARnH. Pointer modification or address generation is either linear or circular, depending on the pointer configuration in the status register ST2_55. Note that in rows 2 to 5, ARn is incremented (or decremented) by 1 if a 16-bit data operation or a 1-bit register-bit operation is used. For 32-bit data operation

TABLE 4.15   AR Indirect-Addressing Operands in DSP Mode

	Operand	Pointer modification
1	`*ARn`	ARn is not modified
2	`*ARn+`	ARn is incremented by 1 (or 2) after access
3	`*ARn−`	ARn is decremented by 1 (or 2) after access
4	`*+ARn`	ARn is incremented by 1 (or 2) before access
5	`*−ARn`	ARn is decremented by 1 (or 2) before access
6	`*(ARn+T0/AR0)`	ARn is incremented by T0 (or AR0) after access
7	`*(ARn−T0/AR0)`	ARn is decremented by T0 (or AR0) after access
8	`*ARn(T0/AR0)`	ARn is not modified; T0 (or AR0) is used as offset
9	`*(ARn+T0B/AR0B)`	Increment is done for bit-reversal addressing
10	`*(ARn−T0B/AR0B)`	Decrement is done for bit-reversal addressing
11	`*(ARn+T1)`	ARn is incremented by T1 after access
12	`*(ARn−T1)`	ARn is decremented by T1 after access
13	`*ARn(T1)`	ARn is not modified; T1 is used as offset
14	`*ARn(#K16)`	ARn is not modified; K16 is used as offset
15	`*+ARn(#K16)`	K16 is added to ARn before access

or 2-bit register-bit operation, ARn is updated by 2. In addition, in rows 6 to 10, T0 is used if the C54CM bit is zero. Otherwise, AR0 is used in order to make the code compatible with the TMS320C54x. In rows 14 and 15, the capital K16 means 16-bit signed constant, where k16 used in absolute-addressing mode represents a 16-bit unsigned constant. Compared with Table 4.6, we show that C55x processors support more flexible indirect-addressing operations.

Example 4.19

The following instructions use AR indirect-addressing mode:

(A) MOV   *AR4-,T2
(B) MOV   dbl(*AR4+),pair(T2)
(C) MOV   *-AR4,T2
(D) MOV   *(AR4+T0),T2
(E) MOV   *AR4(T0),T2
(F) MOV   *AR4(#8),T2
(G) MOV   *+AR4(#8),T2

In instruction (A), the CPU loads the word at the address pointed to by XAR4 into T2. After being used for the address, AR4 is decremented by 1. In instruction (B), the CPU loads two words at the address pointed to by XAR4 and loads the following (if the value of AR4 is an even number) or the preceding (if the value of AR4 is an odd number) address into T2 and T3, respectively. After being used for the addresses, AR4 is incremented by 2. In instruction (C), AR4 is decremented by 1 before being used for the address. In instruction (D), the CPU loads the word pointed to by XAR4 into T2. After being used for the address, AR4 is incremented by the value in T0. In instruction (E), the CPU loads the word at address (XAR4+T0) into T2, and AR4 is not modified. In instruction (F), the CPU loads the word at the address equal to the value of XAR4+8 into T2, and AR4 is not modified. In instruction (G), XAR4 first is added to the constant 8, and then the CPU loads the data at the memory pointed to by XAR4 into T2.

To address MMRs with indirect-addressing modes, we load the correct data-space address into the pointer. For example, the following instruction

```
AMOV    #AC0L,XAR1
```

loads the address of the low word of AC0 into XAR1. This instruction initializes XAR1 as the pointer that can be used to access the low word of AC0 using indirect addressing.

To access register bits with indirect-addressing modes, we use indirect operands to access register bits if the instruction has the syntax element Baddr. First, make sure the pointer contains the correct bit number. For example, the instruction

```
MOV    #3,AR1
```

initializes AR1 as the pointer to access bit 3 of a register. Then, the following instruction

```
BSET    *AR1+,AC2
```

sets bit 3 of AC2. After being used for the address, AR1 is incremented by 1.

To address I/O space with indirect-addressing modes, we use the `port()` qualifier. For example,

```
MOV    port(*AR1),T2
```

loads the value at the I/O space pointed to by AR1 into T2. Note that the indirect operands `*ARn(#K16)`, `*ARn+(#K16)`, `*CDP(#K16)`, and `*CDP+(#K16)` cannot be used for accessing I/O space. In addition, the delay operation cannot be used for accessing I/O space.

Dual AR Indirect-Addressing Mode   Dual AR indirect-addressing mode supports two data-memory accesses through the use of two extended auxiliary registers that contain addresses. This mode may be used for executing an instruction that accesses two data-memory words or for executing two instructions in parallel. For example, the following instruction adds two operands to ACx:

```
ADD    Xmem,Ymem,ACx
```

For executing two instructions in parallel, both instructions must access a single memory value such as Smem or Lmem. For example,

```
MOV       Smem,dst
|| AND    Smem,src,dst
```

The operands listed in rows 1, 2, 3, 6, 7, 8, 11, and 12 of Table 4.15 can be used for dual AR indirect-addressing mode.

CDP Indirect-Addressing Mode   CDP indirect-addressing mode uses CDP to point to data. For accessing data space, the 16 LSBs are supplied by CDP, and the 7 MSBs are supplied by CDPH. For accessing a register bit, a bit number is contained in CDP. For accessing I/O space, CDP contains a 16-bit I/O address. The operands available for this addressing mode are `*CDP`, `*CDP+`, `*CDP-`, `*CDP(#K16)`, and `*+CDP(#K16)`. For example,

```
MOV    *CDP+,T2
```

In this instruction, the CPU loads the value pointed at by XCDP (CDPH:CDP) into T2. After being used for the address, CDP is incremented by 1.

Coefficient Indirect-Addressing Mode   Coefficient indirect-addressing mode uses the same address generation process as the CDP indirect- and dual AR indirect-addressing modes for accessing three memory operands per cycle. Two of these operands (Xmem and Ymem) are accessed with the dual AR indirect-addressing mode, and the third operand (Cmem) is accessed with the coefficient indirect-addressing mode. For example, the following syntax

```
MPY          Xmem,Cmem,ACx
:: MPY       Ymem,Cmem,ACy
```

allows two multiplications to be performed in parallel. The operand (Cmem) is common to both multiplications, while Xmem and Ymem are other values in the multiplication. An example of detailed parallel instruction is given as follows:

```
MPY          *AR0,*CDP+,AC0
:: MPY       *AR1,*CDP+,AC1
```

In this parallel instruction, the CPU multiplies the value at address XAR0 by the coefficient at address XCDP and stores the product to AC0. At the same time, the CPU multiplies the value at address XAR1 by the same coefficient at address XCDP and stores the product to AC1. After being used for the address, CDP is incremented only once by 1. This mode mainly is used by instructions including FIR filtering, multiply, multiply-add, multiply and subtract, and dual-multiply and add/subtract. Operands available for the coefficient indirect-addressing mode are *CDP, *CDP+, *CDP−, and *(CDP + T0/AR0).

4.4.4 Instruction Set

This section introduces some useful terms, symbols, and abbreviations used in the TMS320C55x mnemonic instruction set, as will as some commonly used instructions. Detailed descriptions of individual instructions are available in the *TMS320C55x DSP Mnemonic Instruction Set Reference Guide* [10].

Introduction

The C55x mnemonic instruction set can be divided into six types of operations: (1) arithmetic, (2) bit manipulation, (3) extended auxiliary register, (4) logical, (5) move, and (6) program-control. Basic C55x mnemonic instructions include ABS (absolute value), ADD (addition), AND (bitwise AND), B (branch), CALL (function call), CLR (clear or assign the value to 0), CMP (compare), CNT (count), EXP (exponent), MAC (multiply-accumulate), MAR (modify contents of auxiliary register), MAS (multiply-subtract), MAX (maximum), MIN (minimum), MOV (copy data), MPY (multiply), NEG (two's complement), NOT (one's complement), OR (bitwise OR), POP (pop from the top of stack), PSH (push to the top of stack), RET (return), ROL (rotate left), ROR (rotate right), RPT (repeat), SAT (saturate), SET (set, assign the value to 1), SFT (shift), SQA (square add), SQR (square), SQS (square subtract), SUB (subtraction), SWAP (swap register), TST (test bit), XOR (bitwise XOR), XPA (expand), and XTR (extract). Mnemonic syntax keywords and operand modifiers are case insensitive. For example, MOV or mov are treated as the same by the assembler.

Some of these basic syntax roots may be modified with mnemonic syntax prefixes (A or B). For example, the MOV instruction with prefix A forms AMOV and occurs in the DAGEN functional unit in the address phase: thus, it cannot be placed in parallel with any instruction that uses dual-addressing mode. For example,

```
AMOV     #x,XAR0      ; copy the address of x to XAR0
```

In addition, mnemonic syntax suffixes include 40 (enable M40 mode), B (borrow), C (carry), CC (conditional), I (enable interrupt), K (multiply has a constant operand), L (logical shift), M (the option of assigning operand to T3), R (round), S (signed shift), U (unsigned), V (absolute value), and Z (delay on the memory operand). Some suffixes can be combined.

Simple instructions are only allowed to stay on the same line. However, parallel instructions using the notation :: to imply parallelism may be split. For example,

```
MPYR40        uns(Xmem),uns(Cmem),ACx
:: MPYR40     uns(Ymen),uns(Cmem),ACy
```

In this parallel instruction, the optional symbol R (or rnd) is applied to the instruction MPY, indicating that rounding is performed in the instruction. The optional 40 keyword sets the status bit M40 to 1 (i.e., treats the accumulator as 40-bit registers) for execution of the instruction.

The optional operand modifier uns means unsigned is applied to the input operand that is zero extended. However, the instruction suffix U (unsigned) is used when the whole operation is affected (MPYMU, CMPU, BCCU). Additional operand modifiers include dbl (access a true 32-bit memory operand), dual (access a 32-bit memory operand for use as two independent 16-bit operands), HI (access the upper 16 bits of the accumulator), LO (access the lower 16 bits of the accumulator), high_byte (access the high byte of the memory location), low_byte (access the low byte of the memory location), pair (dual-register access), rnd (round), and saturate.

Absolute-address Smem is denoted with the @ syntax. Addresses may be assembly-time constants, symbolic link-time constants, or expressions. A valid address is a # followed by a number (e.g., #7), an identifier #x, or a parenthesized expression #(x + 1). However, when an address is used in a data memory address, the address does not need to have a leading #. When used in contexts such as branch targets or absolute-address Smem, addresses generally need a leading #. An optional mmr prefix used in indirect-addressing mode indicates an access to a memory-mapped register. If several prefixes are specified, mmr must be the innermost prefix. For example, uns(mmr(*AR0)) is valid, but mmr(uns(*AR0)) is not.

Parallel Instructions

The C55x supports the execution of two instructions (or operations) in parallel within the same CPU cycle, thus reducing total execution time. C55x instructions use dedicated resources within three independent computation units: PU, AU, and DU. If all parallelism rules are satisfied, two instructions that independently use any two of the independent resources may be placed in parallel. There are three types of parallelism: (1) built-in parallelism, (2) user-defined parallelism, and (3) combined built-in and user-defined parallelism.

Built-In Parallelism   Some instructions perform two different operations in parallel. Built-in parallelism within a single instruction uses double colons :: to separate the two operations. This type of parallelism is also called implied parallelism. For example,

```
MPY         *AR0,*CDP,AC0
:: MPY      *AR1,*CDP,AC1
```

This is a single instruction using a dual MAC. The data referenced by AR0 is multiplied by the coefficient referenced by CDP, and the product is saved in AC0. At the same time, the data referenced by AR1 is multiplied by the same coefficient pointed to by CDP, and the product is saved in AC1.

User-Defined Parallelism   The programmer, or the C compiler, can place two instructions in parallel so that both instructions are executed in a single cycle. Parallel bars || are used to separate the two instructions to be executed in parallel. For example,

```
MPYM        *AR+,*CDP,AC1
|| XOR      AR2,T1
```

In this example, the first instruction performs multiplication in the DU and the second instruction performs a logical operation in the AU; thus, these two instructions can be executed in parallel.

Combined Built-In and User-Defined Parallelism   Some built-in parallel instructions can be combined with user-defined parallelism. In other words, one of the two user-defined instructions may have built-in parallelism. For example,

```
MPY         *AR0,*CDP,AC0
:: MPY      *AR1,*CDP,AC1
|| RPT      CSR
```

Parallelism between two instructions is allowed only if all of the rules are respected. For example, two instructions can be executed in parallel if the length of the parallel instruction does not exceed 6 bytes, if there are no hardware resource conflicts, and if there is parallelism between the three main computation units, AU, DU, and PU.

We may use the following procedure to simplify the process of using user-defined parallelism to produce optimized assembly code:

Step 1. Write and verify the assembly code without the use of user-defined parallelism, but use the available built-in parallelism.

Step 2. Identify potential user-defined parallel-instruction pairs in the code and place them in parallel. Perform this step on critical sections of code first.

Step 3. Run the optimized code using the assembler to validate the parallel-instruction pairs. If there are invalid pairs, refer to the set of parallelism rules to make necessary changes. Make sure all of the parallel pairs are valid before continuing to the next step.

Step 4. Make sure the code still functions correctly.

4.4.5 Programming Considerations

This section introduces some useful methods for optimizing C55x assembly code to use special architectural features such as dual-MAC units, circular addressing, and zero-overhead looping. A detailed discussion of these issues is available in [11].

Efficient Use of Dual-MAC Units

As illustrated in Fig. 4.20, dual-MAC units in the DU receive three operands via three independent data buses: BB, CB, and DB. During a dual-MAC operation, each MAC unit requires two data operands from memory. However, the three data buses are capable of providing at most three independent operands. Thus, both MAC units use the same data value on the BB. Dual-MAC units are capable of implementing two multiply-and-add (or two multiply-and-subtract) operations in one cycle as follows:

$$ACx = ACx + Xmem \times Cmem$$

$$ACy = ACy + Ymem \times Cmem$$

The common operand (Cmem) has to be addressed by XCDP and should be kept in internal memory since BB is not connected to external memory. In addition, Cmem should not reside in the same memory block with respect to the other two operands since the internal DARAM block only allows two accesses in one cycle. For addressing Xmem and Ymem operands, any of the eight auxiliary registers (XAR0-XAR7) can be used. An example of a typical dual-MAC instruction is

```
MAC        *AR2+,*CDP+,AC0
:: MAC     *AR3+,*CDP+,AC1
```

When an algorithm has internal symmetry, it can be exploited for efficient dual-MAC implementation. For example, a symmetric FIR filter has coefficients that are symmetrical with respect to delayed values of the input signal. The detailed implementation of FIR filters using dual-MAC hardware is introduced in Chapter 6. In addition, dual-MAC operations can be applied for complex vector multiply, matrix multiply, block FIR filtering, and multichannel applications.

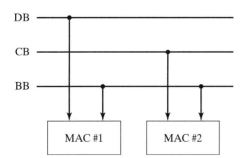

Figure 4.20   Data-bus structure for a dual-MAC operation

Circular Addressing

Circular addressing can be used with any of the indirect-addressing modes. Each auxiliary register ARn and the CDP can be configured independently to be a linear or circular pointer. This configuration is done by setting (circular) or clearing (linear) the linear-/circular-configuration bit (ARnLC for ARn and CDPLC for CDP) in the ST2_55 status register. For example,

```
BSET    AR1LC    ; AR1LC=1, sets AR1 as circular pointer
BCLR    AR0LC    ; AR0LC=0, clears AR0 as linear pointer
```

Five 16-bit circular-buffer-start-address registers (BSA01, BSA23, BSA45, BSA67, and BSAC) define a circular buffer with a start address not bound by any alignment constraint. In addition, three 16-bit circular-buffer-size registers (BK03, BK47, and BKC) specify the number of words (up to 65,535) in a circular buffer or the number of bits within a register. Each buffer-start-address register and buffer-size register is associated with a particular pointer, as summarized in Table 4.16. The value in the pointer (ARn or CDP) acts as an index, selecting words relative to the start address. That is, a 23-bit address has the form ARnH:(BSAxx + ARn) or CDPH:(BSAC + CDP).

The procedure for setting up a circular buffer in data memory is summarized as follows:

Step 1. Initialize the appropriate buffer-size register (BK03, BK47, or BKC). For example, load BK03 with the value buffer_size = 5 for a buffer size of 5 words.

Step 2. Initialize the appropriate configuration bit in ST2_55 to choose circular modification for the selected pointer.

Step 3. Initialize the appropriate extended register (XARn or XCDP) to select a main data page.

TABLE 4.16   Circular-Buffer Registers and Associated Pointers (adapted from [9])

Pointer	ST2_55 bit	Buffer-start-address register	Buffer-size register
AR0	AR0LC	BSA01	BK03
AR1	AR1LC	BSA01	BK03
AR2	AR2LC	BSA23	BK03
AR3	AR3LC	BSA23	BK03
AR4	AR4LC	BSA45	BK47
AR5	AR5LC	BSA45	BK47
AR6	AR6LC	BSA67	BK47
AR7	AR7LC	BSA67	BK47
CDP	CDPLC	BSAC	BKC

Step 4. Initialize the appropriate buffer-start-address register (BSA01, BSA23, BSA45, BSA67, or BSAC). The main data page in ARnH or CDPH (b22-b16) concatenated with the contents of the buffer-start-address register specifies the 23-bit start address of the buffer.

Step 5. Load the selected pointer, ARn or CDP, with a value from 0 to (`buffer_size - 1`).

Note that the buffer must be placed in 1 of the 128 main data pages of data space because we cannot address data across main data pages without changing the value in ARnH or CDPH. In addition, make sure that the absolute value of each offset is less than or equal to (`buffer_size - 1`) if indirect-addressing operands with offsets are used. For a buffer of register bits, we only need to load ARn or CDP. The buffer-start-address register defines the reference bit, and the pointer selects bits relative to the position of that reference bit.

Example 4.20

The circular-buffer size is 3 words, and AR1 is configured as a circular pointer. The following instructions show how to initialize and access a circular buffer:

```
MOV     #3,BK03         ; circular buffer size is 3 words
BSET    AR1LC           ; configure AR1 as a circular pointer
AMOV    #010000h,XAR1   ; AR1H=1, buffer at main data page 1
MOV     #0A02h,BSA01    ; circular buffer start at 010A02h
MOV     #0000h,AR1      ; index AR1 = 0
MOV     *AR1+,AC0       ; loads data at 010A02h to AC0
ADD     *AR1+,AC0       ; adds data at 010A03h to AC0
ADD     *AR1+,AC0       ; adds data at 010A04h to AC0
ADD     *AR1+,AC0       ; adds data at 010A02h to AC0
```

Repeat Loops

The C55x supports three types of repeat (zero-overhead) loops: single repeat, local-block repeat, and block repeat. In addition, the processor also provides BCC (branch on auxiliary register not zero) instruction to implement instruction looping, but requires a five-cycle loop overhead. The selection of the looping method depends on the number of instructions that need to be repeated, the layer of loops that need to be nested, and the methods that need to control the loop counters. The most efficient loops are single repeat and local-block repeat. Single-repeat instructions repeat a single-cycle instruction or two single-cycle instructions that are executed in parallel. Local-block-repeat looping repeats a block of code from the instruction-buffer queue. Since the code already has been fetched and placed in the queue, we can reduce program-memory accesses, pipeline conflicts, power consumption, wait states, and access penalties.

For single-repeat instructions, the number of repetitions, N, is a constant embedded in the repeat instruction. The number N is loaded into RPTC, and the following instruction is executed $N + 1$ times. For some applications, the computed single-repeat register (CSR) is copied to RPTC to specify the number N, which allows the user to compute the loop counter during runtime. CSR is not decremented because RPTC holds the active loop count; thus, CSR needs to be initialized only once. Initializing CSR outside the loop can save execution time. Both RPTC and CSR are 16-bit registers, which allows the following instruction or instruction pair to be repeated up to 65,536 times. For example,

```
RPT      CSR (k8 or k16)
```

This instruction repeats the next instruction or the next two parallel instructions the number of times specified by the contents of CSR + 1, or an immediate value $kx + 1$.

Block-repeat instructions form loops that repeat a block of instructions. The instructions for repeating a block of instructions are

```
RPTBLOCAL    pmad    ; local block repeat
RPTB         pmad    ; block repeat
```

We can create up to two layers of block-repeat loops without any cycle penalty. One block-repeat loop can be nested inside another, creating an inner (level 1) loop and an outer (level 0) loop. In addition, we can use any number of single-repeat loops inside each block-repeat loop.

The C55x registers associated with loops are summarized in Table 4.17. The 16-bit block-repeat counter (BRC0 or BRC1) indicates the number of times to repeat the instruction block after its initial execution. The 24-bit block-repeat start-address register (RSA0 or RSA1) contains the address of the first instruction in the instruction block, and the 24-bit block-repeat end-address register (REA0 or REA1) contains the address of the last instruction in the instruction block. Whenever BRC1 is loaded, BRS1 is loaded with the same value. The contents of BRS1 are not modified during the execution of the inner loop. Each time the inner loop is triggered, BRC1 is reinitialized from the value of BRS1. This feature initializes BRC1 once it is outside the outer loop, thus reducing the time needed for initializing BRC1 at each repetition of the outer loop.

TABLE 4.17   Summary of Block-Repeat Registers

Level 0 (outer) loop registers		Level 1 (inner) loop registers	
Register	Description	Register	Description
BRC0	Block-repeat counter 0	BRC1	Block-repeat counter 1
RSA0	Block-repeat start-address register 0	RSA1	Block-repeat start-address register 1
REA0	Block-repeat end-address register 0	REA1	Block-repeat end-address register 1
		BRS1	BRC1 save register

An example of two layers of block-repeat loops is given as follows:

```
MOV      #(N0-1),BRC0       ; repeat outer loop N0 times
MOV      #(N1-1),BRC1       ; repeat inner loop N1 times
...
RPTBLOCAL  LoopB-1          ; outer loop from here to LoopB
...
RPT      #(N2-1)            ; single repeat next instruction
...
RPTBLOCAL    LoopA-1        ; inner loop from here to LoopA
...
PRT      #(N3-1)            ; single repeat next instruction
...
LoopA:                      ; end of inner loop
...
PRT      #(N4-1)            ; single repeat next instruction
...
LoopB:                      ; end of outer loop
```

4.4.6 Optimization of C Programs

As mentioned in Section 4.1.3, we can optimize the performance of C programs for DSP processors. This section introduces some techniques for writing efficient C code for the C55x architecture. For detailed information on writing C programs, refer to the *TMS320C55x Optimizing C/C++ Compiler User's Guide* [12].

Introduction

The C55x C compiler defines the size of each C data type (signed and unsigned), as summarized in Table 4.18. In general, we use the int data type for fixed-point arithmetic, especially in multiplication. In particular, we use int or unsigned int rather than long for loop counters since the hardware loop counters on the C55x are only 16 bits wide.

A 16-bit multiplication with a 32-bit result is an operation that does not directly exist in the C language, but is performed on DSP-processor hardware. The correct

TABLE 4.18   Size of Each C Data Type Used in C55x Processors

Data type	Size in bits
char	16
short	16
int	16
long	32
long long	40
float	32
double	64

expression for a 16-bit × 16-bit multiplication with 32-bit accumulation is

```
result = (long)a*b;
```

where the variable `result` is defined as `long`, and the input variables a and b are defined as `int`. When this expression is used, the C compiler notices that each operand fits in 16 bits, so it issues an efficient single-instruction multiplication. Note that the same rules also apply for other arithmetic operations. For example, the following expression

```
result = (long)a+b;
```

adds two 16-bit numbers and obtains a full 32-bit result.

The C compiler offers four optimization options ($-$o0, $-$o1, $-$o2, and $-$o3) for transforming C code into assembly-language source code. For CCS users, the optimization level can be selected from **Project** → **Build Options** and by selecting the desired level in the **Opt Level** field. We can use $-$o2/$-$o3 to maximize compiler optimization. In the **Program Level Opt** field, selecting one of the four options enables the $-$pm option that combines source files to perform program-level optimization. In addition, CCS has extensive profiling options that can be used to evaluate the efficiency of C code.

Optimizing C Code

The following C55x-specific optimization techniques can be used to improve the performance of C code:

1. Use loops that efficiently use C55x hardware loops, MAC hardware, and a dual-MAC architecture.
2. Use intrinsics to replace complicated C code.
3. Avoid the modulus operator % in C when simulating circular addressing.
4. Use long accesses to reference 16-bit data in memory.

There are many ways to write efficient loop code in C. For example, avoiding function calls within the body of repeated loops enables the compiler to use very efficient hardware-looping instructions such as single repeat and block repeat. In addition, keeping the loop code small enables the compiler to use local block-repeat instructions.

A loop count is the number of times that a loop executes, and the loop counter is the variable used to count the number of iterations. To generate efficient code, we can use the `int` (or `unsigned int`) data type for the loop counter whenever possible, the `MUST_ITERATE` pragma to eliminate code to skip around loops and help the compiler generate efficient hardware loops, and the $-$o3 and $-$pm compiler options. For example, the following C function computes the sum of elements inside vector a[]:

```
int sum(const short *a, int n)
{
    int sum = 0;        // sum of elements in array a[]
    int i;              // loop counter
```

```
#pragma MUST_ITERATE(1) // at least execute the loop once
   for (i=0; i<n; i++)
   {
       sum += a[i];        // compute the vector sum
   }
   return sum;
}
```

Since we know the loop always executes at least once, we use the expression `#pragma MUST_ITERATE(1)` to communicate this fact to the compiler. Otherwise, the compiler inserts a conditional-branch instruction to check if the loop could execute zero times. The MUST_ITERATE pragma must appear immediately before the loop.

To facilitate the generation of a single-repeat loop with MAC as its only instruction, use the local variable for the product. For example, the following C function implements the inner (dot) product of two vectors, x[] and y[], using Q.15 arithmetic:

```
int innerp(const int *x, const int *y, unsigned int n)
{
    unsigned int i;           // loop index
    long sumProduct;          // sum of products
    for (i=0; i<=n-1; i++)
    {
        sumProduct += (long)x[i]*y[i];
    }
    return (int)((sumProduct>>15)&0x0000FFFFL);
}
```

The result of the MAC operation is first accumulated into a `long` object. The result is then shifted and truncated before the return.

A dual-MAC operation is one of the most important hardware features of the C55x processor. In order for the compiler to generate a dual-MAC operation, the code must have two consecutive MAC instructions that get all of their multiplicands from memory and share one multiplicand. The two operations must not write their results to the same variable or location. For example, the compiler can generate dual-MAC instructions for the following code:

```
int    *a, *b, onchip *c;
long   s1, s2;
...
s1 = s1 + (*a++ * *c);         // first MAC
s2 = s2 + (*b++ * *c++);       // second MAC
...
```

The onchip qualifier informs the compiler that the memory pointed to by the shared dual-MAC operand c is on chip, which is required for dual-MAC operations, as discussed in Section 4.4.5.

The C55x C compiler provides special functions called intrinsics that map directly to C55x instructions. Intrinsics with leading underscores "_" are accessed by being called as C functions. Note that using compiler intrinsics reduces the portability of C code. For more information on using available intrinsics, refer to the *TMS320C55x Optimizing C/C++ Compiler User's Guide* [12]. For example, the following intrinsic

```
long _smac(long src, int op1, int op2);
```

multiplies `op1` and `op2`, shifts the result left by 1 bit, and adds it to `src` to produce a saturated 32-bit result. In addition, the dual-MAC expressions introduced in the previous example can be replaced with the following intrinsics:

```
s1 = _smac(s1, *a++, *c);
s2 = _smac(s2, *b++, *c++);
```

We use intrinsics for experiments in Chapters 6–9.

C55x buses support 32-bit accesses in a single cycle. Thus, treating 16-bit data as `long` reduces the data-movement time from one memory location to another by half. The requirement for long-data accesses is that the data must be aligned on a double- (even-) word boundary using the `DATA_ALIGN` pragma as follows:

```
short      x[10];
#pragma    DATA_ALIGN(x, 2)
```

4.4.7 System Issues

C55x processors are used in many high-performance, power-efficient applications. These DSP applications require on-chip memory and peripherals for reducing system cost. The block diagram of the TMS320C5510 is illustrated in Fig. 4.21.

The architecture of the C55x provides methods to dynamically conserve power through six software-programmable idle domains: CPU, DMA, peripheral, clock generator, instruction cache, and external-memory interface. Each domain can operate normally or can be placed into a low-power idle state when its capabilities are not required. This control provides the user the capability to modify the power consumption of the processor dynamically based on activity.

The C55x architecture provides access to a maximum of 8M words (16M bytes) of memory, including on-chip and external memory, organized as unified program/data space. For specific memory maps and configurations, refer to a processor-specific data sheet such as that shown in Fig. 1.4 for the TMS320C5510.

The total on-chip memory of TMS320C5510 is 352K bytes. The DARAM is located in the byte-address range 000000h-00FFFFh and is composed of eight blocks (8 Kb per block), as summarized in Table 4.19. Each DARAM block can perform two accesses per cycle (two reads, two writes, or a read and a write). The SARAM is located at byte address range 010000h-04FFFFh and is composed of 32 blocks. Each SARAM block can perform one access (one read or one write) per cycle.

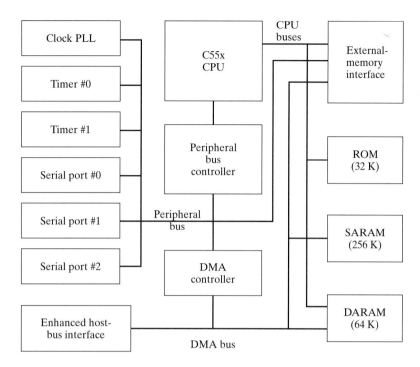

Figure 4.21   Block diagram of the TMS320C5510

TABLE 4.19 Address Range of DARAM Blocks

DARAM block	Byte address range
0	000000h-001FFFh
1	002000h-003FFFh
2	004000h-005FFFh
3	006000h-007FFFh
4	008000h-009FFFh
5	00A000h-00BFFFh
6	00C000h-00DFFFh
7	00E000h-00FFFFh

The C55x processor's external memory interface (EMIF) has been significantly enhanced from that used in the C54x. These improvements include expanded memory options and doubled bandwidth (from 16 bits to 32 bits) to increase performance and reduce power. The C55x extends capabilities to EMIF with a variety of high-speed, low-cost memories such as synchronous-burst static RAM, synchronous dynamic RAM, ROM, Flash, asynchronous static RAM, and dynamic RAM.

All C55x processors use the same CPU architecture, but have different memory configurations and on-chip peripherals. The C5510 supports the following peripherals:

- An EMIF
- A six-channel DMA controller
- Three McBSPs
- A digital PLL clock generator, two timers, and eight configurable general-purpose I/O pins

The EMIF supports a glueless interface from the C55x to a variety of external-memory devices. For each memory type, the EMIF supports 8-bit, 16-bit, and 32-bit accesses for both reads and writes.

The DMA controller transfers data between locations in memory space without CPU intervention. This movement of data to and from internal memory, external memory, and peripherals occurs in the background of the CPU operation. The DMA has six independent, programmable channels, which allow six different contexts of DMA operation. Each of the channels is executed in a time-division multiplexed fashion.

McBSPs are high-speed, full-duplex serial ports that allow direct interfacing with other devices in systems such as other DSP processors and CODECs. McBSPs have the capabilities of double-buffered data registers, which allow a continuous data stream, μ-law and A-law companding, a wide selection of data sizes from 8 to 32 bits, etc.

4.5 TMS320C62x AND TMS320C64x

The TMS320C62x and TMS320C64x are a family of general-purpose, 16-bit fixed-point DSP processors. The C6000 generation is based on a very-long-instruction-word (VLIW) architecture. A floating-point version of the C6000 processor, TMS320C67x, is discussed in Chapter 5. In this section, we focus on the 16-bit fixed-point family of the C6000, including both the C62x and C64x. The C64x family is an extension of the C62x architecture, and the C64x includes many significant enhancements to and more functioning units than the C62x. Its instruction set is also a super-set of the C62x instruction set. The C6000 processor enables advanced imaging, radar and sonar systems, and 3G wireless- and broadband-communications infra-structure applications.

In the following sections, we introduce the architecture, addressing modes, programming considerations, and system issues of C62x and C64x processors. Detailed descriptions of these C6000 processors are available in references [13-16].

4.5.1 Architecture Overview

As shown in Fig. 4.22, C62x and C64x processors have the same core architecture, which consists of two identical fixed-point data paths (A and B) and an instruction-control unit that controls the program fetch, dispatch, decode, and control logic, etc. Each data path has four independent, multipurpose functional units, .Ln, .Sn, .Mn,

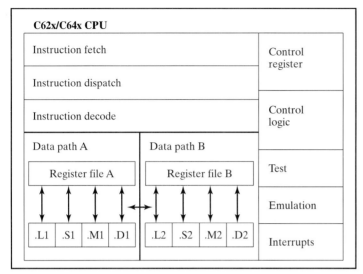

Figure 4.22 TMS320C62x/C64x CPU core

and .Dn, where n = 1 and 2 corresponds to the functional units in data paths A and B, respectively. Therefore, the core CPU consists of eight functional units, which implies that the processor can issue and execute up to eight 32-bit instructions per clock cycle. In general, the cycle time for the C62x is about 4 to 5 nsec, while the C64x processor reduces the cycle time by half to about 1.67 to 2 nsec. Therefore, C62x and C64x processors can achieve 1,600 MIPS operating at 200 MHz, or 3,600 MIPS operating at 600 MHz. This maximum performance can only be achieved when the eight functional units are running eight parallel instructions at their full capacity throughout the program, which is difficult to obtain for many applications.

By using a VLIW architecture, the C6000 processor can achieve a very high level of parallelism, thus avoiding the need for hardware that performs complex instruction scheduling and dispatching at the same time. The scheduling is carried out during compile time by the compiler and the programmer, which greatly simplifies the architecture compared to superscalar-based processors such as Pentium processors. However, the tradeoffs of using a VLIW-based processor are that the compiler is more complex and program memory usage is higher. The latest C64x processor improves on these deficiencies.

There are several new functionalities and enhancements in the new C64x over the C62x, as shown in Fig. 4.23. The C62x processor can perform only two 16 × 16-bit multiplications per cycle using two M units, as compared to four 16 × 16-bit multiplies per cycle in the C64x processor. The dual multipliers per M unit in C64x processors also allow the single instruction multiple data (SIMD) type of multiplication to be performed on 8-bit and 16-bit operands, which supports four 8 × 8-bit or dual 16 × 16-bit multiplications simultaneously. In addition, the C64x processor also supports dot products, extended-precision multiplications, 32-bit rotations, 32-bit bidirectional shifts, and Galois multiplications. Unlike C5000 processors, C6000 processors do not provide guard bits.

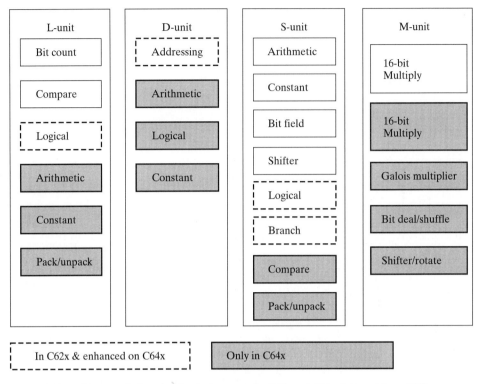

Figure 4.23   Differences in functional units between the C62x and C64x (adapted from [17])

The C62x processor contains a 40-bit ALU in the L unit. The ALU performs 32- or 40-bit arithmetic (with saturated or nonsaturated mode), normalization, and 32-bit logical operations. C64x processors include new instructions for 16-bit dual and 8-bit quad integer SIMD additions and subtractions and for packed-data manipulations (e.g., data selection and swapping). In addition, logical operations such as bitwise logical XOR and AND-NOT are also included.

The D unit in the C62x contains a 32-bit adder/subtractor, which is used for address generation. Two addresses can be generated in one cycle, and both linear and circular addressing can be implemented. Addressing modes are discussed in Section 4.5.4. The D unit can also be used as a general-purpose ALU if address generation is not required. There are several new features in the D unit of C64x processors, including 64-bit load and store, nonaligned load and store, and logical operations like AND, OR, XOR, and AND-NOT.

The S unit contains a 32-bit integer ALU and a 40-bit shifter. With these functional units, both the C62x and the C64x can perform 32-bit arithmetic, logical, bit-field, and 32-bit (or 40-bit) shift operations. In addition, the S unit in the C64x supports dual 16-bit or quad 8-bit compare, packed-data manipulations, new branching, and dual 16-bit or quad 8-bit saturation-addition instructions.

As shown in Fig. 4.22, register files A and B are associated with data paths A and B, respectively. Each register file contains 16 (A0-A15 or B0-B15) 32-bit

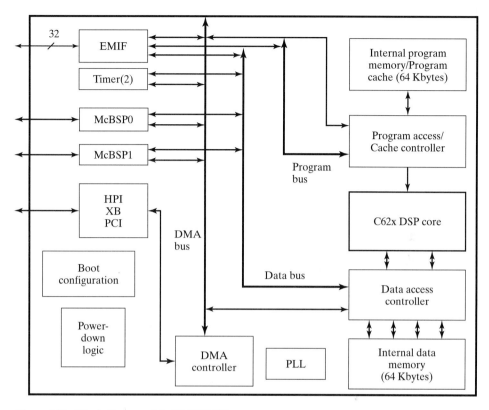

Figure 4.24   Block diagram of the TMS320C62x

general-purpose registers that are used to store data or addresses for the C62x [or 32-bit registers (A0-A31 or B0-B31) for the C64x]. The C6000 processor supports 8- (only in C64x), 16-, 32-, 40-, and 64-bit (only in C64x) data length when loading or storing data to and from memory. Therefore, a pair of adjacent registers must be used to hold the longer wordlength data. In addition, each data path has a cross path for reading operands from the other register file. Figures 4.24 and 4.25 show the internal architecture of C62x and C64x processors, respectively.

4.5.2 Memory Systems

The internal architecture of the C62x/C64x core was described in the previous section. In this section, we examine the memory system of fixed-point C6000 processors. The on-chip memory system of the C62x/C64x is implemented based on the modified Harvard architecture, which provides separate memory for program and data. A 32-bit address bus addresses the program memory, and a 256-bit data bus holds the 8×32-bit instructions. Data memory has two 32-bit address buses and two 32-bit (64-bit for the C64x) data buses. Since the C64x processor has a wider data bus, it can support 64-bit data transfers in addition to the 32-bit and 16-bit transfers in the C62x. It is also important to note that both

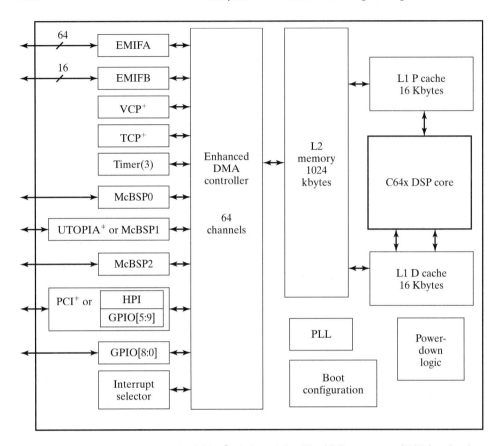

Figure 4.25 Block diagram of TMS320C64x [+ indicates that Viterbi Coprocessors (VCPs) and turbo coprocessors (TCPs) exist only on a C6416 device, and ‡ indicates that the universal test-and-operations physical interface for asynchronous transfer mode (UTOPIA) and PCI peripherals exist only on C6415 and C6416 devices]

program- and data-memory locations are byte addressable, and memory accesses on the C6000 processors must be byte aligned. This issue is further explained in a later section.

Table 4.20 shows the on-chip program and data memory in the C62x and C64x families. The C62x family has an on-chip data and program memory range from 512K bits to 4M bits. The C6211 processor has 32K bits of on-chip level-1 data (L1D) cache and program (L1P) cache. In addition, 512K bits of unified level-2 (L2) program and data cache are provided. C64x processors also use the same two-level cache as the C6211.

On-chip program memory can operate in one of four modes: (1) mapped, (2) cache enable, (3) cache freeze and, (4) cache bypass. Mapped mode, the default mode at reset, allows program code to be fetched from internal program memory. The cache operation is disabled. In cache-enable mode, on-chip program memory becomes direct-mapped cache. During cache freeze, the cache retains its current state and is used to ensure that critical program data is not overwritten in the cache.

TABLE 4.20   On-Chip Data and Program Memory

Processor	Data memory (bit)	Program memory (bit)
C6201	512K	512K
C6202	1M	2M
C6203	4M	3M
C6204	512K	512K
C6205	512K	521K
C6211	32K L1D data cache	32K L1P program cache
	512K L2 cache	
C6411	128K L1D data cache	128K L1P program cache
	2M L2 cache	
C6414	128K L1D data cache	128K L1P program cache
	8M L2 cache	
C6415	128K L1D data cache	128K L1P program cache
	8M L2 cache	
C6416	128K L1D data cache	128K L1P program cache
	8M L2 cache	

In cache-bypass mode, the program is fetched from external memory, and on-chip cache is completely bypassed.

The internal data memory of the C62x processor is divided into several banks for accessing multiple data in different banks simultaneously within a cycle. For example, the internal data memory of the C6201 processor is organized into eight banks, each with 8K bytes of memory. As shown in Fig. 4.26, these memory banks

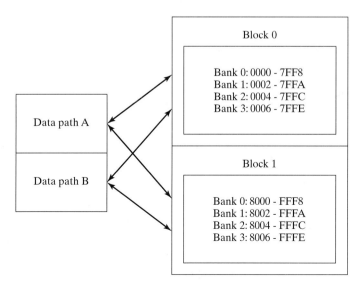

Figure 4.26   On-chip data memory organization of the C6201 processor

are divided into two blocks, each block consisting of four banks. Data paths A and B can access any portion of the internal memory simultaneously without the conflict of using different blocks. Likewise, the DMA controller (not shown in the figure) can also access these blocks.

4.5.3 External-Memory Addressing

In order to access external memory, C6000 processors provide an EMIF. The C62x has one EMIF, while the C64x has two units, EMIFA and EMIFB. The EMIF has a 23-bit external address bus and a 32-bit data bus, which both are multiplexed between program- and data-memory accesses. In contrast, the EMIFA and EMIFB of the C64x have 64-bit and 16-bit data buses, respectively. The memory interface provides a glueless interconnect to a broad range of synchronous and asynchronous external memory, such as RAM, synchronous-burst static RAM, synchronous dynamic RAM, asynchronous static RAM, ROM, flash, etc.

For both C62x and C64x processors, external memory is divided into four spaces: CE0, CE1, CE2, and CE3. In the C62x, only CE1 can be accessed in 8, 16, or 32 bits, while the other three spaces must be accessed in 32 bits. For example, the C6201 processor can access three 128M-bit spaces (CE0, CE2, and CE3) of external memory, which can be either synchronous or asynchronous memory. It further accesses another 32M-bit region (CE1) from asynchronous memory. In the C64x, EMIFA can access memory of 8, 16, 32, or 64 bits, while EMIFB can only access memory of 8 or 16 bits. An additional feature of the C64x is that it supports peripheral data transfer, which allows data to be transferred directly between external devices under the control of the EMIF.

4.5.4 Instruction Set

C6000 processors (C62x, C64x, and C67x) share a common instruction set. All of the instructions valid for the C62x are also valid for the C64x and C67x. However, because the C67x is a floating-point processor, there are some specific instructions for the C67x only. Similarly, the C64x adds functionality to the C62x with some unique instructions. This section introduces the assembly-language instructions for fixed-point C62x and C64x processors.

Assembly-Language Format C6000 processors provide an assembly-language format, as shown in Fig. 4.27. The first two columns contain optional fields, which give the programmer the ability to include options such as executing the current instruction with the previous one in parallel and executing a conditional-branch operation. In the example shown in Fig. 4.27, this instruction is executed only

Optional field	Optional field	Opcode field	Optional field	Operand field with one to four operands	Comment field
| |	[B1]	MPY	.M	A0, A1, A2 ;	multiply

Figure 4.27   Assembly-language format for C6000 processors

if the contents of the B1 register are not zero. The third column is the operation code (opcode) field, which contains the C6000 instruction. The fourth column contains another optional field that indicates which execution unit is used. In this case, the .M unit is used to execute the multiply operation. If the execution unit is not indicated, the compiler automatically assigns an appropriate execution unit. The operand column contains the registers to be used in this instruction. Up to four operands can be specified under this column. Finally, the last column contains an optional comment field. In the example given in Fig. 4.27, the number in A0 is multiplied with the number in A1, and the multiplication result is stored in A2.

TABLE 4.21   Mapping of Instructions to Functional Units (adapted from [15])

.L unit	.M unit	.S unit	.D unit
ABS	MPY	ADD	ADD
ADD	MPYU	ADDU	ADDU
ADDU	MPYUS	ADDK	ADDAB
AND	MPYSU	ADD2	ADDAH
CMPEQ	MPYH	AND	ADDAW
CMPGT	MPYHU	B*disp*	LDB
CMPGTU	MPYHUS	B IRP[+]	LDBU
CMPLT	MPYSU	B NRP[+]	LDH
CMPLTU	MPYHL	B *reg*	LDHU
LMBD	MPYHLU	CLR	LDW
MV	MPYHULS	EXT	LDB (15-bit offset)*
NEG	MPYHSLU	EXTU	LDBU (15-bit offset)*
NORM	MPYLH	MV	LDH (15-bit offset)*
NOT	MPYLHU	MVC[+]	LDHU (15-bit offset)*
OR	MPYLUHS	MVK	LDW (15-bit offset)*
SADD	MPYLSHU	MVKH	MV
SAT	SMPY	MVKLH	STB
SSUB	SMPYHL	NEG	STH
SUB	SMPYLH	NOT	STW
SUBU	SMPYH	OR	STB (15-bit offset)*
SUBC		SET	STH (15-bit offset)*
XOR		SHL	STW (15-bit offset)*
ZERO		SHR	SUB
		SHRU	SUBAB
		SHRL	SUBAH
		SUB	SUBAW
		SUBU	ZERO
		SUB2	
		XOR	
		ZERO	

+: S2 only
*: D2 only

Summary of the Instruction Set   The instruction set of C6000 processors consists of simple reduced instruction-set computer (RISC)-like instructions such as load, store, add, subtract, multiply, shifting, normalization, logical instructions, zero, saturation, and no operation. These instructions are grouped according to their functional unit, as shown in Table 4.21. Note that some of the instructions can be

TABLE 4.22   Additional Instructions to Functional-Unit Mapping in the C64x (adapted from [16])

.L unit	.M unit	.S unit	.D unit
ABS2	AVG2	ADD2‡	ADD2‡
ADD2‡	AVGU2	ADDKPC	AND‡
ADD4	BITC4	AND‡	ANDN
AND‡	BITR	ANDN	LDDW
ANDN	DEAL	BDEC	LDNDW
MAX2	DOTP2	BNOP	LDNW
MAXU4	DOTPN2	BNOP *reg*	MVK‡
MIN2	DOTPNRSU2	BPOS	OR‡
MINU4	DOTPNRUS2	CMPEQ2	STDW
MVK‡	DOTPRSU2	CMPEQ4	STNDW
OR‡	DOTPRUS2	CMPGT2	STNW
PACK2	DOTPSU4	CMPGTU4	SUB2‡
PACKH2	DOTPU4	CMPLT2	XOR
PACKH4	GMPY4	CMPLTU4	
PACKHL2	MPY2	MVK‡	
PACKL4	MPYHI	OR‡	
PACKLH2	MPYHIR	PACK	
SHLMB	MPYIH	PACKH2	
SHRMB	MPYIHR	PACKHL2	
SUB2‡	MPYIL	PACKLH2	
SUB4	MPYILR	SADD2	
SUBABS4	MPYLI	SADDU4	
SWAP2	MPYLIR	SADDSU2	
SWAP4	MPYSU4	SADDUS2	
UNPKHU4	MPYUS4	SHLMB	
UNPKLU4	MPYU4	SHR2	
XOR	MVD	SHRMB	
	ROTL	SHRU2	
	SHFL	SPACK2	
	SMPY2	SPACKU4	
	SSHVL	SUB2‡	
	SSHVR	SWAP2	
	XPND2	UNPKHU4	
	XPND4	UNPKLU4	
		XOR	

32-bits

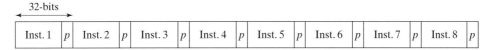

Figure 4.28   Basic format of a fetch packet

carried out in different functional units. For example, the ZERO instruction can be performed in .L, .S and .D units.

In addition to the list in Table 4.21, Table 4.22 shows the new C64x instructions. Some instructions, which are marked as ‡ in Table 4.22, also exist on C62x processors, but have been extended to include additional functional units in the C64x processor.

Instruction Packing

C6000 processors are always fetched in a group of eight instructions at a time. A fetch packet consists of eight instructions with 256 bits (or eight words). Another packet, known as the execute packet, consists of a group of instructions to be executed in parallel. Figure 4.28 shows the basic format of the fetch packet and the use of the *p*-bit (or the LSB of each word) to indicate whether the instruction is executed in parallel with the previous instruction. When the *p*-bit of the instruction word is set to one, the instruction is to be executed in parallel with the previous instruction in the same cycle; otherwise, the instruction is executed in the next cycle.

Example 4.21

Examine the *p*-bit of the fetch packet of the partial-sequential, partial-parallel instructions listed as follows:

```
       Instruction 1   ; instructions 1 and 2 are executed sequentially
       Instruction 2
       Instruction 3   ; instructions 3, 4, and 5 are executed in parallel
    || Instruction 4
    || Instruction 5
       Instruction 6   ; instructions 6, 7, and 8
    || Instruction 7   ; are executed in parallel
    || Instruction 8
```

The *p*-bit in the fetch packet appears as follows:

32 bits

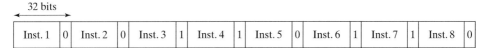

Four execute packets (or four cycles) are required to execute these eight instructions. The first execute packet contains instruction 1; the second packet contains instruction 2; the third packet contains instructions 3, 4, and 5; and the last packet contains instructions 6, 7, and 8. Note that in C62x processors, the last *p*-bit in the fetch packet must be set to zero, which means that every fetch packet starts with a new execute packet. However, C64x processors allow the crossing of boundaries in an execute packet, which increases code-packing density over C62x processors.

TABLE 4.23   Indirect-Address Generation for Load/Store Operations (adapted from [15])

Addressing type	No modification of address register	Pre-increment/ Pre-decrement	Post-increment/ Post-decrement
Register indirect	*R	*++R *−−R	*R++ *R−−
Register relative	* +R[ucst5] * −R[ucst5]	*++R[ucst5] *−−R[ucst5]	*R++[ucst5] *R−−[ucst5]
Base + index	* +R[offsetR] * −R[offsetR]	*++R[offsetR] *−−R[offsetR]	*R++[offsetR] *R−−[offsetR]

R denotes register, ucst5 denotes a 5-bit unsigned constant field, and offsetR denotes the register offset

Addressing Modes

The C6000 has a load/store RISC architecture in which the operands must be loaded into the register file, processed, and stored back into memory. The C6000 processor supports linear- and circular-addressing modes that use an address register as a pointer. Furthermore, the C6000 allows byte addressing, which implies that it provides the flexibility to load/store bytes (8 bits), half words (16 bits), words (32 bits), and double words (64 bits, only in the C64x) based on the given instructions.

We can use the load or store instruction to illustrate the usage of indirect addressing in linear mode. The syntax for indirect load/store address generation is shown in Table 4.23. The three types of indirect addressing are (1) register indirect, (2) register relative, and (3) base + index. As with other TMS320 processors, the indirect address is specified by a symbol * before the register. The pre-increment and pre-decrement of the address register are preceded by ++ and −−, respectively. Similarly, the post-increment and post-decrement of the address register are specified after the register with ++ and −−, respectively. In addition, indirect addressing allows programmers to include a scaled offset to the base address in both the register-relative and base + index modes. The offset is stored in either a register (offsetR) or a 5-bit unsigned constant (ucst5). The bracket [ ] in Table 4.23 indicates that the offset is left-shifted (or scaled) by 0, 1, or 2 for bytes, half words, or word loads, respectively.

Example 4.22

We illustrate several load/store operations, which are performed independently (not sequentially) using indirect-addressing modes based on the memory contents shown in Fig. 4.29.

```
(A)  LDB      *−A4[2], A5
(B)  LDB      *+A4[3], A5
(C)  LDH      *A4++[A1], A5
(D)  LDH      *++A4(2), A5
(E)  LDW      *A4++[1], A5
(F)  LDW      *−−A4[2], A5
```

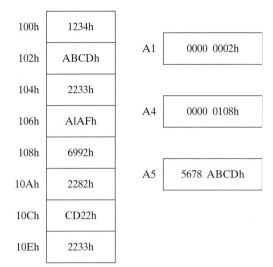

100h	1234h
102h	ABCDh
104h	2233h
106h	A1AFh
108h	6992h
10Ah	2282h
10Ch	CD22h
10Eh	2233h

A1   0000 0002h

A4   0000 0108h

A5   5678 ABCDh

Figure 4.29   Memory and register contents before executing load/store instructions

  (G) STH A5, *A4−−[A1]
  (H) STW A5, *++A4[2]

In instruction (A), the load instruction, LDB, loads the data in memory pointed at by A4 and pre-decremented by 2 (0108h − 2h = 0106h) into the A5 register. Therefore, the 1-byte data AFh is loaded into the A5 register, which becomes FFFF FFAFh. Note that when using a load instruction (e.g., LDB, LDH, and LDW) for a signed number, the destination value is signed-extended. Another set of load instructions, such as LDBU and LDHU, is used to load unsigned numbers to the designated register with the zero-filled final value. Instruction (B) is similar to instruction (A), except that pointer A4 is incremented by 3 to become 010Bh. Since 010B contains 22h, register A5 now contains 0000 0022h.

Instructions (C) and (D) perform a half-word load into register A5. In instruction (C), register A4 points at 0108h, thus loading the half word 6992h into A5, and its contents become 0000 6992h. Register A4 is post-incremented by the value in the A1 register and left-shifted by 1 for the half-word load. Thus, A4 becomes 0108h + 4h = 010Ch. In instruction (D), register A4 is pre-incremented by 2 to become 010Ah, and the half-word value 2282h is loaded into A5. Note that the pre-increment offset is 2 instead of the scaled value of 4, since parentheses ( ) are used to specify the index in this case. Similar operations can be carried out for the LDW in instructions (E) and (F), which load a 32-bit value into the designated register, A5. The result of instruction (E) is A5 = 2282 6992h and A4 = 010Ch. Instruction (F) results in A5 = A1AF 2233h and A4 = 0104h.

Instructions (G) and (H) are used to store the data in a register to a memory location. Similar to load instructions, store instructions (e.g., STB, STH, and STW) can store a byte, a half word, or a word of the register, with an optional offset specified from another register (or 5-bit constant), to the memory. In instruction (G), the half word of register A5, ABCDh, is stored to the memory at address 0108h. The address is pointed at by register A4, which is post-decremented to 0108h − 4h = 0104h. In instruction (H), the entire 32 bits of A5, 5678ABCDh, are stored to the memory address starting from A4 = 0108h + 8h = 0110h.

31	26	25	21	20	16	15	14	13	12	11	10	9	8	7	6	5	4	3	2	1	0
Res		BK0		BK1		B7 mode		B6 mode		B5 mode		B4 mode		A7 mode		A6 mode		A5 mode		A4 mode	

Figure 4.30   Address mode register

TABLE 4.24   Mode-Field Encoding Table

Mode	Addressing option
00	Linear mode
01	Circular mode using BK0
10	Circular mode using BK1
11	Reserved

The C62x only supports two different circular-buffer sizes at any one time, while the C64x supports eight different buffer sizes concurrently. In addition, the C64x includes double-word modulo addressing. Circular addressing can be implemented using any of the registers (A4-A7, B4-B7) as a pointer. These registers can be configured as linear or circular addressing by specifying the mode options in the addressing-mode register, as shown in Fig. 4.30. The mode-field encoding is given in Table 4.24. The block-size fields BK0 and BK1 in the addressing-mode register are used to specify the block size of the circular buffer. Since BK0 and BK1 are 5 bits wide, up to 32 different block lengths for circular addressing are possible. The circular buffer is always byte aligned.

For example, a simple 8-byte circular buffer is set up from 10000000h to 10000007h. A pointer, B4, is used to point to the circular buffer. Assume the current B4 = 10000007h. From Fig. 4.30, the bits under B4 mode and BK1 are set to 10 and 00010, respectively. For example, executing the following instruction

```
LDB        *B4++,B8
```

moves the contents of memory at 10000007h into the B8 register and post increases the pointer B4 to 10000000h. This result occurs because the circular pointer cannot point outside the circular buffer and must wrap to the top of the buffer when reaching the bottom of the buffer.

Conditional Instructions

All C6000 instructions allow conditions to be programmed into them. As shown in Fig. 4.27, square brackets [registers] are used to test whether the value of the register is zero or nonzero. If the value is nonzero, the instruction after the brackets is performed. Only registers A1, A2, B0, B1, and B2 can be used to test the condition in the C62x processor. In the C64x processor, an additional register A0 can be used as a conditional register.

The following example illustrates a simple usage of the conditional instruction:

```
   [A1]    ADD    .L1    A4,A5,A6
|| [!A1]    ADD    .L2    B4,B5,B6
```

The first ADD instruction is conditioned on A1 being non-zero, while the second ADD instruction is conditioned on A1 being zero. Therefore, the preceding instructions are mutually exclusive.

4.5.5 Programming Considerations

In this section, we discuss the pipeline of the C6000 processors, as well as the programming environment and code-efficiency issues.

Pipeline

The pipeline of C62x and C64x processors consists of fetch, decode, and execute stages, which are divided further into the following 11 phases:

A. *Fetch*: The fetch stage is divided into phases 1-4 as follows:
   1. *PG*: The program-address-generation phase computes the next sequential fetch-packet address or branch address.
   2. *PS*: The program-address-send phase sends the program address to memory.
   3. *PW*: The program-address-ready-wait phase waits until either a memory access or a tag compare is completed.
   4. *PR*: The program-fetch-packet-receive phase receives the fetch packet from memory.

B. *Decode*: The decode stage is divided into phases 5-6 as follows:
   5. *DP*: The instruction-dispatch phase separates fetch packets into execute packets.
   6. *DC*: The instruction-decode phase decodes source registers, destination registers, and associated paths.

C. *Execute*: When the instruction enters the execution stage, the CPU executes each instruction within the eight functional units. The number of phases required to complete each instruction varies.

   7-11. *E1-E5*: The execute stage is divided into phases 7-11 (E1-E5). Different types of instructions require different numbers of phases to execute. Most instructions require only one execution phase, E1, and do not have any delay. Others, like the multiply instructions (e.g., MPY and SMPY), require two execution phases, E1 and E2, which implies that a latency of two instruction cycles and a delay of one instruction cycle are introduced in multiply operations.

Latency is the number of cycles between the execution of two consecutive instructions on the same functional unit. Delay is the number of cycles until the result is ready. In data-load instructions (e.g., LDB and LDH), five execution phases (E1 to E5) are required to calculate the memory address, point to the data, read the data, and load it into a register. Thus, a latency and delay of five instruction cycles is introduced in the load instructions. Finally, the branch instruction (B) needs only one execution phase, E1, but reaches its target destinations five cycles later. Therefore, branch instructions have a latency of six instruction cycles.

Programming Environment

There are three methods for writing code for C6000 processors. The first way is writing in C, which is the preferred programming method for C6000 processors. The C compiler for the C6000 has improved tremendously to achieve high efficiency for some applications. In the second method, programmers write code in linear assembly and use the assembly optimizer to schedule the code and take care of pipelining, functional units, and parallelism. The third method is to write the code completely in assembly language, which gives programmers the best efficiency if the code is written optimally, though this process can be very time consuming and tedious.

As with the C5000, the programmer needs to determine the compiler-optimization level when programming in C so that the written C code is able to run fast enough to meet real-time requirements. In C6000 programming, there are several methods for improving the performance of C code by refining the code as follows:

- Using intrinsic functions to replace complicated C code
- Pipelining the instruction manually or using optimizer (similar to using the −o2, −o3 compiler option)
- Using word access to operate on 16-bit data stored in the high and low parts of a 32-bit register

The C6000 compiler provides a set of special functions that map directly to in-lined C6000 instructions. These intrinsic functions are represented with a leading underscore _ and can be accessed easily by being called just like a function call in C. Some commonly used intrinsic functions include _add2, _clr, _ext, _mpy, _sadd, _norm, _sat, _set, and _sub. A complete listing of intrinsic functions can be found in the *TMS320C62x/C67x Programmer's Guide* [13].

Software pipelining is a technique for scheduling instructions in a loop to execute multiple iterations of the loop in parallel, thus adding delay slots and, at the same time, maximizing the usage of functional units. This optimization can be executed automatically using the compiler options −o2 or −o3 or by using the assembly optimizer. Note that the C62x does not support hardware looping and that all loops must be implemented in software. In the C64x processor, hardware-assisted software looping is provided. Two new instructions, BDEC and BPOS, are used to implement nested loops.

Code Efficiency

Loop unrolling is a method for improving the performance of looping in a program. Loop unrolling reduces the iterations of a loop by adding copies of the loop code.

This method increases the number of instructions to be executed in parallel at the expense of increasing code size. A simple example to illustrate loop unrolling is given as follows:

```
; program A: original code without loop unrolling
          MVK   4,B0
loop:
          LDH   *A5++,A0
    ||    LDH   *A6++,A1
          ADD   A0,A1,A2      ; add four times
          SUB   B0,1,B0
    [B0]  B     loop

; program B: code with loop unrolling once
          MVK   2,B0
loop:
          LDH   *A5++,A0
    ||    LDH   *A6++,A1      ; add first two numbers
          ADD   A0,A1,A2
          LDH   *A5++,A0
    ||    LDH   *A6++,A1      ; add the other two numbers
          ADD   A0,A1,A2
          SUB   B0,1,B0
    [B0]  B     loop
```

Program A is the original code with a loop count of 4, while program B reduces the loop count to 2 by repeating the loop instructions twice within the loop. Program B can be extended further by loop unrolling into a complete straight-line code without any loop. However, loop unrolling can only achieve a speed gain up to a certain number of loop unrollings. Beyond that, it may require larger code size without any increase in speed.

When operating on a series of short data, code can be written to take advantage of accessing two short data at a time using word access. This SIMD optimization is an effective way for improving the performance of code. In C64x processors, programmers can take further advantage of the 64-bit register and operate on four short data or two word data at any one time. However, care must be taken in ensuring that the loop runs for an even number of iterations. In addition, the data must be word aligned. Intrinsic functions are available to access this dual (or quad) data.

For a C program to call an assembly routine, a common coding interface is required to pass data from the C program to the assembly routine. The C compiler assigns the registers that are used to store the return value, return address, and passing arguments, as shown in Fig. 4.31. Note that the return value is stored in register A4, which overwrites the first argument from the C program to the assembly code. The return address from the assembly code, then is written to the B3 register.

For example, the following C program, `mainroutine.c`, calls the assembly routine `addroutine.asm`:

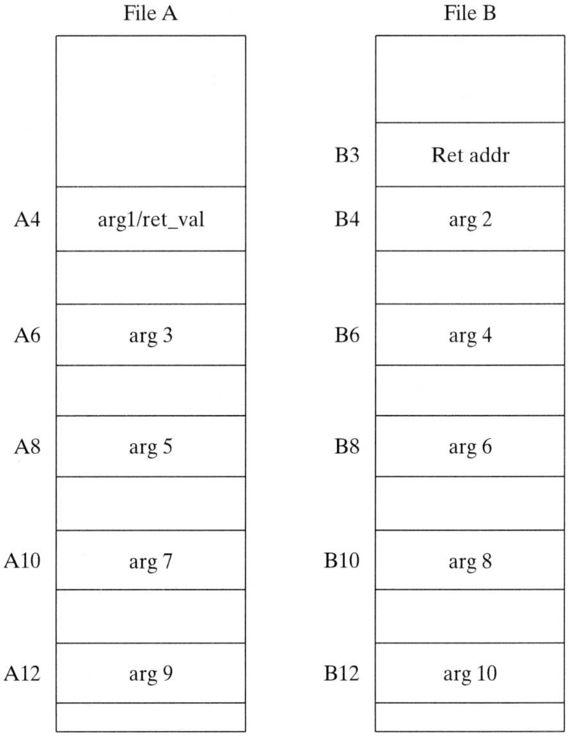

File A   File B

A4	arg1/ret_val	B3	Ret addr
		B4	arg 2
A6	arg 3	B6	arg 4
A8	arg 5	B8	arg 6
A10	arg 7	B10	arg 8
A12	arg 9	B12	arg 10

Figure 4.31   Usage of registers by C compiler

```
;  mainroutine.c
int addroutine(int,int);
int x = 3, y = 4;

void main(void)
{
  y = addroutine(x,y)
}

;  addroutine.asm
   .global         _addroutine
_addroutine:
   add    a4,b4,a4
   b      b3
   nop    5
```

Note that the registers used in the assembly routine and underscore must be written before addroutine to indicate that this variable has been called by the C main function.

The assembly optimizer allows programmers to write assembly code without concerning the pipeline structure, resource (e.g., functional units and registers) allocation, and parallel construct. This simplified assembly code is called linear assembly code, which allows programmers to use symbolic variables just like they do with

C. Programmers just need to insert two assembler directives, .cprog and .endproc, within the code to mark the assembly code to be optimized by the assembly optimizer. (See the example that follows.) Additional features such as passing and returning arguments to and from the assembly code can be easily incorporated into the assembly code as shown in the example. Furthermore, variable names for data and pointers can be assigned automatically to registers by using the .reg assembler directive.

```
_routine        .cprog      pa, pb, count
                .reg        a, b, sum, accsum
                zero        sum
                zero        accsum
loop:
                LDH         *pa++,a
                LDH         *pb++,b
                ADD         a,b,sum
                ADD         sum,accsum,accsum
                SUB         count,1,count
       [count]  B           loop
                .return     accsum
                .endproc
```

4.5.6 System Issues

As shown in Figs. 4.24 and 4.25, the C6000 processor contains several peripherals for interfacing with off-chip devices such as memory, another C6000 processor, a host processor, a CODEC, and serial devices. The general functionalities of these peripherals are explained in Chapter 3. In this section, we only describe the features related to C6000 processors.

The HPI is 16 bits for both C62x and C64x and 32 bits for C64x processors only. Therefore, the HPI in the C64x can operate in two modes: HPI-16 and HPI-32. External devices such as another C6000 processor and host processor can access the internal memory of the C6000 processor via the parallel HPI port. In HPI-16 mode, the host device can read or write to these registers by writing the desired address into the HPI address register and then performing two 16-bit reads or writes to the HPI data register. In HPI-32 mode, a single 32-bit access is possible.

The expansion bus (XB) provides a channel for another processor to access the C6000's internal memory. XB is similar to HPI, but it is 32 bits wide and has the flexibility of operating in both asynchronous (only in HPI) and synchronous modes. These features allow greater flexibility in interfacing to a wider range of devices. The PCI interface also provides standard master/slave 32-bit, 33 MHz PCI operations. In the C6205, the HPI is replaced by the 32-bit PCI interface. All C6000 processors have shared general-purpose input/output (GPIO) pins with unused peripheral pins. In C64x processors, 8 or 16 dedicated GPIO pins are available.

The DMA controller handles the movement of data from one block of memory to another without CPU intervention. The data transfer can be triggered by any interrupt. C62x processors provide a four-channel DMA controller and one extra channel dedicated to HPI. In addition, the C6211 processor has an enhanced DMA capability, which supports up to 16 channels plus 1 channel for the HPI, synchronization, and autoinitialization. Advanced C64x processors support the enhanced DMA and 32 channels that are chainable. The C64x also provides twice the maximum external DMA bandwidth compared to the C62x.

There are two to three full-duplex McBSPs in C62x processors. Each serial port supports wordlengths of 8, 12, 16, 20, 24, and 32 bits and supports up to a 100 Mbps transfer rate. Each McBSP can be programmed independently for internal or external frame synchronization and bit clocking. The McBSP also supports automatic A-law or μ-law companding for 8-bit transfers, thus allowing interoperability with different interface standards. Both C62x and C64x processors support 128 channels through time-division multiplexed data streams. The main difference is that the C62x can activate 32 channels at any one time, compared to the 128 active channels of the C64x processor. UTOPIA is also featured in C6415 and C6416 processors. The peripheral supports the UTOPIA level-2 interface as a slave asynchronous-transfer-mode controller.

The C62x and C64x have two and three 32-bit timers, respectively. Each timer has dedicated input and output pins and an external signal or an internal clock running at 1/4 of the CPU clock rate. The input pin is used for counting external events, while the output pin is used to generate a periodic pulse (or square wave). This output can be used to interrupt the CPU when the count register reaches the timer period.

The PLL within the C6000 processor provides necessary clock scaling. For example, the C6211 processor supports clock-multiplier factors of 1 and 4, whereas the C64x processor has multiplier factors of 1, 6, and 12. Therefore, C6000 processors can generate different output clock signals with reference to the input clock frequency.

Three power-down modes are available in C6000 processors. Power-down-1 mode (IDLE 1) halts the clock, but allows PLL and other peripherals to continue running. The CPU can be awakened through an external or internal interrupt. In power-down-2 mode (IDLE 2), both the core and peripherals are turned off, but the PLL is still running. Only the external interrupt can wake up the processor in this mode. Finally, power-down-3 mode (IDLE 3) shuts down the processor completely. The only way to wake up the processor from IDLE 3 is to reset it.

There are other special peripherals in the latest version of C64x processors. For example, the C6416 processor incorporates additional coprocessors, such as a VCP and a TCP, to speed up wireless voice and data coding. The VCP supports more than 500 voice channels at 8 kbps and allows the user to program the decoder parameters. The TCP supports 35 data channels at 384 kbps and supports turbo coder. With the increased usage of C6000 processors in many different applications, we expect to see more on-chip peripherals in the latest versions of the C6000 family.

4.6 EXPERIMENTS AND PROBLEMS

In this section, we re-examine the experiments described in Section 3.7.3 using C5000 assembly code. The assembly program provides full control of the processors' resources, such as memory, registers, arithmetic modes, and operations, thus resulting in increased coding efficiency. This section introduces the steps of building and executing C54x and C55x assembly programs under the CCS environment. The same linker-command file exp4c54.cmd (or expc55.cmd for the C55x) is used for all of the experiments in this section. Before building the project, users can include the label start from the **Project → Build Option → Linker → Code entry point** to ensure that the PC points to the start of the program.

The assembly program has a flow similar to the intrinsic functions used in Chapter 3 for experiments, except that it provides full control to the programmer in defining registers, defining addressing modes, and setting control registers. For example, the user is allowed to add a long series of numbers with intermediate overflow using assembly. This capability is not possible with the intrinsic function _sadd, which sets the OVM (or SATD in the C55x) status bit to 1 to turn on OVM mode. However, the readability and portability of using assembly instructions is greatly reduced compared with C code. Therefore, we have to use a different assembly program for C55x experiments.

4.6.1 Representing Different Q Formats

In this experiment, we represent a series of numbers in Q.15, Q1.14, and Q2.13 formats using assembly code, as shown in the program listing of exp4c54_1.asm (or exp4c55_1.asm for the C55x). This source file and the linker-command file are added into the project, compiled, and run under the CCS using the steps described in Section 1.7. The exp4c54_1.asm program is listed as follows:

```
                .def    x,y,init_q15,init_q1_14,init_q2_13
x               .usect "vars",12  ; reserve 12 locations for x
y               .usect "vars",3   ; reserve 3 locations for y

                .sect   "table"
init_q15        .int    16384, 8192, 4096, 2048
init_q1_14      .int    8192, 4096, 2048, 1024
init_q2_13      .int    4096, 2048, 1024, 512

start:   ; copy data to vector x using indirect addressing mode
    NOP
copy:
    STM    #x,AR1          ; AR0 pointing to x0
    RPT    #11             ; repeat next instruction 12 times
    MVPD   init_q15,*AR1+  ; copy 12 data samples to x

* mode setting
    .mmregs                ; memory-mapped registers
    SSBX   SXM             ; set sign extension mode
    SSBX   FRCT            ; set operation in fractional mode
    SSBX   OVM             ; set overflow mode
* add data samples

add_q15:
    STM    #x,AR2
    RPTZ   A,#3            ; add 4 samples in Q.15
```

```
          ADD     *AR2+,16,A
          STH     A,*(y)
add_q1_14:
          STM     #x+4,AR2
          RPTZ    A,#3              ; add 4 samples in Q1.14
          ADD     *AR2+,16,A
          STH     A,*(y+1)
add_q2_13:
          STM     #x+8,AR2
          RPTZ    A,#3              ; add 4 samples in Q2.13
          ADD     *AR2+,16,A
          STH     A,*(y+2)
end:
          NOP
          B       end
```

As shown in the program, the C54x instruction SSBX (or BSET for the C55x) is used to set status bits. The fractional mode bit mnemonic FRCT is the same for both the C54x and C55x; however, the saturation mode bit is OVM for the C54x, but is SATD for the C55x. In addition, the sign-extension-mode bit is SXM for the C54x, but is SXMD for the C55x.

In order to verify the assembly code, the project is built, and the executable file is loaded into the C5000 simulator. The detailed step-by-step procedure is given in Section 1.7. Click on the run icon ☒ to execute the program. The user can examine the results in memory by clicking on **View → Memory**. The **Memory Window Options** dialog box allows the user to enter the address for y and set the format to **16-Bit Signed Int**, as shown in Fig. 4.32.

A similar observation and explanation, as reported in Chapter 3, are shown in Fig. 4.33. The results are 30,720 for Q.15 format, 15,360 for Q1.14 format, and 7,680 for Q2.13 format.

As discussed in Section 3.7.3, it is important to interpret the data and results used in the program. The same hex representation can mean different values in different Q formats. For example, the number 16,384 is equal to 0.5 in Q.15 format, but

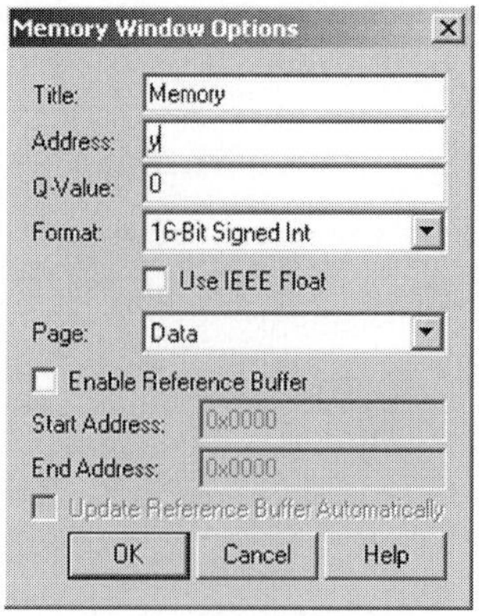

Figure 4.32 Setting of the **Memory Window Options** dialog box to view the result of y

```
0080:   x
0080:    16384    8192    4096    2048
0084:     8192    4096    2048    1024
0088:     4096    2048    1024    512
008C:   y
008C:    30720   15360    7680    0
0090:       0       0        0     0
0094:       0       0        0     0
0098:       0       0        0     0
```

Figure 4.33   Results of the Q.15, Q1.14, and Q2.13 additions

it represents 1 in Q1.14 format and 2 in Q2.13 format. This fact can be more crucial if mixed-Q formats are used in the same program, and care must be taken by the programmer to perform different arithmetic manipulation for using different Q formats in the same program.

4.6.2 Coefficient Quantization

As stated in Section 3.3.2, we can represent filter coefficients in fixed-point DSP processors by multiplying the floating-point coefficients derived from MATLAB with the correct scaling factor. In this experiment, we multiply the filter coefficients listed in Table 3.10 with a scaling factor of 32,768 and store them in the coeff_q15 array. The program performs a series of multiply-accumulate operations on these coefficients with the constant value #7FFFh (0.999969482 in Q.15 format). The final sum of the products is stored in the variable yq15. The C55x assembly program exp4c55_2.asm (or exp4c54_2.asm for the C54x) used for this experiment is listed as follows:

```
            .def      coeff_q15,yq15,init_q15
coeff_q15.usect     "vars",40    ; reserve 40 locations for
                                 ; coefficients
yq15       .usect     "vars",2    ; reserve 2 locations for y

           .sect     "table"
init_q15 .int       -141,-429,-541,-211,322,354,-215,-551
           .int       21,736,332,-841,-871,755,1651,-304
           .int       -2880,-1106,6139,13157,13157,6139,-1106,-2880
           .int       -304,1651,755,-871,-841,332,736,21
           .int       -551,-215,354,322,-211,-541,-429,-141
           .text                 ; create code section
           .def  start       ; label of the beginning of code

   start     ; copy data to vector x using indirect addressing mode
      BCLR   C54CM     ; set C55x native mode
      BCLR   AR1LC     ; set AR1 in linear mode
      BCLR   AR2LC     ; set AR2 in linear mode
      BSET   SXMD      ; set sign extension mode
      BSET   FRCT      ; set operation in fractional mode
      BSET   SATD      ; set saturation mode

   copy
      AMOV   #coeff_q15,XAR1    ; AR1 pointing to x0
      AMOV   #init_q15,XAR2     ; XAR2 pointing to table of data
```

```
        RPT    #39                      ; repeat next instruction 40 times
        MOV    *AR2+,*AR1+              ; copy 40 data to coeff_q15

    mult_add_q15    ; scale the 40 data samples
        AMOV   #coeff_q15,XAR2
        AMOV   #yq15,XAR1
        MOV    #0,AC0
        MOV    #0x7FFF<<16,AC1
        RPT    #39
        MACM   *AR2+,AC1,AC0
        MOV    AC0,dbl(*AR1)

    end
        NOP
        B      end
```

We obtain similar results compared with those shown in Section 3.7.3, which can be verified by viewing the memory of *(yq15) and *(yq15 + 1). The most significant word (bits 31-16) of the AC0 contains the most significant word of the fractional number, while the least significant word (bits 15-0) contains the least significant part of the result. In Q.15 format, only the most significant word needs to be saved, and the least significant word can be ignored. Since multiplication-add operations are performed in Q.15 format, left-shifting by 1 bit of the product is required to get rid of the extra sign bit. This process is done automatically by setting the fractional mode using the instruction BSET FRCT (or SSBX FRCT for the C54x).

Assembly code has the option of turning off the SATD mode for the C55x (or OVM for the C54x) to allow intermediate overflow. When the SATD is cleared to 0, intermediate overflow is allowed. A technique to examine if the result is wrong when adding a long sequence of N signed fractional numbers is to clear the SATD bit to 0 when performing the initial $(N-1)$ additions and to set the SATD bit back to 1 when performing the final addition. Check whether the same result occurs when using the preceding method. Another example of overflow is shown in the next section.

In addition, the intrinsic function _smac (used in Section 3.7.3) using Q.15 fractional arithmetic has restricted the result to be left-shifted by one after every multiplication, which creates a problem for non-Q.15 fractional multiply-add operations. Therefore, assembly code gives the user the flexibility to perform arithmetic with different formats and modes. The user is encouraged to modify the preceding assembly code to represent the filter coefficient using Q.9 and Q.7 formats. Examine and compare the results using different Q formats.

4.6.3 Overflow Handling

This experiment illustrates a case when intermediate overflow is allowed during a series of additions. In the assembly program exp4c54_3.asm (or exp4c55_3.asm for the C55x), a series of numbers is declared as an integer. In the program, we add eight 16-bit integers with the OVM (or SATD for the C55x) bit set to 1, thus producing a saturated 16-bit result. Alternatively, we can clear the OVM bit to 0 (to allow overflow). We then can build the project, load the executable file, and run the code. The result y can be viewed in a memory window. We obtained two results: $y = -16,384$ when OVM = 1 and $y = -9,831$ when OVM = 0. Which is the correct result, and what caused the incorrect result?

```
            .def    x,y,init
    x       .usect  "vars",8        ; reserve 8 locations for x
    y       .usect  "vars",1        ; reserve 1 location for y

            .sect   "table"
    init .int       19660, 19660, 6554, -9830, -32768, -29491, 9830, 6554
                    ; adding these numbers: 0.6+0.6+0.2-0.3-1-0.9+0.3+0.2
            .text      ; create code section
            .def    start   ; label of the beginning of code
    start:    ; copy data to vector x using indirect addressing mode
        NOP
    copy:
        STM     #x,AR1      ; AR0 pointing to x0
        RPT     #7          ; repeat next instruction 8 times
        MVPD    init,*AR1+  ; copy 8 data to x
    add:
        SSBX    SXM         ; turn on sign-extension mode
        ;RSBX   OVM         ; turn off(RSBX) overflow mode
        SSBX    OVM         ; to view the different in the result
        STM     #x,AR2      ; DP pointing to vector x0
        RPTZ    A,#7        ; A = 0, set repeat counter to 8
        ADD     *AR2+,16,A  ; shift 16 bits to left before adding
    write:    ; write the result to memory location y
        STH     A,*(y)      ; y <- A
    end:
        NOP
        B       end
```

We can also check whether the arithmetic results in overflow by observing the OVA status flag, which is displayed in the CPU register window. If the OVA flag is 0, the final result in ACC A is correct; otherwise, the result has overflowed. If ACC B is used, the arithmetic overflow is indicated by the OVB flag. For the C55x, the overflow flags are ACOVn, n = 0, 1, 2, 3 for ACn. In addition, we can use the previous method, which adds the first $(N - 1)$ numbers under OVM = 0 and sets OVM = 1 for the final addition. This method allows intermediate overflow and ensures that the last addition has not overflowed. The user can modify the preceding code to ensure the correct result is obtained.

4.6.4 Scaling and Saturation Mode

This experiment shows the effect of scaling sinusoidal-input values that are larger than the dynamic range allowed in Q.15 format. The original data samples stored in the array init are left-shifted by 16 bits before being stored back to the same memory locations. This operation is carried out using the instruction LD *AR2,16,A, where register AR2 points to the variable x. OVM mode is set using the instruction SSBX OVM to allow saturation. The assembly program `exp4c54_4.asm` (or `exp4c55_4.asm` for the C55x) is listed as follows:

```
        .def    x,s,init
x       .usect  "vars",8; reserve 8 locations for x

        .sect   "table"
init    .int    0,5792,8191,5792,0,-5792,-8191,-5792; value of x
        .text           ; create code section
        .def    start   ; label of the beginning of code
start:     ; copy data to vector x using indirect addressing mode
    NOP
copy:
    STM   #x,AR1       ; AR0 pointing to x0
    RPT   #7           ; repeat next instruction 8 times
    MVPD  init,*AR1+   ; copy 8 data samples to x
scale:
    .mmregs
    STM   #7,BRC
    ;RSBX  OVM         ; OVM is not set, overflow will occur
    SSBX  OVM          ; OVM is set
    STM   #x, AR2
* add the first 4 data using direct addressing mode
    RPTB  done-1       ; block repeat is used to execute
    LD    *AR2,16,A    ; the following 3 instructions 8 times
    SFTA  A,3
    STH   A,*AR2+
done:
end:
    NOP
    B     end
```

In contrast to Section 3.7.3, which requires the intrinsic function to be changed from int type to long type in order to emulate OVM = 0, we just need to clear the OVM to 0 by using the instruction RSBX OVM (or BCLR SATD for the C55x). Figure 4.34 shows the scaled-data samples when OVM mode is turned off (Fig. 4.34a) and turned on (Fig. 4.34b). When clearing the OVM bit to 0, any overflow after left-shifting results in different sign values, as shown in Fig. 4.34(a). When OVM mode is set to 1, data samples are clipped to the maximum positive (#7FFFh) or maximum negative (#8000h) values. Therefore, the polarity of the original signal is maintained, as shown in Fig. 4.34(b).

4.6.5 Rounding

In this experiment, we explore the rounding scheme available in C5000 programming. Several C5000 assembly instructions round results. For example, C54x instructions multiply with rounding (MPYR), multiply-add with rounding (MACR), multiply-subtract with rounding (MASR), etc. As introduced in Sections 4.3.3 and 4.4.4, these instructions end with the letter R.

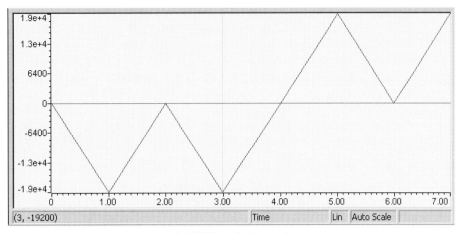

(a) OVM mode clear to 0

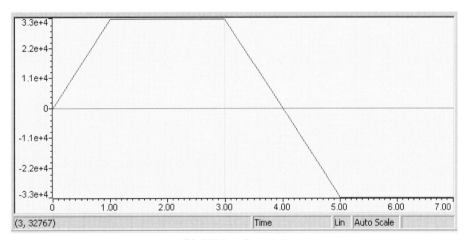

(b) OVM mode set to 1

Figure 4.34   Effects of turning OVM mode (a) off and (b) on

As explained in Section 3.3.4, rounding does not result in undesired bias as compared with truncation, which simply ignores the contribution from the lower word. We can illustrate this fact by using the assembly program `exp4c54_5.asm` (or `exp4c55_5.asm` for the C55x), which multiplies two numbers 0.26 and 0.79, and observing the difference between the results of using the multiply-with-rounding instruction, `MPYR`, and the multiply-without-rounding instruction, `MPY`.

4.6.6 Extracting Guard Bits

This experiment examines a technique for preventing overflow without using saturation mode or scaling. This method uses the guard bits of the accumulator to store overflow values during consecutive additions, as described in Section 3.3.3. In other words, guard bits provide the headroom required for preventing overflow.

In this experiment, we consider the worst case, which adds the most positive number in a signed 16-bit representation (32,767 or #7FFFh) 255 times. The accumulator is also preloaded with #7FFF 0000h before the additions. It is important that overflow mode is turned off to allow overflow into the guard bits. Otherwise, the value of the accumulator is saturated to #7FFF FFFFh and the result is incorrect. The assembly program exp4c54_6.asm (or exp4c55_6.asm for the C55x) used for the experiment is listed as follows:

```
        .def    yl,yh,yg
yl      .usect "vars",1      ; reserve 1 location for AH
yh      .usect "vars",1      ; reserve 1 location for AL
yg      .usect "vars",1      ; reserve 1 location for guard bits
        .text                ; create code section
        .def    start        ; label of the beginning of code
start:
   NOP
add:
   .mmregs
   SSBX   SXM               ; turn-on sign extension
   ;SSBX OVM                ; turn-on overflow mode
   RSBX   OVM               ; turn-off overflow mode (allow overflow)
   LD     #7FFFh,16,A
   RPT    #254
   ADD    #7FFFh,16,A
write:                       ; write the result to memory location y
   STH    A,*(yh)           ; y <- AL
   STL    A,*(yl)           ; yh <- AH
   STH    A,-16,*(yg)       ; yg <- AG
end:
   NOP
   B      end
```

Figure 4.35 shows the result displayed in the CCS memory window. In the case of fractional representation, guard bits can be interpreted as integer bits, with the MSB used as the sign bit. Therefore, the 40-bit accumulator contains 1 sign bit, 8 integer bits, and 31 fractional bits. In our case, the result is interpreted as 255.992, which is close to the actual value of 256. It is also noted that further addition results in overflow. However, it must be stated that we are considering a very extreme case of adding #7FFFh 256 times. In normal operation, this case may never occur, and 8 guard bits are more than sufficient for preventing overflow.

4.6.7 A Special Case of Multiplication

In this final experiment, we investigate a special case in Q.15 multiplication (that involves multiplying −1 by −1). The correct answer should be +1, but +1 cannot be presented in Q.15 format. The closest answer is to saturate the result to the largest positive number 7FFFh (or 0.99996948 in Q.15 format). In order to ensure this is the

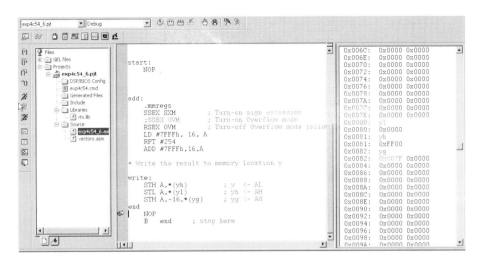

Figure 4.35   Results of carrying out addition using guard bits

case, the OVM is set to 1 before multiplication. The user can refer to the assembly file exp4c54_7.asm (or exp4c55_7.asm for the C55x) to experiment with this special multiplication using OVM (or SATD for the C55x) =1 or 0.

PROBLEMS

PART A

Problem A–4–1

In the C55x assembly program exp1c55.asm given in Section 1.7.1, the first four data are added using direct-addressing mode. Modify that part of program using indirect-addressing mode and a single-repeat instruction. Test the program using CCS and the steps given in Section 1.7.

Problem A–4–2

Redo Problem 1 using a C54x assembly program.

Problem A–4–3

An all-digital PLL with a five-point median filter and a moving-average filter is implemented using the TMS320F2407 assembly program dpll.asm in Section 4.2.6. Implement these two filters using the C55x (or C54x) and test the program on CCS using the same disk file en.dat.

Problem A–4–4

Redo Problem 3 using the C55x (or C54x) assembly program.

Problem A–4–5

Modify the C54x assembly program `exp4c54_1.asm` given in Section 4.6.1 to add the following decimal numbers using Q.15, Q.14, and Q.13 formats: 0.2, 0.4, 0.6, and 0.8. Compare the results from different Q formats.

Problem A–4–6

Redo Problem 5 using the C55x assembly program.

Problem A–4–7

Write a C54x assembly program to implement Eq. (4.2.2), where the coefficients $b_i = \{0.35, 0.725, -0.135, 1.25\}$ and the signal samples are $x(n) = \{0.85, 0.925, 0.215, 0.6\}$. How can we represent the coefficient 1.25, which is bigger than 1? How can we prevent overflow? How can we represent a result that is bigger than 1?

Problem A–4–8

Redo Problem 7 using the C55x assembly program.

Problem A–4–9

Redo Problem 7 using circular buffers for both b_i and $x(n)$.

PART B

Problem B–4–10

Describe the major hardware functional units in all of the fixed-point DSP processors mentioned in this chapter. List them in a table and compare the wordlength and precisions used in each of the functional units.

Problem B–4–11

Most fixed-point DSP processors have a 16-bit × 16-bit multiplier. What is the reason for having a 17-bit × 17-bit multiplier in C5000 processors?

Problem B–4–12

Describe the improvements of the MAC unit from the C2000 family to the C54x and then to the C55x.

Problem B–4–13

Explain how floating-point numbers are emulated in fixed-point processors. The user can refer to the references listed below to find out how floating-point numbers are being programmed in fixed-point processors.

Problem B–4–14

State the number of read and write buses in C2000, C54x, and C55x processors. Why are there more read buses compared to write buses in C5000 processors?

Problem B–4–15

> What are intrinsic functions used for, and why and when are they useful?

Problem B–4–16

> List the number of execution-pipeline stages used in C2000, C5000, and C6000 processors. Do the more stages used in the processor result in faster throughput of the processor?

Problem B–4–17

> Why is zero-overhead looping hardware crucial for DSP algorithms? In single and block repeats, what typical parameters need to be set up before performing these repeats?

Problem B–4–18

> Describe the improvements of looping architectures from the C2000 family to the C54x and then to the C55x.

Problem B–4–19

> Circular buffering is an important feature in a DSP processor. Explain how circular buffers are implemented in C2000, C5000, and C6000 fixed-point processors. How many circular buffers can be specified in these processors at any one time?

Problem B–4–20

> What types of power conservation hardware and features are being employed in the latest fixed-point DSP processors?

Problem B–4–21

> Explain the types of cache used in the latest family of C62x and C64x processors. Why are they useful compared with conventional on-chip memories?

SUGGESTED READINGS

1 Texas Instruments. *TMS320C1x/C2x/C2xx/C5x Assembly Language Tools User's Guide.* SPRU018D, 1995.
2 Texas Instruments. *TMS320F/C24x DSP Controllers, CPU and Instruction Set Reference Guide.* SPRU160C, 1993.
3 Texas Instruments. *TMS320F243/F241/C242 DSP Controllers, Systems and Peripherals Reference Guide.* SPRU276C, 2000.
4 Best, R. E. *Phase-Looked Loops: Design, Simulation, & Applications.* 3rd Ed. New York, NY: McGraw-Hill, 1997.
5 Texas Instruments. *TMS320C54x DSP CPU and Peripherals: Reference Set.* Vol. 1, SPRU131F, 1999.
6 Texas Instruments. *TMS320C54x DSP Mnemonic Instruction Set Reference Set.* Vol. 2, SPRU172B, 1998.
7 Kuo, S. M. and B. H. Lee. *Real-Time Digital Signal Processing: Implementations, Applications and Experiments with the TMS320C55x.* New York, NY: Wiley, 2001.

8 Texas Instruments. *TMS320VC5510 Fixed-Point Digital Signal Processor*. SPRS076, 2000.

9 Texas Instruments. *TMS320C55x DSP CPU Reference Guide*. SPRU371D, 2001.

10 Texas Instruments. *TMS320C55x DSP Mnemonic Instruction Set Reference Guide*. SPRU374E, 2001.

11 Texas Instruments. *TMS320C55x DSP Programmer's Guide*. SPRU376A, 2001.

12 Texas Instruments. *TMS320C55x Optimizing C/C++ Compiler User's Guide*. SPRU281C, 2001.

13 Texas Instruments. *TMS320C62x/C67x: Programmer's Guide*. SPRU198C, 1999.

14 Texas Instruments. *TMS320C6000 Peripherals: Reference Guide*. SPRU 190C, 1999.

15 Texas Instruments. *TMS320C62x/C67x: CPU and Instruction Set: Reference Guide*. SPRU189C, 1998.

16 Texas Instruments. *TMS320C64x: Technical Overview*. SPRU395, 2000.

17 Berkeley Design Technology. Benchmark results for the latest VLIW–based processors, presented at Int. Conf. on Signal Processing Application and Technology, 2000.

5

Floating-Point Digital Signal Processors

In this chapter, we introduce high-performance floating-point DSP processors with better precision, a larger dynamic range, and more on-chip memory. Floating-point processors support both integer and floating-point arithmetic, thus allowing software developers to concentrate on algorithms with minimal concerns about overflow and scaling problems. We use the TMS320C3x and TMS320C67x from Texas Instruments for introducing architectures, instruction sets, the programming, and the application of floating-point DSP processors.

5.1 INTRODUCTION

As discussed in Chapters 3 and 4, 16-bit fixed-point DSP processors such as the TMS320C2000, TMS320C5000, TMS320C62x, and TMS320C64x have limited precision and dynamic range. In fixed-point implementations, DSP processors usually need to scale the signals in order to prevent overflow. However, scaling down signal values reduces the SQNR. The methods for reducing overflow introduced in Chapter 3 must be realized with experiments using real-world data, which is the most critical, difficult, and time-consuming task for fixed-point implementations.

Floating-point arithmetic provides higher precision and a much larger dynamic range. Therefore, floating-point operations support more accurate DSP operations. For example, a 32-bit floating-point processor has 24-bit precision (mantissa) and about 1,530 dB dynamic range (exponent). Floating-point processors support floating-point operations at the speed of processing integers while preventing difficult problems (e.g., overflow, operand alignment, and signal scaling) that commonly occur in fixed-point operations. In many DSP operations such as the FFT and cascade IIR filtering, computed values may increase from one stage to the next. Using fixed-point processors (introduced in Chapter 4) would cause overflow if the incoming numbers are in full scale. In addition, quantization effects also become less of a problem with floating-point arithmetic of 24-bit precision instead of the 16-bit precision in fixed-point processors.

Due to wider instruction word size, floating-point processors support more addressing modes and a higher degree of parallel execution. High-level C language is more efficient on a floating-point processor with a register-based architecture, floating-point instructions, a large address space, powerful addressing modes, a flexible instruction set, and a faster floating-point computation engine. Floating-point arithmetic results in a smooth transition from the simulation environment on a host computer using MATLAB or C programs to a real-time implementation on actual DSP systems. In addition, many DSP algorithms are implemented using floating-point C programs that run on a general-purpose computer. These C programs can be run on floating-point DSP processors with minor modifications. Typically, floating-point DSP processors are used in high-end, high-performance applications such as video conferencing, network packet switching, cellular base stations, radars and sonar, and digital imaging.

TMS320 floating-point DSP processors consist of the C3x, C4x, and C67x. The C3x family is based on the TMS320C30, a 32-bit architecture with 40-bit extended-precision registers. These processors use a modified Harvard architecture with multiple buses for faster throughput. These processors include hardware floating-point computation units (e.g., multiplier, ALU, and shifter), fast memory management with an on-chip DMA controller, a 64-word on-chip instruction cache, and 16M words of address space. The C3x is a 32-bit processor capable of performing integer, logical, and floating-point operations of up to 150 million floating-point instructions per second (MFLOPS). The architecture and programming of the C3x are introduced in Section 5.2.

The TMS320C4x is the first DSP processor designed for parallel processing. The C4x is code compatible with the C3x processor. It includes six communication ports, a six-channel DMA controller, and two independent 32-bit memory interface units. The C40 has six serial ports, while the C44 only has four. As discussed in Section 3.6.7, a C40 can connect directly up to six C40 processors for parallel processing without any glue logic.

The most recent floating-point processor in the TMS320 family is the TMS320C67x, which was introduced in 1997. As discussed in Chapter 4, the C6000 generation is based on a VLIW architecture. This generation includes the floating-point C67x and the fixed-point C62x and C64x introduced in Chapter 4. The instruction set of the C67x is a superset of the C62x that adds floating-point capabilities to six of the C62x processor's eight functional units. Thus, it is code compatible with the fixed-point processor C62x. The processor supports the IEEE floating-point format introduced in Section 3.2.2 and provides up to 1,350 MFLOPS of performance at 225 MHz. The C67x is optimized for high performance and high efficiency in C programming. The TMS320C67x is discussed in Section 5.3.

In addition to the C3x and C67x from Texas Instruments, there are other floating-point DSP processors, such as the DSP9600x from Motorola and the ADSP-21x6x SHARC and ADSP-TS101S TigerSHARC from Analog Devices. The DSP9600x is a 32-bit highly parallel multiple-bus IEEE floating-point processor. The SHARC processor builds on the ADSP-21000 family DSP core to form a complete system-on-a-chip design, adding a dual-ported on-chip static RAM and integrated I/O peripherals supported by a dedicated I/O bus. The TigerSHARC can perform 1,500 MFLOPS at 250 MHz.

The purpose of this chapter is to introduce the architecture, addressing modes, instruction sets, and programming considerations of commonly used floating-point DSP processors. We present some coding examples to illustrate the architecture and instruction set of the processors.

5.2 TMS320C3x

TMS320C3x DSP processors (C30, C31, C32, and C33) are the first generation of 32-bit floating-point processors in the TMS320 family. These processors offer different on-chip memory and peripherals, as summarized in Table 5.1. For example, the TMS320VC33-150 with a 13-ns instruction cycle time can perform 150 MFLOPS and 75 MIPS. To create efficient programs for the C3x, we have to understand the processor architecture, addressing modes, instruction set, and programming considerations. This section focuses on the architecture and the programming of the TMS320C3x and briefly introduces its on-chip peripherals. Detailed information of the processor can be found in the *TMS320C3x User's Guide* [1].

5.2.1 Architecture Overview

The TMS320C3x is based on a Harvard architecture with separate buses for the program and the data; thus, instructions can be fetched while data is accessed. The C3x performs integer, floating-point, and logical operations using two- and three-operand instructions. The C3x's internal buses and special instruction set have the speed and flexibility to execute up to 150 MFLOPS. The C3x can perform parallel multiply and ALU operations on integer or floating-point data in a single cycle. The processor also has a general-purpose register file, a program cache, two dedicated

TABLE 5.1   Summary of On-Chip Memory and Peripherals for TMS320C3x Processors

Processor	RAM (32-bit)	ROM (32-bit)	Serial port	DMA	Timer
TMS320C30	2K	4K	2	1	2
TMS320C31	2K	0	1	1	2
TMS320C32	512	0	1	2	2
TMS320VC33	34K	0	1	1	2

auxiliary-register arithmetic units, and internal dual-access memory. The C3x optimizes execution speed by implementing frequently used operations such as block repeat, circular addressing, and multiplication in hardware.

As shown in Table 5.1, the performance of the C3x is further enhanced through its on-chip memory and memory-mapped peripherals, including a concurrent DMA controller. The block diagram of the C30 is illustrated in Fig. 5.1. Note that the ROM block is available only on the C30. On the C33, there is an additional boot loader, RAM block 2 (16K × 32 bits), and RAM block 3 (16K × 32 bits), which totals 34K × 32 on-chip DARAM to improve its performance. A 64 × 32 bits instruction cache stores repeated (or frequently used) sections of time-critical code, thus greatly reducing off-chip accesses. As a result, the code can be stored in slower, lower-cost off-chip memory. The cache frees external buses from program fetches so that they can be used by the DMA controller or other system components. The cache can operate automatically without user intervention.

The performance of the C3x is enhanced by the internal bus structure and parallelism. The separate program buses, data buses, peripheral buses, and DMA buses support program fetches, data accesses (read and write), and DMA accesses in parallel. The processor uses 24-bit addresses for accessing 32-bit instructions and data words in memory. The 24-bit program-address bus and the 32-bit program-data bus can fetch an instruction word every cycle. The 24-bit data-address buses and the 32-bit data-data bus support two data-memory accesses. The 24-bit DMA address bus and the 32-bit DMA data bus support the DMA controller. These two buses allow the DMA to perform memory accesses in parallel with the memory accesses on the program and data buses.

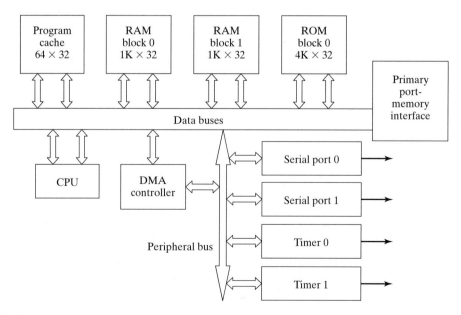

Figure 5.1   Block diagram of the TMS320C30

The most important multiprocessing configuration is the sharing of global memory by multiple processors. In order for multiple processors to access global memory and share data in a coherent manner, the C3x provides five instructions and two external I/O flags that can be configured as input and output pins under software control to support interlocked operations. Through the use of external signals, these instructions guarantee the integrity of communication.

General-purpose applications are greatly enhanced by providing large dual-access memory, a multiprocessor interface, internally and externally generated wait states, a DMA controller supporting concurrent I/O, two timers, two serial ports, and a multiple interrupt structure.

5.2.2 Central Processing Unit

The C3x has a register-based CPU architecture. As illustrated in Fig. 5.2, the CPU consists of two main functional units supported by a register file and four internal buses. Important CPU elements are listed as follows:

1. A 32/40-bit integer and floating-point hardware multiplier
2. A 32/40-bit integer and floating-point ALU
3. A 32-bit barrel shifter
4. An address generation unit that has two auxiliary-register arithmetic units (ARAU0 and ARAU1) and that is supported by eight auxiliary registers (AR0-AR7), index registers (IR0 and IR1), and a block-size register (BK)
5. Eight 40-bit extended-precision registers (R0-R7) that function as accumulators

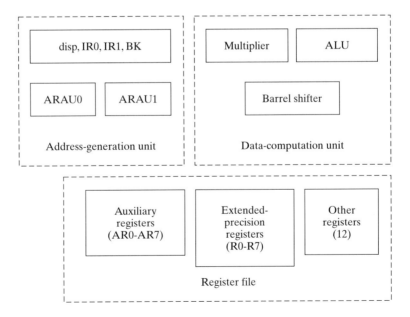

Figure 5.2   Block diagram of the C3x CPU

Computation Unit

The multiplier performs single-cycle multiplications on 24-bit integer and 32-bit floating-point values. When the multiplier performs floating-point multiplication, the inputs are 32-bit floating-point numbers, and the product is a 40-bit floating-point number. For integer multiplication, the inputs are 24 bits and the result is 32 bits. The ALU performs single-cycle operations on 32-bit integer, 32-bit logical, and 40-bit floating-point data, including integer and floating-point conversions. The barrel shifter shifts its input data up to 32 bits left (or right) in a single cycle.

C3x processors can perform parallel multiply and ALU operations on integer or floating-point data in a single cycle. The parallel instruction set includes multiply/add, add/store, load/load, or load/store and is very effective for implementing DSP algorithms.

The internal buses CPU1 and CPU2 carry two data-memory operands from memory to the multiplier, ALU, and register file every machine cycle. The internal-register buses REG1 and REG2 carry two data operands from the register file to the multiplier and the ALU. This bus structure allows parallel operations (multiplication and addition/subtraction) on four operands (integer or floating-point) in a single cycle.

Data-Address-Generation Unit

Two auxiliary-register arithmetic units (ARAU0 and ARAU1) with index registers (IR0 and IR1) can generate two data addresses for fetching two operands in a single cycle. The ARAU operates in parallel with the multiplier and ALU, which support addressing with displacements (disp), circular addressing, and bit-reversal addressing. This unit is discussed further in Section 5.2.3.

A very powerful feature of the TMS320C3x is the block-repeat capability, which allows a block of instructions to be repeated any number of times without overhead for loop control. To support single-repeat and block-repeat operations, the C3x provides the 32-bit repeat-start address register and repeat-end address register, which specify the starting address and the ending address of the loop, respectively. The 32-bit repeat-count register (RC) is used to specify the number of times a block of code is to be repeated when a block repeat is performed. To execute the loop N times, we have to load the number $N - 1$ into the RC. The repeated code runs as straight-line code, but the program size is equal to the loop code. Therefore, the most frequently used loop can be implemented in block-repeat form to reduce execution time and program size.

Register File

The C3x provides 28 registers in a multiport register file that can be accessed by the address-generation units, multiplier, shifter, and ALU. These registers support addressing, integer/floating-point operations, stack management, processor status, block repeats, and interrupts. These registers can be used as general-purpose registers; however, some registers also have special functions. For example, the PC contains the address of the next instruction to be fetched, the status register contains the status of the CPU, and the SP contains the address of the top of the stack.

Addressing Registers   The eight 32-bit auxiliary registers (AR0-AR7) support a variety of indirect-addressing modes for generating 24-bit data addresses. They can be accessed by the CPU and modified by the ARAU. They can also be used as loop counters. In addition, they can be treated as 32-bit general-purpose registers that are modified by the multiplier and ALU. The details of using auxiliary registers for generating 24-bit addresses are discussed in Section 5.2.3.

The eight LSBs of the 32-bit DP is used for direct addressing. Memory space (16M words) is portioned into 256 data pages (64K words per page), and the DP is used as a pointer to the data page. The 32-bit index registers (IR0 and IR1) are used by the ARAU for indexing/incrementing the contents of the auxiliary registers. These index registers make indirect-addressing mode very flexible. Finally, the 32-bit block-size register is used by the ARAU in circular addressing, which is discussed in Section 5.2.5.

Extended-Precision Registers   The eight extended-precision registers (R0-R7) are 40-bit registers that can be used as accumulators for supporting operations on 32-bit integer and 40-bit floating-point numbers. In some applications, these registers provide sufficient temporary storage of internal variables. Performing arithmetic operations with operands stored in these registers significantly increases code efficiency. As illustrated in Fig. 5.3, only the 32 LSBs (b31-b0) are used if the operands are integers (signed or unsigned), and bits 39-32 remain unchanged. Instructions that assume the operands are floating-point numbers use 40 bits (b39-b0), where the eight MSBs (b39-b32) are dedicated to the exponent (exp), and the 32 LSBs (b31-b0) are used for the mantissa (man), with bit 31 as the sign bit (s). The data formats are discussed further in Section 5.2.5.

5.2.3 Memory Organization and Addressing Modes

The C3x uses a 24-bit address bus. Therefore, the processor provides a unified memory space of $2^{24} = 16M$ words (32 bits). The memory maps for the microcomputer and microprocessor modes are similar, except that on-chip ROM is not used in microprocessor mode. The C3x provides a variety of addressing modes to support flexible access of data from memory, registers, and I/O ports.

The C3x supports five groups of addressing modes: general, three operand, parallel, long immediate, and conditional branch. Six types of addressing (register, direct, indirect, short immediate, long immediate, and PC relative) may be used

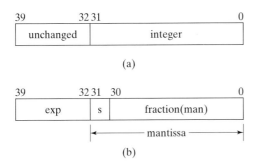

(a)

(b)

Figure 5.3   Extended-precision register formats: (a) integer format and (b) floating-point format

within each group. In addition, the C3x provides two special indirect-addressing modes (circular and bit-reversal) for supporting important DSP algorithms. Circular addressing is required by many algorithms such as convolution and correlation for implementing circular buffers in memory. This issue is discussed in Section 5.2.5. Bit-reversal-addressing mode is very useful in implementing the FFT, which is introduced in Chapter 8.

Memory Organization

The memory space for the C3x is 16M words, which allows program code, tables, coefficients, or data to be stored in either RAM or ROM. As shown in Fig. 5.1, each RAM and ROM block is capable of supporting two CPU accesses in a single cycle. The separate program buses, data buses, and DMA buses support program fetches, data reads and writes, and DMA operations in parallel. For example, the CPU can fetch an instruction and access two data values in one RAM block in parallel with the DMA loading another RAM block in a single cycle.

The memory map of a given processor depends on whether the processor is running in microprocessor mode or microcomputer mode. For example, the C30 memory map in microprocessor mode is illustrated in Fig. 5.4. In both modes, RAM block 0 is located at addresses 809800h–809BFFh; RAM block 1 is located at

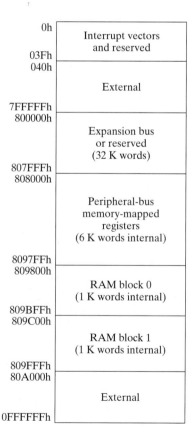

Figure 5.4   TMS320C30 memory map in microprocessor mode

addresses 809C00h–809FFFh; locations from 0h–03Fh consist of interrupt vector, trap vector, and reserved locations; and external memory is located from 80A000h–0FFFFFFh. Locations 040h–7FFFFFh are also accessed over the external memory port. In microprocessor mode, the 4K on-chip ROM or boot loader is not mapped into the C3x memory map. In microcomputer mode, locations 0C0h–0FFFh are mapped to the 4K internal ROM (C30) or the boot loader (C31).

Addressing Types

The C3x provides six types of addressing for accessing data from memory, registers, and instruction words. In register addressing, the operand is a CPU register. Floating-point operations use extended-precision registers as operands. For example,

```
ABSF    R1      ; R1 <- |R1|
```

For integer operations, any register in the register file can be used.

In direct addressing, the data address is formed by the concatenation of the 8 LSBs of the DP with the 16 LSBs of the instruction word, which is specified by @addr, where the symbol @ indicates direct addressing. In this scheme, the DP points to 1 of 256 data pages (0-255), and the 16-bit offset from the instruction word specifies the location inside the 64K-words page. For example,

```
ADD     @01B2Fh,R1 ; R1 <- R1 + (DP:1B2Fh)
```

If DP = 2 Ch, the data at the memory address 2C1B2Fh is added to the data in R1, and the sum is stored back into R1.

The most commonly used addressing mode is indirect addressing. Indirect-addressing mode uses the contents of an auxiliary register (ARn, n = 0, 1, ..., 7) with optional displacement (disp) or an index register (IR0 or IR1) to specify the address of the operand in memory. Only the 24 LSBs of the auxiliary registers (AR0-AR7) and index registers (IR0 and IR1) are used in indirect addressing. A displacement is either an explicit unsigned 8-bit integer (a value from 0-255) contained in the instruction word or a default, 1. The ARAU modifies auxiliary registers in parallel with operations within the data computation unit. The assembler syntax and operations of indirect addressing are summarized in Table 5.2, where addr denotes a memory address, disp represents displacement, IRx represents IR0 or IR1, B specifies a bit-reversed operation, and % or circ() indicates an address updated by circular addressing.

For example, *+AR0(1) means that the memory address is the value of AR0 plus 1 and that AR0 is not updated. The post-increment modifier *ARn++(disp) indicates that after ARn is used as a memory address, its value is incremented by disp. For example, the modifier *++AR0(1) generates the same memory address (AR0)+1, but AR0 is also incremented by 1. Similarly, a double minus −− decrements both the memory address and the register value based on displacement. The displacement may be replaced by an index register IRx (IR0 or IR1), where the value of IRx is used for computing the address or modifying ARn. Modifiers for circular buffers are discussed in Section 5.2.5.

TABLE 5.2 Syntaxes and Operations of Indirect Addressing

Syntax	Operation
*ARn	addr = ARn
*+ARn(disp)	addr = ARn + disp
*−ARn(disp)	addr = ARn − disp
*++ARn(disp)	addr = ARn + disp ARn = ARn + disp
*−−ARn(disp)	addr = ARn − disp ARn = ARn − disp
*ARn++(disp)	addr = ARn ARn = ARn + disp
*ARn−−(disp)	addr = ARn ARn = ARn − disp
*ARn++(disp)%	addr = ARn ARn = circ(ARn + disp)
*ARn−−(disp)%	addr = ARn ARn = circ(ARn − disp)
*+ARn(IRx)	addr = ARn + IRx
*−ARn(IRx)	addr = ARn − IRx
*++ARn(IRx)	addr = ARn + IRx ARn = ARn + IRx
*−−ARn(IRx)	addr = ARn − IRx ARn = ARn − IRx
*ARn++(IRx)	addr = ARn ARn = ARn + IRx
*ARn−−(IRx)	addr = ARn ARn = ARn − IRx
*ARn++(IRx)%	addr = ARn ARn = circ(ARn + IRx)
*ARn−−(IRx)%	addr = ARn ARn = circ(ARn − IRx)
*ARn++(IR0)B	addr = ARn ARn = B(ARn + IR0)

In short-immediate addressing, the operand is a 16-bit immediate value contained in the instruction word. For example,

```
SUBI    2,R0    ; R0 <- (R0) - 2
```

The immediate value, 2, is encoded in the instruction word and is used as an operand for execution.

In long-immediate addressing, the operand is a 24-bit immediate value contained in the instruction word. For example,

```
BR      08000h    ; branch to address 08000h
```

Long-immediate addressing mode is used for program-control instructions such as BR, BRD, and CALL. The 24-bit immediate value usually is specified as a label in assembly code.

PC-relative addressing is used for branching. For example,

```
BR      next    ; branch to the instruction with label "next"
```

The assembler takes a label or address specified by the programmer and generates a displacement. The displacement is stored as a 16-bit (or 24-bit) signed integer in the instruction word. The displacement is added to the PC during the pipeline-decode phase.

Addressing Groups

Instructions that use general addressing modes are general-purpose instructions such as ADD and MPY that use the following syntax:

```
mnemonic   src,dst   ; dst <- src operation dst
```

Four types of addressing (register, short immediate, direct, and indirect) may be used for these instructions. For example,

```
ADDF    *AR4,R0
```

In this instruction, the data in memory pointed to by AR4 is added to the contents of R0, and the sum is stored back into R0.

Two types of addressing (register and indirect) can be used as the three-operand addressing mode. Three-operand instructions with the suffix 3, such as ADDF3 and MPYF3, have the syntax

```
mnemonic   src2,src1,dst   ; dst <- src1 operation src2
```

where src1 and src2 may be registers or indirect operands. For example,

```
ADDF3   R4,R5,R0      ; R0 <- R4 + R5
```

Parallel-addressing mode is indicated by ||, which allows the parallel execution of two instructions in the same cycle. For example,

```
    ADDF3   src2,src1,dst1   ; dst1 <- src1 + src2
||  STF     src3,dst2        ; dst2 <- src3
```

Register and indirect addressing are used for parallel addressing. The limitations are similar to those present in the three-operand mode.

Instructions using conditional-branch addressing mode (e.g., B*cond*, CALL*cond*, and DB*cond*) perform a variety of conditional operations. For example,

```
Bcond    src
```

The condition field (*cond*) specifies the condition that is checked to determine whether to branch to a new location specified by src. A branch is performed if the condition is true. Two types of addressing modes (register and PC-relative) can be used for conditional-branch addressing. If the src operand is expressed in register-addressing mode, the contents of the specified register are loaded into the PC. If the src operand is expressed in PC-relative mode, the assembler generates a displacement. This displacement is stored as a 16-bit signed integer in the 16 LSBs of the branch-instruction word.

5.2.4 Instruction Set

All C3x instructions are encoded in a single word (32 bits), and most instructions require one clock cycle to execute. The instruction set contains 113 instructions organized into the following six functional groups:

1. Load and store
2. Two-operand arithmetic/logical
3. Three-operand arithmetic/logical
4. Program control
5. Interlocked operations
6. Parallel operations

Note that many instruction pairs have the same mnemonic, but use the suffix F to indicate floating-point operations and I to indicate integer operations. For example,

```
ADDF    src,dst    ; src and dst are floating-point numbers
ADDI    src,dst    ; src and dst are signed integers
```

The TMS320C3x supports conditional loads, branches, traps, calls, and returns. The C3x provides 20 condition codes (*cond*) that can be used with any of the conditional instructions, such as B*cond* src. The conditions include signed and unsigned integer comparisons, comparisons to zero, and comparisons based on the status of individual condition flags. Seven conditional flags provide information about the result of arithmetic and logical instructions. Some widely used condition mnemonics, *cond*, are EQ (equal to), NE (not equal), LT (less than), LE (less than or equal), GT (greater than), GE (greater than or equal), Z (zero), NZ (not zero) V (overflow), NV (no overflow), C (carry), NC (no carry), P (positive), N (negative), and NN (nonnegative). For example,

```
BZ     R0 ; R0 -> PC if Z flag = 0
```

In this example, the condition code Z is used with the branch mnemonic B to form the conditional instruction BZ. The CPU tests the Z flag and branches to program memory specified by R0 if Z = 0.

Load-and-Store Instructions

There are 12 load-and-store instructions, including LDF/LDI (load floating-point/integer value), STF/STI (store floating-point/integer value), PUSH (push integer on stack), and PUSHF (push floating-point value on stack). These instructions load a word from memory into a register, store a word from a register into memory, or manipulate data on the system stack. When pushing (or popping) a 40-bit extended-precision register, the high and low part of the register must be pushed or popped separately, thus requiring two instructions. Two instructions (LDF*cond* and LDI*cond*) load data conditionally, which is useful for locating the maximum or minimum value in a data set. For example,

```
LDFZ   R3,R5  ; if condition flag Z = 0, R3 -> R5
```

Two-Operand Instructions

There are 35 two-operand arithmetic and logical instructions for both integer and floating-point operations, such as ABSF/ABSI (absolute value), ADDF/ADDI (add), MPYF/MPYI (multiply), and SUBF/SUBI (subtract). For example,

```
ADDF   src,dst   ; src + dst -> dst
```

The two operands are source (`src`) and destination (`dst`). The source operand may be a memory word, a register, or a part of the instruction word. The destination operand is always a register.

Example 5.1

An input signal from a 16-bit ADC to $x(n)$ and an output signal $y(n)$ to a DAC can be implemented in the following I/O interrupt-service routine:

```
ISR:   PUSH    AR0            ; save registers into stack
       PUSH    R0             ; save both upper and lower
       PUSHF   R0             ; 32-bit of R0

*      Input signal from an ADC
       LDI     @ADCA,AR0      ; AR0 points to ADC
       LDI     *AR0,R0        ; get signal from ADC -> R0
       ASH     -16,R0         ; shift right by 16 bits
       FLOAT   R0,R0          ; convert R0 to floating point
       STF     R0,@XN         ; store R0 at x(n) for processing

*      Output signal to a DAC
       LDF     @YN,R0         ; output y(n) -> R0
       FIX     R0,R0          ; convert R0 to integer value
       LSH     16,R0          ; shift left by 16 bits
       STI     R0,*AR0        ; output signal to DAC
```

```
            POPF    R0                  ; restore upper 32-bit of R0
            POP     R0                  ; restore lower 32-bit of R0
            POP     AR0                 ; restore AR0
            RETI                        ; return to main program
```

In the routine, `LDI @ADCA,AR0` loads the address of the converters (ADC and DAC), represented by the label `ADCA`, into AR0. The number obtained from the ADC is stored in R0 using the instruction `LDI *AR0,R0`. The instruction

```
    FLOAT   src,dst
```

converts the integer operand `src` to it equivalent floating-point value and saves the result into the `dst` register. Similarly, the instruction

```
    FIX     src,dst
```

converts the floating-point number `src` to the nearest integer, and the result is loaded into the `dst` register (Rn).

In Example 5.1, we used the shift instruction supported by the barrel shifter. The arithmetic shift instruction

```
    ASH     count,dst
```

shifts the `dst` operand up to 32 bits. If `count` is greater (or less) than zero, `dst` is left- or right-shifted by the value of the `count` operand. The logical-shift instruction, LSH, is similar to the arithmetic shift, ASH, except that for right-shift operations the high-order bits of the `dst` operand are zero-filled in LSH and sign-extended in ASH.

Three-Operand Instructions

There are 17 three-operand instructions that allow the C3x to read two operands from memory or the CPU register file in a single cycle, process the data, and store the result in a register. Three-operand instructions may have two source operands and a destination operand. A source operand may be a memory word or a register, while the destination is always a register. These instructions include ADDC3 (add with carry), ADDF3/ADDI3 (add), CMPF3/CMPI3 (compare), MPYF3/MPYI3 (multiply), and SUBF3/SUBI3 (subtract). Note that all three-operand instructions can be written without the suffix 3. For example,

```
    ADDI3   R0,R1,R2
```

can be written as

```
    ADDI    R0,R1,R2
```

Program-Control Instructions

There are 16 program-control instructions that affect program flow such as repeat, call, return, and branch instructions. The repeat mode allows a zero-overhead repetition of a single instruction or a block of code. We introduce repeat operations in Section 5.2.5. The C3x supports both standard and delayed branches, as well as several instructions that are capable of conditional operations. These instructions include B*cond*[D] (branch conditionally), BR[D] (branch unconditionally), DB*cond*[D] (decrement and branch conditionally), CALL (call subroutine), CALL*cond* (call subroutine conditionally), RETI*cond* (return from interrupt conditionally), and RETS*cond* (return from subroutine conditionally). The optional suffix D after a standard branch instruction indicates delayed branching. A standard branch instruction, such as B*cond*, is executed in four cycles since pipeline flush occurs, and a delayed branch, such as B*cond*D, can be executed effectively in a single cycle. We discuss delayed branching further in Section 5.2.5.

Parallel-Operations Instructions

Parallel instructions occur in pairs and are executed in parallel in order to support a high degree of parallelism. These instructions offer parallel loading and storing of registers (e.g., LDF‖LDF, LDI‖LDI, LDF‖STF, LDI‖STI, STF‖STF, and STI‖STI), parallel arithmetic operations (e.g., MPYF3‖ADDF3, MPYI3‖ADDI3, MPYF3‖SUBF3, and MPYI3‖SUBI3), or arithmetic/logical instructions used in parallel with a store instruction (e.g., ADDF3‖STF, MPYF3‖STF, and SUBI3‖STI). Each instruction in a pair is entered as a separate source statement. The second instruction must be preceded by two vertical bars ‖. For example,

```
     LDF     *AR2--(1),R1
||STF        R3,*AR4++(IR1)

     ADDI3   *AR0--(IR0),R5,R0
||STI        R3,*AR7

     MPYF3   *AR5++(1),*--AR1(IR0),R0
||ADDF3      R5,R7,R3
```

Example 5.2

We use a C3x assembly program to implement the simple power estimator given in Eq. (2.5.10). In the following assembly code, the value of α is declared using ALPHA, and the precomputed value $(1 - \alpha)$ is represented by MALPHA [3].

```
     LDF     @POWER,R0      ; P^x(n-1) -> R0
     LDF     @MALPHA,R3     ; (1 - alpha) -> R3
     MPYF    R3,R0          ; (1 - alpha) * P^x(n-1) -> R0
     MPYF3   R6,R6,R1       ; x(n) * x(n) -> R1
     LDF     @ALPHA,R2      ; alpha -> R2
     MPYF    R2,R1          ; alpha * x(n) * x(n) -> R1
     ADDF    R1,R0          ; R0 + R1 -> R0
     STF     R0,@POWER      ; R0 -> P^x(n)
```

In this example, the signal $x(n)$ is stored in R6, and the power estimate is saved in the memory location POWER.

5.2.5 Programming Considerations

The C3x provides a complete set of operations that efficiently implement DSP algorithms for real-time applications. In this section, we discuss some important issues regarding software development.

Data Formats

In the C3x, data is organized into three types: signed integers, unsigned integers, and floating-point numbers. The C3x supports short and single-precision formats for signed integers, unsigned integers, and floating-point numbers. It also supports an extended-precision format for floating-point numbers. As shown in Example 5.1, the C3x provides instructions for converting floating point to integer (FIX) and integer to floating point (FLOAT).

Integer Formats   The C3x supports two integer formats: a 16-bit short integer and a 32-bit single-precision integer. The short integer is a 16-bit, two's complement integer for immediate integer operands in the range of -2^{15} $(-32,768) \leq x \leq 2^{15} - 1$ (32,767). For instructions that use integer operands, this number is sign-extended to 32 bits. The range of a single-precision integer x is $-2^{31} \leq x \leq 2^{31} - 1$. As shown in Fig. 5.3, when extended-precision registers (R0-R7) are used as integer operands, only the 32 LSBs are used, and the 8 MSBs (b39-b32) remain unchanged.

Similarly, 16-bit short-unsigned integers have a range of $0 \leq x \leq 2^{16} - 1$. For instructions that use this type of operand, the data is zero-filled to 32 bits. In the single-precision unsigned-integer format, the number is represented as 32-bit data in the range of $0 \leq x \leq 2^{32} - 1$. Similar to the signed integers, only the 32 LSBs are used when R0-R7 is used as unsigned-integer operands.

Floating-Point Formats   As shown in Fig. 5.3, all C3x floating-point formats (short, single precision, and extended precision) consist of three fields: an exponent field (exp), a sign bit (s), and a fraction field (man). The sign bit and fraction field may be considered as one unit called the mantissa field. The exponent is represented in two's complement, while the mantissa field is represented in normalized two's complement.

The 16-bit short floating-point format uses a 4-bit (b15-b12) exponent field and a 12-bit (b11-b0) mantissa field. The range of the numbers is $-2.56 \times 10^2 \leq x \leq 2.5594 \times 10^2$. The 32-bit single-precision floating-point format uses an 8-bit (b31-b24) exponent field and a 24-bit (b23-b0) mantissa field. The range of the numbers is $-3.4028236 \times 10^{38} \leq x \leq 3.4028234 \times 10^{38}$. Finally, in the extended-precision format, floating-point numbers use an 8-bit (b39-b32) exponent field and a 32-bit (b31-b0) mantissa field. The range of the numbers is $-3.4028236691 \times 10^{38} \leq x \leq 3.4028236683 \times 10^{38}$.

In order to achieve the highest possible efficiency in hardware implementation, the TMS320C3x uses a floating-point format that differs from the IEEE-754 standard introduced in Section 3.2.2. For some real-time applications, the input signal is obtained from the ADC, as shown in Example 5.1. The C3x converts the integer into its floating-point format, processes this signal, converts the output to integer format, and then sends it to the DAC. In this stand-alone DSP system, it does not matter what kind of floating-point format the processor uses. However, if the input signal samples are floating-point numbers from data files or other computers that use IEEE floating-point format, conversion between the different formats is required. Two versions of the conversion routines are given in the *TMS320C3x User's Guide* [1].

Subroutines and Software Stack

The CALL and CALL*cond* instructions cause the SP to increment and store the contents of the PC (which is pointing to the instruction following the call instruction) onto the top of the system stack. At the end of the subroutine, the RETS*cond* instruction performs a conditional return, which means that the top of the stack is popped to the PC if the condition is true. The programmer has to initialize the arguments of the subroutine appropriately in the main routine before the subroutine is called. The CALL src instruction transfers control to the section of program memory at the long-immediate address src, which contains the subroutine for execution, and then returns to the calling routine by executing the RETS*cond* instruction at the end of the subroutine.

At the beginning of the subroutine, the programmer has to save the critical registers used by the subroutine using PUSH instructions, as shown in the code in Example 5.1. The saved registers are stored on the system stack. Before the last RETS*cond* instruction, the programmer restores the used registers using the POP instruction. It is important to note that the pop sequence is ordered according to the reversed-push sequence; the first one pushed in should be the last one popped out.

The C3x uses the software stack in data memory whose location is determined by the contents of SP, which always points to the last value pushed on the stack. The stack is used not only during the subroutine call instructions CALL and CALL*cond*, and interrupt operations for saving the program-counter address, but also inside the subroutine (or ISR) as temporary storage for the registers to be used by that routine. There are two additional instructions, PUSHF and POPF, for saving and restoring floating-point numbers in the 40-bit extended-precision registers R0-R7. As shown in Example 5.1, using PUSH and PUSHF on the same register saves the lower 32 and upper 32 bits, and POPF followed by POP recovers this extended-precision number.

Repeat Operations

For many DSP algorithms, most of the execution time is spent in the inner kernels of code. Using the repeat mode allows these time-critical sections of code to be executed efficiently. The TMS320C3x provides two instructions for supporting zero-overhead looping: RPTS (repeat a single instruction) and RPTB (repeat a block of code).

The RPTS instruction loads all registers and mode bits necessary for the repeat of the single instruction. For example,

```
RPTS    src
```

repeats the instruction following RPTS a number of times specified by src without any penalty for looping. The advantage of the repeat operation is that the repeated instruction is fetched only once; thus, the buses are available for accessing operands. It is important to note that the RPTS instruction is not interruptible; in other words, interrupt requests are not be served during repeat operations.

Example 5.3

The sum of the products of two arrays a[i] and b[i] of length 512 can be computed, and the result is stored in R0 using the following code [1]:

```
     LDI     @addra, AR0          ; AR0 points to vector a[i]
     LDI     @addrb, AR1          ; AR1 points to vector b[i]
     LDF     0.0, R0              ; clear (zero) R0
     MPYF    *AR0++(1), *AR1++(1), R1   ; compute the first product
     RPTS    511                  ; repeat the following parallel
                                  ; instruction 512 times
     MPYF3   *AR0++(1), *AR1++(1), R1   ; compute next product
  || ADDF3   R1, R0, R0           ; accumulate the previous one
     ADDF    R1, R0               ; accumulate the last product
     ...
```

The instruction RPTS 511 loads the immediate value 511 into RC; as a result, the parallel instruction MPYF3‖ADDF3 is executed 512 times. Note that we have to compute the first product before the repeat operations since the parallel instruction accumulates the previous product. At the end of the repeat loop, we have to add the last product into R0.

The RPTB instruction repeats a block of code a specified number of times. The RC must be loaded before the RPTB instruction is executed. The typical setup of a block-repeat operation is illustrated as follows:

```
           LDI     N, RC        ; N -> RC, repeat block code N+1 time
           RPTB    endLoop      ; repeat a block of code to endLoop
           ...                  ; instructions to be repeated
 endLoop   ...                  ; the last instruction in repeat loop
```

Since the block-repeat mode modifies the PC, the last instruction in the block cannot be an instruction (e.g., B*cond* or BR) that also modifies the PC.

Three registers are associated with block-repeat instructions. The repeat-start address register (RS) holds the address of the first instruction, and the repeat-end address register (RE) holds the address of the last instruction of the block of code to be repeated. The RC contains a value one less than the number of times the block of code is to be repeated. In repeat mode, the CPU compares the contents of the RE with the PC after the execution of each instruction. If they match and the RC is nonnegative, the RC is decremented and the PC is loaded with the RS. Note that the start- and end-address registers, RS and RE, are automatically set from the code; thus, the programmer only needs to set the RC for the correct operation of the repeat modes. That is, we must load the value N into the RC before the RPTB instruction if we want to repeat a block of code $N + 1$ times.

Example 5.4

The following code uses block repeat to find a maximum of 128 floating-point numbers in the buffer [1]:

```
        ...
        LDI     127, RC        ; 127 -> RC, repeat block code 128 times
        LDI     @addr, AR0     ; AR0 points to the beginning of data buffer
        LDF     *AR0++(1), R0  ; initialize the first number as MAX in R0
        RPTB    loop           ; repeat the following block of code to loop
        CMPF    *AR0++(1), R0  ; compare the next number with MAX in R0
loop    LDFLT   *-AR0(1), R0   ; if greater, load new MAX into R0
```

Note that the CMPF src, dst instruction performs the dst-src operation. The result is not loaded into dst register R0, thus supporting nondestructive compares.

Block repeats can be nested by saving and restoring the registers RS, RE, and RC, as well as the status register that controls the repeat-mode status. Because there is only one set of repeat registers, it is necessary to save these registers before entering an inner loop. Therefore, it may be more efficient to implement a nested loop using the traditional method of assigning a register as a counter followed by a delayed branch (see Section 5.2.5) rather than using the nested repeat-block approach.

Circular Addressing

To implement circular addressing, the C3x uses the 32-bit BK to specify the size of the circular buffer. By labeling the MSB that is equal to 1 in the BK as bit $M(M \leq 15)$, the address of the top of the circular buffer can be found by setting bits M through 0 of the selected auxiliary register, ARn, to 0. The address immediately following the bottom of the circular buffer is formed by concatenating bits 31 through $M + 1$ of the ARn with bits M through 0 of the BK. A circular buffer of size R must start on a K-bit boundary, where $K \geq M + 1$ is an integer that satisfies $2^K > R$. For example, a 31-word circular buffer must start at an address whose 5 ($K = 5$) LSBs are 0, and the value 31 must be loaded into the BK.

In circular addressing, the index refers to the M LSBs of the selected ARn, and the step is the quantity added to or subtracted from the ARn. The step (unsigned integer) must be less than or equal to the block size. The ARn used to specify an address within a circular buffer must point to an element in the circular buffer the first time the buffer is addressed.

As listed in Table 5.2, the circular pointer can be modified by using *++ARn(disp)%, *−−ARn(disp)%, *++ARn(IRx)%, or *−−ARn(IRx)%. The modulus operator % indicates circular addressing. The ARAU automatically creates the desired circular buffer. After the ARn reaches the bottom or higher address of the buffer, it wraps back to the top address of that buffer. Circular buffers are very useful for implementing FIR filtering and correlation and are discussed in Chapter 6.

Delay Branches

The C3x offers three types of branch instructions: standard, delayed, and conditional. Standard branches (e.g., repeats, calls, returns, and traps) empty the pipeline before performing the branch, thus taking four cycles for execution. A standard branch should be avoided whenever possible since a delayed branch (conditional or unconditional) creates single-cycle branching.

Delayed branches (e.g., BcondD, BRD, and DBcondD) do not empty the pipeline, but they do guarantee that the next three instructions in the pipeline are executed before the PC is modified by the branch instruction. As a result, the branch requires only a single cycle, thus making the speed of the delayed branch very close to that of the block repeats. The only limitation is that none of the three instructions following a delayed branch can be a branch (standard or delayed), call, return, repeat, trap, or idle instruction.

Every delayed branch has a standard branch counterpart that is used when a delayed branch cannot be used. Sometimes a branch is necessary in the flow of the program, but fewer than three instructions can be placed after a delayed branch. For faster execution, it is still advantageous to use a delayed branch by inserting a NOP instruction. For example,

```
        ...
        LDF     *+AR1(1), R2    ; load data in memory to R2
        BGED    next            ; if R2 >= 0, branch to next after
                                ; executing following 3 instructions
        LDFZ    R2, R1          ; if R2=0, R2 -> R1
        ADDF    3.0, R1         ; R1+3.0 -> R1
        NOP                     ; dummy instruction
        MPYF    1.5, R1         ; continue here if R2 < 0
        ...
next    LDF     R1, R3          ; continue here if R2 >= 0
```

In this example, the three instructions (LDFZ, ADDF, NOP) following a delayed branch (BGED) are executed. The tradeoff is more instruction words (because of using the dummy instruction NOP) for faster execution.

5.2.6 System Issues

As shown in Table 5.1, the C3x provides two timers, a serial port (the C30 has two), and an on-chip DMA controller (the C32 has two) for supporting DSP systems. The DMA is used to perform memory and I/O operations without interfacing with the operation of the CPU. It is possible to interface the C3x processor to slow external memory and peripherals such as ADCs, DACs, and CODECs without reducing the computational throughput of the CPU. The result is improved system performance and reduced system cost.

Peripherals

All C3x peripherals are controlled through memory-mapped registers on dedicated peripheral buses that are composed of a 24-bit address bus and a 32-bit data bus. As shown in Fig. 5.1, C30 peripherals include two timers and two serial ports. The two timers are general-purpose 32-bit timer/event counters with two signaling modes. The two bidirectional serial ports are independent, and each can be configured to transfer 8, 16, 24, or 32 bits of data. The general operation of timers is introduced in Section 3.6.2, while the functions of serial ports are discussed in Section 3.6.3.

Timers   C3x timers are general-purpose, 32-bit timer/event counters. Timers can be used to signal the processor or the external devices at specified intervals or to count external events. For example, with an internal clock, the timer can be used to signal an external ADC to start a conversion or to interrupt the DMA controller to begin a data transfer. With an external clock, the timer can count external events and interrupt the CPU after a specified number of events.

Each timer has three 32-bit memory-mapped registers. The global-control register determines the operating mode of the timer, monitors the timer status, and controls the function of the timer. The period register specifies the timer's signaling frequency. Finally, the counter register contains the current value of the incrementing counter.

Serial Ports   The C30 has two independent bidirectional serial ports (the C31, C32, and C33 only have one). We can configure each port to transfer 8, 16, 24, or 32 bits of data simultaneously in both directions. Each serial port has eight memory-mapped registers. A global control register controls the global functions of the serial port and determines the operating mode, and two control registers control the functions of the six serial-port pins. In addition, there are three receive/transmit timer registers (timer control, timer counter, and timer period), a data-transmit register, and a data-receive register.

When the data-transmit register is loaded, the transmitter loads the word into the transmit-shift register, and the bits are shifted out one by one. The word is not loaded into the shift register until the shifter is empty. When the data-transmit register is loaded into the transmit-shift register, it sets a transmit-ready bit to specify that the buffer is available to receive the next data. When the serial port receives data, the receiver shifts the bits one by one into the receive-shift register. After the specified numbers of bits are shifted in, the data-receive register is loaded from the receive-shift register, and the receive-ready bit is set. The data-receive register must be read to allow new data in the receive-shift register to be transferred to the data-receive register.

Direct Memory Access Controller   As introduced in Section 3.6.4, the on-chip DMA controller can read from or write to any location in the memory space without interfering with the operation of the CPU. Therefore, the C3x can interface to slow external memories and peripherals without reducing throughput to the CPU. As illustrated in Fig. 5.5, the DMA controller contains its own address generators, source and destination registers, and transfer counter. Dedicated DMA address and data buses minimize conflicts between the CPU and the DMA controller. A DMA operation consists of a block or single-word transfer to or from memory.

A DMA transfer consists of two operations: a data read from a memory location and a data write to a memory location. The operation of the DMA is controlled

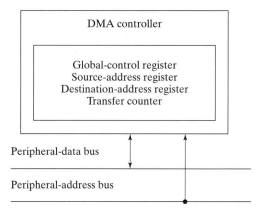

Figure 5.5   Block diagram of
the DMA controller

by the DMA global control, source address, destination address, and transfer-counter registers. The global-control register controls the operations and indicates the status of the DMA. The source (for a read) and destination (for a write) addresses can be incremented, decremented, or synchronized at the end of the corresponding memory access using specified bits. The transfer-counter register controlled by a 24-bit counter (counts down) supports the size of the block-data transfer.

Each DMA memory transfer consists of read data from the address specified by the DMA source register and write data to the address specified by the DMA destination register. At the end of a DMA read, the source address is modified as specified by the dedicated bits in the global-control register. At the end of a DMA write, the destination address is modified, and the transfer counter is decremented. DMA needs a single cycle to read and write for on-chip memory and peripherals, while off-chip reads and writes require two cycles.

Interrupt Operations

The C3x supports multiple internal and external interrupts, which can be used for a variety of applications. External interrupts on the RESET, INT0, INT1, INT2, and INT3 pins are synchronized internally. Once synchronized, the interrupt input sets the corresponding interrupt-flag register bit if the interrupt is active (enabled). Four CPU registers contain bits used to control interrupt operations: status register, CPU/DMA interrupt-enable register, CPU interrupt-flag register, and DMA global-control register. The C3x allows the CPU and DMA to respond to and process interrupts in parallel.

Interrupts on the C3x are prioritized and vectored. In microprocessor mode, interrupt vectors contain the addresses of ISRs that start execution when an interrupt occurs. A block diagram that describes the flow of an ISR is given in Section 3.4.4. In CPU interrupt processing, the corresponding interrupt flag in the IF register is cleared, and interrupts are disabled globally. The current PC is pushed onto the top of the stack. The interrupt vector is fetched and loaded into the PC, and the CPU starts executing the first instruction in the ISR. The interrupt-acknowledge instruction can be used to signal that an interrupt has been serviced.

It is important to save the processor context during the execution of an ISR. As shown in Example 5.1, the context must be saved before the execution of the ISR and restored after the routine is finished. The PC is automatically pushed onto the stack when processing a subroutine call or an interrupt. Important information in other C3x registers such as the status, auxiliary, or extended-precision registers must be saved using PUSH or PUSHF instructions.

5.2.7 An Application: Pseudo-Random Number Generator

Two basic techniques can be used for generating random numbers. The first one is the lookup-table method using a set of stored random samples, and the second one is based on the random-number generation algorithm. Both techniques generate a pseudo-random number sequence that repeats itself after a finite period; therefore, the sequence is not truly random for all time. Since the algorithm approach is suitable for floating-point processors, we introduce this algorithm and its implementations with C and C3x assembly programs in this section.

Algorithm

The linear-congruential sequence generator generates integer numbers recursively as

$$x(n + 1) = [Jx(n) + 1] \bmod M, \qquad (5.2.1)$$

where

$$J = 4K + 1 \text{ and } M = 2^L. \qquad (5.2.2)$$

The parameters K and L are integers such that $M > J$. Equation (5.2.2) guarantees that the period of the sequence in Eq. (5.2.1) is of full-length M.

The sequence tends to be more random for large values of J. A good set of parameters is $M = 2^{20} = 1{,}048{,}576$, $J = 4(511) + 1 = 2{,}045$, and the seed $x(0) = 12{,}357$. We can normalize the nth random sample as

$$r(n) = \frac{x(n) + 1}{M + 1} \qquad (5.2.3)$$

so that the random samples $r(n)$ are greater than 0 and less than 1.

Implementations

The pseudo-random number generator described in Eqs. (5.2.1) and (5.2.3) can be implemented in the C function uran.c, which is listed as follows:

```
static  long  xn = (long)12357;// seed x(0) = 12357
float uran()

{
  float ran;                  // random number r(n)
  xn = (long)2045*xn + 1L;    // x(n+1)=2045*x(n)+1
```

```
    xn -= (xn/1048576L)*1048576L;// x(n+1) mod M
    ran = (float)(xn + 1L)/(float)1048577;// normalization
    return(ran);                    // return r(n) to main function
}
```

The random numbers generated by this C function are distributed uniformly between 0 and 1, with a mean of 0.5 and a variance of $\sqrt{1/12}$. As discussed in Chapter 2, subtracting 0.5 can obtain a zero-mean sequence from the generated random numbers.

The section of the C3x assembly implementation (uran_c3x.asm) of the preceding C program is listed as follows [3]:

```
LDI     SEED,R4         ; load SEED of white noise generator
STI     R4,@WN          ; R4 -> WN
LDI     @WN,R0          ; x(n-1) -> R0
MPYI    @S2045,R0       ; 2045 * R0 -> R0
ADDI    1,R0            ; R0 + 1 -> R0
LDI     R0,R1           ; R0 -> R1
FLOAT   R1,R1           ; convert R1 to floating point
MPYF    @S1048576,R1    ; R1 * (1/1048576) -> R1
FIX     R1,R1           ; convert R1 to integer value
MPYI    @SS1048576,R1   ; R1 * 1048576 -> R1
SUBI    R1,R0           ; R0 - R1 -> R0
STI     R0,@WN          ; store new x(n) for next iteration
ADDI    1,R0            ; R0 + 1 -> R0
FLOAT   R0,R0           ; convert R0 to floating point
MPYF    @S1048577,R0    ; R0 * (1/1048577) -> R0
SUBF    @M05,R0         ; R0 - 0.5 -> R0 = r(n)
```

The parameters used in the program are defined at the beginning of the main program as follows:

```
SEED        .set    12357        ; seed for white noise generator
S2045       .int    2045         ; scaling factor 2045
S1048576    .float  9.5367432e-7 ; scaling factor 1/1048576
SS1048576   .int    1048576      ; scaling factor 1048576
S1048577    .float  9.5367341e-7 ; scaling factor 1/1048577
M05         .float  0.5          ; constant 0.5
```

Note that the parameters S1048576 = 1/1048576 and S1048577 = 1/1048577 are computed by the programmer at programming time. This technique effectively replaces the time-consuming computation of division by simple multiplication in real-time processing.

5.3 TMS320C67x

The TMS320C67x is a family of 32-bit floating-point DSP processors. Its architecture is based on a VLIW architecture, which is similar to the fixed-point

TMS320C62x and TMS320C64x processors described in Section 4.5. The TMS320C67x extends the TMS320C62x instruction set to support floating-point arithmetic. Therefore, the C67x processor is upward compatible with the C62x processor, but it is not compatible with the C64x. The C67x instruction set with added floating-point capability is a superset of the C62x fixed-point instruction set. Therefore, C62x instructions can run on the C67x processor. The C67x processor is particularly useful in applications that require high precision, a large dynamic range, and intensive computation, such as radar, sonar, 3-D graphics, wireless base stations, digital scriber loops, and medical imaging.

Similar to the C62x processor, the C67x processor can execute up to eight instructions per clock cycle. The early version of the C67x family (such as the C6701) operates at a maximum of 167 MHz clock rate, with a 1.9 V for core supply and 3.3 V for I/O supply. Some versions of C67x processors (such as the C6711 and C6712) reduce the clock rate to 100 MHz and eliminate on-chip peripherals to reduce cost.

5.3.1 Architecture Issues

The C67x processor has two floating-point data paths, a program control unit, a DMA controller, on-chip memory, and program and data memory interfaces, as shown in Fig. 5.6. The two data paths support 64-bit data for 32-bit single-precision and 64-bit double-precision floating-point arithmetic. Each data path includes a set of four execution units (.L, .S, .M, and .D), a general-purpose register file, and paths for moving data between memory and registers. The C67x floating-point processor can execute eight instructions in one cycle, which includes two single-precision multiplications or one double-precision floating-point multiplication. In addition, it can perform 16×16-bit and 32×32-bit integer multiplications.

The C67x's bus structure has been modified from the C62x to allow 64-bit data load in order to access double-precision floating-point operands. However, the processor only stores 32-bit data. There are two independent register files (A and B), each containing 16 32-bit general-purpose registers for storing addresses and data. A pair of adjacent registers can be used to hold 64-bit data for supporting 64-bit floating-point arithmetic. The time required to perform double-precision floating-point arithmetic is much longer compared with single-precision arithmetic. This multiple-cycle, double-precision floating-point arithmetic restricts the usage of cross paths, resulting in slower throughput compared with single-precision arithmetic. A cross path is used to transport data between data paths A and B. However, only one cross-path communication can be used at any one time. The C67x processor also supports conversion between fixed- and floating-point formats. In order to control the operations, control registers are used to configure the rounding mode in each unit. Status bits at each unit also indicate overflow or underflow, divide by zero, not a number, etc.

The on-chip memory system is based on a modified Harvard architecture, which provides separate program and data-memory spaces. Two data paths are connected to the data memory, and each path has a 32-bit address bus and a 64-bit data bus. Program memory has a 32-bit address bus and a 256-bit data bus. The latest generation of the C67x has the ability to configure on-chip memory as level-1 (L1) and

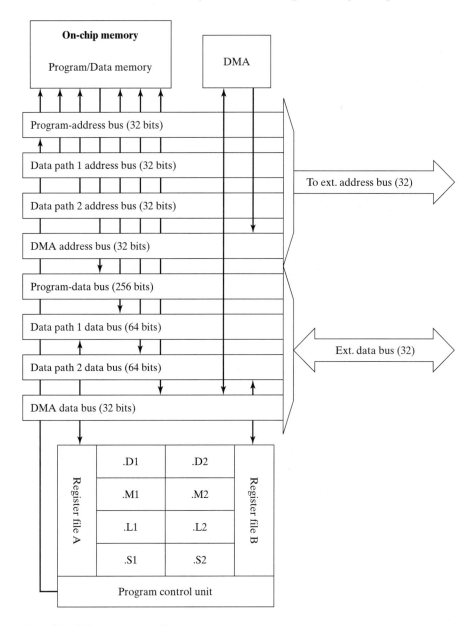

Figure 5.6 C67x processor architecture

level-2 (L2) program and data cache. Similar on-chip cache organization can be observed in the C62x and C64x processors shown in Fig. 4.24 and Fig. 4.25, respectively. The only exception is that a 64-bit (instead of 32 bits in the C62x) read path is used between the C67x core and its L1 data cache.

The C67x processor has one EMIF, which provides a 23-bit address bus and a 32-bit data bus. These buses are multiplexed between program and data accesses.

TABLE 5.3   On-Chip Memory and Peripherals for C67x Processors

Processors	On-chip memory		Peripherals and external-memory interface
	Data memory	Program memory	
C6701	16K × 32	16K × 32	A four-channel DMA, a 16-bit HPI, two BSPs, two timers, and a 32-bit EMIF
C6711	32K bits L1 cache	32K bits L1 cache	A 16-channel enhanced DMA, a 16-bit HPI, two BSPs, two timers, and a 32-bit external memory interface
	512K bits unified L2 cache		
C6712	32K bits L1 cache	32K bits L1 cache	A 16-channel enhanced DMA, a 16-bit HPI, two serial ports, two timers, and a 16-bit EMIF
	512K bits unified L2 cache		
C6713	4K bytes L1 cache	4K bytes L1 cache	A 16-channel enhanced DMA, a 16-bit HPI, two McBSPs, two timers, and a 32-bit EMIF
	64K bytes L2 cache and 192K bytes L2 SRAM		

When interfaced to a high-speed synchronous-burst static RAM via the DMA controller, the peak external-memory bandwidth is around 167 millions words (32 bits) per second for a 167 MHz processor. Other external memory such as asynchronous static RAM supports lower memory bandwidth. Table 5.3 summarizes the data, program, cache, and external-memory interface available on C67x processors. Note that the C6712 has a 16-bit EMIF as compared to the other C67x processors, which have a 32-bit EMIF. Therefore, the C6712 is not suitable for floating-point data transfer due to the bottleneck of accessing external memory.

The C67x processor also contains similar peripherals to those contained by the C62x and C64x processors described in Section 4.5.6. The last column of Table 5.3 lists the peripherals available on different C67x processors. Note that the C6000 processor does not offer dedicated I/O to facilitate multiprocessor connections.

5.3.2 Instruction Set

As discussed in Section 4.5.4, the TMS320C67x processor not only supports all of the instructions available for the C62x, but it also uses additional instructions that include 32-bit integer multiply, double-word load, and floating-point operations (e.g., addition, subtraction, and multiplication). The instruction set for the C62x was introduced in Section 4.5.4; thus, this section only introduces additional C67x-specific instructions.

Compared with other floating-point DSP processors such as the TMS320C3x, the C67x instruction performs fewer operations due to its architecture. The C67x

TABLE 5.4 TMS320C67x Instructions and Functional-Unit Mapping (adapted from [6])

.L unit	.M unit	.S unit	.D unit
ADDDP	MPYDP	ABSDP	ADDAD
ADDSP	MPYI	ABSSP	LDDW
DPINT	MPYID	CMPEQDP	
DPSP	MPYSP	CMPEQSP	
INTDP		CMPGTDP	
INTDPU		CMPGTSP	
INTSP		CMPLTDP	
INTSPU		CMPLTSP	
SPINT		RCPDP	
SPTRUNC		RCPSP	
SUBDP		RSQRDP	
SUBSP		RSQRSP	
		SPDP	

processor also requires intensive memory usage since the execution packet may consist of up to eight 32-bit instructions. In order to reduce memory accesses, the C67x processor uses several techniques such as variable-size execution packets, conditional execution, and instruction packing.

As mentioned before, the C67x instruction set is a superset of the C62x. The C67x processor provides extra instructions, as listed in Table 5.4. These new instructions can be classified into the following groups:

1. Single- and double-precision floating-point addition, subtraction, and absolute value
2. Single- and double-precision floating-point multiplication, 32×32-bit integer multiplication with a 64-bit product, and 32×32-bit integer multiplication with a 32-bit product
3. Double-word load and address calculation
4. Single- and double-precision floating-point compare
5. Single- and double-precision floating-point reciprocal estimates, reciprocal square root, fixed-point and floating-point conversions, and single-precision and double-precision floating-point conversions

The C67x processor has the same parallel-move capability as the C62x processor. In addition, the C67x processor supports 64-bit double-word loads, but it only allows 32-bit stores. Double words can be loaded at one access in the C67x processor using the LDDW (double-word load) instruction. In this case, the 32 LSBs are loaded into the even-address register, and the 32 MSBs are loaded into the next higher odd-address register.

The C67x and other C6000 processors do not support hardware looping; thus, loops must be implemented using software. Furthermore, there is no bit-reversal addressing. Similarly, the C67x processor does not provide a hardware stack. A software stack can be implemented by using any general-purpose register as a stack pointer. Push and pop operations can be realized using load and store instructions with post- or pre-increment (or decrement).

5.3.3 Pipeline Architecture

The C67x pipeline architecture is a superset of the C62x processor. The C67x pipeline adds 5 new execute stages (E6-E10), resulting in a 16-stage pipeline (in contrast to the 11-stage pipeline in C62x/C64x processors). As described in Section 4.5.5, the fetch phase includes the PG, PS, PW, and PR stages, and the decode phase is divided into the DP and DC stages. The fetch and decode phases of the C67x pipeline are identical to those of the C62x. When the instruction enters the execution phase, the CPU executes each instruction within the eight functional units. Some floating-point instructions in the C67x processor require additional delay slots (E2-E10), which comprise the additional cycle delay after the E1 stage of the pipeline.

Instruction latency is defined as the total cycles for a given instruction to be executed, which are equal to the number of delay slots (summarized in Table 5.5) plus one. For example, the six delay slots in the ADDDP instruction produces a result after a latency of 7 cycles. Functional-unit latency (FUL) is defined as the number of cycles for a given instruction to hold onto that functional unit. All of the C62x and C64x instructions require one cycle of FUL, which means that a new instruction can be started on the functional unit at each cycle. However, there are five double-precision floating-point instructions in the C67x processor that require two to four FUL cycles, as listed in the last five rows of Table 5.5. For example, the instruction ADDDP needs two FUL cycles, which only allows a new instruction to use the functional unit two cycles later.

TABLE 5.5 Delay Slots and Latency Cycles of C67x Instructions (adapted from [6])

Instructions	Delay slots
ABSSP, ADDAD, CMPEQSP, CMPGTSP, CMPLTSP, RCPSP, RSQRSP	0
ABSDP, RCPDP, RSQRDP, SPDP	1
CMPEQDP, CMPGTDP, CMPLTDP	2
ADDSP, DPINT, DPSP, DPTRUNC, INTSP, INTSPU, MPYSP, SPINT, SPTRUNC, SUBSP	3
INTDP, INTDPU, LDDW	4
ADDDP, SUBDP	6 (FUL = 2)
MPYI MPYID, MPYDP	8 (FUL = 4) 9 (FUL = 4)

5.3.4 Programming Considerations

Writing correct and efficient assembly code for C67x processors can be a very challenging task due to the complex architecture and deep pipeline. Therefore, programming in C is highly recommended for the C67x. As introduced in Section 4.5.5, the user may write the code in linear assembly (with an `.sa` extension), which is assembly code that has not allocated registers. The assembly optimizer performs the tasks of assigning registers, inserting NOP instructions automatically, and using loop optimization before passing the code to the assembler and linker. In addition, using intrinsics in C code can further enhance the efficiency of the program. A table of C6000 C compiler intrinsics can be found in the *TMS320C62x/C67x Programmer's Guide* [7]. Extensive discussion on writing parallel assembly code can be found in textbooks [8, 9].

Similar to the C62x, the C67x processor uses the same optimization methods, such as parallel optimization, filling delay slots, loop unrolling, and SIMD optimization. The optimization techniques for the C6000 processor were introduced in Section 4.5.5. The SIMD optimization is further enhanced in the C67x processor with its long data path (64 bits). For example, the C67x processor can perform LDDW, which reads 64 bits of data into a register pair. It can read two words or four short words, thus accessing two single-precision floating-point data at a time. C67x instructions also perform two 32×32-bit multiplications or four 16×16-bit multiplications simultaneously. These SIMD instructions double the performance of the C62x processor.

5.3.5 Real-Time Implementations

The Embedded Target for the TI TMS320C6000 DSP Platform [10] is software developed by The MathWorks for the TMS320C67x processor. This tool integrates MATLAB, Simulink, and Real-Time Workshop with the Texas Instruments eXpressDSP tools. In the integrated environment, Simulink and MATLAB are used to develop and verify DSP designs using blocksets and M-files. Once the design has been validated, the Real-Time Workshop is used to generate C code, and Texas Instruments development tools such as CCS with C compiler, assembler, and linker are used to build and load the executable file for the targeted C67x processors. A more in-depth overview of MATLAB tools is given in Appendix A and in references [10–13].

Currently, the embedded target for the C6000 only supports the C6701 evaluation module and the C6711 DSK from Texas Instruments. Other C6000 development boards that support communication over JTAG can also use the real-time data-exchange functional block to develop code that can run on those boards. Detailed information on Texas Instruments development tools can be found in references [14–17].

A 3-Band Graphic Equalizer

In this section, we use an example of designing and implementing a 3-band graphic equalizer to illustrate the integrated tools. As shown in Fig. 5.7, the input signal $x(n)$ is processed by a 3-band graphic equalizer. The equalizer consists of a lowpass filter

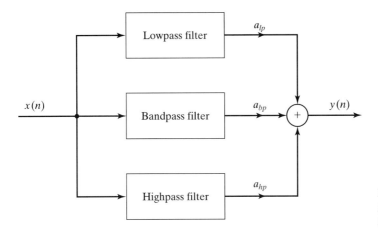

Figure 5.7   Block diagram of a 3-band graphic equalizer

with the cutoff frequency F_{lp} = 5 kHz, a bandpass filter with the cutoff frequencies F_{bp1} = 5 kHz and F_{bp2} = 10 kHz, and a highpass filter with the cutoff frequency F_{hp} = 10 kHz. The amplitude of the output signal from each filter is adjusted by the corresponding gain factor a_{lp}, a_{bp}, or a_{hp}. These tuned outputs are then summed to form the equalized signal $y(n)$.

Building a Simulink Model

The equalizer can be easily designed using Simulink by first clicking on the icon ▥ in the MATLAB window. Once the Simulink **Library Browser** is opened, the user can click on the icon ▢ to open a new worksheet for the Simulink design. Under **DSP Blockset**, select **Filtering** → **Filter Design** → **Digital Filter Design** to select the filter design block. Drag three **Digital Filter Design** blocks into the workspace to design and implement the filters shown in Fig. 5.7. Double-click on the **Digital Filter Design** blocks to design the lowpass, bandpass, and highpass filters by entering the parameters for these filters. In this experiment, the filter-design specifications are:

1. Use the equiripple FIR-filter design technique with minimum order
2. Set the sampling frequency to 48 kHz
3. Specify the passband ripple and stopband ripple to be less than 1 dB and 50 dB, respectively, for all of the filters

A more detailed explanation of the FIR-filter design is given in Chapter 6, and an example of using Simulink is given in Appendix A.3.

Once the filters have been designed, the user can add three **Gain** blocks, a **Sine Wave** block, and a **Spectrum Scope** block into the worksheet and complete the design by connecting these blocks, as shown in Fig. 5.8. The user can also open the `geq.mdl` file to see the parameters specified in each block.

Before running the simulation, the user must specify the simulation parameters by clicking on **Simulation** → **Simulation parameters**, as shown in Fig. 5.9. The simulation can be run by clicking on the ▶ icon. A window that shows the spectrum of the

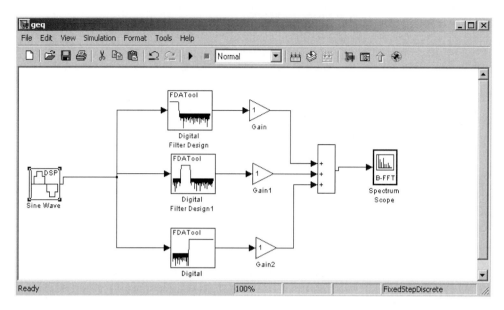

Figure 5.8 Design and implementation of a 3-band graphic equalizer using Simulink

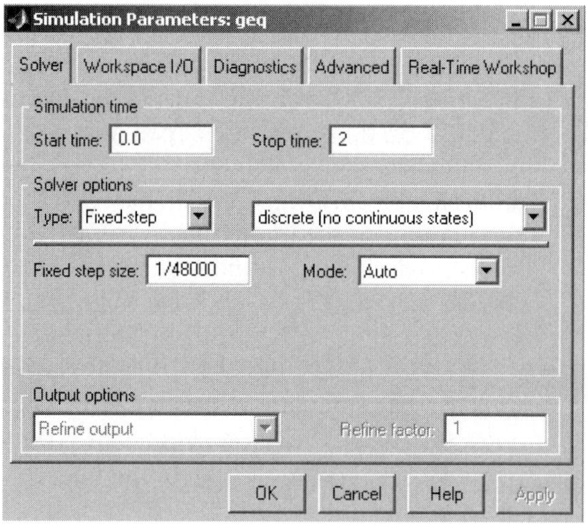

Figure 5.9 Setting simulation parameters

output signal $y(n)$ then is displayed. We can change the input signal to different frequencies and vary the gain values to alter the response of the graphic equalizer.

Real-Time Processing

Once the algorithm is working, the next step is to implement the designed equalizer on a C6000 DSK board for real-time processing. In this section, the C6711 DSK is

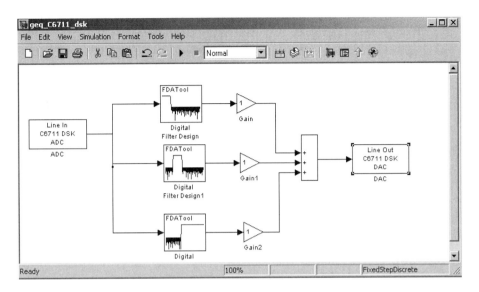

Figure 5.10   Implementation of a 3-band graphic equalizer on the C6711 DSK

used to illustrate the real-time implementation. The **Sine Wave** block and the **Spectrum Scope** block are replaced by the **Line In C6711 DSK ADC** block and **Line Out C6711 DSK DAC** block, respectively, in the modified Simulink model `geq_C6711_dsk.mdl` (Fig. 5.10). The user can double-click on these blocks to specify the ADC and DAC specifications. Note that the C6711 DSK only operates at a fixed sample rate of 8 kHz. Therefore, the previous filter designs based on a sampling rate of 48 kHz must be altered accordingly.

After all of the parameters are confirmed, the user can specify the simulation parameters for real-time implementation on the C6711 DSK. Select **Simulation →** **Simulation parameters** to open the **Simulation Parameters** dialog box. Click on **Solver** and change **Stop time** to **inf** in order to run this equalizer in real time infinitely. Click on **Real-Time Workshop** under the **Simulation Parameters** dialog box to specify the appropriate versions of the **System target file** and **Template makefile**, as shown in Fig. 5.11. With the `ti_c6000.tlc` and `ti_c6000.tmf` configurations, the user can generate real-time executable code and download it into the C6711 DSK by clicking on **Build & Run**. The template make file and the block parameters are combined to form the target make file for the C6711 DSK. This make file invokes the Texas Instruments cross-compiler to build an executable file, which automatically downloads the code into the C6711 DSK using the parallel port and runs the equalizer program.

For real-time experiments, a signal generator (or CD player) and spectrum analyzer (or speakers) are attached to the input and output ports of the DSK, respectively. The user can also analyze real-time signal processing by altering the settings of the blocks and performing the build-and-run operation again. This tool provides a quick way of prototyping the code in the target C67x platform.

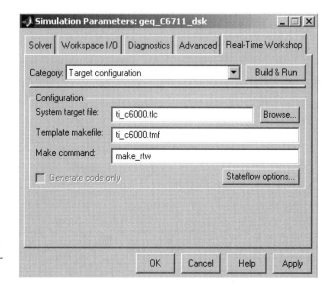

Figure 5.11   Simulation parameters for Real-Time Workshop running on the C6711 DSK

5.4 EXPERIMENTS AND PROBLEMS

In this section, we use Simulink and Fixed-Point Blockset to experiment with the differences between floating-point and fixed-point arithmetic. A brief introduction to the important features and usage of Simulink and Fixed-Point Blockset is given in Appendix A and in reference [18]. In this section, we use Fixed-Point Blockset to explore the details of fixed-point and floating-point arithmetic that mimic the performance of fixed- and floating-point DSP processors. The main objectives of these experiments are to understand the dynamic range and precision used in fixed- and floating-point processors and to prevent overflow when performing arithmetic in fixed-point processors. In the last experiment, we show how to use optimized DSP functions in C5000 DSPLIBs from Texas Instruments [19, 20] for a given application.

5.4.1 Addition and Subtraction

In this section, we examine the differences between using fixed-point and floating-point formats for performing addition and subtraction. A Simulink file ex5_sum.mdl is developed for this experiment. Simply type ex5_sum in the MATLAB command line, and the Simulink model is opened, as shown in Fig. 5.12. This simulation compares Q.15 addition in the top part of Fig. 5.12 and single-precision floating-point addition in the middle part of Fig. 5.12 with the double-precision floating-point addition in the bottom part of Fig. 5.12. The Q.15 numbers are converted from the double-precision numbers using sfix(16) and 2^{-15} as the output data type and output scaling, respectively. The single-precision numbers are formed using float('single') as the output data type. It is shown that the errors for using Q.15 and single-precision additions in this example are $\sim 3.3 \times 10^{-5}$ and $\sim 4.7 \times 10^{-8}$,

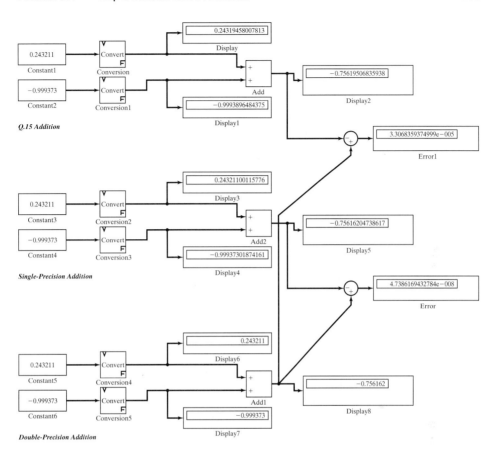

Figure 5.12   Simulink file that shows additions in Q.15, single-precision, and double-precision floating-point formats

respectively. The difference in errors is due to the precision used in different data formats.

The user can run the Simulink file ex5_sum.mdl to compare the quantization error of adding two numbers using the preceding fixed-point 16-bit and 32-bit arithmetic against the double-precision floating-point arithmetic. Modify ex5_sum.mdl, shown in Fig. 5.12, to perform the following tasks:

Step 1. Change the wordlength of the data and addition from Q.15 to Q.7. Observe and note the precision lost compared to the Q.15 arithmetic.

Step 2. Convert the precision back to Q.15 and change the first number from 0.243211 to 1.243211. Double-click on the **Conversion** block, and make sure that the option **Saturate to max or min when overflows occur** is checked. Run the simulation, and note that in the Q.15 addition the first number has changed to 0.99996948242 instead of the correct 1.243211, which produces a significant error compared to single-precision floating-point addition. Consider how to solve this problem, and consider what happens when the saturation option is unchecked.

Step 3. Use the same set of numbers as in the preceding example, and change the fixed-point number format to Q1.14 by specifying the output scaling to 2^{-14}. What happens to the fixed-point addition? The reason the result is now correct is that we have increased the dynamic range from $[+0.99996948242, -1]$ in Q.15 format to $[1.99993896484, -2]$ in Q1.14 format, which allows 1.243211 to be represented correctly.

Step 4. Check the saturation option. Numbers that exceed the dynamic range are clipped to the maximum or minimum values. Otherwise, the result from the arithmetic operation is allowed to overflow into the opposite number range, thus producing a significant overflow error. As stated in previous chapters, overflow may cause severe problems in fixed-point arithmetic and must be prevented. The reason overflow does not happen in floating-point numbers is the use of a much wider dynamic range. The user can enter the MATLAB commands `realmin` and `realmax` in order to determine the dynamic range of double-precision floating-point values used in MATLAB.

Step 5. Use guard bits to allow overflow in fixed-point DSP processors. Change the second number from -0.999373 to 0.999373 in the preceding example and perform the addition. The result is incorrect in Q.15 addition. Double-click on the **Add** block, and change the output data type to `sfrac(15,2)`, which specifies a 15-bit signed fractional data type with 2 guard bits. The wordlength is 17 bits and extends the dynamic range to $[-3.999, +4]$.

Step 6. Examine the rounding methods used in number representations. Rounding is required when converting a fixed-point number from a higher resolution to a lower resolution. The details of rounding methods were explained in Section 3.7.1. Double-click on the **Convert** block and change the **Round toward** option to **zero**, **nearest**, **ceiling**, or **floor**. Which rounding method has the smallest error, and which has the largest?

Step 7. Compare 16-bit floating-point addition with 16-bit fixed-point addition to determine that floating-point arithmetic does not always produce more accurate results. Using the preceding Simulink file, change constant numbers from 0.243211 to 100.243211 and from 0.99996948242 to -100.99996948242. Since both numbers have magnitudes larger than 1, Q.15 number format cannot be used. Instead, these numbers are represented in Q7.8 fixed-point format, which is specified in the **Conversion** blocks as `sfix(16)` and 2^{-8} for **output data type** and **output scaling**, respectively. The dynamic range of `sfix(16)` uses an output scaling of 2^{-8}, which is $[+127.99609375, -128]$.

These numbers can also be specified as 16-bit floating-point numbers with four exponential bits by using `float(16,4)` in the floating-point **Conversion** block. The floating-point number representation in MATLAB is given by

$$X = (-1)^s \times 2^{\exp-(\exp1-1)} x1.man$$

where $\exp = 2^{\exp\_bit}$, $\exp1 = 2^{\exp\_bit-1}$, and man is the mantissa bit. In this case, exp_bit must be specified as equal and above 4 in order to cater to the number range higher than ±100. In fact, the dynamic range of float(16,4) is ±512 × 1.99951171875. The details can be found in the file ex5_sum1.mdl.

5.4.2 Multiplication and Division

In this experiment, we investigate the effects of multiplication in fixed-point and floating-point arithmetic using the Simulink model ex5_mult.mdl shown in Fig. 5.13. Similar to the example for addition in the previous section, we highlight the differences of finite-wordlength arithmetic using different formats. The user can run the

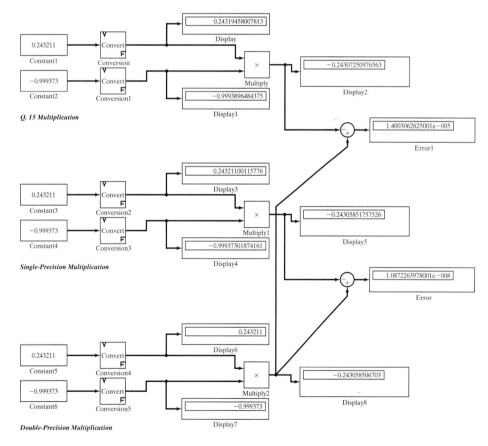

Figure 5.13   Simulink model showing multiplication using Q.15, single-precision, and double-precision floating-point formats

Simulink file and note the differences as compared to double-precision floating-point multiplication.

Modify the `ex5_mult.mdl` model shown in Fig. 5.13 for the following tasks:

Step 1. Reduce the precision of Q.15 multiplication to Q.8 multiplication. Note the difference in the results.

Step 2. Test a special case involving Q.15 fractional multiplication. As mentioned in Chapter 3, the result of Q.15 fractional multiplication is always within the number range $[+0.99996948242, -1]$. A special case involves multiplying -1 by -1, which produces $+1$ and thus cannot be represented exactly in Q.15 format. This special case can be easily tested in the Simulink by replacing both numbers with -1 and enabling the **Saturation on Integer overflow** option in the **Multiply** block. What happens to the result? What happens when the **Saturation** option is not turned on?

Step 3. Test the fact that requiring twice the wordlength for holding the product of the multiplication of two numbers, which is very important for integer multiplication, is not an issue for Q.15 arithmetic since the result is always bounded within ± 1. A simple modification of `ex5_mult1.mdl` can be used to illustrate this fact. Change the single-precision floating-point multiplication to 16-bit integer multiplication. Convert the fractional numbers to integers by using the conversion formula `floor(32768*fraction_number)`, where `floor` truncates the multiplication result. Connect these two numbers to the **Multiply** block, which is set to either `sint(16)` (result is stored in 16 bits) or `sint(32)` (result is stored in 32 bits). The integer result is scaled back to its fractional equivalent by dividing by a constant of 2^{15} or 2^{30}, which is specified in block `Constant8` for a 16-bit or 32-bit result. Examine the results using `sint(16)` and `sint(32)`, and note the requirements in performing integer multiplication.

Step 4. Change the number to be multiplied from 0.243211 to 10.243211 and from -0.999373 to -10.999373. Since the multiplication results are around 100, we can use a 16-bit fixed-point representation `sfix(16)` with an output scaling of 2^{-8} to represent all fixed-point arithmetic blocks. In addition, we can use a 16-bit floating-point representation `float(16,4)` to represent all 16-bit floating-point multiplication blocks. These modifications are carried out and saved in the new file `ex5_mult2.mdl`. Run this simulation file and observe the differences between the 16-bit fixed-point arithmetic and the double-precision floating-point arithmetic.

Step 5. Note that division is seldom used in DSP processors. If division is required, it can be implemented by using multiplication of the inversed value. Unlike multiplication, division does not require twice the wordlength for the result. A Simulink example is given in `ex5_div.mdl`.

5.4.3 Finite-Impulse Response Filtering

This section examines the FIR filter using Q.15 and single-precision formats as given in the ex5_fir.mdl file shown in Fig. 5.14. A detailed description of FIR filters is given in Chapter 6. In this example, two sinusoids at 2 and 80 Hz are mixed before being quantized to Q.15, single- and double-precision floating-point formats. These quantized signals are used as input to the corresponding FIR filters using the same format. Run the simulation and observe the results. Because this FIR filter is a low-pass filter, the 80 Hz sinewave is attenuated and only the 2 Hz sinewave is left.

Click on the **FixPt GUI** block after simulation has been completed. The **FixPt GUI** block shown in Fig. 5.15 is the **Fixed-Point Blockset** interface tool for the Simulink file. The interface tool displays the block names, minimum and maximum simulation values, data types, and scaling factors. In addition, an error message shows the occurrence of saturation or overflow. In this case, overflow has occurred in the **FIR2 filter** block, which uses Q.15 implementation for fixed-point FIR filtering. Double-click on the **Output Scope** block to view the results of the filtering.

The reason for overflow is due to the multiply-add operations, which produce results that exceed the dynamic range of Q.15 numbers. A method for preventing overflow is to insert a scaling factor before the FIR filter. Based on the discussions given in Chapter 3 on scaling, what is a suitable scaling factor for preventing overflow? Verify that overflow is under control after proper scaling. What happens if a Q.7 format (8 bits) is used to represent the FIR coefficients? Is scaling using a smaller factor required?

5.4.4 Infinite-Impulse Response Filtering

This experiment uses IIR filters (direct-form I and II) implemented in different precisions to examine the same filtering problems given in previous section. Open the file ex5_iir.mdl, as shown in Fig. 5.16, and run the simulation. Click on the **Comparison Scope** block and observe the filter output.

Right-click on the **Fixed-Point Direct Form I Implementation** block and select **Fixed-Point Settings** to view the precision and the data format used in the fixed-point IIR filter. All numerator coefficients are set to sfix(32) with 31 fractional bits, while

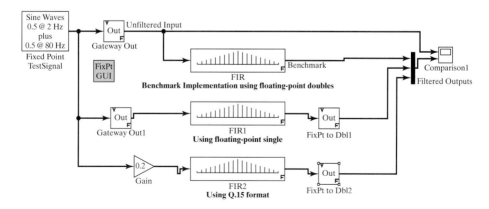

Figure 5.14   Simulink using Fixed-Point Blockset to implement a quantized FIR filter

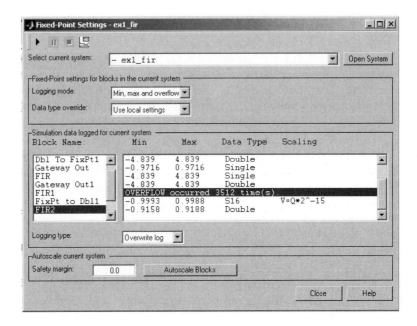

Figure 5.15   Fixed-Point Blockset interface window

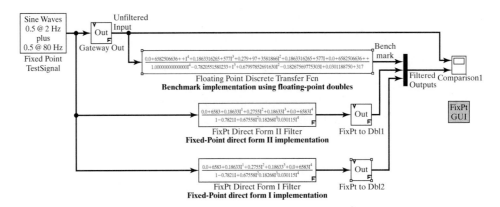

Figure 5.16   Fixed-point implementation of IIR filters

the denominator coefficients are set to sfix(32) with 30 fractional bits. The input to the filter is in Q.15 format, and addition is performed in Q1.30 format. Check the precision and data format used in the direct-form II IIR filter.

The user can also look at the internal structure of the direct-form I IIR filter by right-clicking on the **Fixed-Point Direct Form I Implementation** block and selecting the **Look Under Mask** option to open the detailed block diagram of the IIR filter. Notice that the feedforward section is performed before the feedback section. This order is opposite to that in the direct-form II implementation, which performs feedback before feedforward. Check the internal structure used in the direct-form II IIR filter and observe the differences compared with direct-form I. A detailed description on the structures of IIR filters is given in Chapter 7.

5.4.5 Using DSPLIB

The C55x (or C54x) DSPLIB is an optimized DSP function library for C programming on C55x (or C54x) processors. It includes more than 50 assembly-optimized routines that can be called from either C or assembly programs. Source code is provided to allow the user to modify the functions to match specific needs. These routines are installed in CCS and can also be downloaded from the Web. By using these ready-to-use DSP functions, the user can achieve a faster execution speed and shorten application development time. We use these routines for experiments in later chapters for FIR filtering, IIR filtering, FFT, and adaptive filtering. In this section, we briefly introduce this library and show how to use it for DSP applications.

C5000 DSPLIB functions operate on the Q.15 data type. A Q.15 operand is represented by a `short` data type (16 bits) that is predefined as DATA in the `dsplib.h` header file. A Q.31 operand is represented by a `long` data type (32 bits) that is predefined as LDATA. DSPLIB functions typically operate over vector operands for greater efficiency; however, they may be used to process short arrays or scalars with longer execution times. Complex elements are stored in a real-imaginary format.

In order to use these optimized routines in C, we have to include `dsplib.h`, which is located in `c:\ti\c5500\dsplib\include\` (or `c:\ti\c5400\dsplib\include` for the C54x), and link the code with the DSPLIB object-code library `55xdsp.lib` (or `54xdsp.lib` for the C54x), which is located in `c:\ti\c5500\dsplib\` (or `c:\ti\c5400\dsplib\` for the C54x). DSPLIB provides functions in eight functional categories: (1) FFT, (2) filtering and convolution, (3) adaptive filtering, (4) correlation, (5) math, (6) trigonometric, (7) miscellaneous, and (8) matrix. The convention for describing the arguments for each individual function is summarized in Table 5.6. The user can find examples on how to use every function in DSPLIB. For specific

TABLE 5.6   DSPLIB Function Arguments

Argument	Description
x, y	Input-data vectors
r	Output-data vector
nx, ny, nr	Size of vectors x, y, r
h	Filter-coefficient vector
nh	Size of vector h
DATA	Short Q.15 number
LDATA	Long Q.31 number
ushort	Unsigned short number

functions, refer to the *TMS320C55x* (or *TMS320C54x*) *DSP Library Programmer's Reference* [19, 20].

In this experiment, we use the following random-number-generation function for generating 128 random samples:

```
ushort rand16(DATA *r, ushort nr)
```

The C program `rangen.c` is listed as follows:

```
#include  "math.h"
#include  "dsplib.h"
#pragma   DATA_SECTION(xn, "signal_out");

ushort    N=128;    // length of vector
DATA      xn[128];  // signal buffer

void main()
{
    rand16(xn, N);  // call random number generator
}
```

The `DATA_SECTION` pragma allocates space for `symbol` in a section named `section_name`, which is useful if the user wants to link data objects into an area separate from the .bss section. The syntax for the pragma in C is:

```
#pragma DATA_SECTION(symbol,"section name")
```

This program is added into a new project with the linker-command file and the object-code library `55xdsp.lib` (or `54xdsp.lib` for the C54x). Compile, load, and run the code using CCS. The user must include the search path for `dsplib.h` in the CCS build option by clicking on **Project** → **Build Options . . .** → **Compiler** →

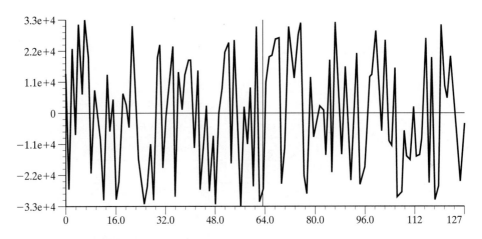

Figure 5.17   Random signal generated by `rand16`

Preprocessor... and inserting the path `c:\ti\c5500\dsplib\include\` (or `c:\ti\c5400\dsplib\include` for the C54x) in **Include Search Path**. The 128 random numbers generated are shown in Fig. 5.17.

PROBLEMS

PART A

Problem A–5–1

Use the Fixed-Point Blockset and Simulink to implement a sixth-order, direct-form II IIR filter using the MATLAB function `butter(6,0.4)`. In addition, implement the same IIR filter using cascade structures with (a) high- to low-gain cascade (i.e., the gain of the biquads is in descending order) and (b) low- to high-gain cascade (i.e., the gain of the biquads is in ascending order). The preceding filters are implemented in 16-bit fixed-point precision. Modify the existing Simulink file `ex5_iir.mdl` by replacing the existing IIR filters with these new sets of IIR filters. Run the simulation and observe the difference using a Spectrum Scope. Also, benchmark the results against the results of using a double-precision floating-point IIR filter. Comment on the effects of ordering the gain of the biquad in ascending and descending order.

Problem A–5–2

Modify the C program `rangen.c` given in Section 5.4.5 to generate 256 samples of sinewave by calling the following C55x assembly-optimized function in the C55x DSPLIB:

```
sine(x,r,NX);
```

Use CCS to test the mixed C-and-ASM program, and plot the generated waveform.

Problem A–5–3

Repeat Problem 2 for the C54x processor.

Problem A–5–4

Modify the C program `rangen.c` given in Section 5.4.5 to implement a floating-point random-number generator by calling the Q.15-to-float conversion function in the C54x (or C55x) DSPLIB.

Problem A–5–5

Convert Problems 2 and 3 into a floating-point sinewave generator using the Q.15-to-float conversion function in the C54x (or C55x) DSPLIB, and examine any difference in the fixed-point and floating-point representations.

Problem A–5–6

Use the Embedded Target for the Texas Instruments TMS320C6000 DSP platform to design, simulate, and implement the following applications. These applications can be implemented in both floating-point (single-precision) and fixed-point (Q.15) representations. To perform a fixed-point implementation and simulation, use the C62x DSP Library, which contains a set of C62x DSP blocksets. These blocksets can be run on the C6711 DSK since the C67x processor is a superset of the C62x processor. Examine the difference in results when implementing the following applications in C67x and C62x processors:

(a) A 5-band IIR filter-based graphic equalizer. The sampling frequency used is 8 kHz, and these 5 bands are [0–500 Hz], [500–1,000 Hz], [1,000–2,000 Hz], [2,000–3,000 Hz], and [3,000–4,000 Hz]. Use any suitable IIR filter-design method, and select suitable passband and stopband ripples and cutoff frequencies.

(b) A simple notch (bandstop) filter to cancel the 60 Hz power-line interference in a speech signal using a sampling frequency of 8 kHz. Use either an FIR or IIR filter-design method.

(c) A simple FFT spectrum analyzer that takes in a block of 512 input data samples, after which FFT is performed on this block of data and the magnitude of the frequency components is displayed. The subsequent data block is being overlapped by 50% from the previous one. The sampling frequency is 8 kHz, and a sinewave can be used to test the performance.

Problem A–5–7

Use Simulink and Fixed-Point Blockset as shown in Section 5.4.2 to extend the multiplication block to mimic the MAC hardware in DSP processors. In a typical 16-bit DSP processor, a 16-bit × 16-bit multiplier and a 40-bit adder are provided. A 40-bit accumulator, which includes 8 guard bits, is used to hold the result of the MAC unit.

Problem A–5–8

Use Simulink and Fixed-Point Blockset to investigate the roundoff error in IEEE-754 single-precision floating-point arithmetic and in Q8.23 fixed-point arithmetic when performing addition and multiplication.

PART B

Problem B–5–9

State the advantages and disadvantages of using floating-point DSP processors rather than fixed-point DSP processors.

Problem B–5–10

List some of the hardware functional units that are common features of floating-point DSP processors. Are there any major hardware differences when they are compared with fixed-point DSP processors?

Problem B–5–11

Compare the dynamic range and precision of IEEE-754 single- and double-precision floating-point formats. How does the IEEE-754 single-precision floating-point format compare with the 32-bit fixed-point format in terms of precision and dynamic range?

Problem B–5–12

Block floating-point format derives the exponent from the largest number in a data array. This single exponent is then associated with the entire array and controls its scale. Block floating-point format commonly is used in audio coding to reduce the amount of bits for transmission and storage. If 20 floating-point (IEEE-754 single-precision) numbers are grouped with one common exponential term, how much reduction can be gained by using block floating-point format?

Problem B–5–13

Some floating-point DSP processors use floating-point formats other than the IEEE-754 floating-point format. For example, the C3x processor uses its own floating-point format, as described in Section 5.2.5. How do C3x devices talk to C67x devices that use the IEEE-754 format?

Problem B–5–14

Compare and contrast the architecture and instruction set differences between C3x and C67x processors.

Problem B–5–15

List the number of execution pipeline stages used in C3x and C67x processors.

Problem B–5–16

The C67x processor uses the IEEE-754 floating-point format, while the C3x processor uses its own floating-point format, as shown in Section 5.2.2. The C3x floating-point number representation is expressed as $x = \{(-2)^s + (.\text{man})\} \times 2^{\text{exp}}$, where exp is in two's complement. Devise a conversion scheme showing how data derived from the C67x and C3x can be exchanged. What are the smallest and largest numbers to be represented in C3x floating-point format?

Problem B–5–17

Denormalization is the process of converting a floating-point number to its fixed-point equivalent. It is the inverse operation of normalization, which is used in fixed-point processors to convert fixed-point numbers to floating-point numbers. Explain the steps used in performing denormalization in DSP processors.

SUGGESTED READINGS

1 Texas Instruments. *TMS320C3x User's Guide*. SPRU031D, 1994.
2 Texas Instruments. *TMS320VC33 Digital Signal Processor*. SPRS087B, 2000.
3 Kuo, S. M. and D. R. Morgan. *Active Noise Control Systems: Algorithms and DSP Implementations*. New York, NY: John Wiley, 1996.
4 Chassaing, R. *Digital Signal Processing: Laboratory Experiments Using C and the TMS320C31 DSK*. New York, NY: John Wiley, 1999.
5 Sorensen, H. V. and J. P. Chen. *A Digital Signal Processing Laboratory Using the TMS320C30*. Upper Saddle River, NJ: Prentice Hall, 1997.

6 Texas Instruments. *TMS320C62x/C67x CPU and Instruction Set Reference Guide*. 1998.

7 Texas Instruments. *TMS320C62x/C67x Programmer's Guide*. 1999.

8 Dahnoun, N. *Digital Signal Processing Implementation Using the TMS320C6000 DSP Platform*. Upper Saddle River, NJ: Prentice Hall, 2000.

9 Kehtarnavaz, N. and M. Keramat. *DSP System Design Using the TMS320C6000*. Upper Saddle River, NJ: Prentice Hall, 2001.

10 The MathWorks. *Embedded Target for the TMS320C6000 DSP Platform: For Use with Real-Time Workshop User's Guide*. 2002.

11 The MathWorks. *Using MATLAB*. Version 6, 2000.

12 The MathWorks. *Simulink: User's Guide*. Version 4.1, 2000.

13 The MathWorks. *Real-Time Workshop: For Use with Simulink, User's Guide*. Version 4.1, 2000.

14 Texas Instruments. *Real-Time Data Exchange (RTDX)*. White paper, SPRY012, 1998.

15 Texas Instruments. *Code Composer Studio: Getting Started Guide*. SPRU509C, 2001.

16 Texas Instruments. *TMS320C6201/6701 Evaluation Module User's Guide*. SPRU269C, 1998.

17 Texas Instruments. *TMS320C6000 Code Composer Studio Tutorial (Rev. C)*. SPRU301C, 2000.

18 The MathWorks. *Fixed-Point Blockset: For Use with Simulink Users's Guide*. 2002.

19 Texas Instruments. *TMS320C54x DSP Library Programmer's Reference*. SPRU422b, 2001.

20 Texas Instruments. *TMS320C55x DSP Library Programmer's Reference*. SPRU422F, 2002.

6

Finite-Impulse Response Filtering

FIR filtering was introduced in Chapter 2. Chapters 4 and 5 showed that the DSP processor architecture, including two independent memory blocks with associate buses, hardware MAC unit(s), zero-overhead looping, circular buffers, and special MAC instructions, is optimized to support FIR filtering. In this chapter, we introduce FIR-filter structures, characteristics, designs, and implementations. We take a practical approach that avoids the use of a lot of mathematics. We use MATLAB's FDATool for designing filters and quantizing filter coefficients for finite-precision implementations. Finally, we implement the designed FIR filters using both fixed-point and floating-point DSP processors. An important application of FIR filters for multirate DSP systems is also introduced in this chapter.

6.1 FINITE-IMPULSE RESPONSE FILTERS

As discussed in Section 2.3, there are two major types of digital filters: FIR and IIR filters. IIR filters are discussed in Chapter 7. In this chapter, we focus on FIR filters. The major advantages of FIR filters are:

    1. *Linear phase*: FIR filters can have an exact linear-phase response, resulting in a constant group delay over the frequency range of interest. Therefore, no phase distortion is introduced by the filter.

2. *Guaranteed stability*: FIR filters are always stable due to nonrecursive realization.

3. *Fewer finite-precision errors*: FIR filters are less sensitive to finite-wordlength effects such as coefficient quantization errors and roundoff noise.

4. *Efficient implementation*: FIR filters can be implemented efficiently using DSP processors with hardware MACs, circular addressing, and special instructions for FIR filtering.

The main disadvantage of FIR filters is that they may require a much higher-order filter (more coefficients) than IIR filters with comparable performance, thus resulting in longer delays and higher computational costs and memory requirements. In addition, there is no direct conversion from the well-known analog filter design to the digital FIR filter. Optimization of FIR-filter designs requires computer programs for intensive computations.

The phase and group-delay characteristics of FIR filters are generally better than those of IIR filters. For applications where wave shape is important, an FIR filter with good phase characteristics is usually a good choice. However, for narrow-band, sharp cutoff filters where phase is not important, IIR filters may be better than FIR filters.

6.1.1 Definitions

As defined in Section 2.3.2, the I/O equation for an FIR filter of length L (or order $L - 1$) can be expressed as

$$y(n) = b_0 x(n) + b_1 x(n - 1) + \cdots + b_{L-1} x(n - L + 1)$$
$$= \sum_{i=0}^{L-1} b_i x(n - i)$$
$$= \mathbf{b}^T \mathbf{x}(n), \tag{6.1.1}$$

where T denotes the transpose of the vector, $\mathbf{b}$ is the filter coefficient vector $(L \times 1)$ defined as

$$\mathbf{b} = [b_0\, b_1\, \cdots\, b_{L-1}]^T, \tag{6.1.2}$$

and the signal vector (array or buffer) at time n is defined as

$$\mathbf{x}(n) = [x(n)\, x(n - 1) \cdots x(n - L + 1)]^T. \tag{6.1.3}$$

The design of FIR filters determines the coefficients $\{b_i, i = 0, 1, \ldots, L - 1\}$ needed to achieve the desired filter characteristics with as few coefficients as possible.

Equation (6.1.1) shows the FIR-filtering equivalent to an inner (or dot) product of vectors or to a linear convolution of the two sequences. An FIR filter is a non-recursive filter, which generates its output by simply scaling previous input samples by filter coefficients and then summing the weighted inputs. The coefficients are constants determined by using the filter-design process. However, the signal vector $\mathbf{x}(n)$ is a time function, which needs to be updated every sampling period when the newest sample $x(n)$ is available.

By taking the z-transform of both sizes of Eq. (6.1.1) and using the time-shift property, we have

$$Y(z) = b_0 X(z) + b_1 z^{-1} X(z) + \cdots + b_{L-1} z^{-L+1} X(z) = X(z) \sum_{i=0}^{L-1} b_i z^{-i}.$$

By rearranging the terms, we obtain the transfer function of the FIR filter expressed as

$$B(z) = \frac{Y(z)}{X(z)}$$

$$= b_0 + b_1 z^{-1} + \cdots + b_{L-1} z^{-L+1} = \sum_{i=0}^{L-1} b_i z^{-i}. \tag{6.1.4}$$

Thus, the FIR filter has a transfer function that is a polynomial in z^{-1}. This FIR filter has $L - 1$ zeros and poles. Because the poles are located at the origin $z = 0$, they do not affect the behavior of the filter. Only the zeros in the z-plane determine the frequency-response characteristics. Therefore, an FIR filter is also called an all-zero filter.

6.1.2 Filter Characteristics

The characteristics of the filter in the time domain can be described by the impulse response of the filter, which is the output (response) of the filter when the input is an impulse function. Applying the unit-impulse function $\delta(n)$ defined in Eq. (2.1.3) to the input of the FIR filter defined in Eq. (6.1.1), the output of the filter can be obtained as

$$h(n) = \begin{cases} b_n, & n = 0, 1, \ldots, L - 1 \\ 0, & n < 0 \text{ and } n \geq L \end{cases}. \tag{6.1.5}$$

Therefore, the impulse response of the FIR filter is identical to the coefficient vector defined in Eq. (6.1.2), which consists of a finite number (L) of coefficients.

The characteristics of a filter in the frequency domain can be described by using the frequency response (or transfer function) of the filter. As discussed in Section 2.4, the transfer function given in Eq. (6.1.4) can be evaluated on the unit circle to obtain the frequency response expressed as

$$B(\omega) = B(z)|_{z=e^{j\omega}} = \sum_{i=0}^{L-1} b_i e^{-j\omega i}. \tag{6.1.6}$$

The frequency response is a function of the continuous-frequency variable ω in radians per second and is a periodic function with period 2π. Thus, it is only necessary to compute the frequency response over a 2π interval. Because ω is a continuous variable, the DFT can be applied to evaluate the frequency response at certain frequencies. The DFT of the impulse response $h(n)$ of length N results in N samples of $B(\omega)$ at frequencies $2\pi k/N$, $k = 0, 1, \ldots, N - 1$.

The frequency response defined in Eq. (6.1.6) is a complex function. Thus, it can be expressed further as the polar form

$$B(\omega) = |B(\omega)|e^{j\phi(\omega)}, \tag{6.1.7}$$

where $|B(\omega)|$ is the magnitude response and $\phi(\omega)$ is the phase response. Both the magnitude and phase responses are periodic functions with period 2π. If the filter has real coefficients, the real part of the frequency response is an even function, and the imaginary part is an odd function. Therefore, the magnitude and phase responses are the even and odd functions, respectively, expressed as

$$|B(\omega)| = |B(-\omega)| \text{ and } \phi(\omega) = -\phi(-\omega). \tag{6.1.8}$$

By exploiting these results, we only need to compute the response from 0 to π for FIR filters with real-valued coefficients.

Filters can be categorized by their magnitude-response characteristics. There are four common filter types: lowpass, highpass, bandpass, and bandstop. We define these filters later when we introduce the filter-design specification. In addition, there are allpass, comb, arbitrary (multipassband) filters, etc.

When the input signal is a sinewave of amplitude A at frequency ω_0, as defined in Eq. (2.1.6), the filter output is still a sinusoidal signal expressed as

$$y(n) = A|B(\omega_0)| \sin[\omega_0 n + \phi(\omega_0)]. \tag{6.1.9}$$

This steady-state sinusoidal response shows that the input sinusoidal signal changes its amplitude according to the magnitude response and shifts the phase according to the phase response of the filter.

If the input signal consists of several frequency components, the amount of time delay for each frequency component at the filter output is expressed as

$$\tau_p(\omega) = \frac{-\phi(\omega)}{\omega}, \tag{6.1.10}$$

which is called the phase delay of the filter. It is important to note that the phase delay is different from the group delay, $T_d(\omega) = [-d\phi(\omega)/d\omega]$, defined in Eq. (2.4.9), which represents the delay of the envelope (average time delay) of the signal. Phase distortion occurs when different frequency components have different phase delays through a filter. A filter with a linear phase has a constant phase (or group) delay. In contrast, a filter with a nonlinear phase produces a phase distortion.

Example 6.1

Given an FIR filter with coefficient vector $\mathbf{b} = [0.1\ 0.3\ 0.5\ 0.3\ 0.1]^T$, the transfer function of the filter is given as

$$B(z) = 0.1 + 0.3z^{-1} + 0.5z^{-2} + 0.3z^{-3} + 0.1z^{-4}.$$

The frequency response is expressed as

$$
\begin{aligned}
B(\omega) &= 0.1 + 0.3e^{-j\omega} + 0.5e^{-2j\omega} + 0.3e^{-3j\omega} + 0.1e^{-4j\omega} \\
&= e^{-2j\omega}\lfloor 0.1e^{2j\omega} + 0.3e^{j\omega} + 0.5 + 0.3e^{-j\omega} + 0.1e^{-2j\omega} \rfloor \\
&= e^{-2j\omega}\lfloor 0.5 + 0.3(e^{j\omega} + e^{-j\omega}) + 0.1(e^{2j\omega} + e^{-2j\omega}) \rfloor \\
&= e^{-2j\omega}[0.5 + 0.6\cos\omega + 0.2\cos 2\omega].
\end{aligned}
$$

The magnitude response is obtained as

$$|B(\omega)| = |0.5 + 0.6\cos\omega + 0.2\cos 2\omega|,$$

and the phase response is

$$\phi(\omega) = -2\omega.$$

The preceding example shows that the FIR filter with symmetric coefficients has a linear phase. The phase (or group) delay of this filter is two samples, which is a constant for all frequencies. Thus, signals of different frequencies have the same delay as they pass through the filter without phase distortion. This important fact is discussed further in Section 6.1.3.

MATLAB provides the impz, freqz, grpdelay, and zplane functions for analyzing digital filters. For example, the command

```
[h, t] = impz(b, 1, L);
```

computes L samples of the impulse response and stores them in the output vector h, the input vector b consists of L FIR-filter coefficients, and t = [0:L - 1]' is the time index. The impz function with no output arguments plots the impulse response using stem(t, h). The functions freqz, grpdelay, and zplane are introduced in Chapter 2.

Example 6.2

For the FIR filter defined in Example 6.1, the following MATLAB script exmp6_2.m can be used to plot the pole-zero diagram, magnitude and phase responses, and group delay:

```
b = [0.1 0.3 0.5 0.3 0.1];    % define an FIR filter
zplane(b, 1); pause;          % plot pole-zero diagram
freqz(b, 1); pause;           % plot magnitude & phase spectra
grpdelay(b, 1)                % plot group delay
```

The pole-zero diagram and the magnitude and phase responses are shown in Fig. 6.1. From the MATLAB results, we observed that the given FIR filter has a linear phase; thus, it has a constant group delay of 2.

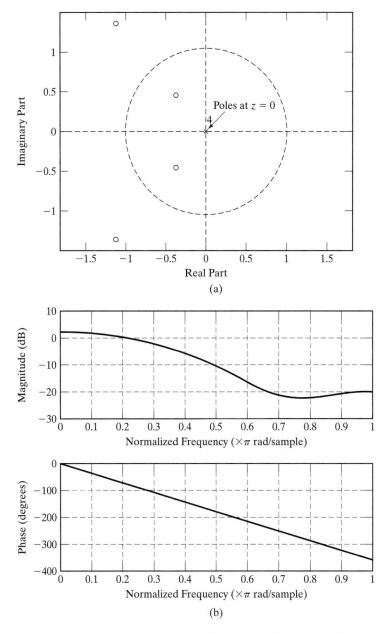

Figure 6.1 Characteristics of an FIR filter: (a) Pole-zero diagram and (b) magnitude and phase responses

6.1.3 Filter Structures

For a given transfer function, there are different ways to implement a digital filter. These various realizations are represented with signal-flow diagrams called filter structures. Many factors determine the selection of a particular structure for a particular application, including programming considerations, regularity of the hardware implementation, sensitivity of quantizing coefficients, and noise introduced by quantization of the signal. For example, for the given filter transfer function in Eq. (6.1.4) or the I/O equation given in Eq. (6.1.1), the signal-flow diagram of the FIR filter is illustrated in Fig. 6.2. It represents the direct calculation of Eq. (6.1.1) and is called a direct structure, a tapped-delay line, or a transversal filter. The direct structure is very easy to program and can be efficiently implemented with most DSP processors, which have an architecture and instructions tailored for FIR filtering.

As introduced in Chapter 2, the box labeled z^{-1} in Fig. 6.2 represents a unit delay. When the filter is implemented with tapped-delay-line hardware, the box z^{-1} corresponds to a physical-delay element. However, if the filter is implemented on a DSP processor or software, the delay unit corresponds to the storage of the signal value in memory rather than to the physical-delay device.

The transpose structure of the FIR filter is illustrated in Fig. 6.3, which implements exactly the same I/O equation given in Eq. (6.1.1). However, the transpose structure leads to a very different program, where the partial sums feed into succeeding stages. Thus, the transpose structure is more sensitive to roundoff noise for computing filter output. Implementation of the direct structure defined in Fig. 6.2 and the transpose structure defined in Fig. 6.3 on the TMS320 fixed-point processor is given in reference [1], which shows that the direct form can take advantage of the special instruction and has less quantization noise.

A high-order FIR filter can also be realized as a cascade of several second-order filters. However, the direct structure shown in Fig. 6.2 can take better advantage

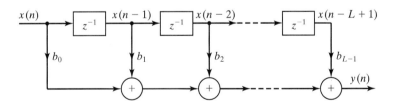

Figure 6.2   Direct structure of an FIR filter

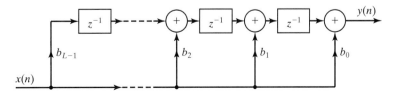

Figure 6.3   Transpose structure of an FIR filter

312 Chapter 6 Finite-Impulse Response Filtering

of DSP processor architectures. Therefore, a direct structure is preferred for FIR filtering.

As shown in Examples 6.1 and 6.2, when the filter coefficients are symmetric about the midpoint, the filter has a linear phase. By assuming L is an odd number and defining $M = (L - 1)/2$, the frequency response of the FIR filter given in Eq. (6.1.6) can be rewritten as

$$B(\omega) = \sum_{i=0}^{L-1} b_i e^{-j\omega i} = e^{-j\omega M} \sum_{i=0}^{L-1} b_i e^{j\omega(M-i)}. \tag{6.1.11}$$

When the impulse response is symmetric about M, equation (6.1.11) can be rewritten as

$$B(\omega) = e^{-j\omega M} \left[b_M + 2\sum_{i=1}^{M} b_{M-i} \cos(\omega i) \right], \tag{6.1.12}$$

which results in the linear phase being expressed as

$$\phi(\omega) = -M\omega. \tag{6.1.13}$$

Therefore, the phase (or group) delay of this linear-phase FIR filter is a constant M over the frequency band of interest. This property preserves the wave shape of signals in the passband without phase distortion. Note that a filter with phase response $\phi(\omega) = c - M\omega$ also has a linear-phase response, where c is a constant.

Based on the filter length L and the symmetric characteristic, there are four possible types of linear-phase FIR filters:

Type 1: Filter length L is an odd number (M is an integer), and the impulse response is even symmetric about its midpoint M. That is,

$$b_i = b_{L-i-1}, \quad i = 0, 1, \ldots, M. \tag{6.1.14}$$

Type 2: Filter length L is an even number (M is not an integer), and the impulse response is even symmetric about M.

Type 3: Filter length L is an odd number, and the impulse response is odd symmetric (antisymmetric) about M. That is,

$$b_i = -b_{L-i-1}, \quad i = 0, 1, \ldots, M - 1, \tag{6.1.15}$$

and $b_M = 0$.

Type 4: Filter length L is an even number, and the impulse response is odd symmetric.

For example, the signal-flow diagram of a Type 2 symmetric FIR filter is illustrated in Fig. 6.4. The I/O equation is expressed as

$$y(n) = \sum_{i=0}^{L/2-1} b_i[x(n - i) + x(n - L + 1 + i)]. \tag{6.1.16}$$

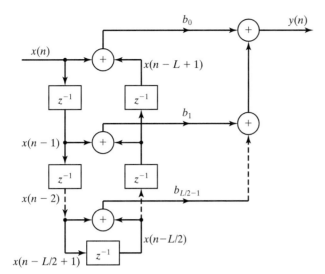

Figure 6.4 Signal-flow diagram of a Type 2 symmetric FIR filter

By comparing this equation with the direct form given in Eq. (6.1.1), it can be seen that the symmetric filter requires half the number of multiplications. It only requires half of the memory for storing filter coefficients. However, the computational advantage in Eq. (6.1.16) is lost in the more complicated indexing of data in the signal buffer in most DSP processors. This fact is shown in Sections 6.3 and 6.4 for DSP implementations.

Some important characteristics for analyzing or designing linear-phase FIR filters are discussed in [1]. For example, a Type 3 or Type 4 filter has $|B(0)| = 0$, which is undesirable for a lowpass filter. A Type 2 or Type 3 filter has $|B(\pi)| = 0$, which is unsuitable for a highpass filter. In addition, a Type 3 or Type 4 filter gives a constant $90°$ phase shift, which is desired for a differentiator or a Hilbert transformer. These important features are summarized in Table 6.1.

TABLE 6.1 Summary of Linear-Phase FIR Filters

Filter type	Filter length (L)	Symmetric type	Magnitude response	Phase shift	Comment
Type 1	Odd	Even			
Type 2	Even	Even	$|B(\pi)| = 0$		Unsuitable for highpass filter
Type 3	Odd	Odd	$|B(0)| = 0$ $|B(\pi)| = 0$	$90°$	Unsuitable for lowpass or highpass filter
Type 4	Even	Odd	$|B(0)| = 0$	$90°$	Unsuitable for lowpass filter

6.1.4 Filter Designs

Digital-filter design involves the computation of filter coefficients to approximate a desired frequency response. The design process starts from the filter specification and continues to the implementation of the desired filter in the following five steps:

Step 1. *Specification*: A desired frequency response is determined by specifying filter characteristics in a frequency domain.

Step 2. *Design criteria*: A measure of the quality of the filter is chosen by selecting design criteria.

Step 3. *Realization*: A class of filter is chosen by selecting the filter type, structure, and length of the filter. An algorithm is then selected to calculate the desired filter coefficients.

Step 4. *Quantization*: Quantization involves quantizing filter coefficients for a given wordlength, analyzing the finite-precision effects, and optimizing the quantized filter to reduce quantization errors.

Step 5. *Implementation*: Implementation involves verifying the designed filter using simulation and implementing it in software (or hardware).

Some FIR-filter design methods may be simple such as the Fourier technique, which is introduced later. However, no single method is optimal for every filter characteristic. Filter coefficients can be generated using some reasonable equations. However, using an optimization technique to obtain an optimum filter demands intensive computation. In addition, the major concern for implementing filters on fixed-point DSP processors for real-time applications is the limited precision and dynamic range, as discussed in Chapter 3. The theory for analyzing finite-wordlength effects is too complicated to compute by hand; thus, we usually rely on computer software for designing digital filters.

A filter-design software package, such as MATLAB with the Signal Processing Toolbox and Filter Design Toolbox, not only saves time when designing a filter quickly, but also supports the design of a quantized filter for a given wordlength. We can explore many possibilities such as filter characteristics, structures, and different wordlengths quickly using the software package.

Filter Specifications

As mentioned earlier, a filter specification is usually based on the desired magnitude response. For example, the specification for a lowpass filter is illustrated in Fig. 6.5. The passband is defined as the frequency range over which the input signal is passed with approximate unit gain. Thus, the passband is defined by $0 \leq \omega \leq \omega_p$, where ω_p is called the passband-cutoff (edge) frequency. The width of the passband is usually called the bandwidth of the filter. For a lowpass filter, the bandwidth is ω_p. The input-signal components that lie within the stopband are attenuated to a level that effectively eliminates them from the output signal. The stopband is defined by $\omega_s \leq \omega \leq \pi$, where ω_s is called the stopband-edge frequency. The transition band is defined as $\omega_p < \omega < \omega_s$, which is between the passband and the stopband. In this region, the filter magnitude response typically

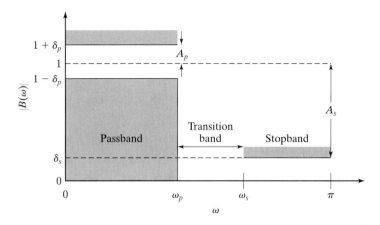

Figure 6.5   Lowpass filter specification

makes a smooth transition from the passband-gain level to that of the stopband. The width of the transition band is $\omega_s - \omega_p$. In general, the steeper the desired filter rolloff (i.e., the narrower the transition band), the larger the number of coefficients that is required. The passband ripple (deviation) is specified by δ_p, which is the maximum deviation from unit gain within the passband. The stopband ripple (attenuation) is defined by δ_s.

Digital-filter specifications are usually given in dB defined as follows:

$$A_p = 20 \log_{10}(1 + \delta_p) \text{ dB} \tag{6.1.17}$$

and

$$A_s = -20 \log_{10}(\delta_s) \text{ dB.} \tag{6.1.18}$$

For example, suppose that $\delta_p = 0.1$ and $\delta_s = 0.01$. From Eq. (6.1.17), we have $A_p = 20 \log_{10}(1.1) = 0.828$ dB. From Eq. (6.1.18), we have $A_s = -20 \log_{10}(0.01) = 40$ dB. Note that the negative sign in Eq. (6.1.18) indicates A_s is the minimum stopband attenuation, while A_p is the passband gain above 0 dB ($\delta_p = 1$). Different definitions of A_p, such as

$$A_p = 20 \log_{10}(1 - \delta_p) \text{ dB} \tag{6.1.19}$$

and

$$A_p = 20 \log_{10}\left(\frac{1 + \delta_p}{1 - \delta_p}\right) \text{ dB,} \tag{6.1.20}$$

are used by different DSP books.

The lowpass filter passes low-frequency components to the output while attenuating high-frequency components. Conversely, the highpass filter permits

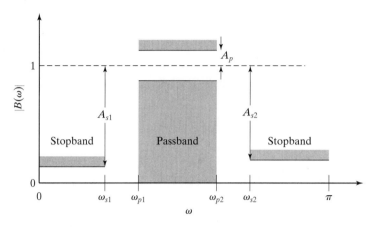

Figure 6.6   Bandpass-filter specification

high-frequency components to appear at the output while effectively attenuating low-frequency components. The highpass filter specification is similar to the low-pass filter given in Fig. 6.5, except that the stopband is at the low-frequency range from 0 to ω_s, while the passband is at the high-frequency range from ω_p to π.

The bandpass filter rejects both high- and low-frequency components. The specification for a bandpass filter is illustrated in Fig. 6.6. The passband is defined by $\omega_{p1} \leq \omega \leq \omega_{p2}$, where ω_{p1} and ω_{p2} are the lower and upper passband-edge frequencies, respectively. The lower stopband is defined by $0 \leq \omega \leq \omega_{s1}$, and ω_{s1} is the lower stopband-edge frequency. The upper stopband is defined by $\omega_{s2} \leq \omega \leq \pi$, and ω_{s2} is the upper stopband-edge frequency. The passband ripple is specified by A_p, while stopband ripples are defined by A_{s1} and A_{s2} for lower and upper stopbands, respectively.

The bandstop filter rejects an intermediate band of frequencies while passing high- and low-frequency components. The bandstop filter specification is similar to the bandpass filter given in Fig. 6.6, except that the stopband is at the middle-frequency range from ω_{s1} to ω_{s2}, the lower passband is at the low-frequency range from 0 to ω_{p1}, and the upper passband is at the high-frequency range from ω_{p2} to π.

Design Criteria

The filter-design error is the difference between the desired response and the filter's actual response. Three criteria are generally used in FIR-filter design:

1. **Least square (LS) approximation,** which is the average of the squared error in the frequency response between the desired and designed filters
2. **Chebyshev approximation,** which is the maximum error over specified regions of the frequency response
3. **Butterworth approximation,** which is the approximation to the desired response based on a Taylor-series approximation

Most useful filter-design algorithms are based on one (or a combination) of these three approximations.

Some useful filter-design algorithms are the Fourier (window) method, the frequency-sampling method, and the Parks–McClellan (optimal) algorithm. The frequency-sampling method is fast and simple, but it gives the least control over the overall frequency response. The Fourier design has a frequency response with oscillations or overshoots that may be undesirable, and the window functions are used to control these effects. An important advantage of the window method is its simplicity, and the major disadvantage is its lack of flexibility to control passband and stopband ripples. The Parks–McClellan algorithm generates a linear-phase FIR filter and minimizes the Chebyshev error, but the design can be slow without using computer-design software. With the availability of efficient filter-design packages such as MATLAB, the Parks–McClellan algorithm is widely used for most practical applications. In the following section, we use the Fourier method to explain some basic concepts, options, and tradeoffs of designing FIR filters.

Fourier Design Method

For a given desired frequency response $B(\omega)$, the filter's impulse response can be obtained by computing the inverse Fourier transform

$$h(i) = \frac{1}{2\pi} \int_{-\pi}^{\pi} B(\omega)e^{j\omega i}\, d\omega. \tag{6.1.21}$$

The impulse response possibly has an infinite length. However, we only can implement the filter with a finite length of impulse response. An FIR filter is designed by choosing L coefficients to best match the impulse response. That is,

$$b_i = \frac{1}{2\pi} \int_{-\pi}^{\pi} B(\omega)e^{j\omega i}\, d\omega, \quad i = 0, 1, \ldots L - 1. \tag{6.1.22}$$

It is clear that the filter's frequency response becomes closer to the desired response as more coefficients are used.

Finite-length filters that are designed using the Fourier technique have the following two deficiencies due to the truncation of the IIR to a finite length:

1. The frequency response has significant oscillations (ripples) in both the passband and stopband. The effect of undesired ripples and overshoots is called Gibb's phenomenon. The period of the ripples decreases as the filter length increases.
2. The frequency response has a slower transition (wider transition band) than the desired response. The width of the transition band decreases as the filter length increases.

Some applications are more sensitive to the amplitude and period of the ripples, and some are more sensitive to the width of the transition band.

The truncate operation is equivalent to multiplying the impulse response $h(i)$ by a rectangular window of length L. The truncated filter has coefficients b_i, which are the products of the Fourier coefficients $h(i)$ given in Eq. (6.1.21) and the rectangular-window coefficients $w(i)$. That is,

$$b_i = w(i)h(i), \tag{6.1.23}$$

where

$$w(i) = \begin{cases} 1, & i = 0, 1, \ldots, L-1 \\ 0, & \text{otherwise} \end{cases} \tag{6.1.24}$$

for the rectangular window.

The time domain multiplication given in Eq. (6.1.23) is equivalent to the convolution in the frequency domain expressed as

$$B(\omega) = \frac{1}{2\pi} \int_{-\pi}^{\pi} W(\omega - \chi)H(\chi)\, d\chi. \tag{6.1.25}$$

Thus, the ideal window is a delta function $W(\omega) = \delta(\omega)$ such that $B(\omega) = H(\omega)$, which requires the rectangular window given in Eq. (6.1.24) to have an infinite length. Otherwise, $W(\omega)$ has the classic sinc(x) shape. The rippling and wide-transition regions of FIR filters are generated by this convolution. Thus, the most effective method for reducing these effects is to design a finite-length window such that the resulting impulse response decays toward zero smoothly instead of using the rectangular window defined in Eq. (6.1.24). The window functions supported by MATLAB are Hamming, Hann, Blackman, Kaiser, Bartlett, Chebyshev, etc. MATLAB also contains a window design and analysis tool, which is introduced in Section 8.4. The best window for most applications is the Kaiser window, which has a parameter to adjust the compromise between the overshoot reduction and the transition bandwidth.

Example 6.3

Consider an ideal lowpass filter with cutoff frequency ω_c. This filter has magnitude 1 at frequencies below ω_c and magnitude 0 at frequencies from ω_c to π. From Eq. (6.1.21), we have

$$h(i) = \frac{1}{2\pi} \int_{-\omega_c}^{\omega_c} e^{j\omega i}\, d\omega = \frac{\omega_c}{\pi} \text{sinc}\left(\frac{\omega_c}{\pi} i\right).$$

The MATLAB script `exmp6_3.m` designs lowpass filters of order 50 using rectangular and Hamming windows. As shown in Fig. 6.7(a) in linear scale, a Hamming window effectively reduces passband ripples. However, the width of the transition band is increased, as shown in Fig. 6.7(b) in dB scale. The transition width of the filter is determined by the width of the main lobe of the window, while the side lobes of the window produce ripples in both the passband and the stopband. In general, efforts to reduce the amplitude of the ripples increases the transition bandwidth and vice versa. Fig. 6.7(b) also shows that the Hamming window improves the stopband attenuation by more than 20 dB.

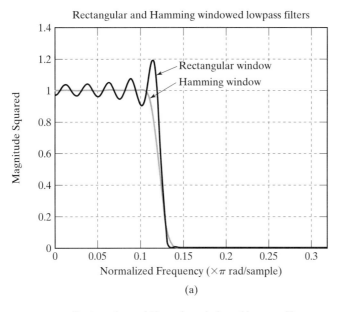

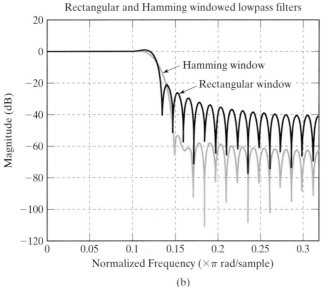

Figure 6.7 Magnitude responses of lowpass filters designed using the rectangular and Hamming windows: (a) linear scale and (b) dB scale

As shown in Fig. 6.7, the windowed FIR filter shows that the ripples vary in amplitude. The largest ripple occurs near the band edge. The principle of equiripple-filter design is to even out the ripple amplitudes to achieve a better approximation of an ideal filter response. The Parks–McClellan algorithm and the Remez exchange algorithm result in equiripple filters.

MATLAB Functions

MATLAB functions for designing FIR filters in the Signal Processing Toolbox are summarized in Table 6.2. These functions design Type 1 or Type 2 linear-phase filters by default. Both the `firls` and `remez` functions design Type 3 and Type 4 linear-phase FIR filters using a `hilbert` or `differentiator` flag. The toolbox also includes the function `intfilt` for designing interpolation filters, which is introduced in Section 6.5.

The Signal Processing Toolbox is data oriented. The user creates separate variables for each parameter required to characterize a given filter type. In addition, filter coefficients are represented in double-precision floating-point format. We cannot design single-precision floating-point or fixed-point filters using this toolbox. To design and analyze quantized filters, we have to use the object-oriented Filter Design Toolbox to convert the double-precision filter to a quantized filter. This toolbox, which is discussed in Section 6.2 and Appendix A.2, also provides the new FIR-filter design methods `firlpnorm` and `gremez`.

6.1.5 Finite-Wordlength Effects

Filter-design algorithms assume infinite precision, and filter design tools use a very high precision (such as double-precision floating-point format in MATLAB) to represent filter coefficients. However, it is often necessary to represent filter coefficients using fewer bits, such as 16 bits in most fixed-point DSP processors, thus leading to lower filter performance.

As discussed in Chapter 3, finite-wordlength effects may be divided into two categories: (1) Quantization errors in representing signal and coefficient values using a finite number of bits, and (2) arithmetic errors such as roundoff noise and overflow. In this section, we focus on fixed-point implementations since most practical applications use 16-bit, fixed-point DSP processors, and fixed-point arithmetic requires careful consideration.

Coefficients of the FIR filter must be quantized to N bits. Thus, the filter actually implements the I/O equation

$$y(n) = \sum_{i=0}^{L-1} b_i' x(n - i), \tag{6.1.26}$$

TABLE 6.2 Summary of MATLAB Functions for Designing FIR Filters

Method	MATLAB functions
Windowing	`fir1, fir2, kaiserord`
Multiband	`firls, remez, remezord`
Constrained LS	`fircls, fircls1`
Arbitrary response	`cremez`
Raised cosine	`firrcos`

where b_i' represents quantized filter coefficients. Coefficient quantization errors cause the frequency response of the quantized filter to deviate from the desired response. The frequency response $B'(\omega)$ with quantized coefficients may be considered as the sum of the ideal response $B(\omega)$ and the frequency response of the error filter $B_e(\omega)$, which is expressed as

$$B_e(\omega) = \sum_{i=0}^{L-1} [b_i' - b_i] e^{-j\omega i}. \tag{6.1.27}$$

It is clear that the main effects of coefficient quantization are a possible increase in the passband ripple and a reduction in stopband attenuation. An objective of finite-precision design is to limit the magnitude of $B_e(\omega)$ so that the frequency response of the actual filter meets the specification.

The addition of the error filter may limit the maximum attenuation in the stopband, thus allowing additional signal components to be passed as the output of the filter. For example, the maximum value of the error-filter response is bounded by

$$20 \log_{10} |B_e(\omega)| \leq (20 \log_{10} L - 6N) \text{ dB}. \tag{6.1.28}$$

This equation shows that the coefficient-quantization error is related to the number of bits used to represent filter coefficients and to the length of the FIR filter. The bound in Eq. (6.1.28) is very conservative. For example, when $L = 100$ and $N = 16$, the maximum error is -56 dB. A straightforward solution is to use more bits to represent filter coefficients. However, an optimization procedure can be used to determine the best quantized coefficients rather than simply rounding the coefficients to N bits. Optimization techniques, such as mixed-integer programming algorithms for obtaining coefficients, are supported by MATLAB, which is discussed in Section 6.2.3.

The direct implementation of an FIR filter is shown in Fig. 6.2. However, $(L - 1)$ summing junctions are realized by a single accumulator. Roundoff noise occurs by rounding out the less significant bits. It is good practice to accumulate the sum of the products in a double-precision accumulator. In this case, the value of the sum is reduced to the original wordlength N only after the output $y(n)$ is computed, thus reducing roundoff noise. This technique is used in Section 6.3 for a fixed-point implementation.

Another important issue in implementing FIR filters with fixed-point DSP processors is the overflow problem, which occurs when partial sums or the final filter output exceeds the given dynamic range of the DSP system. Methods for preventing overflow when computing Eq. (6.1.1) include saturation mode, guard bits, and scaling the signal values and/or filter coefficients, as introduced in Chapter 3.

6.1.6 Implementations of Finite-Impulse Response Filters

When the desired filter is implemented on a digital system, the signal and coefficient values are represented with a finite numbers of bits. As discussed in Chapter 1, the conversion of the signal from continuous time to discrete time is called sampling.

The discrete-time signal is then converted to a digital signal using quantization. The combination of sampling and quantization is called analog-to-digital conversion. Similarly, after an FIR filter has been designed, the filter must be realized by digital hardware or by a program that implements the I/O equation of the filter. We implement digital filters on fixed-point and floating-point DSP processors in Sections 6.3 and 6.4, respectively.

As discussed in Section 6.1.1, the transversal filter generates its output $y(n)$ by performing linear convolution (or inner product). The MATLAB function `filter` can be used for FIR filtering as follows:

```
y = filter(b, 1, x);
```

This function filters the input signal in vector x with the FIR filter given by the coefficient vector b and stores the generated output signal in vector y. FIR filtering can also be implemented using the function y = conv (b, x), which computes the convolution of sequences b and x with an appropriate number of zeros appended to the input.

The implementation of Eq. (6.1.1) can be written in C language as

```
y[n] = 0.0;
for (i=0; i<L; i++)
    y[n] + = b[i]*xn[i];
```

where b[i] represents the i-th filter coefficient b_i, and xn[i] represents the signal sample $x(n - i)$ in the delay buffer.

The FIR filtering given in Eq. (6.1.1) can also be implemented in a C function `fir.c` as follows:

```
float fir(x, w, ntap)

float *x, *w;       // x is signal vector, w is coefficient vector
int ntap;           // length of filter
{
  float yn = 0.0;   // output of FIR filter, y(n)
  int i;            // loop index
  for (i=0; i<ntap; ++i){
    yn += w[i]*x[i];// FIR filtering
  }
  return(yn);       // return y(n) to the calling function
}
```

This function can be called by the main function as

```
y[n] = fir(xn, b, L);   // FIR filtering
```

where xn and b are arrays defined in the main function for signal and coefficient buffers, respectively.

To compute the output continuously such as in real-time operations, the signal buffer has to be refreshed after obtaining the output $y(n)$. The signal buffer can be refreshed by shifting the data in the buffer one position to the right, as shown in Fig. 6.2, where the last element $x(n - L + 1)$ is discarded and the new data is placed into the first location, labeled as $x(n)$. The C function for refreshing the signal buffer, datamov.c, is listed as follows:

```
void  datamov(x, L, input)

float *x;                 // signal vector
int L;                    // length of vector
float input;              // newest input sample
{
  int i;                  // loop index
  for (i=L-1; i>0; --i){
    x[i] = x[i-1];        // shift old data x(n-i), i = 1, 2, ... L-1
  }                       // such as x(0) -> x(1), etc.
  x[0] = input;           // insert new data x(n) into x(0)
}
```

This function can be called by the main function as

```
datamov(xn, L, ADC);      // update signal buffer
```

where ADC is the memory location that contains the newest sample.

Updating the signal buffer requires intensive operations when the filter length L is large. By comparing the contents of current buffer at time n with the new buffer at time $n + 1$, we observe that $L - 1$ samples are the same, except the oldest sample $x(n - L + 1)$ has been replaced by the new sample. Therefore, instead of moving data in the linear buffer, we can update the data buffer in a circular fashion, as shown in Fig. 6.8. At the current time n, the data pointer points at $x(n)$. After accessing that data for computing the filter output, the pointer is incremented by one and thus points at the next sample, $x(n - 1)$. This process is repeated L times in a clockwise direction for accessing the data $x(n - i), i = 0, 1, \ldots, L - 1$ in the buffer. It is important to note that the pointer is not incremented after the last data, $x(n - L + 1)$, is accessed; thus, the pointer still points at the end of the buffer and

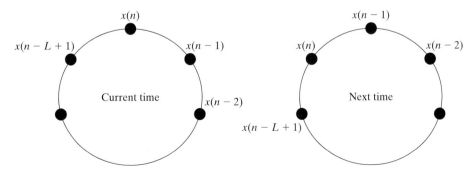

Figure 6.8   Circular buffer at the current time (left) and at the next time (right)

waits for the next iteration. At the next iteration, $n + 1$, the new sample is placed in the location pointed at by the pointer. This new sample replaces the oldest sample in the previous iteration and is called $x(n)$, as shown in the right side of Fig. 6.8. The rest of the $L - 1$ samples are still at the same physical locations of the buffer (memory); however, the old $x(n)$ (at the left) now has become $x(n - 1)$ (at the right), the old $x(n - 1)$ (at the left) has now become $x(n - 2)$ (at the right), etc.

This efficient circular technique is supported by circular-addressing mode, which in turn is supported by most of the modern DSP processors introduced in Chapters 4 and 5. In addition, the FIR-filter coefficients $\{b_0 \, b_1 \, \cdots \, b_{L-1}\}$ are also placed in a circular buffer. This placement can save the pointer-update operation, since at the end of the current iteration the coefficient pointer automatically wraps around from the end of the buffer that contains b_{L-1} to the beginning of the buffer that contains b_0. We discuss the implementation of FIR filters using circular buffers in Sections 6.3 and 6.4.

6.2 DESIGN OF FINITE-IMPULSE RESPONSE FILTERS USING MATLAB

MATLAB contains a set of functions for the design, analysis, and implementation of FIR filters. In this section, we use FDATool, which incorporates the Filter Design Toolbox and Signal Processing Toolbox, to design, analyze, and quantize FIR filters for practical applications. This powerful tool integrates the conventional command-line filter-design functions under the graphic window environment. A brief introduction of FDATool and related toolboxes is given in Appendix A, and detailed information can also be found in references [11, 12].

6.2.1 Introduction to the Filter Design and Analysis Tool

We can start FDATool by typing

```
fdatool
```

in the MATLAB command window. FDATool opens the default design mode, as illustrated in Fig. 6.9. FDATool contains an interactive GUI, which allows the user to specify the following filter-design parameters:

1. Filter type can be lowpass, highpass, bandpass, bandstop, multiband, differentiator, Hilbert transformer, arbitrary magnitude, arbitrary phase, raised cosine, etc.
2. Filter design methods can be equiripple, least square, and windows for FIR filters. Note that filter-design methods that are not available for the selected filter type (as summarized in Table 6.1) are removed from the **Design Method** menu.
3. Filter order is either user specified or the minimum order determined by the filter-design method.
4. Window specification is only available for FIR-filter designs using Fourier method. There are 16 windows from which to choose.

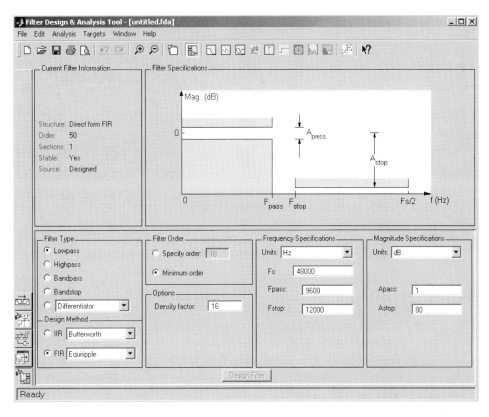

Figure 6.9   The default window of FDATool

5. Frequency specification can be sampling frequency (f_s), passband-edge frequencies, or stopband-edge frequencies, as illustrated in Figures 6.5 and 6.6.
6. Magnitude specification can be passband and stopband ripples in linear or dB units.

A graphic filter-design specification is shown in the upper-right side of the window to guide the design process. The default frequency unit is Hz. However, we can change the unit to kHz, MHz, or a normalized frequency (0 to 1, where 1 corresponds to $\omega = \pi$ or $f = f_s/2$) by selecting the **Units** menu in the **Frequency Specifications** window. Similarly, the default magnitude unit is dB (for A_p and A_s), and we can change it to linear (for δ_p and δ_s) by selecting the **Units** menu in the **Magnitude Specifications** window. The relationships between the linear and dB scales used in MATLAB are defined in Eqs. (6.1.18) and (6.1.20). The upper-left window shows the **Current Filter Information** (e.g., structure, order, stability, sections) for the designed filter.

One useful feature is the **Convert Structure** option from the **Edit** pull-down menu. When this option is selected, a window pops up, as shown in Fig. 6.10. This window allows the user to select the desired filter structure used for realization. For FIR filters, we can select the direct-form structure shown in Fig. 6.2 or the direct-form-transpose structure shown in Fig. 6.3.

Figure 6.10 Window for converting a
filter structure in FDATool

6.2.2 Finite-Impulse Response Filter Design and Analysis

In this section, we use an example to illustrate the process of designing an FIR filter using FDATool. The filter specifications are given as follows:

1. A bandpass FIR filter should be designed using the equiripple design method with minimum order.
2. The frequency specifications in Hz are Fstop1 = 800 Hz, Fpass1 = 1,000 Hz, Fpass2 = 1,300 Hz, Fstop2 = 1,500 Hz, and the sampling frequency f_s = 8,000 Hz.
3. The magnitude specifications for the lower and upper stopband ripples are 50 dB (A_{s1}) and 60 dB (A_{s2}), respectively. The passband ripple is limited to 1 dB (A_p).

In designing a bandpass filter, as shown in Fig. 6.6, we can specify the passband with two frequencies: Fpass1 determines the lower edge of the passband, and Fpass2 specifies the upper edge of the passband. Similarly, Fstop1 determines the upper edge of the first stopband, and Fstop2 specifies the lower edge of the second stopband. Enter the preceding parameters into FDATool, and click on the **Design Filter** button to compute the filter coefficients.

After we have designed the filter, we can analyze the filter characteristics. As shown in Fig. 6.9, we can access the analysis methods by selecting the method listed in the **Analysis** menu or by clicking on the toolbar buttons. For example, by selecting **Magnitude Response** from the **Analysis** menu, the magnitude-response plot is displayed in the upper-right window, as shown in Fig. 6.11. The minimum order of this filter using the equiripple method is 79. The magnitude response allows the user to check if the designed filter meets the given frequency specifications in terms of its ripple tolerances and edge frequencies.

FDATool also provides additional displays for showing the filter's characteristics. These plots include phase response, group delay, impulse response, step response, pole/zero plot, and filter coefficients. The impulse response of an FIR filter displays the coefficients of the designed filter. The pole/zero plot shows the locations of poles

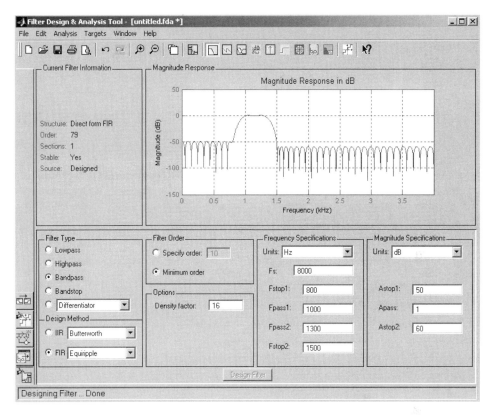

Figure 6.11   Magnitude response and information regarding the designed filter

and zeros. For example, the phase response and impulse response are displayed in Fig. 6.12. The phase response (Fig. 6.12a) shows that the filter has a linear phase within the passband, which is one of the important characteristics of the FIR filter. This linear-phase filter is obtained by having symmetric coefficients, as shown in the impulse response (Fig. 6.12b).

The filter coefficients can be exported to the MATLAB workspace by clicking on the main menu, **FILE → Export**, and a small window pops up, as shown in Fig. 6.13. Besides exporting to the workspace, we can export to the text file or the MAT-file. The coefficients can also be exported to a C header file with the required data type by selecting **FILE → Export to C Header file**. In Section 6.6, we export designed FIR-filter coefficients to a C header file, which can be included in a C program. In addition, we can export the designed filter to SPTool, which was introduced in Chapter 2, by selecting **FILE → Export to SPTool** for filtering the input signal.

6.2.3 Quantization Process and Analysis

We have used FDATool to design an FIR filter with the given specifications. The filter coefficients are represented using double-precision floating-point format,

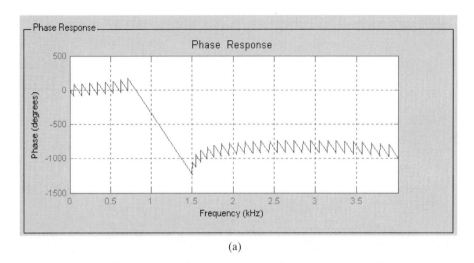

(a)

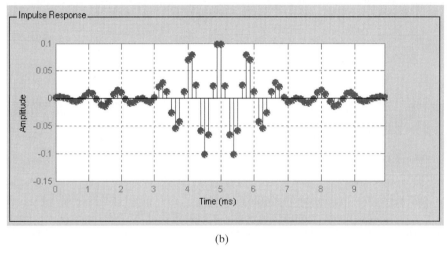

(b)

Figure 6.12   Filters characteristics: (a) phase response and (b) impulse response

Figure 6.13   Window that exports filter coefficients to the MATLAB workspace

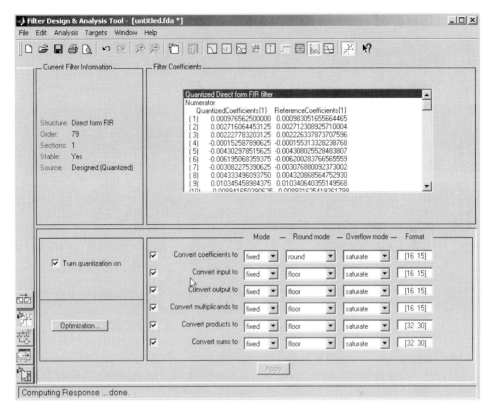

Figure 6.14   Setting quantization parameters window

which requires a 64-bit wordlength. We have to quantize these coefficients either to single-precision floating-point numbers (32 bits) for implementation on the floating-point processors introduced in Chapter 5 or to fixed-point numbers (16 bits or 24 bits) for the fixed-point DSP processors introduced in Chapter 4. To analyze the quantization effects, we quantize the filter coefficients to fixed-point representations.

We can turn on quantization mode by clicking on the **Set Quantization Parameters** icon located at the left side of window shown in Fig. 6.11 and clicking on the **Turn quantization on** box. A new window shown in Fig. 6.14, replaces the previous display. This new window allows the user to set quantization parameters for filter coefficients, input, output, multiplicands, products, and sums. These quantization parameters can be set individually in terms of the Mode, Round mode, Overflow mode, and Format.

The Mode allows the user to choose the following options:

1. **fixed**, which is signed or unsigned fixed-point arithmetic with a wordlength from 2 to 53 bits and a fractional bit from 0 to one less than the given wordlength.

2. **float**, which is user-specified floating-point arithmetic with a wordlength from 2 to 64 bits and an exponent length from 0 to 11 bits.

3. **double**, which is double-precision IEEE-754 floating-point format using 64-bit wordlength with 11 exponent bits.

4. **single**, which is single-precision IEEE-754 floating-point format using 32-bit wordlength with an 8-bit exponent.

There are five arithmetic rounding modes: `ceil` (round up to the next quantized value), `convergent` (round to the nearest quantized value–numbers halfway between two quantized values are rounded up if the LSB is 1), `fix` (round negative numbers up and round positive numbers down to the next quantized value), `floor` (round down to the next quantized value), and `round` (round to the nearest quantized value–numbers halfway between two quantized values are rounded up).

Overflow mode can also be set to either `saturated` or `wrap` on overflow. The data format can be set for fixed and float mode. In fixed-point mode, [N n] specifies the total number of bits (wordlength) by N and the fractional length by n. The default setting for coefficients, input, output, and multiplicands is [16 15] (Q.15). However, the default setting for products and sums is [32 30] (Q1.30), which uses twice the wordlength to simulate a double-precision accumulator in fixed-point DSP processors. In floating-point mode, the format is specified by [N e], where e represents the total number of exponent bits.

We can set these parameters properly to mimic the actual arithmetic operations performed by DSP processors in implementing quantized digital filters. Doing so allows the user to evaluate the effects of representing filter coefficients and signal values using finite wordlength and performing the fixed-point arithmetic accurately.

We examine the quantization effects of the designed FIR filter as follows:

1. Quantize the coefficients, input, output, and multiplicands to Q.15 format. However, represent the products and sums using Q.31 format.

2. Repeat the same experiment using Q.7 format for coefficients, input, output, and multiplicands, but use Q.15 format for products and sums.

The format settings for Q.31, Q.15, and Q.7 are [32 31], [16 15], and [8 7], respectively. We also set the mode, round mode, and overflow mode to `fixed`, `floor`, and `saturation`, respectively.

We can compare the responses of the quantized filter using Q.15 format and the reference filter using double-precision floating-point format. The performance of the fixed-point Q.15 filter is close to that of the reference filter. However, when using the smaller wordlength of Q.7, there is noticeable degradation, as shown by the magnitude responses in Fig. 6.15. The magnitude responses show that the quantized filter has very large stopband ripples (less stopband attenuation) that are unacceptable based on the frequency specifications. We can also evaluate quantization errors in phase, impulse, and step responses by clicking on the corresponding options available from the **Analysis** menu.

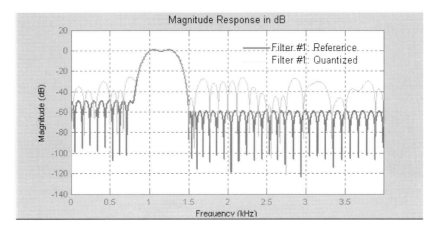

Figure 6.15 Magnitude responses of the reference filter and the quantized filter using Q.7 format

We can view the quantized coefficients by selecting the **Filter Coefficients** option from the **Analysis** menu. The quantized coefficients in Q.7 format and the reference coefficients in double-precision format are displayed side by side, as shown in Fig. 6.16. The symbols +, −, and 0 may appear in the leftmost column of the listing. The symbol + indicates that the quantized coefficient overflowed towards $+\infty$, the symbol − indicates that the quantized coefficient overflowed towards $-\infty$, and the symbol 0 marks the coefficient that has been quantized to zero. In the example of using the reduced-precision Q.7, many coefficients are quantized to zero, as shown in Fig. 6.16, thus degrading the filter performance significantly.

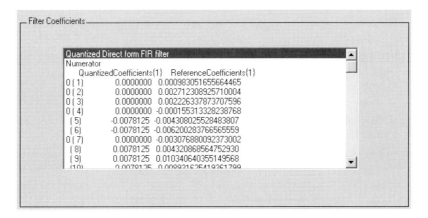

Figure 6.16 List of quantized coefficients vs. reference-filter coefficients

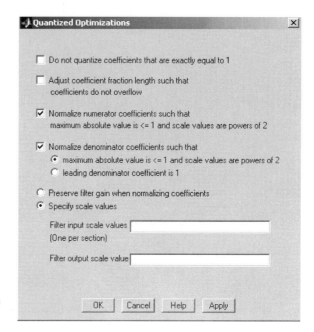

Figure 6.17   **Quantized Optimizations** window

An **Optimization** button is located at the bottom-left corner of Fig. 6.14. Clicking on the button opens the optimization window displayed in Fig. 6.17. This option scales the reference-filter coefficients and quantizes the reference filter again. It also scales the input and output of the filter. We can scale the input and output differently by specifying [input_scale output_scale] or by using a single scalar value for both the input and the output. In our example, we use a scalar value of 1 since all of the coefficients are less than 1. We examine the properties of scale value in more detail in Chapter 7 when we introduce cascade IIR-filter structures.

The latest version of FDATool provides the option of exporting coefficients to the CCS integrated development environment (IDE) and to the XILINX Coefficient (.COE) file. The first option is activated by clicking on **Targets → Export to Code Composer Studio (tm) IDE**. This useful feature allows the user to export filter coefficients directly into the CCS environment as a C header file or to write directly into DSP processor memory in different data types. It also allows the user to select the target DSP platform if more than one DSP board is configured.

In order to implement the designed filter in C, we can export the quantized coefficients to a C header file. This export can be done by selecting **FILE → Export to C Header File**, and a window appears, as shown in Fig. 6.18. The filter coefficients are stored in arrays B of length BL. Table 6.3 shows a C header file generated by FDATool that consists of 80 coefficients represented with signed 16-bit integers. Note that we can change the variable names BL and B in the header file (shown in Table 6.3) by changing the **Numerator** and **Numerator length** fields shown in Fig. 6.18, respectively. We include the coefficient header file in the C main program in Section 6.6 for experiments.

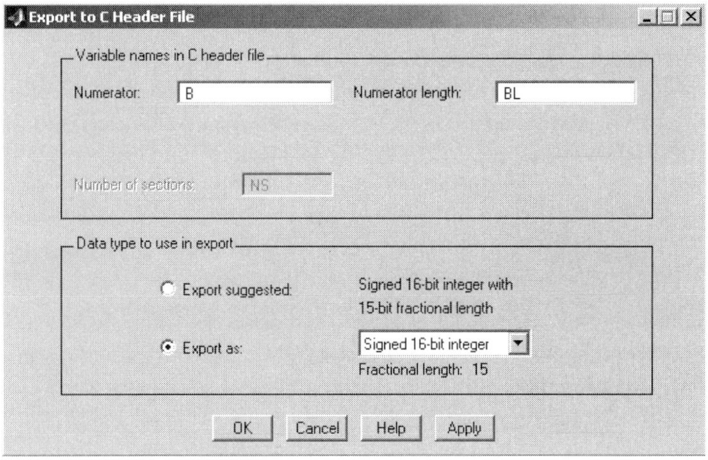

Figure 6.18   Export to C Header File window

TABLE 6.3   Listing of the C Header File

```
/*
* Filter Design and Analysis Tool - Generated Filter Coefficients - C Source
* Generated by MATLAB - Signal Processing Toolbox
*/
/* General type conversion for MATLAB generated C-code   */
#include "tmwtypes.h"
/*
* Expected path to tmwtypes.h
* C:\MATLAB6p5\extern\include\tmwtypes.h
*/
const int BL = 80;
const int16_T B[80] = {
        32,       89,       73,       -5,     -141,     -203,     -101,      142,      339,
       293,      -21,     -378,     -474,     -194,      253,      499,      350,      -34,
      -300,     -256,      -54,        5,     -150,     -235,       72,      665,      949,
       377,     -871,    -1800,    -1382,      399,     2266,     2574,      769,    -1921,
     -3342,    -2197,      755,     3217,     3217,      755,    -2197,    -3342,    -1921,
       769,     2574,     2266,      399,    -1382,    -1800,     -871,      377,      949,
       665,       72,     -235,     -150,        5,      -54,     -256,     -300,      -34,
       350,      499,      253,     -194,     -474,     -378,      -21,      293,      339,
       142,     2101,     -203,     -141,       -5,       73,       89,       32
};
```

6.3 FIXED-POINT IMPLEMENTATIONS

Real-valued FIR filters are commonly used for practical DSP applications. The basic operation for FIR filtering is the MAC operation. In this section, we introduce the implementation of FIR filters using the fixed-point DSP processors introduced in Chapter 4.

6.3.1 Implementation Using the TMS320C2000

In order to produce the fastest possible FIR-filtering routine, signal buffers and coefficient vectors are stored in data RAM. As introduced in Section 4.2, the C2000 has 544 words of fast on-chip DARAM, which are divided into three blocks: B0 (256 words), B1 (256 words), and B2 (32 words). Block B0 is configurable as either data memory or program memory. For FIR filtering, B0 is used to store filter coefficients, and B1 is used as data-buffer memory.

The architecture of the C2000 is optimized to implement FIR filters. After execution of the instruction

```
SETC    CNF      ; configure block B0 as program memory
```

the filter coefficients b_i from B0 (via the program bus) and data $x(n - i)$ from B1 (via the data bus) are available simultaneously for the parallel multiplier. The MACD instruction enables multiply-accumulate, data move, and pointer update to be accomplished in a single cycle, since the filter coefficients are stored in on-chip DARAM. The implementation of the inner product in Eq. (6.1.1) can be made even more efficient with the repeat instruction RPT. Thus, an FIR filter of length L can be implemented as follows:

```
MPY     #0           ; clear P register
LACC    #1,15        ; load rounding offset in ACC
LAR     AR1,LASTAP   ; point to x(n-L+1)
RPT     #L-1         ; repeat next instruction L times
MACD    COEFFP,*-    ; MAC and decrement data pointer
APAC                 ; add the last product to ACC
```

In the preceding code, AR1 is an address register that initially points to $x(n - L + 1)$, and the PC points to the last filter coefficient b_{L-1}. The coefficients are stored in B0, as shown in Fig. 6.19. When the MACD instruction is repeated, the coefficient address

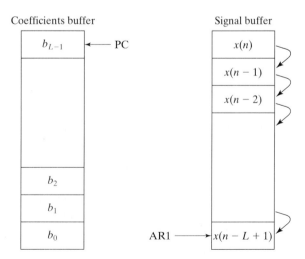

Figure 6.19 Memory layout for FIR filtering using the MACD instruction

TABLE 6.4   Summary of FIR-Filter Routines for the C24x DSP Library [13]

Routine name	Description
FIRFILT_GEN	Generic FIR filter using a linear buffer (looped code)
FIRFILT_ORD10	Tenth-order FIR filter using a linear buffer
FIRFILT_ORD20	Twentieth-order FIR filter using a linear buffer
FIRFILT_CGEN	Generic FIR filter using a circular buffer (looped code)
FIRFILT_CORD10	Tenth order FIR filter using a circular buffer
FIRFILT_CORD20	Twentieth order FIR filter using a circular buffer

contained in the PC is incremented by one during its operation. The MACD in the repeat mode also copies the data in B1 pointed at by AR1 to the next (higher) on-chip RAM location, which is also depicted in Fig. 6.19.

In general, rounding (or truncation) occurs after each multiplication, producing a roundoff-noise component. However, the C2000 processor has a 16-bit multiplier and a 32-bit accumulator, so there is no truncation when summing the set of products in Eq. (6.1.1). All multiplication products are presented and accumulated in full precision, and rounding is performed only after they are all summed. Thus, we get $y(n)$ from the accumulator with only one rounding, which minimizes the roundoff noise in the output $y(n)$.

Texas Instruments provides the C24x Filter Library, which contains the set of FIR routines listed in Table 6.4, which can be called by C programs [13]. FIR filters require a delay buffer to store the past inputs, and the buffer is implemented using two methods: linear buffer and circular buffer. A linear buffer is implemented using the DMOV instruction, and it requires the delay buffer to be placed in the internal DARAM blocks. A circular buffer is implemented using the bit-reversed addressing technique, and it requires appropriate buffer alignment. The C24x Filter Library provides FIR-filter modules for both the generic and fixed-order form. Generic FIR-filter modules are implemented in looped code, and the fixed-order implementation unrolls the loop. Hence, generic FIR filters need a longer execution time than fixed-order FIR filters.

6.3.2 Implementation Using the TMS320C54x

C54x processors have a 17-bit $\times$ 17-bit multiplier and a 40-bit adder to perform a MAC operation in a single cycle. As described in Chapter 4, the CPU of the C54x processor also contains an advanced multiple-bus architecture with one program bus, three data buses, and four associated address buses. These special hardware features support single-instruction repeat, block repeat, block-memory move, and circular buffering, which are required for efficient implementation of FIR filtering. In addition to special hardware and instructions, the C54x processor also allows data and program memory be overlapped by setting the OVLY bit in the PMST register. This setting provides fast access of filter coefficients since both data and program use fast on-chip DARAM.

In the following sections, the methods of configuring memory for data and filter coefficients, circular addressing, and repeat operations are introduced for the C54x in both sample- and block-processing modes. In addition, we also introduce the implementation of symmetric FIR filters.

Sample Processing

In sample-by-sample processing, L-tap FIR filtering is performed when the processor receives a new sample $x(n)$ at each sampling period. This process is repeated for every input sample to the filter.

The first step in the program is to copy the filter coefficients from slow off-chip memory to fast on-chip DARAM. For example, to perform 16-tap FIR filtering, the 16 coefficients (starting at address `init_coeff`) are copied from program memory to data memory (starting at address `coeff`) using the following repeated MVPD instruction:

```
STM    #coeff,AR1        ; AR1 points to on-chip coefficient buffer
RPT    #15              ; repeat next instruction 16 times
MVPD   #init_coeff,*AR1+  ; copy the coefficients
```

The second step is to set up the address pointers to point at the signal and coefficient buffers as follows:

```
STM    #coeff,AR3   ; AR3 points to coefficient buffer
STM    #data,AR2    ; AR2 points to signal buffer
```

The circular buffer can also be set up by initializing the BK register to the length of the FIR filter. In this case, set BK to 16 using the following instruction:

```
STM   #16,BK      ; set circular buffer size to 16
```

If both the coefficient and signal arrays are configured as circular buffers of length L, the starting addresses of these buffers must be properly aligned to the next 2^L boundary that is greater than the value of BK. In this case, the first address must be a multiple of 32 or the last five binary digits of the address are all 0, as shown in Fig. 6.20. The alignment can be set by issuing the command `align 32` in the linker-command file. Note that only one circular-buffer size can be set in the C54x at any given time because of the single BK register. This limitation restricts the implementation of multiple FIR filters with different sizes on the C54x using circular buffers.

The final part of FIR filtering is the execution of MAC instructions in repeat mode as follows:

```
loop:   RPTZ   A,#14              ; repeat next instruction 15 times
        MAC    *AR3+0%,*AR2+0%,A  ; FIR filtering
        MAC    *AR3+0%,*AR2,A     ; the last product
        STH    A,*(y)             ; store y(n)
        PORTW  *(y),out_port
        PORTR  in_port,*AR2
        B      loop
```

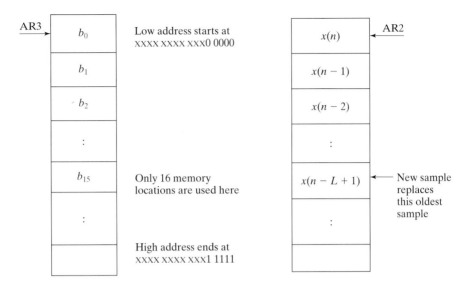

Figure 6.20   Circular buffers for both the coefficient and the signal

The repeat instruction, RPTZ, clears the accumulator A to zero and repeats the first MAC instruction 15 times. The second MAC instruction

```
MAC   *AR3+0%,*AR2,A
```

does not increment the data pointer (AR2) so that it still points at the oldest data sample, $x(n - L + 1)$, which is replaced by the new data sample, $x(n + 1)$, for next iteration, as shown in Fig. 6.20. The final result (filter output) in the upper 16 bits of accumulator A is stored at address y. Note that if the fractional mode is set to one by using the instruction SSBX FRCT, the multiplication result is automatically shifted left 1 bit, and only the upper 16 bits are stored. Furthermore, +0% is the only modifier supported by circular buffering in the dual-operand MAC instruction. Therefore, AR0 needs to be initialized to 1 using the instruction STM #1,AR0 before circular addressing is used. The PORTW and PORTR instructions write the result to out_port and copy the new input sample from in_port, respectively. The FIR-filtering loop is now ready for processing the new input sample for the next iteration.

Block Processing

The sample processing code described in the previous section can be modified for processing a block of signal samples. The principle of block processing was explained in Chapter 3. For example, to process a block of eight samples, the length of the signal buffer equals the number of coefficients (e.g., a 16-tap filter) plus the number of block samples minus 1 (e.g., $16 + 8 - 1 = 23$). Instead of using a single repeat in sample processing, a block repeat (instructions in italic) is used as follows:

```
        STM    #7,BRC           ; block repeat counter set to 7
        STM    #y,AR1           ; AR1 points to output y(n)
        STM    #data,AR2        ; AR2 points to signal buffer
        STM    #coeff,AR3       ; AR3 points to coefficient buffer
        STM    #16,BK           ; initialize circular buffer size to 16
        STM    #1,AR0           ; offset for circular points
        RPTB   blk_loop-1       ; repeats block of code 8 till blk_loop

        RPTZ   A,#15            ; repeat next instruction 16 times
fir:    MAC    *AR3+0%,*AR2+,A  ; FIR filtering
        MAR    *+AR2(#-8)       ; modify AR2 to access the next block
        STH    A,*AR1+          ; store y(n)
blk_loop:   RET
```

For block-repeat operations, the RSA (start address) register is automatically set at the next line of code after RPTB. The REA (end address) register is also specified by the RPTB instruction as the address of the label blk_loop minus 1. The programmer only needs to set the BRC (repeat loop count) register as the loop count minus 1. Note that AR2 no longer is implemented as a circular pointer and must be modified manually using the MAR instruction, which resets the pointer to the start of the next block for subsequent block processing.

Symmetric Finite-Impulse Response Filters

The C54x processor has a special instruction called FIRS for implementing the symmetric FIR filter, as defined in Eq. (6.1.16). This instruction performs both multiply-and-accumulate and the summation of the two signal samples in a single cycle. This instruction can be repeated as follows:

```
    ADD    *AR3+,*AR2-,A       ; add two corresponding samples
    RPTZ   B,#(L/2-1)          ; repeat next instruction L/2 times
    FIRS   *AR3+,*AR2-,#coeff  ; symmetric FIR filtering
```

In the preceding code, AR3 and AR2 are registers that point at the newest data sample, $x(n)$, and the oldest data sample, $x(n - L + 1)$, respectively. As shown in Fig. 6.4 and Eq. (6.1.16), the ADD instruction adds these two data samples and stores the sum in accumulator A. This first step must be done before the repeat of the FIRS instruction is entered. The repeat counter is set to $(L/2) - 1$, as described in Eq. (6.1.16). The FIRS instruction performs the following two operations in parallel $(L/2)$ times:

1. Multiply the previous value in AH (the sum) with the coefficient b_{i-1}, and store the product in accumulator B.
2. Add the next data pair, $x(n - i)$ and $x(n - L + 1 + i)$, store the sum in the AH, and update the pointers as AR3+ and AR2- for the next iteration.

After one complete loop for the symmetric FIR filtering, the data pointers must be reset to the new addresses.

Benchmarks

Texas Instruments provides the *TMS320C54x DSP Library Programmer's Reference* [14], which contains a set of assembly-optimized functions for C programming on C54x processors. In the library, there are five functions, listed in Table 6.5, that are dedicated to FIR filtering. This table also summarizes the cycle count and the code size (in 16-bit words) for using these functions. From the table, we notice that the symmetric FIR-filter function, firs2, has no advantage over the direct-form FIR-filter function, fir, in terms of execution speed and code size. This finding is in contrast to the previous result in Section 6.1.3 indicating that the symmetric FIR filter just needs half of the multiplications when compared to the direct-form FIR filter. The reason is that the symmetric FIR filter requires extra steps to handle the pointers (two instead of one) for the signal samples. This overhead exceeds the advantage gained by halving the multiplications required by the symmetric FIR filter. Another symmetric FIR-filter function, firs, restricts the coefficients to be stored in the program memory, which results in a faster benchmark cycle. However, the cycle advantage over the direct-form FIR function, fir, only takes place when the number of coefficients, L, exceeds 18 taps. Another advantage in using the symmetric FIR filter is the memory saved as a result of storing only half of the coefficients of the FIR filter.

The DSPLIB also provides functions that support decimation (discarding every $D - 1$ samples) and interpolation (inserting $I - 1$ samples) FIR filters. The details of these interpolation and decimation filters are discussed in Section 6.5.

6.3.3 Implementation Using the TMS320C55x

Efficient implementation of FIR filters on C55x processors requires specialized arrangements that can take advantage of dual-MAC units. As discussed in Chapter 4, the CPU has only three 16-bit data buses, as shown in Fig. 6.21, but the computations of two sums of products require four data values (two filter coefficients and two signal samples). Therefore, the CPU needs a special arrangement for performing two MACs per clock cycle to maximize the throughput of FIR filtering on C55x processors.

TABLE 6.5 FIR-Filter Functions and Benchmarks [14], where N is the Length of the Block Size

FIR function	Description	Cycle count	Code size
fir	Direct-form FIR filter	$4 + N \times (4 + L) + 34$	42
firdec	Decimation FIR filter with a decimating factor D	$(N/D) \times [12 + L + 4(D - 1)] + 86$	67
firinterp	Interpolation FIR filter with an interpolating factor I	$N \times [6 + (I - 1) \times (17 + L/I)] + 88$	74
firs	Symmetric FIR filter	$N \times (16 + L/2) + 35$	56
firs2	Symmetric FIR filter with no restriction on placement of coefficients	$N \times (15 + 2 \times L/2) + 43$	58

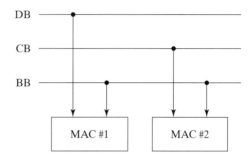

Figure 6.21 Simplified dual-MAC
architecture on the C55x

Sample Processing

In a real-time application of the FIR filter with sample-by-sample processing, one new input is received and stored in $x(n)$, and the signal buffer is updated. The new set of samples is used to compute one output $y(n)$, as described by Eq. (6.1.1). To use the dual-MAC architecture shown in Fig. 6.21 efficiently, we have to interlace the calculations of the current sampling period with those of the next iteration in order to achieve a net performance of two MAC operations per cycle.

To explain this method, the computations of the four-tap FIR filter are expressed as follows:

$$y(n) = b_0 x(n) + b_1 x(n-1) + b_2 x(n-2) + b_3 x(n-3) \qquad (6.3.1)$$

$$y(n+1) = b_0 x(n+1) + b_1 x(n) + b_2 x(n-1) + b_3 x(n-2) \qquad (6.3.2)$$

At current time n, the signal sample $x(n)$ is multiplied by b_0 to obtain product $b_0 x(n)$ for generating $y(n)$, as described in Eq. (6.3.1). This sample value is multiplied by b_1 at the next iteration (at time $n+1$) to obtain $b_1 x(n)$ for generating the next output, $y(n+1)$, given in Eq. (6.3.2). Therefore, we can access this common data $x(n)$, compute $b_0 x(n)$ using one of the MAC units, and precompute $b_1 x(n)$ using the other MAC unit. In this way, these two MAC operations require only three different values from memory; thus, they can be supported effectively by three data buses, as shown in Fig. 6.21. Similarly, we can access the common data $x(n-2)$, compute $b_2 x(n-2)$ using one of the MAC units, and precompute $b_3 x(n-2)$ using the other MAC unit. Note that to obtain $y(n)$, we also need the product terms $b_1 x(n-1)$ and $b_3 x(n-3)$. However, these terms were computed during the previous sampling period at time $n-1$. In this arrangement, two separate running sums are maintained in different accumulators, one with partial products for the current time n and the other with precalculated terms for the next time, $n+1$. At the next time, the precalculated running sum becomes the initial value of the current running sum, and a new precalculated running sum is started from zero. At the end of each sampling period, the current running sum contains the current filter output.

This technique can be applied to any FIR filter with an even number of coefficients. For filters with an odd number of coefficients, the computation groupings become problematic since the last grouping is missing the precalculated term from the previous sampling period. To overcome this problem, we can expand the filter length to the next higher even number by using a zero coefficient for the additional term.

The complete C55x assembly program that implements sample-by-sample FIR filtering using a dual-MAC is available in [15]. The program consists of two parts: a main routine and an ISR. The filtering operation is implemented in the ISR, and part of the computation kernel is listed as follows:

```
      MAC    *(AR0+T0),*(CDP+T0),AC0    ; first dual-MAC
   ::MPY    *(AR1+T0),*(CDP+T0),AC1    ; these 3 instructions are
   ||RPT    #(N_TAP/2-3)               ; executed in parallel
;
      MAC    *(AR0+T0),*(CDP+T0),AC0    ; main computations
   ::MAC    *(AR1+T0),*(CDP+T0),AC1
;
      MAC    *AR0,*CDP+,AC0             ; final dual MAC
   ::MAC    *AR1,*CDP+,AC1
```

In the code, T0 = 2 is the offset for coefficient address pointers. The address pointer AR0 points at b_0, AR1 points at b_1, and the CDP points at the signal buffer. The accumulator AC0 contains the current running sum, and the precalculated running sum is stored in AC1.

Block Processing

As discussed in Chapter 3, block processing assumes we have a block of $N(\geq 2)$ input samples, and FIR filtering is performed to generate N output samples. In this case, we have at least two signal samples; thus, we can compute two sequential filtering operations in parallel. For example, we compute Eqs. (6.3.1) and (6.3.2) in parallel using the dual-MAC units. For the first product of these two equations, one MAC unit computes $b_0x(n)$, while the other MAC unit computes $b_0x(n + 1)$. Since the filter coefficient b_0 is common to both MAC units, these two computations require only three different values from memory: b_0, $x(n)$, and $x(n + 1)$. Thus, the multiplication process can be implemented effectively using the dual-MAC units shown in Fig. 6.21.

After we obtain two output samples, $y(n)$ and $y(n + 1)$, the first MAC unit computes $y(n + 2)$, and the second MAC unit computes $y(n + 3)$ in parallel for the next iteration. In this case, the C55x performs two MAC operations per clock cycle, and we need $N/2$ cycles for computing N output samples. This method can handle filters with either an even or an odd number of coefficients. Since only two new input samples are required at each iteration in computing two output samples, the minimum block size for supporting this technique is $N = 2$.

Block FIR filtering is implemented with C55x assembly code [15], and part of the computation kernel is listed as follows:

```
   ||RPTBLOCAL end_outer            ; the outer loop
;
      MPY    *AR0-,*CDP+,AC0         ; first dual-MAC
   ::MPY    *AR1-,*CDP+,AC1
```

```
  ||RPT    #(N_TAP-3)                      ; single repeat inner loop
;
    MAC    *AR0-,*CDP+,AC0                 ; main computation
  ::MAC    *AR1-,*CDP+,AC1
;
    MAC    *(AR0+T1),*(CDP+T0),AC0         ; final dual MAC
  ::MAC    *(AR1+T1),*(CDP+T0),AC1
end_outer:
    MOV    pair(HI(AC0)),dbl(*AR2+)   ; write both results
```

In the code, $T0 = -(N\_TAP - 1)$ is used to rewind the coefficient pointer CDP, and $T1 = N\_TAP + 1$ is used to rewind the data pointers AR0 and AR1. The accumulator AC0 contains the current filter output $y(n)$, and the next output $y(n + 1)$ is stored in AC1.

Symmetric Finite-Impulse Response Filters

As shown in Fig. 6.4 and Eq. (6.1.16), a symmetric FIR filter that requires only three values $[b_i, x(n - i)$, and $x(n - L + 1 + i)]$ could be implemented effectively using the dual-MAC architecture shown in Fig. 6.21. To support this popular linear-phase filter, the C55x processor provides the following two powerful instructions:

```
FIRSADD   Xmem, Ymem, Cmem, ACx, Acy
FIRSSUB   Xmem, Ymem, Cmem, ACx, Acy
```

The instruction FIRSADD supports Type 1 and Type 2 even symmetric filters, while the instruction FIRSSUB supports Type 3 and Type 4 odd symmetric (antisymmetric) filters.

The FIRSADD instruction performs the following parallel operations:

```
ACy = ACy+(ACx×Cmem)
::ACx = (Xmem<<#16)+(Ymem<<#16)
```

For example, in the following instruction

```
FIRSADD *AR0,*AR1,*CDP,AC0,AC1
```

the signal samples $x(n - i)$ and $x(n - L + 1 + i)$ are pointed at by AR0 and AR1, respectively, and the coefficient b_i is pointed at by the CDP. The sum of $x(n - i)$ and $x(n - L + 1 + i)$ is stored in AC0, and the product is stored in AC1. Note that the coefficients b_i are accessed through the BB bus, which is connected only to the internal memory. Symmetric FIR filtering is implemented using C55x assembly code [15], and part of the computational kernel is listed as follows:

```
; clear AC0 and pre-load AC1 with the sum of the 1st and last inputs
    ADD     *db_ptr1+,*db_ptr2-,AC1

; inner loop
  ||RPT inner_cnt
    FIRSADD  *db_ptr1+,*db_ptr2-,*h_ptr+,AC1,AC0
```

```
; 2nd to last iteration has different pointer adjustment
   FIRSADD   *(db_ptr1-T0),*(db_ptr2+T1),*h_ptr+,AC1,AC0

; last iteration is a MAC with rounding
   MACMR     *h_ptr+,AC1,AC0
```

Similar to Eq. (6.1.16), the antisymmetric FIR filter can be expressed as

$$y(n) = \sum_{i=0}^{L/2-1} b_i[x(n-i) - x(n-L+1+i)]. \qquad (6.3.3)$$

This operation is supported by the FIRSSUB instruction, which performs the following parallel instructions:

```
ACy = ACy+(ACx×Cmem)
::ACx = (Xmem<<#16)-(Ymem<<#16)
```

The symmetric or antisymmetric filter implemented by FIRSADD or FIRSSUB may result in fewer overhead cycles for pointer setup, context save and restore, etc., when compared with the sample-processing and block-processing techniques introduced in previous subsections. In addition, FIRSADD and FIRSSUB may result in lower power consumption since they perform one multiplication and two additions for every two coefficients, whereas the other two methods (sample and block processing) perform two multiplications and one addition.

Texas Instruments provides the TMS320C55x DSPLIB [16], which contains a set of assembly-coded FIR-filtering functions for C programs. These functions are summarized in Table 6.6. Note that N is the block size, with $N = 1$ for sample processing and $N \geq 2$ for block processing, and the code size is in bytes. The last numeric numbers in the cycle-count column represent the overhead for setting the computational kernel.

6.3.4 Implementation Using the TMS320C62x

As explained in Chapter 4, the C62x processor uses the advanced VLIW architecture. It has eight functional units, and each has a multiplier and three ALUs. Thus, up

TABLE 6.6   FIR-Filtering Functions and Benchmarks for the C55x [16], where N is the Length of the Block Size

FIR function	Description	Cycle count	Code size
fir	Direct-form FIR filter	$N \times (2 + L) + 25$	107
fir2	Block FIR filter using a dual MAC	$(N/2) \times [9 + (L - 2)] + 32$	134
firdec	Decimating FIR filter	$(N/D) \times [10 + L + (D - 1)] + 67$	144
firinterp	Interpolating FIR filter	$N \times [2 + I \times (1 + L/I)] + 72$	164
firs	Symmetric FIR filter	$N \times (5 + L/2) + 72$	133

to eight instructions can be executed in a single cycle. Due to the highly efficient C complier available for the C6000 family, code is normally written in C. The C62x also allows the user to write in linear assembly and lets the assembly optimizer handle parallelism and scheduling. In this section, we explore the implementation of a FIR filter using C and linear assembly for C62x processors.

Floating-point C programs for FIR filtering were introduced in Section 6.1.6. FIR filters can be written in C without much knowledge of the C6000 architecture. For example, the fixed-point C code for the FIR filter can be simply written as

```
for (i=0; i<L; i++)
    yout += b[i]*x[i];
```

In fixed-point C code, the variables b and x are declared as short (16 bits), and the output variable yout is declared as int (32 bits).

As explained in Chapter 4, C6000 intrinsics can be used to map special functions, such as MAC operations, directly to C6000 instructions. The use of intrinsics increases code performance. For example, the preceding C code can be modified to include the _mpy intrinsic as follows:

```
for (i=0; i<L; i++)
    yout += ((_mpy(b[i],x[i]))<< 1);
```

Note that the multiplication is performed using Q.15 format; thus, the product must be shifted left 1 bit to get rid of the extra sign bit.

The code can also be written in linear assembly as follows:

```
fir_loop      LDH     *x_pointer,x
              LDH     *b_pointer,b
              MPY     x,b,yout
              SHL     prod,1,prod
              ADD     yout,prod,yout
              SUB     loop_count,1,loop_count
[loop_count]B         fir_loop
```

In the linear assembly code, we let the assembly optimizer optimize the code. The linear assembly code can be further optimized by using a dependency graph, performing resource allocation, and using a scheduling table [17].

As discussed in Section 6.1.6, we have to update the signal buffer when the new sample is available for FIR filtering. As shown in datamov.c, this update can be achieved by shifting the data sample as follows:

```
for (i=L-1; i>0; --i)
    x[i]=x[i-1];
```

As mentioned earlier, shifting data is not efficient, and circular addressing can be used instead of using datamov.c for data shifting. As shown in Fig. 4.30, circular addressing can be set up by writing to the address-mode register (AMR), and the block sizes are specified by the BK0 and BK1 (5-bit) fields of the AMR. Therefore, two circular buffers with different lengths are allowed in C6000 processors. A value L written into the BK register specifies a block size of 2^{L+1} bytes. Circular buffers can be accessed only by registers A4 to A7 and B4 to B7. These registers can be set to linear or circular (using BK0 or BK1), as shown in Table 4.24.

To set up the circular buffer in C for FIR filtering, the following in-line assembly code must be applied:

```
asm("MVKL   4040h,B2")    ; set up A7 and B7 as circular pointers
asm("MVKH   0002h,B2")    ; block size used is 2^3 = 8 in BK0
asm("MVC    B2, AMR")     ; set up the AMR
```

In addition to setting up the AMR, the signal and coefficient buffers have to be aligned on the buffer-size boundary using the #pragma DATA_ALIGN directive. For example an 8-byte boundary can be aligned using

```
#pragma DATA_ALIGN(b,8)
#pragma DATA_ALIGN(x,8)
```

This directive sets A7 to point at the coefficient b[0] and sets B7 to point at the current sample, x[n]. The coefficient and signal buffers are arranged as shown in Fig. 6.22. After the eighth MAC operation, the pointer A7 points at b[0] again, while B7 points back at x[n]. Since the new sample x[n + 1] replaces the old x[n - 7], B7 must be pre-incremented by 1 before placing the new sample.

FIR filtering can be improved further by performing a pair of MAC operations in one cycle using the _mpy and _mpyh intrinsics as follows:

```
for (i=0; i<L/2; i++)
{
    yout += ((_mpy(b[i],x[i]))<<1);
    yout1 += ((_mpyh(b[i],x[i]))<<1);
}
yout_tot = yout+yout1;
```

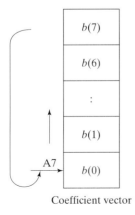
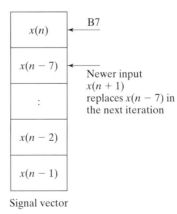

Figure 6.22 Allocation of coefficients and signal samples in circular buffers

TABLE 6.7   FIR-Filtering Functions in the C62x DSPLIB [18], where N is the Length of the Block Size

FIR function	Description	Cycle count	Code size (bytes)
DSP_fir_gen	FIR filter	$[9 + 4 \times \text{ceil}(L/4)] \times \text{ceil}(N/2) + 18$	640
DSP_fir_r4	FIR filter (Radix 4)	$(8 + L) \times N/2 + 14$	544
DSP_fir_r8	FIR filter (Radix 8)	$L \times N/2 + 28$	544
DSP_fir_sym	Symmetric FIR filter	$(3 \times L/2 + 10) \times N/2 + 20$	416
DSP_fir_cplx	Complex FIR filter	$2 \times L \times N + 20$	384

The variables x, h, yout, and yout1 now are specified as int (32 bits). In this way, two MAC operations can be executed in parallel, thus cutting execution time in half. Further enhancement can be achieved in C64x and C67x processors, which can access 64-bit data. This enhancement is explained in the following sections.

Texas Instruments provides the C62x DSPLIB [18], which contains a set of assembly-optimized C functions. By calling these time-critical functions, both the code development time and the program execution time can be reduced significantly. The FIR-filtering functions available in the library are listed in Table 6.7. All of the FIR-filter functions, except the symmetric FIR filter, operate on 16-bit data with a 32-bit accumulation. The symmetric FIR filter operates on 16-bit data with a 40-bit accumulator. There are several restrictions on the size of the coefficients and the number of output samples to be computed for different functions. The FIR-filtering functions with radix 4 and radix 8 imply that the inner loop of the FIR filtering is unrolled four and eight times, respectively. It is noted that the symmetric FIR filter does not provide any advantage in terms of execution time, which is mainly due to the overhead in handling pointers for symmetric FIR filtering.

6.3.5 Implementation Using the TMS320C64x

As introduced in Chapter 4, the C64x processor is an extension of the C62x and has the capability of performing four 16-bit $\times$ 16-bit multiplications in a single cycle. The C64x is also object-code compatible with the C62x processor and contains many 8-bit and 16-bit extensions to the instruction set.

We can take advantage of these new features to implement FIR filters. The C64x processor comes with a special intrinsic (DOTP2) for performing dot products between two pairs of 16-bit numbers in two 32-bit registers. For example, a 64-tap FIR filter using the _dotp2 intrinsic is listed as follows:

```
for (i=0; i<32; i++)
    yout +=_dotp2(b[i],x[i]);
```

The variables b, x, and yout are defined as int (32 bits). Since two 16-bit $\times$ 16-bit MAC operations are performed, the loop count is reduced in half.

TABLE 6.8   FIR-Filtering Functions in the C64x DSPLIB [19], where N is the Length of the Block Size

FIR function	Description	Cycle count	Code size
DSP_fir_gen	FIR filter	$[11 + 4 \times \text{ceil}(L/4)] \times \text{ceil}(N/4) + 18$	544
DSP_fir_r4	FIR filter (Radix 4)	$(8 + L) \times N/4 + 9$	308
DSP_fir_r8	FIR filter (Radix 8)	$L \times N/4 + 17$	336
DSP_fir_sym	Symmetric FIR filter	$(10 \times L/8 + 15) \times N/4 + 26$	664
DSP_fir_cplx	Complex FIR filter	$L \times N + 24$	432

In addition, the C64x processor can access 64-bit data (or four 16-bit data) using the LDDW instruction. In order to perform four 16-bit × 16-bit MAC operations per cycle using two _dotp2 functions, the variables b and x are defined as double, while yout still is defined as int. The 64-bit variable can be specified in terms of its upper 32-bit value and lower 32-bit value using the _hi and _lo intrinsics, respectively. Further, the loop count can be reduced in half using the following approach:

```
for (i=0; i<16; i++)
{
    yout+=_dotp2(_lo(x[i]),_lo(b[i]))+_dotp2(_hi(x[i]),_hi(b[i]));
}
```

In addition to the C62x processor, Texas Instruments also provides the C64x DSPLIB [19], which uses the enhanced features of the C64x architecture and instructions and thus achieves a shorter cycle count when compared to that of the C62x. The benchmark for FIR filtering is shown in Table 6.8. We can compare the differences in terms of cycle count and code size between the C62x and C64x processors.

6.4 FLOATING-POINT IMPLEMENTATIONS

Floating-point formats and floating-point processors are described in Chapters 3 and 5. In this section, we implement FIR filters using floating-point DSP processors TMS320C3x and TMS320C67x introduced in Chapter 5.

6.4.1 Implementation Using the TMS320C3x

As discussed in Chapter 5, the C3x provides two data buses for accessing two data at the same time. This architecture supports parallel-data load, store, or one data load with one data store simultaneously. In addition, the hardware multiplier and ALU are separated, thus making them capable of performing one multiplication and one addition (or subtraction) at the same time in a single cycle. With these two combined features, RPTS, and circular addressing, the FIR filtering given in Eq. (6.1.1) can be implemented as follows [6]:

```
   MPYF3  *AR0++(1)%,*AR1++(1)%,R1    ; b[0]*x[n]
   RPTS   L-2                    ; repeat next parallel instruction
                                 ; L-1 times
   MPYF3  *AR0++(1)%,*AR1++(1)%,R1    ; b[i]*x[n-i]
|| ADDF3  R1,R2,R2               ; sum of products b[i]*x[n-i]
   ADDF3  R1,R2,R2               ; sum the last product
```

In the preceding code, the auxiliary registers AR0 and AR1 point at signal and coefficient buffers. Due to the pipeline architecture, the first instruction computes the first product and puts it into R1. The ADDF3 in the parallel instruction adds the previous product in R1 to R2. At the end of the repeat operation, the final product is saved in R1; thus, we need the last ADDF3 instruction to add the product to R2, which contains the filter output $y(n)$. It is important to note that the code uses circular buffers for both signal samples and filter coefficients.

We can also implement FIR filtering in the following assembly routine (c3xfir.asm):

```
FIR  MPYF3  *AR0++(1)%,*AR1++(1)%,R1  ; bL-1 * x(n-L+1) -> R1
     NOP
     LDF    0.0,R2     ; 0.0 -> R2
     RPTS   RC         ; repeat next parallel instructions L-2 times
     MPYF3  *AR0++(1)%,*AR1++(1)%,R1  ; bi * x(n-i) -> R1
||   ADDF3  R1,R2,R2   ; R2 + R1 -> R2
     ADDF3  R1,R2,R2   ; add last product, R1+R2 -> R2 = y(n)
     RETS              ; return to main program
```

This routine can be called by the assembly main program as follows:

```
LDI   @ADDXN,AR1   ; AR1 points to x(n-L+1)
LDI   @ADDBI,AR0   ; AR0 points to bL-1
LDI   L,BK         ; circular buffer size = L
LDI   L-2,RC       ; L-2 -> repeat counter
LDF   @XN,R2       ; x(n) -> R2
STF   R2,*AR1++(1)% ; store x(n) in buffer and update pointer AR1
STI   AR1,@ADDXN   ; save new address of x(n-L+1)
CALL  FIR          ; call FIR filtering routine to compute y(n)
```

To set up circular addressing, initialize BK to filter length L. The locations for the signal and the coefficient should start from a memory location whose address is a multiple of the smallest power of 2 that is greater than L.

6.4.2 Implementation Using the TMS320C67x

The C67x processor is a floating-point version of the C62x processor. It contains many floating-point instructions, which operate on data represented in either IEEE-754 single-precision (32-bit) or double-precision (64-bit) format. In this section, we study the single-precision floating-point implementation of FIR filtering. The variables b and x are specified as single-precision floating-point numbers. These variables are loaded into the registers and are followed by single-precision multiply (MPYSP) and add (ADDSP) instructions. The following loop is repeated until all of the coefficients in the FIR filter are multiplied by the corresponding signal samples:

```
fir_loop       LDW       *x_pointer++,x
               LDW       *b_pointer++,b
               MPYSP     x,b,yout
               ADDSP     yout,prod,yout
               SUB       loop_count,1,loop_count
  [loop_count]B          fir_loop
```

We can enhance the performance of the code further by using the 64-bit data access available on the C67x processor. Instead of loading a word (LDW), the C67x processor can load a double word (LDDW) that represents two 32-bit signal samples (x1:x0) or two 32-bit coefficients (b1:b0). This process is followed by two MPYSP instructions used to perform single-precision multiplication of x1 × b1 and x0 × b0 independently. The partial products are accumulated in the variables yout1 and yout0 for the even and odd elements using two single-precision ADDSP instructions. The detailed program is listed as follows:

```
fir_loop       LDDW      *x_pointer++,x1:x0
               LDDW      *b_pointer++,b1:b0
               MPYSP     x1,b1,yout1
               MPYSP     x0,b0,yout0
               ADDSP     yout1,prod1,yout1
               ADDSP     yout0,prod0,yout0
               SUB       loop_count,1,loop_count
  [loop_count]B          fir_loop
```

6.5 AN APPLICATION: MULTIRATE SIGNAL PROCESSING

In many applications such as interconnecting DSP systems operating at different sampling rates, sampling-frequency changes are necessary. For example, in telecommunication systems that transmit and receive different types of signals (e.g., facsimile, speech, and video), there is a requirement for processing the various signals at different rates. The process of converting a digital signal to a different sampling rate is called sampling-rate conversion. Systems that employ multiple sampling rates are called multirate DSP systems. The purpose of converting digital signals to a new sampling rate is to make the signals easier to process or to achieve compatibility with other systems. The key processing for sampling-rate conversion is lowpass filtering. Either FIR or IIR filters can be used for this application. However, FIR filters are commonly used since they can be implemented by polyphase structures for saving computation.

A sampling rate increased by an integer factor U is called interpolation (or upsample), while a sampling rate decreased by an integer factor D is called decimation (or downsample). Interpolation and decimation allow the digital system to change the sampling rate by integer factors. One of the main applications of decimation is to eliminate the need for high-quality analog anti-aliasing filters. In a DSP system that uses oversampling and decimation, the analog input is first filtered by a simple analog anti-aliasing filter and then is sampled at the higher rate. The decimation filter then reduces the bandwidth of the sampled digital signal to the desired lower rate. The digital-decimation filter provides high-quality lowpass filtering and reduces the cost of using high-quality analog filters.

In this section, we briefly introduce the principles of decimation and interpolation. Detailed treatments on these subjects can be found in [20, 21].

6.5.1 Interpolation

Interpolation is the process of inserting additional samples between successive samples of the original low-rate signal. The interpolated samples are computed by an FIR filter called an interpolation filter. An example of a two-fold ($U = 2$) interpolator, which increases the sampling rate by a factor of two, is illustrated in Fig. 6.23. For a general interpolator of 1:U rate increases, the upsample process inserts ($U - 1$) zeros in between the successive samples of the original signal $x(n)$ of sampling rate f_s; thus, the sampling rate is increased to Uf_s, or the sampling period is reduced to T/U. This intermediate signal $x(n')$ is filtered by a lowpass filter to produce the final interpolated samples $y(n')$.

The simplest lowpass filter is the direct-form FIR filter $B(z)$ introduced in Section 6.1. This FIR filter can be designed to have a linear phase with the specified passband and stopband-edge frequencies and ripples. Design techniques are introduced in Section 6.1. FDATool, introduced in Section 6.2, can be used for designing this interpolation filter. The interpolating filter operates at the high rate of $f_s' = Uf_s$ with the frequency response

$$B(\omega) = \begin{cases} U, & 0 \le \omega \le \omega_c \\ 0, & \omega_c < \omega \le \pi \end{cases}, \tag{6.5.1}$$

where the cutoff frequency is determined as

$$\omega_c = \frac{\pi}{U} \quad \text{or} \quad f_c = f_s'/2U = f_s/2. \tag{6.5.2}$$

Since the insertion of ($U - 1$) zeros spreads the energy of each signal sample over U output samples, the gain of the filter is U in order to compensate for the energy loss of the upsampling process. Figure 6.23 shows that interpolation increases the sampling rate. However, the bandwidth of the interpolated signal is still the same $f_s/2$ as the original signal.

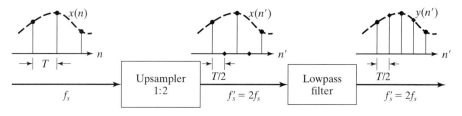

Figure 6.23   Interpolation process, where $U = 2$

Because the upsampler introduces $(U - 1)$ zeros between successive samples of the input signal, only one out of every U input samples to the interpolation filter is nonzero. To implement this filter efficiently, the required filtering operations may be rearranged to operate only on the nonzero samples. Suppose at time n, these nonzero samples are multiplied by the corresponding FIR-filter coefficients $b_0, b_U, b_{2U}, \ldots, b_{L-U}$. At the following time $n + 1$, the nonzero samples are multiplied by the coinciding filter coefficients $b_1, b_{U+1}, b_{2U+1}, \ldots, b_{L-U+1}$. This step can be accomplished by replacing the high-rate FIR filter of length L by U shorter FIR filters $B_m(z), m = 0, 1, \ldots, U - 1$ of length $I = L/U$, which are called polyphase filters operating at the low rate f_s.

Therefore, the interpolation concept shown in Fig. 6.23 is implemented as illustrated in Fig. 6.24, where the set of U polyphase filters are connected in parallel. For each new input sample $x(n)$, U output samples $y_m(n)$ are generated by the corresponding polyphase filters $B_m(z)$. The output of each polyphase filter is computed as

$$y_m(n) = \sum_{i=0}^{I-1} b_{m,i} x(n - i), \quad m = 0, 1, \ldots, U - 1, \tag{6.5.3}$$

where $b_{m,i}$ is i-th coefficient of the m-th polyphase filter $B_m(z)$. The coefficients of the polyphase filter $B_m(z)$ are extracted from the lowpass FIR filter by taking every U-th entry, starting with the m-th entry. That is,

$$b_{m,i} = b_{m+iU}, m = 0, 1, \ldots, U - 1 \text{ and } i = 0, 1, \ldots, I - 1. \tag{6.5.4}$$

As shown in Fig. 6.24, polyphase filters are implemented using the commutative model, which rotates in the counterclockwise direction starting at the top. The interpolated signal $y(n')$ consists of the samples

$$\ldots y_0(n), y_1(n), \ldots y_{U-1}(n), y_0(n + 1), y_1(n + 1), \ldots y_{U-1}(n + 1), \ldots$$

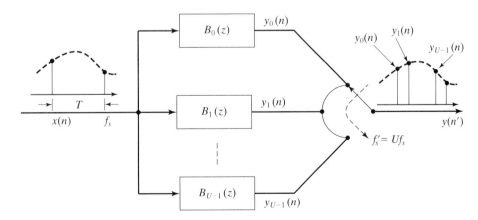

Figure 6.24   Polyphase implementation of an interpolation filter

which means that the m-th interpolated value is computed by the m-th polyphase filter $B_m(z)$. The rotation of the commutator begins with the point at $m = 0$. Thus, the lower-order polyphase filter $B_m(z)$ performs computations at the low sampling rate f_s, and upsampling is achieved from the fact that U output samples $y_m(n), m = 0, 1, \ldots, U - 1$ are generated for each input sample $x(n)$. The computational efficiency of the polyphase filter structure comes from dividing the single L-point FIR filter into a set of smaller filters of length L/U, each of which operates at the lower sampling rate f_s. Furthermore, these U polyphase filters share a single signal buffer of size L/U.

6.5.2 Decimation

As illustrated in Fig. 6.25, decimation of a high-rate signal with sampling rate f_s' by a factor D results in the low-rate $f_s'' = f_s'/D$, which is the reverse process of the interpolation introduced in Section 6.5.1. The downsample by D may be done simply by discarding the $(D - 1)$ samples that are between the low-rate ones. However, decreasing the sampling rate by a factor D reduces the bandwidth by the same factor D. Thus, if the original high-rate signal has frequency components outside the new bandwidth, aliasing occurs. This problem can be solved by filtering the original signal $x(n')$ by the lowpass filter before the downsample process. The cutoff frequency of filter is given as

$$f_c = f_s'/2D = f_s''/2. \tag{6.5.5}$$

This lowpass filter is called the decimation filter. The high-rate filter output $y(n')$ is downsampled to obtain the desired low-rate decimated signal $y(n'')$ by discarding $(D - 1)$ samples for every D samples of the filtered signal $y(n')$.

    The decimation filter operates at the high-rate f_s'. Because only every Dth output of the filter is needed, it is unnecessary to compute output samples that would be discarded. Therefore, the overall computation is reduced by a factor of D. By transposing the interpolator structure in Fig. 6.24, we can obtain a commutative structure for a decimator that is based on the parallel bank of polyphase filters, as illustrated in Fig. 6.26. The commutator rotates in a clockwise direction starting from the top at sampling instant $n'' = 0$. The polyphase filters are related to the original decimation filter as follows:

$$b_{m,i} = b_{m+iD}, m = 0, 1, \ldots, D - 1 \text{ and } i = 0, 1, \ldots, L/D - 1. \tag{6.5.6}$$

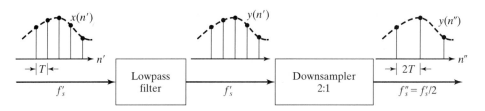

Figure 6.25   Decimation process, where $D = 2$

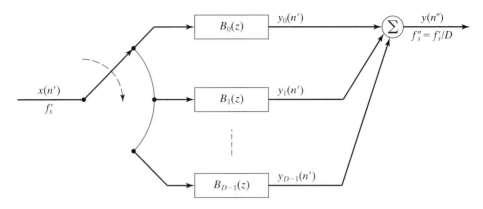

Figure 6.26 Polyphase implementation of a decimation filter

At each sampling period $1/f_s'$, only one polyphase filter, $B_m(z)$, is used to compute the corresponding output, $y_m(n')$. After D outputs are available, the decimated signal is obtained by summing these D outputs as

$$y(n'') = \sum_{m=0}^{D-1} y_m(n'). \tag{6.5.7}$$

6.5.3 Sampling-Rate Conversion

Sampling-rate conversion of a digital signal can be accomplished using two different methods. The first method is to convert the digital signal back into an analog signal using a DAC and then redigitize the analog signal at the desired rate using an ADC with the proper anti-aliasing filter. Quantization and aliasing errors inherent in digital-analog-digital conversion processes degrade the signal further. The second method is to perform the sampling-rate conversion by a rational factor U/D entirely in the digital domain using proper interpolation and decimation. The advantage is that it uses high-quality digital lowpass filtering. We can achieve this digital sampling-rate conversion first by performing interpolation of the factor U and then decimating the signal by the factor D. For example, we can convert digital-audio signals for broadcasting (32 kHz) to professional audio (48 kHz) using a factor of $U/D = 3/2$. That is, we interpolate the 32 kHz signal with $U = 3$ and then decimate the resulting 96 kHz signal with the factor $D = 2$ to obtain the desired 48 kHz. It is very important to note that we have to perform interpolation before decimation in order to preserve desired spectral characteristics. Otherwise, the decimation may remove some of the desired frequency components that cannot be recovered by interpolation.

The principle of sampling-rate conversion can be realized by cascading the interpolator shown in Fig. 6.23 with the decimator shown in Fig. 6.25. Since the lowpass filter in Fig. 6.23 operates at the same rate as the filter in Fig. 6.25 and since they are in cascade, we can combine them using a single lowpass filter, as illustrated in

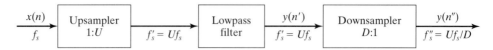

Figure 6.27   Sampling rate conversion by a factor of U/D

Fig. 6.27. The relationship between the three frequencies can be expressed as

$$f'_s = Uf_s = Df''_s. \qquad (6.5.8)$$

The interpolation filter must have the cutoff frequency given in Eq. (6.5.2). The cutoff frequency of the decimation filter is given in Eq. (6.5.5). The frequency response of the combined filter must incorporate the filtering operations for both interpolation and decimation; hence, it should ideally have the cutoff frequency

$$f_c = \frac{1}{2}\min(f_s, f''_s). \qquad (6.5.9)$$

6.5.4 Digital Signal Processing Implementations

The interpolation introduced in Section 6.5.1 can be implemented by the MATLAB function `interp` with the following syntax:

```
y = interp(x, U);
```

The output interpolated vector `y` is `U` times longer than the original input `x`. We can also specify the lowpass filter order `L` and the cutoff frequency `wc` using the following syntax:

```
y = interp(x, U, L, wc);
```

In addition, we can save the coefficients of the FIR filter used for interpolation into vector `b` using the following syntax:

```
[y, b] = interp(x, U, L, wc);
```

The decimation for decreasing the sampling rate of a given sequence can be implemented by the MATLAB function `decimate` with the following syntax:

```
y = decimate(x, D);
```

This function uses an eighth-order lowpass Chebyshev Type I filter by default. We can specify the filter order `L` by using `y = decimate(x, D, L)`. In addition, we can employ the FIR filter by using the following syntax:

```
y = decimate(x, D, 'fir');
```

This command uses a 30-point FIR filter instead of the Chebyshev IIR filter. We can also specify the FIR-filter order by using `y = decimate(x, D, L, 'fir')`.

Example 6.4

Given the speech file `timit_1.asc`, which is sampled by a 16-bit ADC with a sampling rate of 16 kHz, we can use the following MATLAB script (`exmp6_4.m`) to decimate it to 8 kHz:

```
load timit_1.asc -ascii;   % load speech file
soundsc(timit_1, 16000)    % play at 16 kHz
pause;                     % hit 'Enter' at the end
timit1 = decimate(timit_1,2,60,'fir'); % decimation
soundsc(timit1, 8000)      % play the decimated speech
```

We can tell the sound-quality (bandwidth) difference by listening to `timit_1` with 8 kHz bandwidth and `timit1` with 4 kHz bandwidth.

For sampling-rate conversion, we first use the MATLAB function `gcd` to find the conversion factor U/D. For example, to convert an audio signal from CD (44.1 kHz) for transmission using telecommunication channels (8 kHz), we can use the following commands:

```
g = gcd(8000, 44100);   % find the greatest common divisor
U = 8000/g;             % upsample factor
D = 44100/g;            % downsample factor
```

In this example, we obtain $U = 80$ and $D = 441$ since $g = 100$.

The sampling-rate-conversion algorithm shown in Fig. 6.27 is supported by the function `upfirdn` in the Signal Processing Toolbox. This function implements the efficient polyphase-filtering technique. For example, we can use the following command for sampling-rate conversion:

```
y = upfirdn(x, b, U, D);
```

This function first upsamples the signal in vector x with factor U, then filters the resulting signal by the FIR filter given in coefficient vector b, and finally downsamples using factor D to obtain the output store in vector y. The quality of the resampling result depends on the quality of the FIR filter.

Another function that performs sampling-rate conversion is `resample`. For example,

```
y = resample(x, U, D);
```

This function converts the sequence in vector x to the sequence in vector y with the sampling rate ratio U/D. It designs the FIR lowpass filter using `firls` with a Kaiser window. In addition, we may use [y, b] = resample(x, U, D) to obtain the coefficients of the filter used in the resampling process and save them in vector b.

MATLAB provides the function `intfilt` for designing interpolation (and decimation) FIR filters. For example,

```
b = intfilt(U, L, alpha);
```

designs a linear-phase FIR filter with the interpolation ratio 1:U. The bandwidth of the filter is `alpha` times the Nyquist frequency. In addition, the following command

```
b = intfilt(U, L, 'Lagrange');
```

designs an FIR filter that performs Lth-order Lagrange polynomial interpolation.

As shown in Tables 6.5 and 6.6, the decimation filter `firdec` and the interpolation filter `firinterp` are available in both the C54x and C55x DSPLIBs.

6.6 EXPERIMENTS AND PROBLEMS

In this section, we examine the design and implementation of an FIR filter using all of the tools described previously. We follow the entire process, from designing and evaluating the filter using high-level MATLAB tools, to implementing and testing using C programs, to writing assembly code for fixed-point C54x and C55x processors. In order to provide more exercises, we are using a different FIR filter instead of the one designed in Section 6.2.2 to illustrate these procedures.

6.6.1 Design a Bandpass Finite-Impulse Response Filter Using MATLAB

Given a sinewave at 2 kHz that is corrupted by white noise with zero mean and a variance of 0.01. A digital bandpass FIR filter is designed using SPTool to enhance the desired 2 kHz sinewave by reducing the broadband noise. The bandpass filter is designed according to the following design specifications:

1. The bandpass filter has cutoff frequencies around 1,800 Hz and 2,200 Hz.
2. The passband ripple is about 1.5 dB, and stopband attenuation is about 35 dB.
3. The filter is a linear-phase filter.
4. The design method with the lowest order should be used.

A signal that consists of a 2 kHz sinewave corrupted by white noise is generated by the MATLAB script `singen6_6.m` and saved in the file `in.mat`. We can load this file by typing

```
load in.mat
```

Open the SPTool window by typing `sptool` in the MATLAB command window. Click on the **File** button and select **Import...** from the menu. Import the signal variable `in` and its sampling frequency `Fs` from the **Workspace Contents** field to the **Data** and **Sampling Frequency** fields, respectively, and replace `sig1` by `in` in the **Name** field and click on **OK**. Click on the **View** button under the **Signals** column to display the imported signal, as shown in Fig. 6.28. The user can use other tool icons to examine the imported signal. For example, we can listen to the noisy signal by clicking on the play icon ▶ and note that the sinewave is corrupted by the noise.

The next step is to design the bandpass filter that enhances the desired 2 kHz sinewave corrupted by white noise. The **Filters** column in SPTool provides a tool for

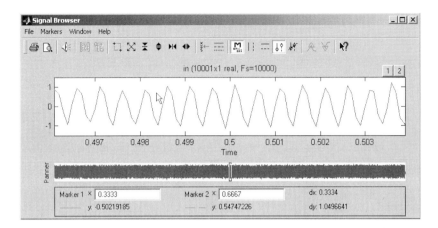

Figure 6.28   **Signal Browser** window in SPTool

filter design. Click on the **New** button to open a new **Filter Designer** window. Design specifications, such as the sampling frequency and the passband- and stopband-edge frequencies, along with their respective passband and stopband ripples for a band-pass filter, are specified as shown in Fig. 6.29. Click on the **Apply** button to complete the design process. Note that the equirriple FIR-filter design method is used to design the lowest order FIR filter that satisfies the specification. Make sure that the

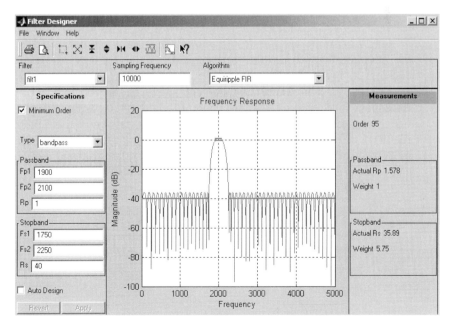

Figure 6.29   **Filter Designer** window for designing a bandpass FIR filter

Minimum Order box is checked to ensure that the filter with the least order is designed.

Note that the **Measurements** column on the right-hand side of Fig. 6.29 reports the final specification of the designed filter. It indicates that an FIR filter of length 95 is designed with a passband ripple of 1.578 dB and a stopband attenuation of 35.89 dB. The **Weight** value indicates the emphasis in minimizing the error in different bands. In this case, it shows that more emphasis (5.75 versus 1) is on minimizing the error in the stopband over the passband. The name of this filter is given as `filt1`. This filter is now ready to be used for filtering the signal in the data file `in`.

We can also design the bandpass filter using FDATool and exporting the filter coefficients from FDATool (**File → Export to SPTool**) to SPTool for verification. We use some powerful features in FDATool at a later time.

From the main SPTool window, select the input signal `in` and the filter `filt1` by highlighting these variables. Click on the **Apply** button in the **Filters** column to activate the **Apply Filter** window, as shown in Fig. 6.30. Change the name of the **Output Signal** from `sig1` to `out` and click on **OK**. The `out` variable appears in the **Signals** column.

View the `out` signal in the Signal Browser and listen to the filtered signal. A cleaner sinewave is heard. We can evaluate the result of the bandpass filter further by examining the spectra of the `in` and `out` signals. The spectra can be examined by selecting the `in` signal and clicking on the **Create** button in the **Spectra** column to open the **Spectrum Viewer** window. At the left-bottom corner of the window, click on **Apply** to generate the power-spectrum-density plot for the `in` signal. Repeat the same process for the **out** signal. To view the spectra of both `in` and `out` signals, select both `spect1` (spectrum for `in`) and `spect2` (spectrum for `out`) by using Shift key, and click on the **View** button in the **Spectra** column to display them, as shown in Fig. 6.31. We can observe that the noise has been reduced by about 35 dB.

After completing the design and verification of the filter, we have to export the filter coefficients into the workspace or a data file. This export can be done by clicking on **File → Export...** to open the **Export from SPTool** window, selecting `Filter: filt1` from **Export List**, and clicking on the **Export to disk** or the **Export to workspace** button. In order to implement the designed bandpass FIR filter in C, we need to represent the coefficients in different formats and export them as a header file that can be included in the C program. Since SPTool cannot do this, we can use FDATool, introduced in Section 6.2, for designing the filter and creating the coefficient header file for experiments.

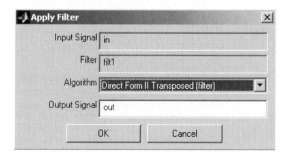

Figure 6.30 Parameters used in the
Apply Filter window

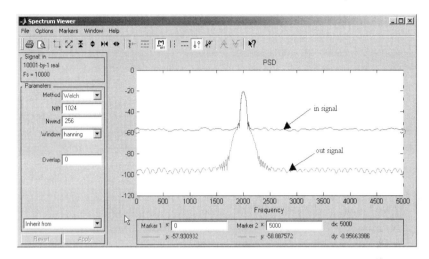

Figure 6.31   **Spectrum Viewer** window showing in and out signals

6.6.2 Implementation Using Floating-Point C

Following the design process described in Section 6.2, open FDATool, enter the same specifications shown in Fig. 6.29, and then click on the **Design Filter** button. The designed filter coefficients are represented with double-precision, floating-point format numbers. Following the quantization process described in Section 6.2.3

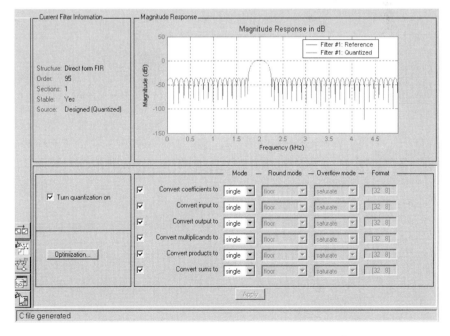

Figure 6.32   Magnitude responses of reference and quantized bandpass filters

and after **Mode** is set to `single` (single-precision, floating-point format), a new window is displayed, as shown in Fig. 6.32. The magnitude responses show that the quantized filter in a single-precision, floating-point representation is almost identical to the reference filter in double-precision, floating-point format.

From the **File** menu, select **Export to C Header file** and click on **OK**. Export the coefficients to the header file `coeff_c_sp.h` by entering the file name and then clicking on **Save**. We can open the `coeff_c_sp.h` file using any text editor and change the variable names of the filter length and coefficient array. Note that this header file requires another header file `tmwtypes.h`, which performs general-type conversion for MATLAB-generated C code.

FIR filtering can be executed on a general-purpose computer by using a C program that inputs a signal from the data file `in.dat` (generated by `singen6_6.m`), performs the FIR filtering, and writes the output signal to the data file `out.dat`. The filter coefficients, input, and output are all represented in 32-bit floating-point format. We can compare the output of this C program with the output obtained by MATLAB. The following floating-point C program `fir_floatpt.c` can also be tested using Visual C/C++:

```c
#include <stdio.h>
#include <stdlib.h>
#include <math.h>
#include "coeff_c_sp.h" /* filter length BL and coefficient buffer
                           buffer B[] is defined in header file */
void main()
{
/*    Define variable arrays, define and clear variables     */
  float yn = 0.0;                // y(n), output from FIR filter
  float xn;                      // x(n)
  float xnbuf[96];               // signal buffer
  int i,j,k;
/*    Declare file pointers     */
  FILE *xn_in;                              // file pointer of x(n)
  FILE *yn_out;                             // file pointer of y(n)
  xn_in = fopen("in.dat","r");     // open file for input x(n)
  yn_out = fopen("out.dat","w");   // open file for output y(n)

/*    Start of main program     */
  for(k=0; k<BL; k++)
    xnbuf[k]=0.0;                           // clear signal buffer
  while ((fscanf(xn_in,"%f",&xn)) != EOF)
  { /*  read in x(n) from data file and processing it   */
    for (i=BL-1; i>0; i--){
      xnbuf[i]=xnbuf[i-1];   // refresh signal buffer
    }
    xnbuf[0]=xn;                  // insert newest sample into buffer
    yn = 0.0;
    for (j=0; j< BL; j++){
      yn += xnbuf[j] * (real32_T)B[j]; // FIR filtering
    }
```

```
        fprintf(yn_out,"%f\n",yn);
    }
    printf("Finish");
    fcloseall();                      // close all opened files
}
```

It is important to note that the filter length `BL` and the coefficient buffer `B[j]` are defined in the header file `coeff_c_sp.h`; thus, they can be used in the main function.

6.6.3 Implementation Using Fixed-Point C

The floating-point C program `fir_floatpt.c` introduced in the previous experiment can be modified to become a fixed-point C program for execution on a general-purpose computer. As discussed in Chapter 3, the coefficients represented in floating-point format must be converted to an N-bit integer using the conversion formula

$$int\_number = round(float\_number \times 2^{N-1}). \qquad (6.6.1)$$

For example, when $N = 16$ is used, the original floating-point number is multiplied by 32,768 and then rounded to the nearest integer.

As described in Section 6.2.3, FDATool can quantize the reference-filter coefficients to 16-bit coefficients in signed-integer (Q.15) format and export the coefficient set to the C header file `coeff_16int.h`. In addition, the input data samples are rounded to a 16-bit integer and saved in the file `in_int.dat` by MATLAB script `singen6_6.m`. As discussed in Chapter 3, the input signals need to be normalized by the largest value in the file (so that $-1 \leq x(n) < 1$), multiplied by 32,768, rounded to 16-bit integers, and saved in an ASCII file. These operations can be done using the following MATLAB commands (in `singen6_6.m`):

```
in_int = round(32768*in./max(abs(in)));
fid = fopen('in_int.dat','w');
fprintf(fid,'%4.0f\n',in_int);
```

The MAC operation in FIR filtering is implemented using a `long` integer (32 bits). The integer output is computed by scaling the accumulated value with a variable `scale` (32,768 for 16-bit fixed-point C). The partial listing of `fir_fixpt.c` is given as follows:

```
yn = 0;
scale = 32768;
for (j=0; j< BL; j++)
{
    yn += (long)(xnbuf[j] * (int16_T)B[j]);
}
fprintf(yn_out,"%d\n",(int)(yn/scale));
```

6.6.4 Implementation Using Fixed-Point C for C5000 Processors

In this experiment, we modify the fixed-point C program `fir_fixpt.c` to run on C54x and C55x processors. The CCS for the C5000 is used to develop and test this fixed-point FIR-filtering program in C. For this experiment, the coefficients of the FIR filter designed by FDATool are exported to a C header file by clicking on **Targets → Export to Code Composer Studio (tm) IDE**. The user can select the target DSP processor by clicking on the **Select target...** button located at the bottom-right corner of the window. The default target is set as the C54x (or C55x) Simulator. Click on **OK**, and save the header file as `coeff_ccs_16int.h`. The user can compare this header file with `coeff_16int.h`, which is generated for a general-purpose computer.

In this experiment, we use the file I/O capability of CCS (introduced in Chapter 2) to test the C code running on C54x and C55x processors. The C program reads a sample from the input data file, uses it as the input signal for the bandpass FIR filter, and displays the filter output via a probe point. The main function enters a loop that calls function `dataIO()` for data input/output and `firproc(input, output)` for FIR filtering. This C program, `fir_fixpt_ccs.c`, is listed as follows:

```
#include <stdio.h>
#include <stdlib.h>
#include <math.h>
#include "coeff_ccs_16int.h"   // filter coefficient header file

int in_buffer[300];
int out_buffer[300];
#define TRUE 1

/* Declare functions */
static int firproc(int *input, int *output);
static void dataIO(void);

void main()
{
/*    Declare file pointers    */
int *input = &in_buffer[0];
int *output= &out_buffer[0];
/*    Start of main program    */
  while(TRUE)
  { /*  read in x(n) from data file and processing it  */
    dataIO();
    firproc(input, output);
  }
}

/* C function firproc for FIR filtering */
static int firproc(int *input, int *output)
{
/*    Define variable arrays, define and clear variables    */
  long yn = 0;              // y(n), output from FIR filter
  int xnbuf[96];            // signal buffer
```

```
      int i,j,k;
      int size = 300;

      for(k=0; k<BL; k++){
        xnbuf[k]=0;            // clear signal buffer
      }
      while(size--){
        for (i=BL-1; i>0; i--){
          xnbuf[i]=xnbuf[i-1];// refresh signal buffer
        }
        xnbuf[0]=*input++;     // insert newest sample into buffer

        yn = 0;
        for (j=0; j< BL; j++){// FIR filtering
          yn += (((long)(int)xnbuf[j] * (long)(int16_T)B[j])>>15);
        }
        *output++ = (int)(yn);
      }
      return(TRUE);
}

/* C function for dataIO */
static void dataIO()
{
   return;
}
```

The C program `fir_fixpt_ccs.c` and the files `vectorsc54.asm` (for the C54x only), `cmdc54.cmd` (or `c55x.cmd` for the C55x), and `rts.lib` (or `rts55.lib` for the C55x) are included in the project. The header files automatically are inserted into the project when the project is built. After successfully compiling and linking the program, we can load the executable file (`.out`) into the simulator using **File** → **Load Program**. The detailed procedure is introduced in Section 2.6.4.

The probe point is a useful tool for algorithm development because it can transfer data from a file to memory `in_buffer` on the target processor or from memory `out_buffer` to a file. It can also update the display window. Probe points and breakpoints can be inserted into the program by placing the cursor at the line `dataIO()` and clicking on the ▧ and ✋ buttons, respectively. For setting file I/O, choose **File** → **File I/O**, and the dialog box **File I/O** displays. From the **File Input** tab, click on the option **Add File**. Select `in_int.dat`, and type in `in_buffer` and 300 for **Address** and **Length**, respectively. Click on **Add Probe Point**, and in the probe point list highlight the line that states `fir_fixpt_ccs.c` line 51 (48 for the C55x) → No Connection. In the **Connect To** field, click on the down arrow and select `in_int_ccs.dat` file. Finally, click on **Replace**, **OK**, **OK** to complete the link between the file and the probe point. Note that we inserted a header (1611 1 0 0 0) on top of the actual data in `in_int.dat` for generating the data file `in_int_ccs.dat`, which can be interpreted by CCS. In addition, add the second breakpoint at the line `*output++ = (int)(yn)` for updating the display window.

The time-domain plots of signal samples `in_buffer` and `out_buffer` can be displayed by choosing **View** → **Graph** → **Time/Frequency** (select **Dual Time** option in

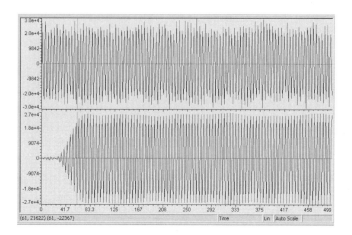

Figure 6.33 Time-domain
input (top) and output
(bottom) signals

Display Type field). Click on **OK** after setting the graph property, and a blank graphical plot is displayed. Make sure that **Float in Main Window** option (available from right click of mouse) is unchecked in order to place the graphic window manually.

After completing the preceding procedure, click on the animate button 🏃 , which executes the program until it reaches a breakpoint. The program is halted, and the display windows are updated. It then continues to execute the program until it reaches another breakpoint. This run-break-resume process continues until it is manually halted. Figure 6.33 shows the time-domain waveforms obtained from in_buffer (upper plot) and out_buffer (lower plot). Note that the output signal is identical to the results of the previous fixed-point C program (in Section 6.6.3) executed on a general-purpose computer.

The efficiency of FIR-filtering code can be profiled in terms of the number of clock cycles. After **Profiler** → **Enable Clock** and **Profiler** → **View Clock** are selected from the menu, the clock window is displayed at the bottom-right corner of the CCS window. Set one breakpoint at the beginning of the code at the C statement

```
for (j=0; j< BL; j++)
```

and the other at the statement

```
*output++ = (int)(yn)
```

Run the program to the first breakpoint by selecting **Debug** → **Run** from the menu. The execution stops at the first breakpoint. Double-click on **Clock Window** to clear the cycle count. Run the program to the second breakpoint, and the **Clock Window** displays the number of cycles (Clock = 3465) that the C code took to execute 96-tap FIR filtering on the C54x. For the C55x experiment, Clock = 2598, which shows that the C55x is faster than the C54x.

The user can compile the C code with the −k option to keep the .asm file generated by the C compiler, which can be done by clicking on **Project** → **Build Options**. In the **Category** window, click on **Assembly**, and check the **Keep generated .asm File (-k)** box. This selection allows the user to compare the differences of handwritten assembly code (which is introduced in Section 6.6.6) with C-compiler-generated assembly code.

6.6.5 Implementation Using Fixed-Point C with Intrinsics

The performance of fixed-point C code can be improved further by including intrinsics for C54x and C55x processors. However, the disadvantage of using intrinsics is the lack of flexibility in porting code between different DSP processors. When using the `long_smac` intrinsic, the integer must first be scaled down before multiplication. Otherwise, the subsequent addition of products may result in overflow and produce a saturated result. In this experiment, we use separate multiplication (`_lsmpy`) and addition (`_lsadd`) intrinsics to implement the MAC operation of the FIR filtering. The kernel code of the C program, `fir_fixpt_intr.c`, is listed as follows:

```
for (j=0; j< BL; j++)
{
   temp = _lsmpy(xnbuf[j],(int16_T)B[j])>>16;
   yn = _lsadd(yn,temp);
}
```

When using intrinsics, we have to include the following statement at the beginning of the C program:

```
#include <intrindefs.h>
```

It is also important to note that the C programs `fir_floatpt.c`, `fir_fixpt.c`, `fir_fixpt_ccs.c`, and `fir_fixpt_intr.c` can be executed on both C54x and C55x simulators. These programs show that C is flexible and allows porting on different DSP processors.

6.6.6 Implementation in Assembly for C5000 Processors

In this section, we implement the FIR filter using both C54x and C55x assembly programs. We first implement complete FIR filtering in C54x assembly code. For the C55x, we implement FIR filtering in assembly routine, which can be called from the main C program.

Implementation on the C54x

The first experiment in this section is the development and testing of assembly code for FIR filtering on the C54x. The input data (`in1_int.dat`) is copied to data memory using the `.copy` directive, while the coefficients are placed in the section labeled as `coeff_table`. Two uninitialized sections, `out` and `coeff`, are declared for 300 output samples and 96 coefficients, respectively, in the internal memory of the C54x processor. The linker-command file `c54.cmd` is used to link these sections to the data and program memories. This assembly code, `fir_fixpt_asm.asm`, is listed as follows:

```
          .def    start
out_buffer  .usect "out",300      ; reserve 300 location for y(n) output
coeff       .usect "coeff",96,1   ; reserve 96 location for coefficients
```

```
            .sect   "coeff_table"
init_coeff  .int    317, 26, -125, -132, 35, 149, 38, -144, -141
            .int    36, 131, 28, -97, -76, 6, 7, -17, 48
            .int    90, -65, -252, -105, 302, 364, -176, -636, -229
            .int    646, 722, -316, -1102, -375, 1034, 1111, -463, -1572
            .int    -517, 1396, 1458, -589, -1955, -627, 1658, 1693, -667
            .int    -2170, -681, 1765, 1765, -681, -2170, -667, 1693, 1658
            .int    -627, -1955, -589, 1458, 1396, -517, -1572, -463, 1111
            .int    1034, -375, -1102, -316, 722, 646, -229, -636, -176
            .int    364, 302, -105, -252, -65, 90, 48, -17, 7
            .int    6, -76, -97, 28, 131, 36, -141, -144, 38
            .int    149, 35, -132, -125, 26, 317

            .sect   "indata"
in_buffer   .copy   "in1_int.dat"       ; copy in1_int.dat to in_buffer
            .mmregs
            .text                       ; create code section

start:
    STM     #0, SWWSR
    SSBX    FRCT                        ; fractional mode
    SSBX    SXM                         ; sign extension mode
    RSBX    OVM                         ; overflow mode
copy:
    STM     #coeff,AR1                  ; AR1 pointing to filter coefficient
    RPT     #95                         ; repeat next instruction 96 times
    MVPD    init_coeff,*AR1+            ; copy coefficients to on-chip memory
* Setup the pointers to point to the coefficient and signal buffers
    STM     #coeff,AR2                  ; AR2 points to coefficient
    STM     #in_buffer,AR3             ; AR3 points to in_buffer
    STM     #out_buffer,AR4            ; AR4 points to out_buffer
    STM     #1,AR0                      ; offset for circular buffer
loop:
    STM     #299,BRC                    ; initializes block repeat counter
    STM     #96,BK                      ; initializes circular buffer size
    RPTB    fir_end-1                   ; block repeat till fir_end
firloop:
    MPY     *AR2+,*AR3+,A               ; first multiply on 1st tap
    RPT     #94                         ; subsequent MAC for 95 taps
    MAC     *AR2+0%,*AR3+,A             ; FIR filtering
    MAR     *+AR3(#-95)                 ; update signal pointer
    STH     A,*AR4+                     ; write result to memory location y
fir_end:
    NOP
```

In the main FIR-filter routine, we set the overflow mode to OFF, fractional mode to ON, and sign-extension mode to ON. The coefficients in coeff_table are originally located in external memory and are copied to the internal memory coeff. After this transfer is completed, address pointers AR2, AR3, and AR4 are set to point at coeff, in_buffer, and out_buffer, respectively. The program then executes the block FIR filtering described in Section 6.3.2. In this FIR-filtering experiment, we treat the entire input sequence as a block. The single-repeat instruction is responsible for the 96-tap FIR filtering, while the block of code is repeated to generate 300 output samples.

As with the procedure given in Section 6.6.4, we create a new project to include the files `fir_fixpt_asm.asm`, `c54.cmd`, and `vectors.asm`. Perform a build operation before loading the executable program into memory by clicking on **File → Load Program . . . .** Use **GEL → C5402_Configuration → CPU_Reset** to reset the program pointer to the start of the program. Set a breakpoint at the end of the program by placing the cursor at the `NOP` instruction, and click on the breakpoint icon ⬛ . Set up a display window to view `in_buffer` and `out_buffer` using the same procedure introduced in previous experiments. In order to ensure that the output data does not contain any previous values, we can fill the memory `out_buffer` with zeros by using **Edit → Memory → Fill** and specify the address as `out_buffer` with a length of 300. Run the program, and the program stops at the breakpoint with an updated display for `out_buffer`.

The final step is to profile the 96-tap FIR filter using assembly code. Set breakpoints at the instructions

```
MPY    *AR2+,*AR3+,A
```

and

```
STH    A,*AR4+
```

Enable and view the clock, as described in previous experiments. Only 99 clock cycles are required for running the assembly-coded FIR filtering. Note that 3,465 cycles are needed for the C program. The user can open the assembly file generated by the C compiler in the CCS and compare the differences between these two assembly programs. We find that the C-compile-generated assembly code uses the stack frequently for passing variables and does not make full use of indirect-addressing features, especially the circular buffer, for efficient implementation.

Implementation on the C55x

In this final experiment, we modify the fixed-point C program `fir_fixpt_ccs.c` developed in Section 6.6.4 for experiments. This C program uses the following loop for FIR filtering:

```
for (j=0; j< BL; j++) {
  yn += (((long)(int)xnbuf[j]*(long)(int16_T)B[j])>>15);
}
```

The new C program `fir_fixpt_C55x.c` calls an assembly routine for FIR filtering. This function is written in C55x assembly (`fir_c55.asm`) and is listed as follows:

```
; This function can be called as follows:
;   yn = fir_c55(xnbuf,BL,B); where
```

```
;      xnbuf - signal vector
;      BL - length of FIR filter
;      B - coefficient vector
       .def  _fir_c55
       .text

_fir_c55
       pshm   ST1_55                  ; save ST1, ST2, and ST3
       pshm   ST2_55
       pshm   ST3_55

       or     #0x340,mmap(ST1_55); set FRCT,SXMD,SATD
       bset   SMUL                    ; set SMUL
       sub    #2,T0                   ; T0=BL-2
       mov    T0,CSR                  ; repeat CSR BL-1 times
       rpt    CSR                     ; repeat the next instruction
       macm   *AR0+,*AR1+,AC0         ; MAC
       macmr  *AR0,*AR1,AC0           ; do the last operation
       mov    hi(AC0),T0              ; save Q15 filtered value

       popm   ST3_55                  ; restore ST1, ST2, and ST3
       popm   ST2_55
       popm   ST1_55

                                      ; return signal buffer index
       ret
       .end
```

As with the procedure given in Section 6.6.4, we create a new project to include the files `fir_fixpt_c55x.c`, `fir_c55.asm`, `c55x.cmd`, and `rts55.lib`. The same C header file `coeff_ccs_16int.h` is included in the main program. After successfully compiling and linking the program, we can load the executable file into a simulator for experiments. Following the identical procedure given in Section 6.6.4, we use the probe point to connect `in_int_ccs.dat` for data file I/O and use Graph to view the filter-output waveform. By comparing the output waveform with the one obtained from the fixed-point C program in Section 6.6.4, we verify that the assembly routine is correct.

To evaluate the efficiency of this mixed C-and-assembly routine, we set one breakpoint at the instruction

```
yn = fir_c55(xnbuf,BL,B);
```

and the other at the instruction

```
*output++ = (int)(yn);
```

Run the program to the first breakpoint by selecting **Debug → Run** from the menu. The execution stops at the first breakpoint. Double-click on **Clock Window** to clear the cycle count. Run the program to the second breakpoint, and **Clock Window** displays the number of cycles (`Clock = 144`) the code took to execute 96-tap FIR filtering on the C55x. Compared with `Clock = 2598` in the fixed-point C program given in Section 6.6.4, it clearly shows that the C55x assembly FIR routine is faster than the C-loop code. Note that due to the overhead of calling the subroutine from the main function, the number of required clock cycles is slightly higher than that of the C54x (99). Also, the benchmark results may vary on different computers.

6.6.7 Implementation Using the C5000 Digital Signal Processing Library

As discussed in Section 6.3, Texas Instruments provides C54x and C55x DSPLIBs that contain C-callable assembly-optimized signal-processing functions [14, 16]. These libraries implement some commonly used DSP algorithms such as FIR and IIR filtering, correlation, FFT, adaptive filter, trigonometry, and matrix operations. Therefore, using these optimized functions shortens the DSP application development time.

The FIR-filtering routines available in the DSPLIB for the C54x and the C55x are listed in Tables 6.5 and 6.6, respectively. In this experiment, we introduce the C program that calls the FIR-filtering routine `fir` in the C5000 DSPLIB. The function `fir.asm` is written in optimized C5000 assembly code. This assembly code is located in `c:\ti\c5400\dsplib\54x_src` (or `c:\ti\c5500\dsplib\55x_src` for the C55x). This function can be called in C as follows:

```
oflag = short fir(DATA *x, DATA *h, DATA *r, DATA *dbuffer,
                  ushort nh, ushort nx)
```

The argument `x` is the pointer to the input-signal vector of `nx` elements, `h` is the pointer to the coefficient vector of size `nh`, `r` is the output-signal vector of `nx` elements (if `r = x` implies in-place computation), `dbuffer` is the signal (or delay) buffer that must be initialized before running this function, and `oflag` is the overflow flag, which is set to 1 if an overflow has occurred in any intermediate or final result and is set to 0 for no overflow.

The C code introduced in Section 6.6.4 can be modified to call this `fir` function. The new C code, `fir_fixpt_dsplib.c`, is listed as follows:

```
#include "math.h"
#include "tms320.h"
#include "dsplib.h"

#include "input.h"                    // input signal file
#include "coeff_ccs_16int_lib.h"      // coefficient file

short i;
DATA   *dbptr = &db[0];

#pragma DATA_SECTION(NUM,".coeffs")
void main(void)
{
  for (i=0; i<NX; i++) r[i] = 0;    // clear output buffer (optional)
  for (i=0; i<NL; i++) db[i] = 0;   // clear signal buffer (a must)

  fir(x, NUM, r, &dbptr, NL, NX);   // FIR filtering
  return;
}
```

Some important features of using C functions in DSPLIBs are summarized as follows:

1. The header files `math.h`, `tms320.h` and `dsplib.h` included in the C code are located in the directory `c:\ti\c5400\dsplib\include\` for the C54x processors (or `c:\ti\c5500\dsplib\include\` for the C55x). The user must include this search path in the CCS build option by clicking on **Project →
Build Options . . . → Compiler → Preprocessor . . .** and inserting the path in **Include Search Path**.
2. The `54xdsp.lib` (or `55xdsp.lib` for the C55x) is a standard memory-model library and must be added into the project. This file is located at `c:\ti\c5400\dsplib` for the C54x (or `c:\ti\c5500\dsplib` for the C55x).
3. Since the `fir` function uses the Q.15 data type DATA, the coefficient and the input signal must be declared as this data type. These changes are done in the input-signal file `input.h` and coefficient file `coeff_ccs_16int_lib.h`, which are included in the program.
4. The C54x user can also insert an assembly instruction `asm(" STM #0, SWWSR")` at the beginning of the program for C54x experiments, which sets the simulator to operate at a zero-wait state.

In the program, FIR filtering is operated in block mode, which results in a more efficient implementation. This fact can be verified by benchmarking the block-processing cycle count using different block sizes. The user can modify the C program to benchmark the code using a block size of 300, 75, and 1 (sample mode). For example, the results for the C54x experiments are listed in Table 6.9. It is noted that the `fir` function executes faster in block mode than in sample mode. However, a longer data block reduces the cycle count only slightly, but at the cost of requiring more memory to store more block samples.

The benchmarks for cycle count and program memory are given in Table 6.5 (for the C54x). The cycle count is stated as $[4 + N \times (4 + L)]$ cycles plus 34 overhead cycles, which can be verified against the results in Table 6.9. The program memory required for the `fir` function is 42 words, and the data memory needed is 2×96 words for the coefficient and delay-line buffers.

The accuracy of this C program can be verified by comparing the filter output with the previous fixed-point C-program output. The user can check whether any overflow occurs in the FIR filtering by examining the `oflag` variable in the CCS

TABLE 6.9 Benchmark of a 96-Tap FIR Filter Using Different Block Sizes with the C54x `fir` Function to Complete 300 Input Samples

Number of samples per block, N	Cycle count
300	31,569
75 (four blocks to complete 300 samples)	31,765
1 (sample mode)	57,910

Watch Window. The user can also compare the performance of the symmetric FIR-filter function `firs` with the `fir` function.

Additional hands-on experiments using FIR filters are given in Appendix B. The additional applications include (1) implement a 3-D audio processor for headset listening, (2) implement a 5-band graphic equalizer for audio signals, and (3) implement a two-channel analysis-synthesis filterbank. The user can refer to the MATLAB and Simulink files available on the companion CD and the Web for more details of these exercises and can work through these hands-on experiments progressively to obtain the final solution using the CCS.

PROBLEMS

PART A

Problem A–6–1

Design an FIR filter using FDATool without symmetric coefficients.

Problem A–6–2

Design a lowpass FIR filter (using FDATool) with the frequency specification Fpass = 4,000 Hz, Fstop = 4,100 Hz, Fsample = 48,000 Hz, passband ripple = 0.5 dB, and stopband ripple = 50 dB. Compare the designed FIR filter using the Equiripple, Least-squares, Window, Maximally Flat, Least Pth-norm, and Constrained Equiripple design methods in terms of the magnitude response and the order of the filter.

Problem A–6–3

Following the use of the Equiripple method in Problem 2, use FDATool to quantize the filter coefficients, input, output, multiplicands, products, and sums to the IEEE 754 single-precision floating-point format. Observe any significant changes in responses compared with the reference responses using double precision.

Problem A–6–4

Continuing from Problem 3, use FDATool to quantize the filter coefficients, input, output, and multiplicands using Q.15 format and quantize the products and sums using Q.31 format. Observe any significant changes in responses compared with the single-precision floating-point implementation given in Problem 3.

Problem A–6–5

Use MATLAB (default by double-precision floating-point arithmetic) as a reference to compare the precision of the FIR filter implemented in floating-point C (single precision), fixed-point C (16-bit integer), and fixed-point assembly (16-bit integer) running on C5000 processors.

Problem A–6–6

Implement FIR filtering using a mixed C-and-assembly program for the C54x processor. The C routine sets up the data samples and coefficients before calling the assembly-coded FIR-filter routine. The C5000 DSPLIB contains an optimized FIR-filter routine. Repeat the preceding FIR filtering by calling on the optimized FIR-filter routine `fir`, and profile the difference in cycle count.

Problem A–6–7

Implement an FIR filter on the C55x using complete assembly code. Compare its performance with the mixed C-and-assembly code given in Section 6.6.6.

Problem A–6–8

Implement FIR filtering using the mixed C-and-assembly program `fir.asm` in the C55x DSPLIB. The C routine sets up the data samples and coefficients before calling the assembly-coded FIR-filter routine. Profile the cycle count with the result given in Section 6.6.6.

Problem A–6–9

Modify the assembly routine `fir_c55.asm` for using dual-MAC units. Compare its performance with the original `fir_c55.asm` that uses a single MAC only.

Problem A–6–10

Implement a symmetric FIR filter on the C55x using the mixed C-and-assembly routine for FIR filtering.

Problem A–6–11

Use FDATool to design the lowpass filter for sampling-rate conversion shown in Fig. 6.27. This filter is designed for converting a 32 kHz input signal into a 48 kHz output signal. Implement the sampling-rate conversion using the C54x (or C55x) DSPLIB.

Problem A–6–12

Design a lowpass FIR filter using the MATLAB function `fir1`, which is an FIR-filter design based on the window method. The designed filter should meet the following filter specifications:

(a) Cutoff at normalized frequency (with reference to the Nyquist frequency) 0.35.
(b) Normalized transition band = 0.1.
(c) Passband ripple at 0.5 dB.
(d) Stopband attenuation at 40 dB.

Perform the following steps:

(a) Select a suitable window function to achieve the desired stopband ripple, and determine the order of the filter based on Table 8.2 (in Chapter 8).
(b) Plot the frequency responses (magnitude and phase) of the designed filter, and verify that the filter specification has been met.

(c) Comment on the distribution of zeros in the pole-zero plot.

(d) Determine the cutoff frequency of the lowpass filter if the sampling frequency is set at 16 kHz.

(e) Determine the advantage of using a normalized frequency when designing the filter.

(f) Convert the preceding lowpass filter to a highpass filter with the same cutoff frequency, and comment on the distribution of zeros.

(g) Refer to Table 6.1 and suggest a suitable linear-phase FIR-filter type used for lowpass and highpass filters.

Problem A–6–13

Design a bandpass FIR filter with normalized cutoff frequencies at $F_{c1} = 0.35$ and $F_{c2} = 0.65$ using the MATLAB function `fir1` and Table 8.2. The lower and upper transition bands are specified at 0.1 and 0.15, respectively. The specified stopband attenuations are specified at 40 dB and 50 dB for lower and upper stopband, respectively, and the passband ripple is restricted to 1 dB. Next, design a bandstop FIR filter with the same cutoff frequencies and transition band, a stopband attenuation of 50 dB, and a passband ripple of 1 dB. Again, refer to Table 6.1 and suggest a suitable linear-phase FIR-filter type used for bandpass and bandstop filters. Verify the frequency responses of the designed bandpass and bandstop filters.

PART B

Problem B–6–14

A comb filter can be described by the following I/O equation:

$$y(n) = x(n) + x(n - L).$$

(a) Find the poles of the filter.

(b) Compute the frequency response of the filter, and sketch its magnitude response.

(c) Determine a suitable value of L to remove a narrowband component at 500 Hz and its odd harmonics at 1,500 Hz, 2,500 Hz, 3,500 Hz, and 4,500 Hz with a sampling rate of 10,000 Hz.

Problem B–6–15

Show that the transpose structure of the FIR filter given in Fig. 6.3 is defined by the I/O equation expressed in Eq. (6.1.1).

Problem B–6–16

Show that the frequency response of a digital filter is a periodic function of frequency ω with period 2π. That is, $B(\omega) = B(\omega + 2\pi)$.

Problem B–6–17

The frequency response of a Type 1 linear-phase FIR filter is given in Eq. (6.1.12).

(a) Compute the magnitude response of the filter.
(b) Redo Step (a) when L is even (Type 2).

Problem B–6–18

Redo Problem 17 for antisymmetric linear-phase FIR filters (Type 3 and Type 4).

Problem B–6–19

Rewrite the I/O given in Eq. (6.1.16) for Type 1, Type 3, and Type 4 linear-phase FIR filters.

Problem B–6–20

A C3x DSP processor used in real-time signal-processing applications has a clock cycle time of 50 nsec. One of the instruction is MPYF3 ‖ ADDF3 using circular addressing, which performs floating-point multiplication (of filter coefficient and input data) in parallel with floating-point addition in a single clock cycle. The length of the FIR filter is L. Assume that we need $L + 1$ clock cycles to compute a single output sample. Determine the maximum bandwidth of the signal that can be filtered with an FIR filter of length $L = 128$ in real time. If an audio signal $x(t)$ is sampled at 48 kHz, what is the maximum length of the FIR filter that can be performed in real time?

Problem B–6–21

In a multichannel surround sound system, the sampling rate for all of the five channels is 48 kHz. Data sample from each channel is quantized to 16 bits and processed by an L-tap FIR filter. The number of MACs per sample for implementing a single channel of an equalizer is 200 MACs per sample using an FIR filter. Determine the MIPS rating required for the equalizer using the FIR filter. What is the maximum length of the FIR filter that may be used to filter the audio signal in real time using C54x and C55x processors? What is the memory required to store the filter coefficients and signal vectors?

Problem B–6–22

A 100-MIPS DSP processor is used to perform lowpass filtering and highpass filtering with cutoff frequencies of $\pi/4$ rad/second and a transition width of 0.053π in real time. The stopband ripples for both filters are 0.01. Using the window method to design these FIR filters, determine a suitable window function and the order of the FIR filters. Assume FIR filtering takes $(L + 10)$ instruction cycles to compute one output sample for an FIR filter of length L. Estimate the processing time required for computing the lowpass and highpass FIR filters. Comment on whether real-time processing is possible if the sampling frequency is 96 kHz. If block processing is carried out, the FIR filter routine takes $5 + N \times (5 + L)$ instruction cycles, where N is the number of samples per block. Comment on the block size required to make block processing more efficient than sample processing. What are the disadvantages of using block processing?

Problem B–6–23

Based on Tables 6.5 and 6.6, compare the differences in the cycle counts when performing FIR filtering using different FIR functions on C54x and C55x processors. Explain the cycle count improvement using the C55x processor.

Problem B–6–24

Based on Tables 6.7 and 6.8, compare the differences in the cycle count when performing FIR filtering using different FIR functions on C62x and C64x processors. Explain the cycle count improvement using the C64x processor.

SUGGESTED READINGS

1 Parks, T. W. and C. S. Burrus. *Digital Filter Design*. New York, NY: Wiley, 1987.
2 Oppenheim, A. V., R. W. Schafer, and J. R. Buck. *Discrete-Time Signal Processing*. 2nd Ed. Upper Saddle River, NJ: Prentice Hall, 1999.
3 Proakis, J. G. and D. G. Manolakis. *Digital Signal Processing: Principles, Algorithms, and Applications*. 3rd Ed. Upper Saddle River, NJ: Prentice Hall, 1996.
4 Mitra, S. K. *Digital Signal Processing: A Computer-Based Approach*. 2nd Ed. New York, NY: McGraw-Hill, 2001.
5 Kuo, S. M. and B. H. Lee. *Real-Time Digital Signal Processing: Implementations, Applications and Experiments with the TMS320C55x*. New York, NY: Wiley, 2001.
6 Kuo, S. M. and C. Chen. "Implementation of Adaptive Filters with the TMS320C25 or the TMS320C30." Chapter 7 in *Digital Signal Processing Applications with the TMS320 Family*, vol. 3, edited by P. E. Papamichalis. Upper Saddle River, NJ: Prentice Hall, 1990.
7 Orfanidis, S. J. *Introduction to Signal Processing*. Upper Saddle River, NJ: Prentice Hall, 1996.
8 Van de Vegte, J. *Fundamentals of Digital Signal Processing*. Upper Saddle River, NJ: Prentice Hall, 2002.
9 Ifeachor, E. C. and B. W. Jervis. *Digital Signal Processing: A Practical Approach*. Upper Saddle River, NJ: Prentice Hall, 2002.
10 Grover, D. and J. R. Deller. *Digital Signal Processing and the Microcontroller*. Upper Saddle River, NJ: Prentice Hall, 1999.
11 The MathWorks. *Filter Design Toolbox User's Guide*. Version 2.1, 2000.
12 The MathWorks. *Signal Processing Toolbox for Use with MATLAB*. Version 5, 2000.
13 Texas Instruments. *TMS320F24x Filter Library: Module User's Guide*. 2002.
14 Texas Instruments. *TMS320C54x DSP Library Programmer's Reference*. SPRU422b, 2001.
15 Alter, D. M. *Efficient Implementation of Real-Valued FIR Filters on the TMS320C55x DSP*. Application Note, Texas Instruments, SPRA655, 2000.
16 Texas Instruments. *TMS320C55x DSP Library Programmer's Reference*. SPRU422F, 2002.

17 Dahnoun, N. *Digital Signal Processing Implementation: Using the TMS320C6000 DSP Platform*. Upper Saddle River, NJ: Prentice Hall 2000.

18 Texas Instruments. *TMS320C62x DSP Library Programmer's Reference*. SPRU402a, 2002.

19 Texas Instruments. *TMS320C64x DSP Library Programmer's Reference*. SPRU565a, 2002.

20 Crochiere, R. E. and L. R. Rabiner. *Multirate Digital Signal Processing*. Upper Saddle River, NJ: Prentice Hall 1983.

21 Vaidyanathan, P. P. *Multirate Systems and Filter Banks*. Upper Saddle River, NJ: Prentice Hall 1993.

7

Infinite-Impulse Response Filtering

This chapter introduces IIR filters. In this chapter, we study some commonly used IIR-filtering structures, design processes, and implementations on both fixed-point and floating-point DSP processors. In addition, important practical issues such as quantization effects in different IIR-filter structures, fixed-point implementations, and scaling issues are examined using FDATool. An application that uses IIR filters for DTMF generation and detection is presented at the end of the chapter. A more detailed discussion of the theoretical aspects of IIR filters can be found in textbooks [1–4].

7.1 INFINITE-IMPULSE RESPONSE FILTERS

FIR-filter characteristics were introduced in Chapter 6 and were briefly compared with IIR filters. In general, IIR filters have the following characteristics:

1. *Nonlinear phase*: IIR filters have a nonlinear-phase response over the frequency of interest. Therefore, group delay varies at different frequencies and results in phase distortion.
2. *Stability issue*: IIR filters are not always stable due to their recursive realization. Therefore, a careful design approach is needed to ensure that all of the poles of an IIR filter lie inside the unit circle to guarantee a stable filter, especially for fixed-point implementations.

3. *Finite-precision errors*: IIR filters are very sensitive to finite-wordlength effects. Therefore, the cascade of second-order IIR sections is commonly used to reduce the impact of accumulated quantization errors.

4. *Computation saving*: IIR filters can be implemented with a lower order to achieve a compatible level of performance when compared with the FIR filter. Therefore, an IIR filter needs lower computation loads (in terms of multiplication and addition) as compared with an FIR filter. However, the IIR filter is more complicated to implement on DSP processors and requires longer code to program both the feedforward and feedback sections in the cascade structure.

5. *Sharper frequency response*: IIR filters can produce a sharper cutoff as compared with FIR filters. The poles of the IIR filter greatly contribute to the frequency response.

In summary, IIR filters must be designed and implemented with greater care as compared with FIR filters. Several design methods and software tools are available for designing stable IIR filters. A commonly used filter-design approach is to convert the well-known classical analog filter to the equivalent digital IIR filter.

7.1.1 Definitions

Equation (6.1.1) defines the difference equation of an FIR filter of length L (or order $L - 1$). The IIR filter extends the feedforward section of the FIR filter by augmenting it with a feedback section. The difference equation of an IIR filter of order $(L - 1)$ and $(M - 1)$ for feedforward and feedback sections, respectively, is expressed as

$$
\begin{aligned}
y(n) &= b_0 x(n) + b_1 x(n - 1) + \cdots + b_{L-1} x(n - L + 1) - a_1 y(n - 1) \\
&\quad - a_2 y(n - 2) - \cdots - a_{M-1} y(n - M + 1) \\
&= \sum_{i=0}^{L-1} b_i x(n - i) + \sum_{m=1}^{M-1} -a_m y(n - m) = \mathbf{b}^T \mathbf{x}(n) + \mathbf{a}^T \mathbf{y}(n - 1),
\end{aligned}
\tag{7.1.1}
$$

where $\mathbf{b}$ and $\mathbf{a}$ are the filter-coefficient vectors defined as

$$
\mathbf{b} = [b_0\, b_1\, \cdots\, b_{L-1}]
\tag{7.1.2a}
$$

$$
\mathbf{a} = [-a_1\, -a_2\, \cdots\, -a_{M-1}].
\tag{7.1.2b}
$$

The signal vectors at time n are defined as

$$
\mathbf{x}(n) = [x(n)\, x(n - 1)\, \cdots\, x(n - L + 1)]^T
\tag{7.1.3a}
$$

and

$$
\mathbf{y}(n - 1) = [y(n - 1)\, y(n - 2)\, \cdots\, y(n - M + 1)]^T.
\tag{7.1.3b}
$$

There are several key differences between the FIR filter and the IIR filter. As discussed in Chapter 6, the FIR-filter output is the combination of only the weighted input samples, but the output of the IIR filter contains the weighted past outputs as well as the inputs. Due to the nature of output feedback, there is an inherent one-sample delay in the feedback section. As shown in Eq. (7.1.1), the IIR filter per-forms two inner (or dot) products of vectors, one for the feedforward section between vectors **b** and **x**, and the other for the feedback section between vectors **a** and **y**. The design of the IIR filter determines two sets of coefficients, $\{b_i, i = 0, 1, \ldots, L - 1\}$ and $\{a_m, m = 1, 2, \ldots, M - 1\}$, to meet a given specifica-tion. In addition, the IIR filter is recursive in computation, which results in an infi-nite impulse response. Therefore, the IIR filter must be designed with special care in order to prevent any growing or oscillation of the impulse response that can lead to an unstable filter.

By taking the z-transform of both sides of Eq. (7.1.1) and using the time-shift property, we have

$$
\begin{aligned}
Y(z) &= \left(b_0 + b_1 z^{-1} + \cdots + b_{L-1} z^{-L+1}\right) X(z) \\
&\quad - \left(a_1 z^{-1} + a_2 z^{-2} + \cdots + a_{M-1} z^{-M+1}\right) Y(z) \\
&= X(z) \sum_{i=0}^{L-1} b_i z^{-i} - Y(z) \sum_{m=1}^{M-1} a_m z^{-m}.
\end{aligned}
\tag{7.1.4}
$$

By rearranging the terms, we obtain the transfer function of an IIR filter expressed as

$$
\begin{aligned}
H(z) &= \frac{Y(z)}{X(z)} \\
&= \frac{\sum_{i=0}^{L-1} b_i z^{-i}}{1 + \sum_{m=1}^{M-1} a_m z^{-m}} = \frac{b_0 + b_1 z^{-1} + \cdots + b_{L-1} z^{-L+1}}{a_0 + a_1 z^{-1} + \cdots + a_{M-1} z^{-M+1}},
\end{aligned}
\tag{7.1.5}
$$

where $a_0 = 1$ is a dummy coefficient that is not used in computing filter output, as shown in Eq. (7.1.1). It is important to note that the coefficients of the feedback coefficients $\{a_m\}$ are sign-negated when compared with the difference equation given in Eq. (7.1.1).

Therefore, the IIR filter can be expressed as the ratio of polynomials in z^{-1}, with $(L - 1)$ zeros and $(M - 1)$ poles. The roots of the numerator and the denom-inator polynomials determine zeros and poles, respectively. In order to ensure a sta-ble filter, all of the poles must be placed inside the unit circle on the z-plane. The IIR filter provides greater flexibility in filter design since both the poles and zeros con-tribute to the frequency response.

7.1.2 Filter Characteristics

As discussed in Chapter 6, the coefficient vector of the FIR filter is identical to the impulse response of the filter. However, the impulse response of the IIR filter is not identical to the coefficient vector. The length of the impulse response for the IIR fil-ter can be infinity; thus, the filter is called the infinite-impulse-response filter.

The transfer function given in Eq. (7.1.5) can be evaluated on the unit circle to obtain the frequency response expressed as

$$H(\omega) = H(z)|_{z=e^{j\omega}} = \frac{\sum_{i=0}^{L-1} b_i e^{-j\omega i}}{1 + \sum_{m=1}^{M-1} a_m e^{-j\omega m}}. \tag{7.1.6}$$

The frequency response is a function of the continuous-frequency variable ω and is a periodic function with the period equal to 2π. Therefore, we only need to evaluate the frequency response from 0 to 2π or from $-\pi$ to π. The DFT can be applied to evaluate the frequency response $H(\omega)$ at equally-spaced frequency points $\omega_k = 2\pi k/N, k = 0, 1, \ldots, N-1$.

The magnitude response $|H(\omega)|$ and the phase response $\phi(\omega)$ are periodic functions with period 2π. If the IIR filter has real-valued coefficients, the real part of the frequency response is an even function, and the imaginary part is an odd function. As such, the magnitude and phase responses are even and odd functions, respectively, expressed as

$$|H(\omega)| = |H(-\omega)| \quad \text{and} \quad \phi(\omega) = -\phi(\omega). \tag{7.1.7}$$

Therefore, we only need to compute the magnitude and phase responses from 0 to π for IIR filters with real coefficients.

The FIR filter with symmetric coefficients has a linear-phase response. The IIR filter, however, is not able to guarantee a linear-phase response, which also implies that the IIR filter is not able to produce a constant phase (or group) delay.

Example 7.1

Given an IIR filter with coefficient vector $\mathbf{b} = [1 \quad 0.5]$ and $\mathbf{a} = [0.5 \quad -0.2]$, as defined in Eq. (7.1.2), the transfer function of the filter is given as

$$H(z) = \frac{1 + 0.5z^{-1}}{1 - 0.5z^{-1} + 0.2z^{-2}}$$
$$= 1 + z^{-1} + 0.3z^{-2} - 0.05z^{-3} - 0.085z^{-4} - 0.0325z^{-5} + 0.0007z^{-6} + \ldots$$

This function shows that the impulse response $\{1, 1, 0.3, -0.05, \ldots\}$ is no longer related to the coefficient vectors $\mathbf{b}$ and $\mathbf{a}$.

We can use the MATLAB functions impz, freqz, grpdelay, and zplane to analyze the preceding transfer function. The script example7_1.m is listed as follows:

```
b = [1 0.5]          % coefficients of the numerator
a = [1 -0.5 0.2]     % coefficients of the denominator
impz(b,a); pause;    % impulse response of the H(z)
zplane(b,a); pause;  % pole-zero plot
freqz(b,a); pause;   % frequency response
grpdelay(b,a)        % group delay
```

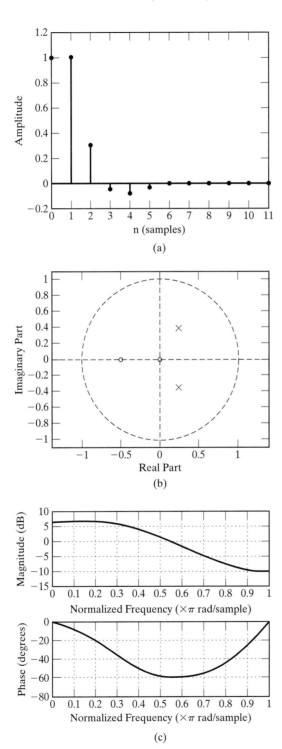

Figure 7.1 Characteristics of an IIR filter: (a) Impulse response, (b) pole-zero diagram, and (c) magnitude and phase responses

The impulse response, pole-zero diagram, and magnitude and phase responses are shown in Fig. 7.1. From these results, we observe that the impulse response of a stable IIR filter converges (decays) to a very small value as time increases. The pole-zero plot shows both the pole and zero locations. For a stable IIR filter, all poles must lie inside the unit circle. From the magnitude response shown in Fig. 7.1(c), the filter is a lowpass filter. However, the phase is nonlinear within the passband; thus, the group delay is no longer a constant over the passband.

7.1.3 Filter Structures

There are many different structures for realizing digital IIR filters, and we need to understand and determine a suitable structure for a particular application. The factors to be considered include errors in quantizing filter coefficients, noise introduced by fixed-point arithmetic, propagation of quantization errors, computational load, memory usage, and programming consideration and flexibility. In this section, we examine the following commonly used IIR filter structures:

1. Direct form I
2. Direct form II

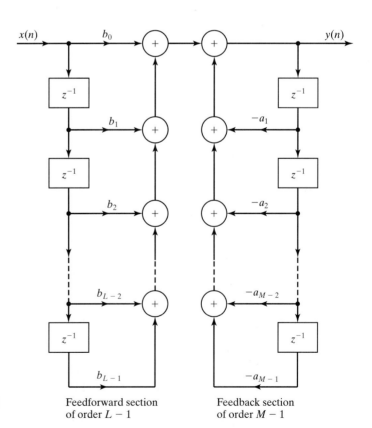

Figure 7.2 Direct-form I realization of IIR filters

Feedforward section of order $L - 1$

Feedback section of order $M - 1$

3. Transposed direct form I
4. Transposed direct form II
5. Cascade form
6. Parallel form

The direct-form I IIR structure is illustrated in Fig. 7.2. It consists of a cascade of two sections: feedforward and feedback. The feedforward section is the same structure as the FIR filter. The feedback section weighs and combines past outputs into the system. The computational requirements for the direct-form I structure are $(L + M - 1)$ multiplications and $(L + M - 2)$ additions per output sample. The filter needs $(L + M - 2)$ delay units (memory locations) for storing intermediate input and output samples. Furthermore, we need to store the $(L + M - 1)$ filter coefficients $\{b_i, i = 0, 1, 2, \ldots, L - 1\}$ and $\{a_m, m = 1, 2, \ldots, M - 1\}$.

By connecting the feedback section before the feedforward section, we derive a different structure known as the direct-form II IIR filter, which is shown in Fig. 7.3 (See the problems at the end of this chapter for details). This structure has an advantage over the direct-form I structure in terms of reducing the memory requirement. The number of delay elements equals the maximum of $\{L - 1, M - 1\}$. A filter structure with the least memory requirement is called a canonical structure. Therefore, the direct-form II is also called the canonical form of the IIR filter. However, the computational load remains the same as that in the direct-form I structure.

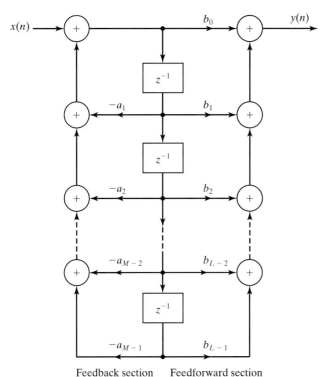

Feedback section   Feedforward section

Figure 7.3 Direct-form II realization of an IIR filter, where $L = M$

Different structures can be derived from the direct-form I and II realizations by using the transposition theorem. This theorem simply states that the input-output properties of the transfer function remain unchanged after the following operations:

1. Reverse the direction of all branches
2. Change branch nodes into summing nodes, and change summing nodes into branch nodes
3. Interchange the input and output

For example, we use the transposition theorem to convert the direct-form II realization shown in Fig. 7.3 into the transposed IIR filter shown in Fig. 7.4. Note the changes of the signal flow, adder nodes, and branching nodes in the transposed IIR filters. By setting $a_i = 0$ for $i = 1, 2, \ldots, M - 1$, we obtain the transpose structure of the FIR filter shown in Fig. 6.3.

Both direct-form and transposed-form IIR filters are very sensitive to quantization errors because quantization errors in the feedback section are fed back and accumulated in the filter, which becomes severe when the filter order is high. A more robust structure for reducing quantization errors is to break down the high-order IIR filter into a cascade of second-order IIR filters (or biquads). The general expression of cascading K biquads to form a $2K$-order IIR filter is

$$H(z) = G \prod_{k=1}^{K} H_k(z) = G \prod_{k=1}^{K} \frac{b_{k0} + b_{k1}z^{-1} + b_{k2}z^{-2}}{1 + a_{k1}z^{-1} + a_{k2}z^{-2}}, \qquad (7.1.8)$$

where G is a gain. An example of cascading two biquads to form a fourth-order IIR filter is shown in Fig. 7.5. Cascading a single first-order IIR filter with a series of

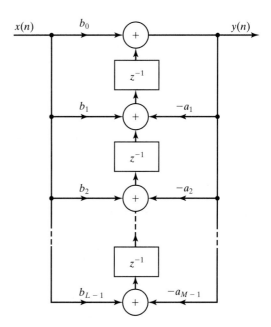

Figure 7.4   Transposed direct-form II IIR filters

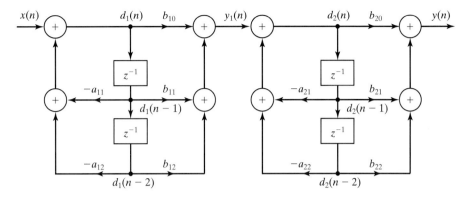

Figure 7.5   A fourth-order IIR filter realized by cascading two biquads

biquads can also form an odd-order IIR filter. For example, we can form a seventh-order IIR filter using a single first-order IIR filter followed by three biquads.

MATLAB provides several functions in the Signal Processing Toolbox to convert a high-order IIR filter from its pole-zero form into second-order sections. The following function

```
[sos, G]=zp2sos(z,p,c);
```

computes the overall gain G and the $K \times 6$ coefficient matrix `sos` containing the coefficients of each second-order section. The coefficient matrix is expressed as

$$\text{sos} = \begin{bmatrix} b_{10} & b_{11} & b_{12} & a_{10} & a_{11} & a_{12} \\ b_{20} & b_{21} & b_{22} & a_{20} & a_{21} & a_{22} \\ \vdots & \vdots & \vdots & \vdots & \vdots & \vdots \\ b_{K0} & b_{K1} & b_{K2} & a_{K0} & a_{K1} & a_{K2} \end{bmatrix}_{K \times 6}, \qquad (7.1.9)$$

where each row corresponds to the numerator and denominator coefficients of each biquad.

Alternatively, MATLAB provides a function that converts its numerator and denominator coefficients directly to the second-order sections as follows:

```
[sos, G]=tf2sos(b,a);
```

In addition to the cascade structure, a high-order IIR filter can also be broken down into a parallel connection of second-order (or first-order) IIR filters. The general expression for implementing a $2K$-order IIR filter as a parallel connection of K second-order IIR filters is obtained by the partial-fraction expansion expressed as

$$H(z) = G + \sum_{k=1}^{K} H_k(z) = G + \sum_{k=1}^{K} \frac{b_{k0} + b_{k1}z^{-1}}{1 + a_{k1}z^{-1} + a_{k2}z^{-2}}. \qquad (7.1.10)$$

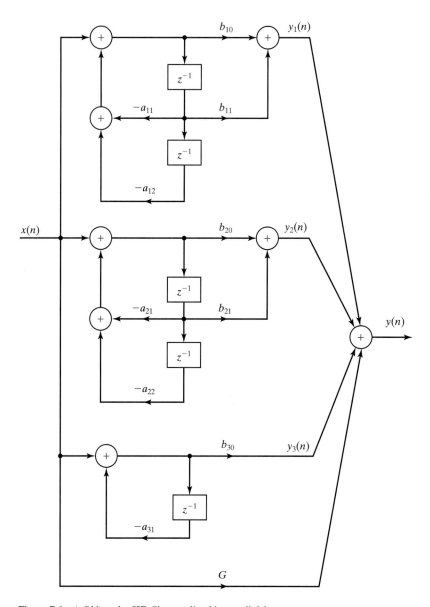

Figure 7.6   A fifth-order IIR filter realized in parallel form

An example of implementing a fifth-order IIR filter using three parallel sections of lower order is shown in Fig. 7.6.

The parallel structure can be obtained by using the MATLAB function

```
[r,p,c]=residuez(b,a)
```

which converts the transfer function to the partial-fraction expansion. The vector r contains the residues, p contains the pole locations, and c contains the direct terms.

Note that $H(z)$ may also be expanded as a sum of first-order sections expressed as

$$H(z) = G + \sum_{k=1}^{P} \frac{A_k}{1 - \alpha_k z^{-1}}. \tag{7.1.11}$$

Because the coefficients A_k and α_k generally have complex values, this structure is difficult to implement for practical applications.

7.1.4 Stability of Infinite-Impulse Response Filters

As discussed in Chapter 6, an FIR filter is always stable. An IIR filter, however, is stable if and only if all of the poles are located inside the unit circle. That is, the absolute value of all of the roots (poles) of the denominator polynomial must be less than 1 (i.e., $|p_m| < 1, m = 1, 2, \dots, M - 1$). Since an IIR filter is normally implemented in a cascade (or parallel) of first- and second-order sections, we can only consider the stability of these two filters.

The transfer function of a first-order IIR filter is expressed as

$$H(z) = \frac{1}{1 + az^{-1}}. \tag{7.1.12}$$

This filter is stable if $|a| < 1$, marginally stable (oscillation) if $|a| = 1$, and unstable if $|a| > 1$.

A second-order IIR filter has the transfer function

$$H(z) = \frac{1}{1 + a_1 z^{-1} + a_2 z^{-2}} = \frac{1}{(1 - p_1 z^{-1})(1 - p_2 z^{-1})}, \tag{7.1.13}$$

where $a_1 = -(p_1 + p_2)$ and $a_2 = p_1 p_2$. To ensure stability, the poles must lie inside the unit circle (i.e., $|p_1| < 1$ and $|p_2| < 1$). Therefore, we obtain

$$|a_2| = |p_1 p_2| < 1, \tag{7.1.14}$$

and the condition for a_1 can be derived from the Schur–Cohn stability test [3] as

$$|a_1| < 1 + a_2. \tag{7.1.15}$$

The stability conditions described in Eqs. (7.1.14) and (7.1.15) can be viewed as the two-dimensional plot shown in Fig. 7.7. A stability triangle is formed in the a_1–a_2 plane that bounds all of the coefficients (a_1, a_2) inside the triangle for a stable second-order IIR filter.

7.1.5 Finite-Wordlength Effects

When implementing an IIR filter in MATLAB, the double-precision floating-point format is used by default for representing filter coefficients and arithmetic operations.

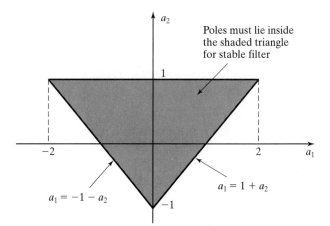

Figure 7.7 Stability triangle for a second-order IIR filter

However, most practical implementations use 16-bit fixed-point processors, which may lead to the degradation of filter performance. The reduced wordlength introduces quantization errors, roundoff noise, and overflow. These errors may circulate in the feedback loops of IIR filters and may get amplified and accumulated. The accumulation of these errors is more severe in the direct-form structures compared with the cascade and parallel structures. In this section, we examine the fixed-point implementation of IIR filters.

As shown in Eq. (3.3.2), the variance of the quantization noise is $2^{-2B}/3$, where B is the wordlength. By assuming that this noise is white noise, we can analyze the power of the quantization noise generated by the IIR filter. The total noise variance can be expressed as

$$\sigma^2_{total} = \frac{2^{-2B}}{3}\frac{1}{2\pi}\int_{-\pi}^{\pi}|H(\omega)|^2\,d\omega, \qquad (7.1.16)$$

which can also be computed using a z-transform as

$$\sigma^2_{total} = \frac{2^{-2B}}{3}\frac{1}{2\pi j}\oint z^{-1}H(z)H(z^{-1})\,dz. \qquad (7.1.17)$$

When the poles are near $z = \pm 1$ (i.e., a narrow-band lowpass or highpass filter), the second-order IIR filter produces more noise. In practice, the cascade-structure filter is recommended for the implementation of high-order narrow-band IIR filters that have closely clustered poles.

Another important issue in implementing the IIR filter using a fixed-point DSP processor is the overflow problem, which is more significant when compared with that of the FIR filter. In an FIR filter, the overflow only causes an error at the output. In an IIR filter, the overflow error is fed back and affects many subsequent output samples. Methods such as saturation mode, guard bits, and scaling the signal and/or coefficients can be used to overcome the effect of overflow, as introduced in Chapter 3.

Another potential problem with using an IIR filter is the possibility of generating limit cycles. This phenomenon was illustrated in Section 3.7.1 in the case of a small-scale limit cycle due to the effects of rounding. Truncation is commonly recommended for eliminating the small-scale limit cycle. Other remedies include using more bits to increase precision and moving the poles away from the unit circle. A large-scale limit cycle can also occur due to overflow. A possible solution is to use the saturation mode provided in most DSP processors.

It is therefore vital to check the stability and finite-wordlength effects of the quantized IIR filter before implementing it on fixed-point DSP processors. FDATool can be used to evaluate the effect of quantizing filter coefficients and arithmetic operation results. An example is given in Section 7.3, which introduces the process of designing and evaluating IIR filters using different structures and precisions.

7.2 DESIGN AND IMPLEMENTATION OF INFINITE-IMPULSE RESPONSE FILTERS

Since analog-filter design is a well-developed technique, a systematic approach in designing a digital IIR filter is to design an analog IIR filter and then convert it into the equivalent digital filter. An alternate approach is to design a digital IIR filter directly in digital domain using the algorithmic design procedure, which solves a set of linear or nonlinear equations. The former approach is used to design typical lowpass, highpass, bandpass, and bandstop filters. The latter approach is used to design filters with arbitrary frequency responses or constraints, which have no analog-prototype filters. We briefly explain the principles of these two design techniques in this section.

The analog-filter specification is very similar to that of the digital filter given in Section 6.1.4. However, there are some minor differences in the definition, as illustrated in Fig. 7.8. The analog frequency variables Ω_p and Ω_s are the

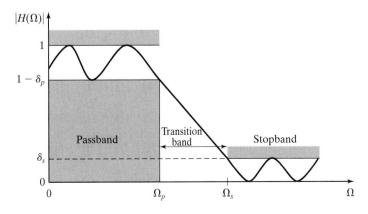

Figure 7.8 Analog lowpass filter specification

passband and stopband edge frequencies, respectively. The passband specification is expressed as

$$1 - \delta_p \leq |H(\Omega)| \leq 1, \quad |\Omega| \leq \Omega_p, \tag{7.2.1}$$

and the stopband specification is

$$|H(\Omega)| \leq \delta_s, \quad |\Omega| \geq \Omega_s. \tag{7.2.2}$$

7.2.1 Analog-to-Digital Filter Design

An analog lowpass filter is first designed based on commonly used analog filters such as Butterworth, Chebyshev I and II, and elliptic filters. It is followed by either (1) performing frequency transformations in the analog domain and then applying analog-to-digital (or s-plane to z-plane) mapping or (2) applying analog-to-digital mapping before performing the frequency transformation in the digital domain. The frequency transformation is the process of converting a lowpass filter into another lowpass, bandpass, highpass, or bandstop filter with different cutoff frequencies using some mapping functions. The frequency transformation can be carried out in either an analog or a digital domain.

An ideal lowpass filter with sharp transition can be obtained by finding a polynomial approximation to the desired magnitude response. This polynomial can be interpreted as a rational function using the Butterworth, Chebyshev, and elliptic filters.

The Butterworth filter is an Mth-order all-pole filter with a squared magnitude response expressed as

$$|H(\Omega)|^2 = \frac{1}{1 + (\Omega/\Omega_c)^{2M}}, \tag{7.2.3}$$

where Ω_c is the 3-dB cutoff frequency defined as $|H(\Omega_c|^2 = 1/2$. The magnitude response of a lowpass Butterworth filter is shown in Fig. 7.9. The magnitude

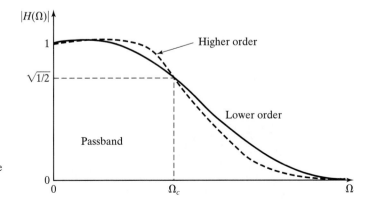

Figure 7.9 Magnitude response of a Butterworth lowpass filter

response is decreased monotonically. Furthermore, the magnitude response is relatively flat over the passband and stopband at the expense of a slower rolloff in the transition region from Ω_p to Ω_s. However, increasing the filter order can reduce the transition bandwidth. Among the different analog filters, the Butterworth filter has the slowest rolloff, thus generally requiring a higher filter order to achieve the same level of rolloff. The phase response is almost linear in the passband and becomes nonlinear around the cutoff frequency.

The required order for the Butterworth filter to satisfy a certain stopband attenuation δ_s at frequency Ω_s can be derived by substituting $\Omega = \Omega_s$ into Eq. (7.2.3), which results in

$$M = \frac{\log_{10}[(1/\delta_s^2) - 1]}{2\log_{10}(\Omega_s/\Omega_p)}. \tag{7.2.4}$$

An interesting feature of the Butterworth filter is the distribution of its poles, s_m, $m = 0, 1, \ldots, 2M - 1$. These poles are distributed uniformly on the unit circle in the s-plane at intervals of π/M radians, and only those poles in the left-half plane are selected for a stable filter. The transfer function in terms of these stable poles is expressed as

$$H(s) = \frac{1}{(s - s_1)(s - s_2)\cdots(s - s_M)}$$

$$= \frac{1}{s^M + a_{M-1}s^{M-1} + \cdots + a_1s + 1}, \tag{7.2.5}$$

where the coefficients $\{a_m\}$ are real numbers.

Chebyshev filters come in two types (Type I and Type II). Chebyshev Type I filters permit ripples in the passband, while Type II filters allow ripples in the stopband. By allowing ripples (or errors) in either the passband or stopband, the Chebyshev filter is able to produce a steeper rolloff compared with the Butterworth filter with the same order. The magnitude responses of Chebyshev Type I and Type II filters are shown in Fig. 7.10.

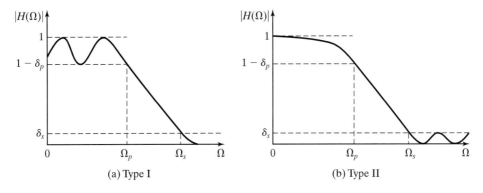

(a) Type I   (b) Type II

Figure 7.10   Magnitude responses of Chebyshev lowpass filters: (a) Type I and (b) Type II

The squared magnitude response of the Chebyshev Type I filter is expressed as

$$|H(\Omega)|^2 = \frac{1}{1 + \varepsilon^2 T_M^2(\Omega/\Omega_p)},\qquad (7.2.6)$$

where $\varepsilon^2 = [(1 - \delta_p)^{-2} - 1]$ is the parameter that determines the amplitude of the passband ripples, and $T_M^2(\Omega/\Omega_p)$ is the Chebyshev polynomial defined as

$$T_M(x) = \begin{cases} \cos(M\cos^{-1}x) & |x| \le 1 \\ \cosh(M\cosh^{-1}x) & |x| > 1 \end{cases}.\qquad (7.2.7)$$

The magnitude of this function oscillates between ± 1. Therefore, the Type I filter has passband ripples that oscillate between 1 and $1/(1 + \varepsilon^2)$. The Type I filter is an all-pole filter with ripples in the passband and monotonic in the stopband. Similar to the transition width of the Butterworth filter, the transition width of Type I filter becomes narrower as the order of filter increases, but at the expense of more ripples (or oscillations) in the passband.

The squared magnitude response of the Chebyshev Type II filter is expressed as

$$|H(\Omega)|^2 = \frac{1}{1 + \varepsilon^2[T_M(\Omega_s/\Omega_p)/T_M(\Omega_s/\Omega)]^2},\qquad (7.2.8)$$

where $\varepsilon^2 = [\delta_s^{-2} - 1]$ is the parameter that controls the amplitude of stopband ripples. Unlike the Type I filter, the Type II filter has a monotonic passband and equiripple stopband and has both poles and zeros. When the order of the Chebyshev Type II filter increases, the number of stopband ripples increases, but the transition width decreases. In general, both Type I and Type II filters require a lower order to meet the same specification when compared with the Butterworth filter, but Chebyshev filters have a poorer phase response.

An elliptic filter has a squared magnitude response expressed as

$$|H(\Omega)|^2 = \frac{1}{1 + \varepsilon^2 U_M^2(\Omega/\Omega_p)},\qquad (7.2.9)$$

where $U_M(\Omega/\Omega_p)$ is a Jacobian elliptic function and can be evaluated using series expansions. The magnitude response of the elliptic filter is shown in Fig. 7.11.

The elliptic filter has the steepest rolloff among the analog filters and requires the least order for achieving a certain specification. However, passband and stopband ripples are present in the elliptic filter, which is similar to the FIR filter produced by the Parks–McClellan algorithm described in Section 6.1.4. The phase response of the elliptic filter is extremely nonlinear in the passband, especially near the cutoff frequency. Therefore, the elliptic filter design is not desired for applications that require good phase response.

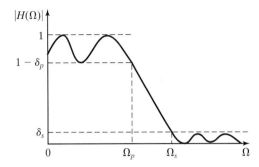

Figure 7.11 Magnitude response of elliptic lowpass filters

TABLE 7.1 IIR-Filter Characteristics Designed by Different Methods

Design method	Magnitude response	Phase response	Filter order
Butterworth	Flat magnitude response in both the passband and stopband; slow rolloff	Approximate linear phase	Requires a higher order
Chebyshev Type I	Shaper transition than Butterworth, but more ripples in the passband	Linearity between Butterworth and elliptic	Order between Butterworth and elliptic
Chebyshev Type II	Shaper transition than Butterworth, but more ripples in the stopband	Linearity between Butterworth and elliptic	Order between Butterworth and elliptic
Elliptic	Equiripple in both passband and stopband, sharpest transition	Highly nonlinear phase	Minimum order

A summary of different analog filters is listed in Table 7.1, which provides a useful guide in selecting a filter-design method to meet the application's requirements.

Frequency Transformation

After designing an analog lowpass filter, we can perform an analog-to-analog frequency transformation and then apply a suitable s-plane to z-plane mapping to get the digital IIR filter, or we can perform the s-plane to z-plane mapping before using the digital-to-digital frequency transformation to derive the desired digital filter.

Analog-to-analog frequency transformation methods are summarized in Table 7.2, which shows the transformation of an analog lowpass filter with cutoff frequency Ω_c to (1) another lowpass filter with a new cutoff frequency Ω_c', (2) a highpass filter with cutoff frequency Ω_c', (3) a bandpass filter with edge frequencies Ω_l and Ω_u, or (4) a bandstop filter with edge frequencies Ω_l and Ω_u. An example of converting a first-order Butterworth lowpass filter $H_{lp}(s) = 1/(s+1)$ to different filters is shown in the last column of Table 7.2.

The second approach is to apply a digital-to-digital frequency transformation after the analog lowpass filter has been converted to a digital filter using the

TABLE 7.2   Analog-to-Analog Frequency Transformations and Examples [3]

Transformation	Mapping	Example
Lowpass $\rightarrow$ Lowpass	$s \rightarrow \dfrac{\Omega_c}{\Omega'_c} s$	$H'_{lp}(s) = \dfrac{\Omega'_c}{\Omega_c s + \Omega'_c}$
Lowpass $\rightarrow$ Highpass	$s \rightarrow \dfrac{\Omega_c \Omega'_c}{s}$	$H'_{hp}(s) = \dfrac{s}{s + \Omega_c \Omega'_c}$
Lowpass $\rightarrow$ Bandpass	$s \rightarrow \Omega_c \dfrac{s^2 + \Omega_l \Omega_u}{s(\Omega_u - \Omega_l)}$	$H'_{bp}(s) = \dfrac{s(\Omega_u - \Omega_l)}{\Omega_c[s^2 + (\Omega_u - \Omega_l)s + \Omega_u \Omega_l]}$
Lowpass $\rightarrow$ Bandstop	$s \rightarrow \Omega_c \dfrac{s(\Omega_u - \Omega_l)}{s^2 + \Omega_u \Omega_l}$	$H'_{bs}(s) = \dfrac{s^2 + \Omega_u \Omega_l}{s^2 + \Omega_c(\Omega_u - \Omega_l)s + \Omega_u \Omega_l}$

TABLE 7.3   Digital-to-Digital Frequency Transformations [3]

Transformation	Mapping	Design parameters
Lowpass $\rightarrow$ Lowpass	$z^{-1} \rightarrow \dfrac{z^{-1} - \alpha}{1 - \alpha z^{-1}}$	$\alpha = \dfrac{\sin[(\omega_c - \omega'_c)/2]}{\sin[(\omega_c + \omega'_c)/2]}$
Lowpass $\rightarrow$ Highpass	$z^{-1} \rightarrow \dfrac{z^{-1} + \alpha}{1 + \alpha z^{-1}}$	$\alpha = -\dfrac{\cos[(\omega_c + \omega'_c)/2]}{\cos[(\omega_c - \omega'_c)/2]}$
Lowpass $\rightarrow$ Bandpass	$z^{-1} \rightarrow -\dfrac{z^{-2} + [2\alpha\beta/(\beta + 1)]z^{-1} + (\beta - 1)/(\beta + 1)}{[(\beta - 1)/(\beta + 1)]z^{-2} + [2\alpha\beta/(\beta + 1)]z^{-1} + 1}$	$\alpha = \dfrac{\cos[(\omega_u + \omega_l)/2]}{\cos[(\omega_u - \omega_l)/2]}$ $\beta = \cot[(\omega_u - \omega_l)/2]\tan(\omega_c/2)$
Lowpass $\rightarrow$ Bandstop	$z^{-1} \rightarrow \dfrac{z^{-2} + [2\alpha/(\beta + 1)]z^{-1} + (1 - \beta)/(\beta + 1)}{[(1 - \beta)/(\beta + 1)]z^{-2} + [2\alpha/(\beta + 1)]z^{-1} + 1}$	$\alpha = \dfrac{\cos[(\omega_u + \omega_l)/2]}{\cos[(\omega_l - \omega_u)/2]}$ $\beta = \tan[(\omega_u - \omega_l)/2]\tan(\omega_c/2)$

s-plane-to-z-plane mapping (as discussed in the next section). Digital-to-digital frequency mapping equations are listed in Table 7.3. Note that ω_c is the cutoff frequency of the original lowpass filter, and ω'_c, ω_l, and ω_u are the new cutoff, lower cutoff, and upper cutoff frequencies, respectively.

Analog-to-Digital Mapping

The design of the digital IIR filter requires the transformation of the analog filter $H(s)$ into its equivalent digital filter, $H(z)$. A mapping function in the form of

$s = f(z)$ is used to map the analog filter into the digital filter. The mapping function must satisfy the following properties:

1. A stable analog filter must be mapped to a stable digital filter. Therefore, all of the poles in the left-half s-plane must be mapped to the poles inside the unit circle in the z-plane.
2. The $j\Omega$ axis in the s-plane is mapped onto the unit circle $|z| = 1$ in the z-plane.
3. The analog-to-digital frequency mapping should be one to one to preserve the frequency response of the analog filter.

Two commonly used mapping techniques, impulse invariant and bilinear, are introduced in this section.

Impulse-Invariant Method   The impulse-invariant method starts from the inverse Laplace transform of analog transfer function $H(s)$ to obtain its impulse response. For example, if $H(s) = A/(s - s_k)$, the impulse response $h(t) = L^{-1}[A/(s - s_k)] = Ae^{s_k t}$. The impulse response is then sampled with sampling period T to obtain $h(nT) = h(t)|_{t=nT} = Ae^{s_k nT}$. We finally derive the transfer function of this digital system by taking the z-transform of $h(nT)$ expressed as

$$H(z) = \sum_{n=0}^{\infty} h(nT)z^{-n} = \sum_{n=0}^{\infty} Ae^{s_k nT}z^{-1} = \frac{A}{1 - e^{s_k T}z^{-1}}. \qquad (7.2.10)$$

By comparing the original analog filter with the digital filter, we have

$$\frac{A}{s - s_k} \rightarrow \frac{A}{1 - e^{s_k T}z^{-1}}. \qquad (7.2.11)$$

It can be easily seen that the transformation function is given as $z = e^{s_k T}$, where s_k is the pole of the analog filter.

The transformation can also be applied to a higher-order IIR filter with p distinct poles by partial-fraction expansion as

$$H(s) = \sum_{k=1}^{p} \frac{A_k}{s - s_k}. \qquad (7.2.12)$$

By applying the transformation to each fractional term, we have

$$\sum_{k=1}^{p} \frac{A_k}{s - s_k} \rightarrow \sum_{k=1}^{p} \frac{A_k}{1 - e^{s_k T}z^{-1}}. \qquad (7.2.13)$$

The sampling process results in the scaling of the magnitude response of the digital filter by T. Therefore, the magnitude response of the digital filter defined in Eqs. (7.2.10) and (7.2.13) must be amplified by T to approximate the same magnitude response of the analog filter.

The mapping function $z = e^{s_k T}$ has a linear relationship between the digital frequency ω and the analog frequency Ω. The analog frequency Ω varies from 0 to

π/T (or from 0 to $-\pi/T$) along the $j\Omega$-axis and is mapped onto ω from 0 to π (or 0 to $-\pi$) around the unit circle. However, this frequency mapping is not one to one since Ω from $\pm k\pi/T$ to $\pm(k + 1)\pi/T$ for $k = 1, 2, \ldots$ are also mapped onto the same digital frequency ω from 0 to $\pm\pi$. Therefore, this mapping produces an aliasing effect and is only suitable for bandlimited filters, such as lowpass and bandpass filters, where the magnitude response of the filter is negligible for frequency that exceeds half of the sampling frequency (π/T).

Bilinear Transformation   A better and more commonly used s-plane to z-plane mapping is the bilinear transformation defined as

$$s = \frac{2}{T}\frac{1 - z^{-1}}{1 + z^{-1}}. \tag{7.2.14}$$

The bilinear transformation is a rational function that maps the left-half s-plane into the internal of the unit circle in the z-plane. It is a one-to-one mapping that maps the entire frequency along the $j\Omega$-axis from $-\infty$ to ∞ onto the unit circle ($-\pi$ to π) only once. This result of a nonlinear relationship between analog frequency and digital frequency is expressed as

$$\omega = 2\tan^{-1}\left(\frac{\Omega T}{2}\right). \tag{7.2.15}$$

This relationship is commonly called the frequency-warping function and is plotted in Fig. 7.12. It compresses the infinite-analog frequency into the digital frequency limited by $\pm\pi$. Therefore, care must be taken when designing the digital filter using bilinear transformation in order to preserve the magnitude response of the analog

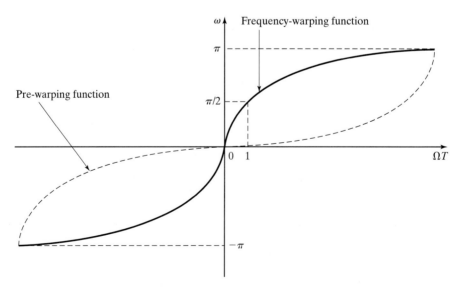

Figure 7.12   Frequency warping and pre-warping functions

filter. A usual approach is to pre-warp the critical frequencies (such as edge frequencies) using the pre-warp function

$$\Omega_c = \frac{2}{T}\tan\left(\frac{\omega_c}{2}\right), \tag{7.2.16}$$

where ω_c is the critical digital frequency. The process of pre-warping is carried out before the use of analog frequency in the analog-filter design.

Example 7.2

Design a first-order digital lowpass filter with a 3 dB cutoff frequency at 0.3π by applying a bilinear transformation to a given analog Butterworth filter with a transfer function $H(s) = 1/(1 + s)$.

   (1) Pre-warp the digital frequency 0.3π to the analog frequency

$$\Omega_c = \frac{2}{T}\tan(0.3\pi/2) = 1.019/T.$$

   (2) Substitute the pre-warped cutoff frequency into the Butterworth filter transfer function to obtain

$$H(s) = 1/(1 + s/\Omega_c) = 1/(1 + sT/1.019).$$

   (3) Apply the bilinear transformation $s = (2/T)(1 - z^{-1}/1 + z^{-1})$ to the transfer function $H(s)$ in (2) to obtain

$$H(z) = H(s)\big|_{s=\frac{2}{T}\frac{1-z^{-1}}{1+z^{-1}}} = \cfrac{1}{1 + 1.963\cfrac{1 - z^{-1}}{1 + z^{-1}}} = 0.337\frac{1 + z^{-1}}{1 - 0.325z^{-1}}.$$

Notice that T does not contribute to the designed filter; thus, the user can simplify the bilinear transformation by selecting $T = 2$. The user can use the MATLAB function `freqz` to verify that the preceding digital IIR filter satisfies the given frequency specification.

7.2.2 Algorithmic Filter Design

Besides using the analog filter to derive the digital IIR filter, an IIR filter can also be designed directly in digital domain by solving a set of equations. Given the impulse response $h_d(n)$ of the desired filter, the algorithmic IIR filter design approximates a causal IIR filter

$$H(z) = \frac{\sum_{i=0}^{L-1} b_i z^{-i}}{1 + \sum_{m=1}^{M-1} a_m z^{-m}} = \sum_{n=0}^{\infty} h(n)z^{-n}. \tag{7.2.17}$$

We can compute a set of parameters $\{b_i, a_m\}$ and substitute them into Eq. (7.2.17) to approximate the desired impulse response [i.e., $h(n) \approx h_d(n)$].

There are several methods for computing parameters $\{b_i, a_m\}$. One method is the Prony method [1], which is based on the least-square approach to find the parameters by minimizing the least-square error

$$E = \sum_{n=0}^{N} |h_d(n) - h(n)|^2. \tag{7.2.18}$$

The denominator coefficients $\{a_m\}$ are solved first, which in turn allows the numerator coefficients $\{b_i\}$ to be calculated. The computed $\{a_m\}$ and $\{b_i\}$ result in the filter's impulse response that closely matches the first N points of the desired impulse response. However, the Prony method is not able to extrapolate the impulse response beyond N points exactly, and there are some means to approximate the extended impulse response by minimizing the norm of the error in a normal equation, as explained in [1].

7.2.3 MATLAB Functions for Infinite-Impulse Response Filter Design

In this section, we introduce some MATLAB functions in the Signal Processing Toolbox for designing digital IIR filters. These functions are summarized in Table 7.4.

The analog-to-digital IIR filter design method uses the first four methods listed in Table 7.4 in a two-step approach. The first step is to find the minimum order that satisfies the given specification by using `buttord`, `cheb1ord`, `cheb2ord`, or `ellipord` for Butterworth, Chebyshev Type I, Chebyshev Type II, or elliptic filter design, respectively. These functions calculate the smallest order and edge frequencies that are used by the corresponding functions `butter`, `cheby1`, `cheby2`, or `ellip` for designing IIR filters.

TABLE 7.4 Summary of MATLAB Functions for Designing IIR Filters

Filter-design method	MATLAB functions
Butterworth	butter, buttord
Chebyshev Type I	cheby1, cheb1ord
Chebyshev Type II	cheby2, cheb2ord
Elliptic	ellip, ellipord
Algorithmic approach	prony, stmcb, lpc
Approximating a piecewise linear-magnitude response	yulewalk
Design filter from a given complex-frequency response	invfreqz, invfreqs

Example 7.3

Design a digital lowpass IIR filter with passband and stopband edge frequencies of 0.3π and 0.4π, respectively, using both the Butterworth and elliptic design methods. The passband and stopband ripples are 1 dB and 50 dB, respectively.

The Butterworth lowpass IIR digital filter is designed by using the following MATLAB script (example7_3.m):

```
[nb,Wnb] = buttord(0.3,0.4,1,50);    % estimate filter order
[bb,ab]= butter(nb,Wnb);             % design the Butterworth filter
fvtool(bb,ab);
```

In the code, the filter-visualization tool (fvtool) is used for analyzing digital filters. Filter characteristics such as passband and stopband ripples and phase response can be viewed using the fvtool function.

The elliptic lowpass IIR digital filter is computed using the following script:

```
[ne,Wne] = ellipord(0.3,0.4,1,50);   % estimate filter order
[be,ae]= ellip(ne,1,50,Wne);         % design the elliptic filter
fvtool(be,ae);
```

Note that the input to the ellip function requires the specification of both passband and stopband ripples, which is not required in the butter function. The ellip function has this specification because the elliptic filter produces both passband and stopband ripples.

The functions butter and ellip combine three steps in designing a digital IIR filter: the design of an analog filter, the analog-to-analog transformation, and the use of bilinear transformation to obtain the coefficients of the digital filter.

Example 7.4

Given an impulse response derived from the desired system, we can obtain the numerator and denominator coefficients of the digital IIR filter using an algorithmic approach, such as the prony function. For example, an impulse response of the elliptic filter is derived as follows (example 7_4.m):

```
[b,a] = ellip(4,1,50,.3);    % design the desired IIR filter
h = impz(b,a,30);            % get impulse response of filter
```

The prony function inputs the desired impulse response h followed by the orders of the numerator and denominator of the IIR filter. We can use the prony function to design an IIR filter as follows:

```
[bb,aa] = prony(h,4,4);   % 4th order numerator and denominator
h1 = impz(bb,aa,30);      % get impulse response of filter
```

We compare the first 30 samples of impulse response h1 derived by the prony function with the original impulse response h. These two impulse responses match, so we have designed a set of IIR-filter coefficients {bb,aa} from the given impulse response h. This method is useful in applications such as accurately modeling a physical system using an IIR filter from the given (or measured) impulse response of system (e.g., room impulse response).

The Signal Processing Toolbox provides an additional set of functions (`buttap`, `cheb1ap`, `cheb2ap`, `ellipap`, and `besselap`) for designing analog lowpass IIR filters. These functions return a set of poles, zeros, and gains of an analog filter with a normalized cutoff frequency of 1 rad/sec. After the prototype lowpass filter has been designed, a set of functions (`lp2lp`, `lp2bp`, `lp2bs`, and `lp2hp`) is used to perform analog-to-analog frequency transformation. Finally, there are two functions (`bilinear` and `impinvar`) that transform the analog filter to a digital filter. These three sets of functions can be used in sequence to design a digital IIR filter. A quicker approach is to use the combined functions, as shown in Example 7.3.

In addition to designing and analyzing a quantized IIR filter, FDATool provides three advanced IIR filter-design methods: `iirlpnorm`, `iirlpnormc`, and `iirgrpdelay`. These functions design IIR filters that meet the arbitrary shape of the frequency response and achieve a certain group delay requirement.

7.2.4 Implementation of Infinite-Impulse Response Filters

In this section, we implement digital IIR filters using MATLAB and C programs. We implement digital IIR filters on fixed-point and floating-point DSP processors in Sections 7.4 and 7.5, respectively.

As discussed in Section 6.1.6 on the implementation of FIR filters using MATLAB, we can implement the IIR filter by using the `filter` function as follows:

```
y = filter(b,a,x);
```

This function filters the input signal in the vector x with the IIR filter given by the coefficient vectors b (numerator) and a (denominator). The output (filtered) signal is placed in the vector y.

The default structure used in MATLAB is the transposed direct-form II (canonical) IIR filter. Note that the first element of vector a is normalized to $a_0 = 1$ with the `filter` function. The `filter` function also allows specification of the initial and final states of the filter as

```
[y,zf] = filter(b,a,x,zi);
```

where zi and zf are the initial and final states of the filter, respectively.

The direct-form I IIR filter can be implemented in the C function iir1.c as follows:

```
float iir1(x,y,b,nb,a,na)

float *x, *y;          // x is input vector, y is output vector
float *b, *a;          // numerator and denominator coefficients
int nb, na;            // length of coefficient vectors
{
  float yn = 0.0;      // output of IIR filter, y(n)
  float yn1 = 0.0;     // temporary storage for feedforward section
  float yn2 = 0.0;     // temporary storage for feedback section
  int i,j;             // loop index
```

```
    for (i=0; i<nb; ++i){  // feedforward section
      yn1 += b[i]*x[i];
    }

    for (j=1; j<na; ++j){  // feedback section
      yn2 += a[j]*y[j];
    }
      yn = yn1-yn2;          // combine two sections
      return(yn);            // return y(n) to calling function
  }
```

This function can be called in the main function as

```
  y[n]= iir1(xn,yn,b,Lb,a,La);    // IIR filtering
```

where xn and yn are arrays defined in the main function for feedforward and feed-back signal vectors, and b and a are coefficient vectors with the lengths Lb and La, respectively.

The direct-form II IIR filter structure can also be implemented in the C function iir2.c as follows:

```
  float iir2(xn,w,b,nb,a,na)

  float *xn, *w;    // x is input signal, w is signal (delay) vector
  float *b, *a;     // numerator and denominator coefficients
  int nb, na;       // length of coefficient vectors
  {
    float yn = 0.0; // output of IIR filter, y(n)
    float w1 = 0.0; // temporary storage for feedback section
    int i,j;                // loop index

    for (j=1; j<na; ++j){ // feedback section
      w1 += a[i]*w[i];
    }
    w1 = xn-w1;
    w[0] = w1;
    for (i=0; i<nb; ++i){ // feedforward section
      yn += b[i]*w[i];
    }
    return(yn);             // return y(n) to calling function
  }
```

Similarly, this function can be called in the main function as

```
  y[n]= iir2(xn,w,b,Lb,a,La);               //IIR filtering
```

where xn is the newest input sample, and w is the signal (or delay) buffer defined in the main function.

In order to compute the IIR filter in real time, the signal buffers (x and y in direct-form I) or intermediate buffer (w in direct-form II) have to be refreshed after obtaining the output $y(n)$. As with FIR filtering, the last element of the buffer is discarded, and new data is placed at the first location of the buffer. The C function datamov.c, listed in Section 6.1.6 for FIR filtering, can be used to update signal buffers in the IIR filter. Circular buffering, described in Section 6.1.6, can also be used in direct-form IIR structures. However, because an IIR filter typically has lower order and is usually implemented in a cascade of biquads, circular buffering may not be necessary in IIR filtering. The implementation of IIR filters using cascade structures in C for experiments is introduced in Section 7.7.

7.3 DESIGN OF INFINITE-IMPULSE RESPONSE FILTERS USING MATLAB

In this section, we use FDATool to design, analyze, quantize, and implement IIR filters. We use the identical specification used in Chapter 6 to design a bandpass filter; thus, we can compare the characteristics between FIR and IIR filters. However, we are not going to cover the detailed steps in designing the IIR filter using FDATool. Refer to Section 6.2.1, Appendix A.2, and references [10, 11] for more information on the usage of FDATool.

7.3.1 Infinite-Impulse Response Filter Design and Analysis

We use the same specification given in Section 6.2.2 for designing a bandpass filter. Figure 7.13 shows the main window of FDATool used to design a direct-form II

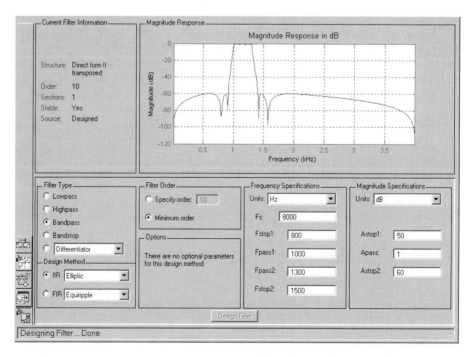

Figure 7.13   FDATool shows the design specifications and magnitude response

transposed IIR filter using the elliptic design method. Note that the order of the designed IIR filter is 10; thus, there are 11 feedforward coefficients b_i and 11 feedback coefficients a_m. As shown in Fig. 6.11, the FIR filter requires 80 coefficients to satisfy the same specification. Therefore, the IIR filter achieves a significant reduction in computational requirements and thus is normally used when limited computational resources are available in the DSP processor.

Other important characteristics of IIR filters can be observed by examining the plots shown in Fig. 7.14. The phase response shown in Fig. 7.14(a) is nonlinear in the passband since the elliptic design method results in a highly nonlinear phase. A better phase response (one that is closer to linear phase) can be obtained by using the Butterworth design method, which is introduced in Section 7.7. Figure 7.14(b) shows that the impulse response gradually converges to a small value; thus, the filter is stable. The group delay shown in Fig. 7.14(c) is not constant throughout the passband, which is due to the nonlinear-phase response. An IIR filter with nonconstant group delay causes different delays at different frequency components. This phase distortion may not be tolerant in some applications that are sensitive to phase. Figure 7.14(d) shows that five complex-conjugate pole pairs are placed closely at the passband. They are all inside the unit circle; thus, this IIR filter is stable.

7.3.2 Quantization Process and Analysis

This section presents the quantization properties of the direct-form II transposed IIR filter. We can turn on quantization mode by clicking on the **Set Quantization Parameters** icon ⬛ and checking the **Turn quantization on** box. We then apply two different fixed-point data formats for different experiments:

1. Q.31 format is applied to coefficients, input, output, multiplicands, and Q.63 format is applied to sums and products.
2. Q.15 format is applied to coefficients, input, output, multiplicands, and Q.31 format is applied to sums and products.

In the first experiment, we examine the magnitude response after selecting the quantization parameters shown in Fig. 7.15. We immediately notice that the quantized magnitude response using the Q.31 and Q.63 formats does not satisfy the required frequency specification. Even poorer results are observed when using the Q.15 and Q.31 formats in the second experiment. Therefore, this direct-form IIR filter cannot be implemented on fixed-point processors using preceding fixed-point formats.

As discussed in Section 7.1.3, quantization errors can be greatly reduced if a cascade of lower-order (normally second-order) sections is used. The cascade IIR filter reduces the accumulation of quantization errors by limiting the quantization-error growth within smaller sections. Therefore, we need to factorize the long polynomial in the direct-form IIR filter into second-order IIR filter sections.

FDATool is able to convert an IIR filter from a direct form to a cascade form. The conversion is activated by selecting the menu **Edit → Convert to Second-Order Section** ..., and the **Order and Scale SOS** window shown in Fig. 7.16 appears. The **Scale** methods include L_2 and L_∞ norms. If **Direction** is set to **Up**, the second-order sections

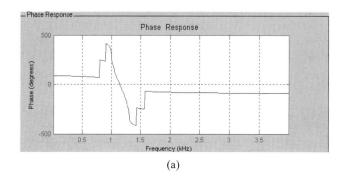

(a)

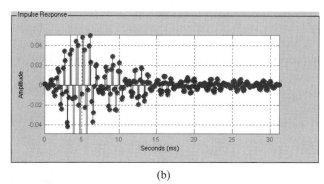

(b)

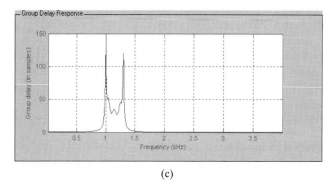

(c)

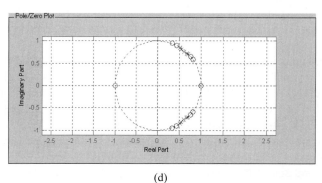

(d)

Figure 7.14   Filter characteristics:
(a) phase response, (b) impulse
response, (c) group delay, and
(d) pole-zero plot

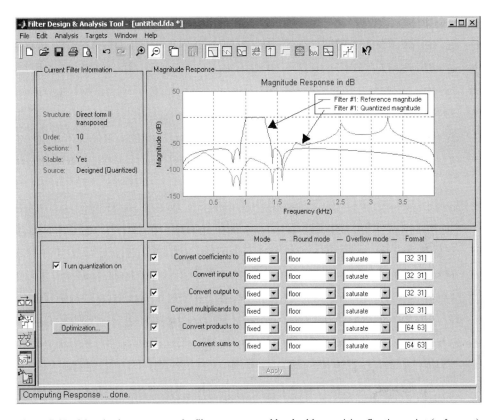

Figure 7.15   Magnitude responses of a filter represented by double-precision floating-point (reference) and fixed-point formats

Figure 7.16   Window for converting to a second-order section IIR filter structure

are ordered from the smallest gain (with poles closest to the origin) to the largest gain (with poles closest to the unit circle). The order is reversed if **Direction** is set to **Down**. The zeros always are paired with the poles closest to them. We can examine the effects of different scaling methods and directions. Note that the **Turn quantization on** box must be unchecked before exploring different scaling method and direction.

The improvement is significant when implementing an IIR filter using second-order sections. Figure 7.17 shows that the magnitude response of the cascade IIR filter

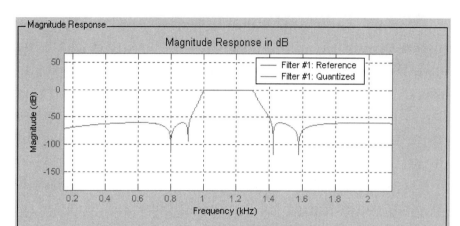

Figure 7.17   Magnitude response of IIR filters with a cascade of biquads

is almost identical to that of the reference filter. It shows that the cascade of biquads is able to reduce quantization errors and produces an overall filter response close to that of the reference filter that uses double-precision, floating-point arithmetic. Therefore, choosing the right filter structure significantly improves the performance of the IIR

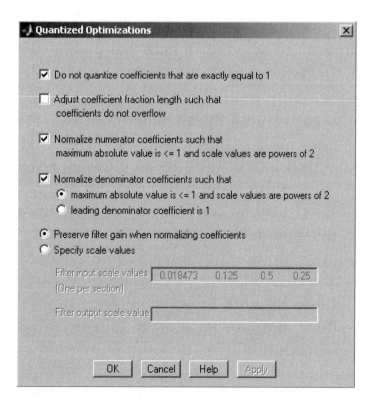

Figure 7.18   **Quantized Optimizations** window

filter. In this example, the tenth-order direct-form IIR filter is converted to a cascade of five second-order IIR filters.

The user can also optimize the quantization of the cascade IIR filter by clicking on the **Optimization**... button. The **Quantized Optimizations** window appears as shown in Fig. 7.18, which displays different features of the quantization process. The selection can determine the coefficient quantization and scaling effects. Table 7.5 summarizes the meanings and implications of each option. It is important to note that scaling may be necessary to prevent arithmetic overflow. A two-step process of normalization and scaling is used for the practical implementation of the IIR filter.

After converting the direct-form filter to the cascade of biquads, the filter-input scale value automatically is updated to a vector of $[0.018473 \quad 0.125 \quad 0.5 \quad 0.25 \quad 1]$. As a result, the input to each section is scaled by the corresponding value in the vector, as

TABLE 7.5   Options for Quantized Optimizations

Option	Meaning and implication
Do not quantize coefficients that are exactly equal to 1	This option eliminates multiplication by 1. Quantization error thus is eliminated
Adjust the coefficient fraction length so that coefficients do not overflow	This option adjusts the fraction length to represent coefficients, which may add complexity to programming
Normalize numerator coefficients such that the maximum absolute value is ≤ 1 and the scale values are a power of 2	This option performs the normalization and scaling of **b** such that all $\lvert b_i \rvert < 1$, as well as the scaling, can be implemented easily via shifting
Normalize denominator coefficients such that the following criteria are met: – The maximum absolute value is ≤ 1, and the scale values are a power of 2 – The leading denominator coefficient is 1	This option performs the normalization and scaling of **a** such that the following criteria are met: – All $\lvert a_m \rvert < 1$, as well as the scaling, can be implemented easily via shifting – $a_0 = 1$, which only occurs after normalization and scaling
Preserve filter gain when normalizing coefficients	This option ensures the filter gain after normalization is the same
Specify the following scale values: – Filter-input scale values (one per section) – The filter-output scale value	This option specifies the following scaling values: – The scaling factor applied to each filter section. If the scaling value is scalar, apply scaling to the input of the filter. If it is vector, apply scaling at the input of each section – Optional scaling at the filter output

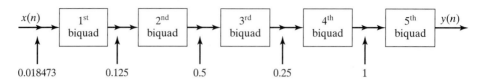

Figure 7.19   Scaling applied to the input of each biquad section

shown in Fig. 7.19. In this case, the output of the last (fifth) biquad is not scaled. An exercise in calculating scaling factors and scaled coefficients for each of the biquads without using FDATool is given in the Problems section at the end of this chapter.

The user can export the coefficients of the biquads into the MATLAB workspace by selecting **File → Export**… from the menu. A window appears to specify the destination and the variable name. The latest version of FDATool also provides a new option for exporting coefficients to SPTool by clicking on the **File → Export to SPTool**. In addition, IIR filter coefficients can be exported to a C header file by selecting **File → Export to C Header file**. The header file can be opened and modified using any text editor.

7.4 FIXED-POINT IMPLEMENTATIONS

In this section, we introduce the implementation of IIR filters using fixed-point DSP processors described in Chapter 4. Similar to its FIR counterpart, IIR filtering is also a MAC-based operation. Usually, a lower-order IIR filter is required to achieve the same performance as compared with a higher-order FIR filter. Therefore, a lower MIPS is required in the implementation of an IIR filter. However, the fixed-point programming of the IIR filter is not as straightforward as that of the FIR filter. As discussed previously, direct-form implementation of the IIR filter is extremely sensitive to coefficient quantization and thus is not recommended for fixed-point processors. Most practical implementations use the cascade of biquads to reduce quantization errors. In general, the use of scaling to prevent overflow, the use of cascade structure, and even the use of double precision to prevent instability are critical in fixed-point IIR filtering. These factors result in more complicated programming than the FIR filtering introduced in Chapter 6.

When implementing a second-order, direct-form II IIR filter, as shown in Fig. 7.5, the feedback section of the IIR filtering operation is performed before executing the feedforward section. The kth biquad can be implemented using the following difference equations:

$$d_k(n) = y_{k-1}(n) - a_{k1}d_k(n-1) - a_{k2}d_k(n-2), \qquad (7.4.1)$$
$$y_k(n) = b_{k2}d_k(n-2) + b_{k1}d_k(n-1) + b_{k0}d_k(n), \qquad (7.4.2)$$

and

$$d_k(n-1) \rightarrow d_k(n-2) \text{ and } d_k(n) \rightarrow d_k(n-1) \qquad (7.4.3)$$

for $k = 1, 2, \ldots, K$, where the filter input $x(n) = y_0(n)$, and the output $y(n) = y_K(n)$.

The first equation performs the feedback section of the direct-form II IIR filter, and the result is stored in $d_k(n)$. The second equation completes the IIR filtering by computing the feedforward section and placing the result in $y_k(n)$. The last equation refreshes the delay buffer by shifting one location in the memory buffer [i.e., $d_k(n-1)$ is shifted to $d_k(n-2)$, and then $d_k(n)$ is shifted to $d_k(n-1)$].

7.4.1 Implementation Using the TMS320C2000

As shown in Eqs. (7.4.1) and (7.4.2), second-order IIR filtering can be implemented using five multiply-add operations. The C24x Filter Library [12] provides two assembly subroutines, IIR5B1Q16 and IIR5B1Q32, which can be called from either the C or assembly main program. The former subroutine implements IIR filtering using a 16-bit signal buffer, while the latter subroutine uses a 32-bit buffer. Multiply-add operations are implemented using the MPY and LTA instructions. The kernel of IIR filtering is listed as follows:

```
biqd:                    ; kth biquad computation where k=1:K
    ; Feedback section
    LT      *-,AR3       ; AR3->ak2, TREG=dk(n-2), AR4->dk(n-1)
    MPY     *+,AR4       ; AR4->dk(n-1),PREG=dk(n-2)*ak2, AR3->ak1
    LTA     *+,AR3       ; AR3->ak1, TREG=dk(n-1),AR4->dk(n-2),
                         ; ACC=in+dk(n-2)*ak2
    MPY     *+,AR4       ; AR4->dk(n-2), PREG=dk(n-1)*ak1,AR3->bk2
    LTA     *-,AR0       ; AR4->dk(n-1), ACC=in+dk(n-2)*ak2+d1(n-1)
                         ; *ak1=dk(n)
    LPH     *+           ; AR4->dk(n-2), PREGH=dk(n-2)
    SPH     *-,AR0       ; AR0=FP, AR4->dk(n-1), dk(n-2)=dk(n-1)
    SACH    *,(16-QFORMAT),AR3   ; AR3->bk2, *FP=dk(n)
    ; Feedforward section
    LACL    #0           ; ACC=0
    MPY     *+,AR4       ; AR4->dk(n-1), PREG=dn(k-2)*bk2, AR3->bk1
    LTD     *,AR3        ; AR3->bk1, TREG=dk(n-1), dk(n-2)=dk(n-1),
                         ; AR4->dk(n-1)
                         ; ACC=dn(k-2)*bk2
    MPY     *+,AR0       ; AR0->dk(n), PREG=dk(n-1)*bk1, AR3->bk0
    LTA     *,AR3        ; ARP=AR3, AR3->bk0, TREG=dk(n)
                         ; ACC=dn(k-2)*bk2+dk(n-1)*bk1
    MPY     *+,AR4       ; ARP=AR4, AR4->dk(n-1), PREG=bk0*dk(n),
                         ; AR3->ak2+1
    APAC                 ; ACC=dn(k-2)*bk2+dk(n-1)*bk1+bk0*dk(n)
    MPY     #1           ; PREGL=dk(n)
    SPL     *-,AR5       ; ARP=AR5,AR4->dk+1(n-2),dk(n-1)=dk(n)
    BANZ    biqd,*-,AR4  ; for next k biquad
```

In the code, AR3 is a coefficient pointer that points at the coefficient buffer, and AR4 is a pointer that points at the delay buffer. The organization of these two buffers is illustrated in Fig. 7.20. Note that in order to use the LTA instruction, the coefficients a_{k1} and a_{k2} of feedback section are negated as shown in Eq. (7.1.2b).

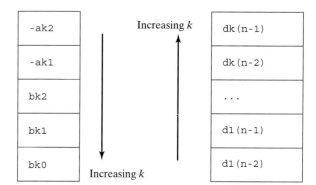

Figure 7.20 Memory arrangement for the coefficient and delay buffers

7.4.2 Implementation Using the TMS320C54x

The MAC unit in the C54x processor can be used to implement both the feedback and feedforward sections of the IIR filter in sequence. The delay elements and the coefficients of the first biquad are arranged in the internal memory of C54x, as shown in Fig. 7.21. It is noted that the input signal $x(n)$ (xin) and the first element in the delay line $d(n)$ can share the same memory location.

The C54x assembly program is listed as follows:

```
          LD     #x0,DP
          SSBX   FRCT
; Feedback section
IIR:   PORTR in_port,xin    ; input to xin
          LD     @xin,16,A    ; shifting the input left by 16 bits
          LD     @dn_1,T      ; store d(n-1) into T register
          MAC    @a11,A       ; A' = -a11*d(n-1)+xin
          LD     @dn_2,T
          MAC    @a12,A       ; A = A'-a12*d(n-2)
          STH    A,@xin       ; replace xin by partial result d(n)
; Feedforward section
          MPY    @b12,A       ; A'' = b12*d(n-2)
          LTD    @dn_1        ; load d(n-1) into T & d(n-2)
          MAC    @b11,A       ; A' = A''+b11*d(n-1)
          LTD    @xin         ; load d(n) into T & d(n-1)
          MAC    @b10,A       ; A = A' + b10*d(n)
          STH    A,@xin       ; result placed in memory @xin
          PORTW @xin,out_port  ; output result to out_port
          B      IIR          ; repeat IIR filtering for next input
```

There are several differences between IIR and FIR programming on the C54x processor. As introduced in Section 6.3.2, FIR filtering needs a long series of multiply-add operations; thus, a single-repeat instruction RPT is very efficient in repeating these MAC operations. The IIR filter is usually implemented as a cascade of biquads, and block repeat is used to implement each section in sequence. Furthermore, scaling

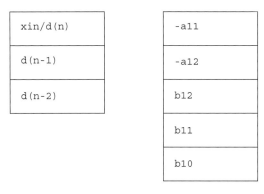

Figure 7.21   Memory arrangement for delay elements and coefficients for a single biquad

TABLE 7.6   IIR-Filter Functions and Benchmarks for K Biquads [13]

IIR function	Description	Cycle count	Code size
iir32	Double-precision IIR filter using direct-form II	$N \times (12 + 56 \times K) + 62$	110
iircas4	Cascaded IIR direct-form II using four coefficients per biquad	$N \times (11 + 5 \times K) + 40$	50
iircas5	Cascaded IIR direct-form II using five coefficients per biquad	$N \times (11 + 6 \times K) + 40$	51
iircas51	Cascaded IIR direct-form I using five coefficients per biquad	$N \times (13 + 8 \times K) + 44$	58

must be carried out at different sections to prevent overflow. Therefore, more instructions are required for IIR filtering.

The TMS320C54x DSPLIB [13] provides a set of assembly-optimized IIR-filtering functions in both sample- and block-processing modes. As listed in Table 7.6, these functions operate on Q.15 and Q.31 operands. The iir32 function operates on 32-bit coefficients and data in a delay buffer, which implies that double-precision storage is required for filter coefficients and intermediate signal elements. The penalty paid for using the iir32 function is an increase of cycle count by more than 12 times and the doubling of the code size over other single-precision IIR-filter functions. The other three IIR functions (iircas4, iircas5, and iircas51) are used to implement the cascade of single-precision, second-order IIR filters. The four-coefficient function (iircas4) assumes that the first feedforward coefficient $b_0 = 1$ and thus provides less flexibility as compared with the five-coefficient function. All these functions output the status of overflow and do not include overflow prevention in the subroutine. Therefore, the user must use a suitable scaling scheme to avoid overflow. A detailed implementation using these DSP functions is given in Section 7.7.7.

7.4.3 Implementation Using the TMS320C55x

As with the C54x, the TMS320C55x DSPLIB [14] provides a set of assembly-optimized IIR-filtering functions that can be called by C programs. As listed in Table 7.7,

TABLE 7.7   C55x IIR Functions and Benchmarks of Using K Biquads [14]

IIR function	Description	Cycle count	Code size (bytes)
iir32	Double-precision IIR filter using direct-form II	$N \times (7 + 31 \times K) + 77$	203
iircas4	Cascaded IIR direct-form II using four coefficients per biquad	$N \times (2 + 3 \times K) + 44$	122
iircas5	Cascaded IIR direct-form II using five coefficients per biquad	$N \times (5 + 5 \times K) + 60$	126
iircas51	Cascaded IIR direct-form I using five coefficients per biquad	$N \times (5 + 8 \times K) + 68$	154

the iir32 function operates on 32-bit coefficients and data to implement the double-precision (32-bit) IIR filter. The IIR functions (iircas4, iircas5, and iircas51) implement a cascade of single-precision (16-bit) biquads. The four-coefficient function, iircas4, assumes that the first feedforward coefficient $b_0 = 1$ and thus provides less flexibility when compared with the five-coefficient function, iircas5. The last function, iircas51, computes the cascade of K biquads, which are implemented using direct-form I. As discussed in Section 7.1.3, this realization requires more memory locations and cycles. Therefore, the most suitable IIR routine is iircas5, which is discussed further in this section.

The assembly subroutine iircas5 realizes the cascade of direct-form II biquads, as shown in Fig. 7.5. This function can be called from the C program as

```
short oflag = iircas5(DATA *x, DATA *h, DATA *r, DATA *dbuffer,
ushort nbiq, ushort nx);
```

The argument x is passed in XAR0, which is the pointer to the input vector with nx real elements. The argument h is passed in XAR1, which is the pointer to the filter-coefficient vector with the format {ak1 ak2 bk2 bk0 bk1} where k is the biquad index. Note that the feedback coefficients a_{k1} and a_{k2} do not need to be negated (as in C2000 and C54x implementations) since the MASM instruction is used for multiply-subtract operations. The argument r is passed in XAR2, which is the pointer to the output vector of nx real elements. The argument dbuffer is passed in XAR3, which is the pointer to the delay buffer of length $2K$. The number of biquads making up the filter is nbiq, which is passed in T0. The number of elements in the input and output vectors nx is passed in T1. The output variable oflag is returned in T0, which indicates if data overflow occurred in an intermediate or final result.

The kernel of IIR-filtering code is listed as follows [14]:

```
; Setup circular addressing
    MOV    T0,T3    ; compute 2*nbiq
```

```
       SFTS   T3,#1
       MOV    @T3,BK03  ; initialize AR0-3 circular buffer size (2*nbiq)
       BSET   AR3LC     ; initialize AR3 = circular (dbuffer)
       ADD    #1,AR3    ; adjust AR3 to buffer start
       MOV    @AR3,BSA23 ; initialize AR2-3 circular start addr:
                        ; dbuffer(1)
       MOV    *AR5,AR3  ; initialize AR3 offset to buffer start
       SUB    #1,T1,T3  ; compute nx-1
       MOV    T3,BRC0   ; initialize outer loop counter (nx-1)
       MOV    XAR1,XAR7 ; reinitialize coefficient buffer pointer
       SUB    #1,T0,T3  ; initialize inner loop counter (#biquads-1)
       MOV    T3,BRC1
       MOV    #SCALE,T2  ;scaling factor
; Kernel - XAR0: x[] input, XAR1: h[] coefficients,
;          XAR2: r[] result
;          XAR3: dbuffer[], XAR7: reinitialize XAR1

    RPTBLOCAL   loop1-1          ; outer loop: process a new input
    MOV    *AR0+<<#16,AC0        ; HI(AC0) = x(n)

    ||RPTBLOCAL  loop2-1         ; inner loop: process a bi-quad
    MASM   T3=*(AR3+T0),*AR1+,AC0  ; AC0 -= a1*d(n-1)
    MASM   T3=*AR3,*AR1+,AC0     ; AC0 -= a2*d(n-2)
    MPYM   *AR1+,T3,AC0          ; AC0 = b2*d(n-2)
    ::MOV  HI(AC0<<T2),*AR3      ; d(n) replaces d(n-2)
    MACM   *(AR3+T0),*AR1+,AC0   ; AC0 += b0*d(n)
    MACM   *AR3+,*AR1+,AC0       ; AC0 += b1*d(n-1), input to next
                                 ; biquad
loop2:
    MOV    XAR7,XAR1             ; reinitialize coefficient pointer
    MOV    rnd(HI(AC0)),*AR2+    ; store result to output buffer

loop1:                          ; signal overflow
    MOV    #0,T0
    XCC    check1,overflow(AC0)
    MOV    #1,T0
check1:                         ; save last index value
    MOV    AR3,*AR5             ; save index to dbuffer[0]
```

Note that the second MAC unit is idle in computing IIR filtering in the routine. It is a challenging job to efficiently use the dual-MAC unit for a cascade of second-order sections.

7.4.4 Implementation Using the TMS320C62x

As described in Chapter 4, the TMS320C62x processor is object-code compatible with the TMS320C64x. In this section, we study the programming of an IIR filter that is applicable for both C62x and C64x processors. In the next section, we exploit special features of the C64x processor for IIR filtering. The C6000 C compiler has very high efficiency as compared with optimized assembly code. Therefore, the C code introduced in Section 7.2.4 may be modified for practical applications.

The direct-form II, second-order IIR filter can be implemented using the following C program with compiler intrinsics:

```
// Feedback section
mult1=_mpy(dn_2,a12) >> 15;        // d(n-2)*a12
mult2=_mpy(dn_1,a11) >> 15;        // d(n-1)*a11
dn=xin - (short)(mult1+ mult2)     // d(n)=xin+d(n-2)*a12+d(n-1)*a11

// Feedforward section
mult3=_mpy(dn_2,b12) >> 15;        // d(n-2)*b12
mult4=_mpy(dn_1,b11) >> 15;        // d(n-1)*b11
mult5=_mpy(dn, b10)  >> 15;        // d(n)*b10
yout=(short)(mult3+mult4+mult5);   // y(n)

// Data shifting
dn_2 = dn_1;                       // d(n-1) -> d(n-2)
dn_1 = dn;                         // d(n)  -> d(n-1)
```

The variables `mult1`, `mult2`, `mult3`, `mult4`, and `mult5` are declared as 32-bit (`int`) variables. The delay elements, input, and output (`dn`, `dn_1`, `dn_2`, `xin`, and `yout`) are specified as 16-bit (`short`) variables.

In the program, we use the intrinsic function

```
int _mpy(int src1, int src2)
```

which multiplies two 16-bit integers (`src1` and `src2`) and produces a 32-bit result. The result is right-shifted 15 bits and saved in a 32-bit (`int`) variable. The results in feedback and feedforward sections are converted to `short` before storing to the `dn` and `yout` variables, respectively. Finally, data shifting is carried out before injecting the new input sample. This second-order IIR-filter code can be extended to implement a cascade IIR filter.

Linear assembly can be used to improve the efficiency of the IIR routine. The allocation of registers and scheduling of instructions are carried out by the compiler. Unlike C5000 processors, the instruction set of the C62x does not contain any complex instructions such as MAC. In order to implement the MAC operation for IIR filtering, a series of simple instructions for load, multiply, add, and shift are used as follows:

```
; Feedback section
MPY    a12,dn_2,mult1
MPY    a11,dn_1,mult2
ADD    mult1,mult2,dn
SHR    dn,15,dn
SSUB   x,dn,dn

; Feedforward section
MPY    b12,dn_2,mult3
MPY    b11,dn_1,mult4
MPY    b10,dn,mult5
ADD    mult3,mult4,sum
ADD    mult5,sum,sum
SHR    sum,15,yout
```

An IIR-filtering routine (`DSP_iir`) is available in the TMS320C62x DSPLIB [15]. The cycle count is $5N + 30$, and the code size is 384 bytes for a biquad IIR function.

7.4.5 Implementation Using the TMS320C64x

The C64x processor extends several fixed-point operations of the C62x processor. The packed-data instructions in the C64x processor are used effectively in FIR filtering, but may not produce significant improvement when running the cascade IIR filter on C64x processors. The reason for this lack of improvement is that second-order IIR filter is commonly used in the cascade structure, and the amount of computation per biquad is limited. However, the code written for the C62x processor can also be used for the C64x processor.

We can also examine the differences between the IIR filter routine in the C62x DSPLIB with the one given in the C64x DSPLIB [16]. The cycle count is $4N + 21$, and the code size is 276 bytes. When compared with the C62x-based IIR filter function, there is only a slight reduction in cycle count on the C64x. The savings in code size is about 100 bytes for the C64x IIR filter routine.

7.5 FLOATING-POINT IMPLEMENTATIONS

In this section, we implement IIR filtering on TMS320C3x and TMS320C67x floating-point DSP processors.

7.5.1 Implementation Using the TMS320C3x

In this section, we implement one biquad on the TMS320C3x since IIR filters are usually realized as a cascade of second-order sections. The difference equations for the single biquad are defined in Eqs. (7.4.1) and (7.4.2). The memory organization for these two equations is illustrated in Fig. 7.22. The signal buffer is implemented as a circular buffer; thus, the address for the start of the value dk(n) must be a multiple

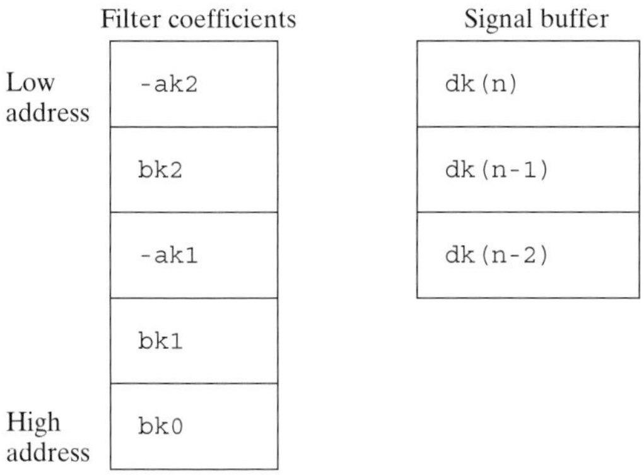

Filter coefficients	Signal buffer
Low address — -ak2	dk(n)
bk2	dk(n-1)
-ak1	dk(n-2)
bk1	
High address — bk0	

Figure 7.22 Data memory organization for a single biquad

of 4. That is, the last 2 bits of the beginning address must be 0. The block-size register BK must be initialized to 3.

An implementation of a single biquad using the TMS320C3x assembly routine
is listed as follows [17]:

```
        .global    IIR1

IIR1    MPYF3 *AR0,*AR1,R0              ;ak2*dk(n-2) -> R0
        MPYF3 *++AR0(1),*AR1-(1)%,R1   ;bk2*dk(n-2) -> R1

        MPYF3 *++AR0(1),*AR1,R0        ;ak1*dk(n-1) -> R0
 ||     ADDF3 R0,R2,R2                 ;ak2*dk(n-2)+x(n) -> R2

        MPYF3 *++AR0(1),*AR1-(1)%,R0   ;bk1*dk(n-1) -> R0
 ||     ADDF3 R0,R2,R2                 ;ak1*dk(n-1)+ak2*dk(n-2)+
                                       ; x(n)->R2

        MPYF3 *++AR0(1),R2,R2          ;bk0*dk(n) -> R2
 ||     STF   R2,*AR1++(1)%            ;store dk(n) and point to
                                       ; dk(n-1)

        ADDF  R0,R2                    ;bk1*dk(n-1)+bk0*dk(n) -> R2
        ADDF  R1,R2,R0                 ;bk2*dk(n-2)+bk1*dk(n-1)+
                                       ; bk0*dk(n)->R0

        RETS                           ; return to main program
   .end
```

The registers used by this IIR1 subroutine are AR0 for the address of the filter coefficient; AR1 for the address of the signal buffer; R2, which contains an input sample;
and BK, which equals 3. The register that contains the result is R0. This routine
needs 11 cycles for execution, and the code size is 8 words.

In most practical applications, the IIR filter contains $K(>1)$ biquads. The
TMS320C3x assembly code for this general case is available in [17]. This more general subroutine requires 17 words for the code and $6K + 17$ cycles for execution.

7.5.2 Implementation Using the TMS320C67x

The C67x has a set of instructions for performing single-precision, floating-point
multiplication (MPYSP) and single-precision, floating-point addition (ADDSP) and subtraction (SUBSP). The user can simply replace MPY, ADD, and SUB in the C62x's linear
assembly code with the preceding instructions. Furthermore, the data can be loaded
in 32 bits using the LDW instruction.

Double-precision, floating-point (64-bit) data can be stored as two 32-bit
data in a register pair. The 32 LSBs are loaded in the even register, while the 32
MSBs (containing the sign bit) are loaded into the next odd register. A set of
double-precision instructions (MPYDP, ADDDP, and SUBDP) is used to handle double-
precision arithmetic.

7.6 AN APPLICATION: DUAL-TONE MULTIFREQUENCY GENERATION AND DETECTION

The IIR filter can be employed in the generation and detection of the DTMF signal used by a touchtone phone. Besides being used as the dialing tone, the DTMF signal can also be used for selecting options in telephone-servicing systems.

The touchtone keypad is shown in Fig. 7.23. The keypad numbers can be encoded with combinations from the low-frequency group (697, 770, 852, and 941 Hz) and the high-frequency group (1,209, 1,336, 1,477, and 1,633 Hz). These DTMF tone frequencies are carefully selected to prevent interference with the speech signal. For example, the digit "5" is represented by a combination of $f_L = 770$ Hz in the low-frequency group and $f_H = 1,336$ Hz in the high-frequency group. Therefore, two tones are generated and combined to uniquely define a particular key on the keypad. Note that the last column containing the frequency of 1,633 Hz is omitted on most phone sets.

7.6.1 Dual-Tone Multifrequency Tone Generation

The generation of dual tones can be implemented by connecting two sinewave generators in parallel and adding the outputs to form the DTMF signal as

$$y(n) = \sin\left(\frac{2\pi f_L n}{f_s}\right) + \sin\left(\frac{2\pi f_H n}{f_s}\right), \tag{7.6.1}$$

where f_s is the sampling frequency, which commonly is set to 8 kHz for telecommunications.

Sinewave can be generated using a look-up table, which stores the sinusoidal values. However, the memory used in the look-up table may be significant. Sinewave can also be generated using the polynomial-approximation technique, but

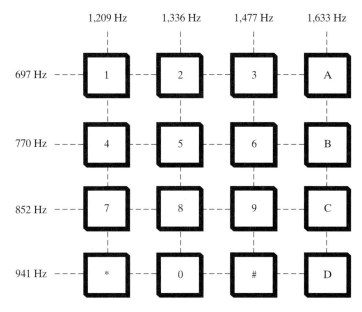

Figure 7.23 Matrix telephone keypad

this technique generally requires intensive computation. In this section, we introduce a simple IIR filter for generating a sinewave.

If the impulse response of an IIR filter is a desired sinewave, it is expressed as

$$h(n) = \sin(\omega_k n), \qquad (7.6.2)$$

which is the filter output when an impulse function $\delta(n)$ is used as input. The filter output in z-domain is expressed as follows [9] (see the Problems section later in this chapter):

$$Y(z) = H(z)X(z) = \frac{\sin \omega_k z^{-1}}{1 - 2 \cos \omega_k z^{-1} + z^{-2}}. \qquad (7.6.3)$$

Since the z-transform of $\delta(n)$ is 1, we have $X(z) = 1$. The transfer function of an IIR filter is expressed as

$$H(z) = \frac{\sin \omega_k z^{-1}}{1 - 2 \cos \omega_k z^{-1} + z^{-2}} = \frac{b_0 + b_1 z^{-1} + b_2 z^{-1}}{1 + a_1 z^{-1} + a_2 z^{-1}}. \qquad (7.6.4)$$

Therefore, this biquad requires only two coefficients,

$$b_1 = \sin \omega_k = \sin\left(\frac{2\pi f_k}{f_s}\right) \text{ and } a_1 = -2 \cos \omega_k = -2 \cos\left(\frac{2\pi f_k}{f_s}\right), \quad (7.6.5)$$

since $b_0 = b_2 = 0$ and $a_2 = 1$. This IIR filter is driven to oscillation mode by placing the poles on the unit circle.

Therefore, an efficient approach to produce the DTMF signal is to use a pair of biquad filters as shown in Fig. 7.24, where $\omega_L = 2\pi f_L/f_s$ and $\omega_H = 2\pi f_H/f_s$. We only need to store eight sets of biquad coefficients (i.e., a total of 16 coefficients). These coefficients can be calculated by substituting the frequency values into Eq. (7.6.5), and the results are summarized in Table 7.8 for the sampling rate of 8 kHz. The technique described in Fig. 7.24 requires an impulse signal $\delta(n)$ to excite the parallel IIR filters. An alternate approach is to set up a proper initial state in the IIR filter, and the filter oscillates without any external excitation. See the problems at the end of this chapter to see how this method is implemented.

The MATLAB function impz allows the user to generate a sinusoidal signal based on the IIR filter. For example, to generate a tone for key 5, the low and high frequencies are 770 Hz and 1,336 Hz, respectively. The coefficients for the IIR filters that generate these tones are given in Table 7.8. The following script (dtmf_gen.m) generates and plays the DTMF tones:

```
Fs = 8000; N = 400;                      % generate a 50 msec tone
b_low = [0 0.56856 0]; a_low = [1 -1.64528 1];
b_high = [0 0.86707 0]; a_high = [1 -0.99638 1];
[H1,T] = impz(b_low, a_low,N,Fs);        % low frequency tone
[H2,T] = impz(b_high, a_high,N,Fs);      % high frequency tone
HT = H1+H2;                              % dual tones
soundsc(HT);                             % play the DTMF tones
save tx_data_5 HT;                       % save in data file
```

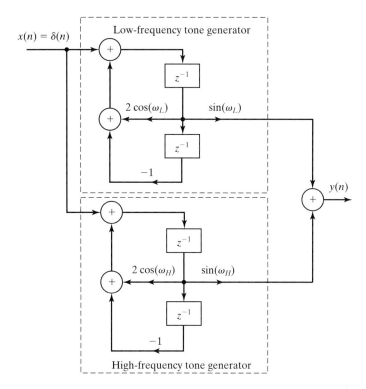

Figure 7.24   A DTMF generator using two IIR filters connected in parallel

TABLE 7.8   Coefficients of IIR Filters for Different DTMF Frequencies

Frequency (Hz)	b_1	$-a_1/2$
697	0.52049	0.85387
770	0.56856	0.82264
852	0.62033	0.78434
941	0.67359	0.73910
1,209	0.81315	0.58205
1,336	0.86707	0.49819
1,477	0.91682	0.39931
1,633	0.95875	0.28427

When implementing the DTMF tone generator in Q.15 format, there is a problem with representing coefficient a_1 since it is larger than 1. There are two solutions to this problem:

1. A different Q format can be used to extend the dynamic range. For example, Q1.14 can be used to represent coefficient values that range from -2 to 1.999.

2. If Q.15 format is used, filter coefficients can be scaled down to less than 1 be-fore implementation, with the tradeoff of reducing the gain of filter. This method was described in Chapter 3 and in Section 7.3.2. In Table 7.8, the coef-ficients a_i are scaled down by half to fit in the Q.15 format.

The DTMF signal must meet some timing requirements for the duration and spacing of tones. The maximum data rate for a touchtone signal is 10 digits per second. For a 100 msec time slot, the duration of the actual tone must be at least 45 msec, but less than 55 msec. The remaining period of 100 msec is for the quiet period. The next section examines the implementation of the DTMF tone detector.

7.6.2 Dual-Tone Multifrequency Tone Detection

The DTMF tone detector detects a valid tone pair and verifies the correct tone duration. DTMF tones have been used to set up calls and to control features in call forwarding, etc. Therefore, a robust DTMF detector is needed in the presence of line interference, noise, and even speech.

A simple approach for detecting a sinewave is to perform spectrum analysis using DFT. A detailed description of DFT and its properties is given in the next chapter. If the received signal is denoted as $x(n)$, the DFT of N samples of $x(n)$ can be computed as

$$X(k) = \sum_{n=0}^{N-1} x(n) W_N^{kn}, \tag{7.6.6}$$

where $k = 0, 1, 2, \ldots, N - 1$, and $W_N^{nk} = e^{-j\frac{2\pi kn}{N}} = \cos(2\pi kn/N) - j\sin(2\pi kn/N)$ is the twiddle factor. Since the DTMF detector only needs to compute eight differ-ent $X(k)$ values corresponding to the DTMF frequencies defined in Fig. 7.23, a more efficient way to calculate $X(k)$ is by using the Goertzel algorithm. The Goertzel algorithm consists of a group of bandpass (second-order IIR) filters that extract the selected frequency information.

Consider the property of the twiddle factor

$$W_N^{-kN} = e^{j\frac{2\pi kN}{N}} = 1. \tag{7.6.7}$$

Multiplying both sides of Eq. (7.6.6) by W_N^{-kN}, we have

$$X(k) = W_N^{-kN} \sum_{n=0}^{N-1} x(n) W_N^{kn} = \sum_{n=0}^{N-1} x(n) W_N^{-k(N-n)}. \tag{7.6.8}$$

We can relate the above DFT to the bandpass filter by defining a sequence

$$y_k(n) = \sum_{m=0}^{N-1} x(m) W_N^{-k(n-m)} = x(n) * W_N^{-kn} u(n), \tag{7.6.9}$$

which is the convolution of a finite-duration sequence $x(n)$ with the sequence $W_N^{-kn}u(n)$. Therefore, the impulse response of the filter is

$$h_k(n) = W_N^{-kn}u(n). \tag{7.6.10}$$

By comparing Eq. (7.6.8) and Eq. (7.6.9), $X(k)$ is equal to the output of filter $y_k(n)$ at time $n = N - 1$. Therefore, instead of computing the DFT, we can simply calculate the filter output at time $n = N - 1$.

Taking the z-transform of Eq. (7.6.10), we obtain

$$H_k(z) = \frac{1}{1 - W_N^{-k}z^{-1}}, k = 0, 1, \ldots, N - 1. \tag{7.6.11}$$

This filter has a pole on the unit circle at $w_k = (2\pi k/N)$, and N parallel filters can be used to implement the N-point DFT. In the DTMF detection, only eight filters are needed, and k can be determined as

$$k \approx \frac{Nf_k}{f_s}. \tag{7.6.12}$$

For example, if the sampling frequency $f_s = 8,000$ Hz, $f_k = 697$ Hz, and $N = 256$, we obtain $k \approx 23$. Figure 7.25 shows the Goertzel filters for the DTMF tone detector.

The transfer function reveals that the coefficient W_N^{-k} is complex, thus requiring complex multiplication and addition. Replacing the complex coefficient with the real coefficient can reduce the computational load. This reduction can be realized by multiplying both the numerator and the denominator of $H_k(z)$ in Eq. (7.6.11) by $(1 - W_N^k z^{-1})$ to form

$$H_k(z) = \frac{1 - W_N^k z^{-1}}{1 - 2\cos\left(\dfrac{2\pi k}{N}\right)z^{-1} + z^{-2}}. \tag{7.6.13}$$

The block diagram of the transfer function in Eq. (7.6.13) is shown in Fig. 7.26. When compared with the DTMF tone generator shown in Fig. 7.24, we notice the close similarity. Therefore, the code written for the DTMF generator can be reused for DTMF detection.

From Fig. 7.26, real coefficients are used in the feedback section, and the recursive computation of $y_k(n)$ is expressed as

$$d_k(n) = x(n) + 2\cos\left(\frac{2\pi f_k}{f_s}\right)d_k(n - 1) - d_k(n - 2),$$

$$n = 0, 1, \ldots, N - 1. \tag{7.6.14}$$

However, the feedforward section of Eq. (7.6.13) contains a complex variable W_N^k. Fortunately, the feedforward section (or nonrecursive part) needs to be computed

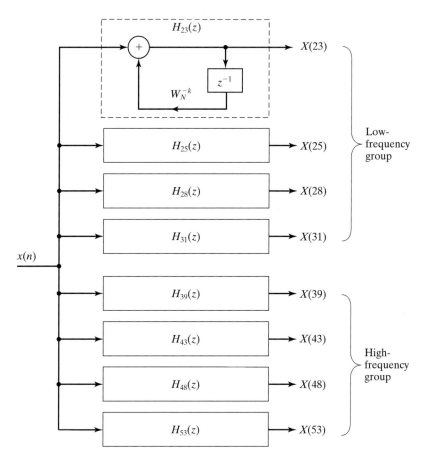

Figure 7.25   Block diagram of Goertzel filters for DTMF detection, where $N = 256$ and $f_s = 8,000$ Hz

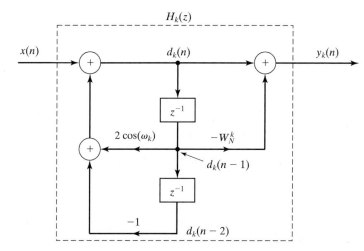

Figure 7.26   Block diagram of the Goertzel algorithm without the complex coefficient in the denominator

only once for every N samples at time $N - 1$ as

$$X(k) = y_k(N - 1) = d_k(N - 1) - W_N^k d_k(N - 2). \qquad (7.6.15)$$

The complex computation can be eliminated by computing a magnitude square of $X(k)$ for tone detection. The squared magnitude is computed as

$$|X(k)|^2 = d_k^2(N - 1) - 2\cos\left(\frac{2\pi k}{N}\right) d_k(N - 1) d_k(N - 2) + d_k^2(N - 2).$$
$$(7.6.16)$$

This equation does not contain any complex coefficients.

The two equations used for the computation of tone detection are Eq. (7.6.14), which needs to be computed for $n = 0, 1, \ldots, N - 1$, and Eq. (7.6.16), which only needs to be computed once at time $n = N - 1$.

MATLAB provides the function `goertzel` for implementing Goertzel filters. The following MATLAB script (`dtmf_goertzel.m`) shows an example of the detection of the DTMF tone generated from the previous section:

```
kL = [23 25 28 31];   % index for low frequency group
kH = [39 43 48 53];   % index for high frequency group
load tx_data_5;       % digit "5" tones generated in previous code
% use Goertzel filter to detect the tone pair
mag_low = abs(goertzel(HT(1:256), kL)).^2;    % low group
[y_low,index_low]= max(mag_low);
mag_high = abs(goertzel(HT(1:256), kH)).^2;   % high group
[y_high, index_high] = max(mag_high);
% select the keypad
key=['1' '2' '3' 'A'; '4' '5' '6' 'B'; '7' '8' '9' 'C'; '*' '0' '#'
    'D'];
fprintf('The key detected is %c',key(index_low, index_high));
```

There are some problems in using Goertzel filters for DTMF tone detection. One problem involves the precise tuning of the center frequencies of Goertzel filters to DTMF tones using a single order N in the DFT computation. Therefore, only approximation of DTMF frequencies is computed, as shown in Eq. (7.6.12). Another problem is the high sensitivity to overflow and roundoff noise because the poles of the filters are on the unit circle. Input must be scaled down to avoid overflow. An alternate approach in DTMF tone detection is to use two groups of four band-pass filters. The first four bandpass filters are tuned to the frequencies in the low-frequency group, while the next four are tuned to the high-frequency group. This method allows the individual bandpass filter to improve tuning to its center frequency, and the filter can be designed to have its poles inside the unit circle, thus resulting in a stable filter. See the Problems section at the end of this chapter for a detailed implementation.

Once Goertzel filtering is computed, the DTMF tone detection enters a tone/pause decision mode. When the pause is detected, DTMF tone detection is disabled, and the decoder has to wait for the completion of the pause period. When the tone is detected, the DTMF detector must identify the DTMF tone presence and execute the tone-validation test. The first test is to check the strength of the possible tone pair, which must exceed a certain threshold to qualify as a valid DTMF tone pair. The second test is of the spectral difference between the row tone and the column tone, where the row frequencies are typically less attenuated than the column frequencies due to the lowpass characteristics of the telephone line. This difference in attenuation is commonly known as forward twist. In contrast, reverse twist occurs when the row frequency is more attenuated than the column frequency. A typical recommendation is that forward twist should not exceed 4 dB and reverse twist should not exceed 8 dB. The third test is to compare the squared magnitude within the low- and high-frequency groups. The fourth test extends the Goertzel filters to reject speech signals that have harmonics close to the frequencies of DTMF tones. Because speech signals contain a significant amount of second harmonic energy, eight more Goertzel filters are implemented to detect the second harmonics of the DTMF tones (such as 1,540 Hz and 2,672 Hz for digit 5) and to reject the signals when the second harmonics component exceeds a certain threshold. The final test is of the validity of the time-duration specification.

7.7 EXPERIMENTS AND PROBLEMS

In this section, we experiment with the complete design and implementation processes for IIR filtering using MATLAB tools, C compilers, and the CCS for C5000 processors. We use the same bandpass filter specification given in Section 6.6 to illustrate IIR filter design in this section.

7.7.1 Design Using MATLAB

Similar to the process described in Section 6.6.1, a signal consisting of a 2 kHz sinewave corrupted by white noise is first created using the MATLAB script singen6_6.m and saved in in.mat. Load the file in.mat into the workspace, open SPTool, and import the signal variable in and its sampling frequency.

The next step is to design a bandpass IIR filter using the **Filters** column in SPTool to satisfy the same specifications shown in Fig. 6.29. Note that we use exactly the same specification for designing an IIR filter in this section as that used for the FIR filter designed in Section 6.6.1.

Click on the **New** button to create a new IIR bandpass filter, as shown in Fig. 7.27. The elliptic IIR filter with equiripple is selected to design an IIR filter with the least order. However, the drawback is its highly nonlinear-phase response. The user either can specify the order of the IIR filter or let the algorithm compute the least order by clicking on **Minimum Order**. Note that the order (3) stated in SPTool means the number of biquads and that the correct filter order is 6. The designed IIR filter only requires a 6th-order elliptic IIR filter compared with the 95th-order equiripple FIR filter shown in Fig. 6.29. We also design a Butterworth filter with the same specification, but require six biquads

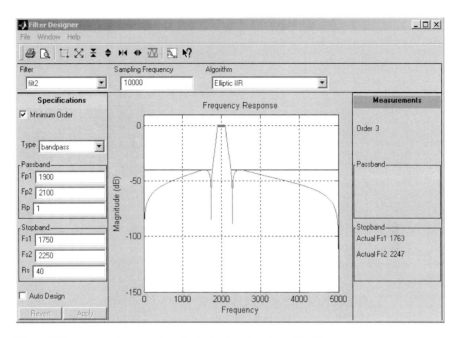

Figure 7.27   **Filter Designer** window for designing a bandpass IIR filter

(12th order). Alternatively, the user can also try to design the bandpass filter by using FDATool and exporting the filter coefficients from FDATool to SPTool for verification.

The name of the Butterworth filter is given as `filt1`, and the name of the elliptic filter is `filt2`. These filters are now ready to filter the noisy signal `in`. In the main window of SPTool, select the input signal `in` and the filter `filt1` by highlighting these variables. Click on the **Apply** button in the **Filters** column to activate the **Apply Filter** window. Change the name of the output signal to `out1` and click on **OK**. Repeat this process for the elliptic filter `filt2`.

View the filtered-output signals in **Signal Browser** and listen to these signals. We can verify the performance of the filter further by examining the spectra of the input and output signals. As shown in Fig. 7.28, the noise has been greatly attenuated by the IIR filters. In Fig. 7.28, the result obtained by the 6th-order elliptic IIR filter is compared with that of the 12th-order Butterworth IIR filter. It shows that the elliptic IIR filter is able to satisfy the same specification and achieve a similar performance using a lower-order (6) filter when compared with the Butterworth IIR filter of order 12. However, the drawback of using the elliptic IIR filter is its highly nonlinear phase compared with the Butterworth IIR filter as shown in Fig. 7.29.

We can use FDATool to design the IIR filter and create the coefficient header file for C programs. Following the design process described in Section 7.3, open FDAtool, enter the specification as shown in Fig. 7.30, and click on **Design Filter** at

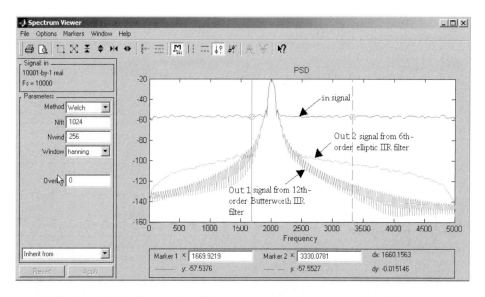

Figure 7.28   Spectra of the input and output signals from Butterworth and elliptic IIR filters

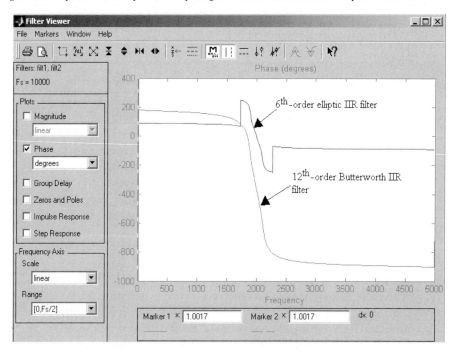

Figure 7.29   Phase responses of Butterworth and elliptic IIR filters

the bottom of window. The designed filter coefficients are represented with double-precision floating-point numbers. Follow the quantization process introduced in Section 7.3.2, and set **Mode** to `single` to obtain the single-precision floating-point coefficients for the floating-point C program that is used in the next section. In

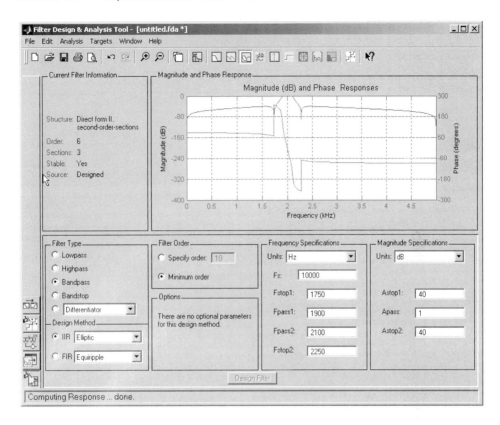

Figure 7.30   Design of an IIR bandpass filter using FDATool

practice, it is more stable to implement the IIR filter using fixed-point arithmetic in cascade form. FDATool can convert the sixth-order IIR filter into a cascade of three biquads if the user clicks on **Edit → Convert to Second-Order Sections**. Click on **File → Export to C Header**, and name the new header file `iircascade_sp.h`. The coefficients are stored in 32-bit, single-precision floating-point format.

7.7.2 Implementation Using Floating-Point C

IIR filtering can be implemented on a general-purpose computer using the C program that inputs a signal from the data file `in.dat` generated by `singen6_6.m`, performs IIR filtering, and writes the output data to the file `out.dat`. The filter coefficients, input, and output samples in the program are all represented by 32-bit floating-point numbers. We can compare the output of this floating-point C program with the output obtained by SPTool in Section 7.7.1. The program `iir_floatpt.c` can be run using a C compiler such as Visual C/C++ on the general-purpose computer. This program is listed as follows:

```
#include <stdio.h>
#include <stdlib.h>
```

```c
#include <math.h>
#include "iircascade_sp.h"     // coefficient header file

void main()
{
  /* Define variable arrays, define and clear variables */
  const int section = 3;
  const int tap = 3;
  float out = 0.0;               // y(n), output from IIR filter
  float xn;
  float delay[9] = {0.0, 0.0, 0.0, 0.0, 0.0, 0.0, 0.0, 0.0, 0.0};
  float gain = 0.004204816817972;
  int j;
  /* Declare file pointers */
  FILE *xn_in;                   // file pointer of x(n)
  FILE *yn_out;                  // file pointer of y(n)
  xn_in = fopen("in.dat","r");   // open file for input x(n)
  yn_out = fopen("out.dat","w"); // open file for output y(n)

  /* Start of main program */
  while ((fscanf(xn_in,"%f",&xn)) != EOF)
  { // read in x(n) from data file and processing it
    out=xn*gain;
    for (j=0; j< section; j++) //IIR filtering
    { // feedback section
      out=out-(delay[1+(j*section)]*(real32_T)DEN[j][1]);
      delay[(j*section)]=out-(delay[2+(j*section)]*
        (real32_T)DEN[j][2]);
      // feedforward section
      out=delay[(j*section)]+(delay[1+(j*section)]*
        (real32_T)NUM[j][1]);
      out=out+(delay[2+(j*section)]*(real32_T)NUM[j][2]);
      // refresh signal buffers
      delay[2+(j*section)]=delay[1+(j*section)];
      delay[1+(j*section)]=delay[(j*section)];
    }
    fprintf(yn_out,"%f\n",out); // output signal y(n)
  }
  printf("Finish");             // mark completion
  fcloseall();                  // close all opened files
}
```

A detailed signal-flow diagram that illustrates IIR filtering in `iir_floatpt.c` is shown in Fig. 7.31. The output file can be loaded into the MATLAB workspace and plotted with the following commands:

```
load out.dat -ascii;
plot(out); title('Output from IIR filter');
```

Thus, we can compare the output from the C program with that obtained by SPTool.

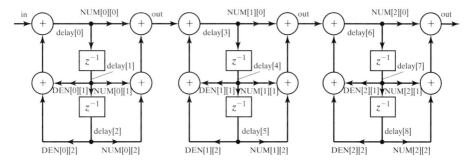

Figure 7.31   An IIR filter consists of a cascade of three biquads

7.7.3 Implementation Using Fixed-Point C

The floating-point C program introduced in the previous section can be modified to a fixed-point C program `iir_fixpt.c` for execution on a general-purpose computer. The input signal samples are first converted from a 32-bit floating-point precision to 16-bit integers using the conversion formula given in Eq. (6.6.1). This conversion is done in `singen6_6.m` and saved in the file `in_int.dat`.

As described in Section 7.3.2, FDATool can quantize reference filter coefficients to 16-bit coefficients in Q.15 format. The filter coefficients are then exported to the C header file `iircascade_16int.h`. The MAC operation in IIR filtering is being implemented using a `long` (32-bit) integer. Finally, the integer output is computed by scaling the accumulated value with a variable `scale` (32,768). Part of the kernel computation of IIR filtering in `iir_fixpt.c` is listed as follows:

```
for (j=0; j<section; j++)
{ // feedback section
  out=out-(((long)(int)delay[1+(j*section)]*(long)(int16_T)
    (DEN[j][1]))>>15);
  delay[(j*section)]=(int)(out-
    (((long)(int)delay[2+(j*section)]*(long)(int16_T)DEN[j][2])>>15));
  // feedforward section
  out=(long)(int)delay[(j*section)]+(((long)(int)delay
    [1+(j*section)]*(long)(int16_T)NUM[j][1])>>15);
  out=out+(((long)(int)delay[2+(j*section)]*(long)
    (int16_T)NUM[j][2])>>15);
  // refresh signal buffer
  delay[2+(j*section)]=delay[1+(j*section)];
  delay[1+(j*section)]=delay[(j*section)];
}
```

Only the output variable `out` is assigned as a `long` variable; the rest of the variables are declared as `int` variables. For a fixed-point (16-bit) implementation, we need to scale back the product after every multiplication in order to prevent the results from overflowing. Note that the final output is multiplied by two to obtain the same dynamic range as the input signal.

7.7.4 Implementation Using Fixed-Point C for C5000 Processors

In this experiment, we modify the fixed-point C program used in the previous section so that it can be executed on C54x and C55x processors. The CCS for the C5000 is used to develop and test the modified fixed-point C program. For this experiment, the coefficients of the IIR filter designed by FDATool are exported to a C header file by clicking on **Targets → Export to Code Composer Studio™ IDE**. The default target is set as the C54x (or C55x) Simulator, and the header file is saved as `iircoeff_ccs_16.h`.

Similar to the experiments for FIR filtering given in Chapter 6, we use the file I/O to test the C code running on C5000 processors in this experiment. The partial listing of C code `iir_fixpt_ccs.c` is given as follows:

```
for (j=0; j<section; j++)
{ // feedback section
  out = out - (long)((int)(((long)delay[1+(j*section)]*
  (long)(int)DEN[j][1])>>15));
  delay[(j*section)]=(int)(out-(long)((int)(((long)delay
    [2+(j*section)]*(long)(int)DEN[j][2])>>15)));
  // feedforward section
  out = (long)(delay[(j*section)]+(int)(((long)delay[1+
  (j*section)]*(long)(int)NUM[j][1])>>15));
  out = out+(long)((int)(((long)delay[2+(j*section)]*
  (long)(int)NUM[j][2])>>15));
  // refresh signal buffer
  delay[2+(j*section)] = delay[1+(j*section)];
  delay[1+(j*section)] = delay[(j*section)];
}
```

The C program `iir_fixpt_ccs.c` is included in the project with the files `vectorsc54.asm` (for the C54x only), `cmdc54.cmd` (or `c55x.cmd` for the C55x), and `rts.lib` (or `rts55.lib` for the C55x). The coefficient header file `iircoeff_ccs_16.h` is inserted automatically into the project. After successfully compiling and linking the program, we can load the executable file into the simulator using **File → Load Program**.

Similar to the FIR-filtering experiments, we use probe points to transfer data from the data file `in_int_ccs.dat` to the memory `in_buffer`. Probe points and breakpoints can be inserted into the program by placing the cursor at the line `dataIO()` and clicking on the 🖎 and 🖑 buttons, respectively. Choose **File → File I/O**, and click on **Add File**. Select `in_int_ccs.dat` and type `in_buffer` and `300` in the **Address** and **Length** boxes, respectively. Click on **Add Probe Point**. In the **Connect To** field, click on the down arrow, select the `in_int_ccs.dat` file, and finally click on **Replace** to complete the link between the data file and the probe point. Add in another breakpoint at the line `output++ = (int)(out*8)` for updating the display plot.

The time-domain plots of `in_buffer` and `out_buffer` can be displayed by choosing **View → Graph → Time/Frequency** and setting the display-property dialog. Click on the animate button 🐾, which executes the program until it reaches a breakpoint. The program halts, and the display windows are updated. Figure 7.32 shows the time-domain plots obtained by displaying the `in_buffer` (upper plot) and the `out_buffer` (lower plot).

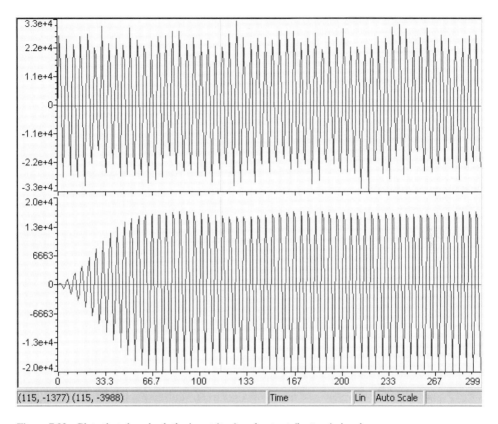

Figure 7.32   Plots that show both the input (top) and output (bottom) signals

The efficiency of the IIR filter code can be profiled in terms of the number of clock cycles. Select **Profiler** → **Enable Clock** and **Profiler** → **View Clock** from the menu. Set one breakpoint at the instruction for (j = 0; j < section; j++) and the other at the instruction output++ = (int)(out*8). Run the program, and the execution stops at the first breakpoint. Double-click on **Clock Window** to clear the cycle count. Run the program to the second breakpoint, and **Clock Window** displays the number of cycles (Clock = 582 for the C54x and Clock = 489 for the C55x) that the C code took to execute the sixth-order IIR filter. Compared with the clock cycles needed for FIR filters (3,465 cycles for the C54x and 2,598 cycles for the C55x), the IIR filter can achieve a similar performance with a much lower computational load. It is interesting to compare the response times of the FIR filter (Fig. 6.33) and IIR filter (Fig. 7.32).

7.7.5 Implementation Using Fixed-Point C with Intrinsics

The fixed-point C code given in Section 7.7.4 can be modified to use the intrinsics of C5000 processors. In this experiment, we use the lsmpy (multiplication) and lsadd (addition) intrinsics to implement the multiply-add operation of IIR filtering. The procedure for testing the C program iir_fixpt_intr.c is identical to the one used in Section 7.7.4. A part of the kernel program is listed as follows:

```
for (j=0; j<section; j++)
{
  // feedback section
  temp = _lsmpy(delay[1+(j*section)],(int16_T)DEN[j][1])>>16;
  out = _lssub(out,temp);
  temp = _lsmpy(delay[2+(j*section)],(int16_T)DEN[j][2])>>16;
  delay[(j*section)] = (int)(_lssub(out,temp));

  // feedforward section
  temp = _lsmpy(delay[1+(j*section)],(int16_T)NUM[j][1])>>16;
  out = _lsadd((long)delay[(j*section)],temp);
  temp = _lsmpy(delay[2+(j*section)],(int16_T)NUM[j][2])>>16;
  out = _lsadd(out,temp);

  // updates signal buffer
  delay[2+(j* section)] = delay[1+(j* section)];
  delay[1+(j* section)] = delay[(j* section)];
}
```

7.7.6 Implementation Using Assembly for C5000 Processors

In this section, we implement the IIR filter on C5000 processors using an assembly program. We first implement IIR filtering on the C54x using an assembly program. We then implement IIR filtering in subroutines using C55x assembly code, which can be called by a fixed-point C program similar to the one given in Section 7.7.4.

Implementation on the C54x

This experiment develops and tests the C54x assembly code that implements the IIR filtering. The input data is physically copied into data memory using the .copy directive, while the coefficients are loaded into the coeff_table section. Two uninitialized sections, out and coeff, are reserved for saving 300 samples of output data and three sets of second-order IIR-filter coefficients in internal memory. The user can also refer to other assembly code (using direct-addressing mode) written for the IIR filter in Section 7.4.2. The assembly code iir_fixpt_asm.asm is shown as follows:

```
            .def    start
out_buffer  .usect "out",300     ; reserve 300 locations for y
coeff       .usect "coeff",12,1  ; reserve 12 locations for
                                 ; coefficients
            .sect  "coeff_table" ; 12 coefficients for 3 biquads
init_coeff  .int   -30677, 19644,-32768,0 ; -DEN[2],
                                          ; -DEN[1],NUM[2],NUM[1]
            .int   -31829, 16158,32767,-9027   ; section 2
            .int   -31866, 23707,32767,-30370 ; section 3

            .sect  "indata"
in_buffer   .copy  "in1_int.dat" ; copy in1_int.dat to in_buffer
```

```
                   .sect "delay"
delay_line    .int  0,0,0,0,0,0,0,0,0 ; reserve 9 locations for
                                      ; signal buffers of 3 sections
            .mmregs
            .text                     ; code section
start:
   SSBX  FRCT                         ; turn on fractional mode
   SSBX  SXM                          ; turn on sign extension mode
   RSBX  OVM                          ; turn on overflow mode
copy:
   STM   #coeff,AR4                   ; AR1 pointing to coeff
   RPT   #11                          ; repeat next instruction 12 times
   MVPD  init_coeff,*AR4+             ; copy the init_coeff to coeff
* Setup the pointers to point to the coeff and data
   STM   #in_buffer,AR1               ; setup the pointer for in_buffer
   STM   #out_buffer,AR2              ; setup the pointer for out_buffer
loop:
   STM   #299,BRC                     ; block repeat for 300 times
   RPTB  iir_end-1
iirloop:
   STM   #coeff,AR4                   ; setup the pointer for coeff
   MPY   *AR1+,#17,A
   STM   #delay_line+2,AR3            ; setup the pointer for delay line
   STM   #2,AR6                       ; number of sections-1
feedback:                            ; feedback section
   MAC   *AR4+,*AR3-,A
   MAC   *AR4+,*AR3-,A
   STH   A,*AR3
   MAR   *+AR3(#2)
feedforward:                         ; feedforward section
   MPY   *AR4+,*AR3-,A
   MAC   *AR4+,*AR3,A
   DELAY *AR3-
   MAC   *AR3,#32767,A
   DELAY *AR3
   MAR   *+AR3(#5)
   BANZ  feedback,*AR6-
   STH   A,3,*AR2+                    ; write result to memory location y
iir_end:
   NOP
```

The coefficients coeff_table, located in the external memory, are first transferred to the internal memory coeff. After the transfer, the pointers AR2, AR3, and AR4 are set to point at out_buffer, coeff, and delay_line, respectively. The coefficients stored in this assembly code are arranged as $\{-a_{k2}, -a_{k1}, b_{k2}, b_{k1}\}$, where k is the section index.

In this IIR-filtering program, we treat the entire input sequence as a block. The single-repeat loop is responsible for the three biquads, while the block repeat generates 300 output samples. The delay instruction is used to update the delay line of the IIR filter instead of using a circular buffer. This instruction moves the data pointed at by the AR register to the next higher memory location.

We can create a new project, similar to the procedure given in Section 6.6.6, to include the files `iir_fxpt.asm`, `c54.cmd`, `vectors.asm`, and `rts.lib`. Perform a complete build before loading the program into memory. Use **GEL** → **C5402_Configuration** → **CPU_Reset** to reset the program pointer to the start of the program. Set a breakpoint at the end of the program by placing the cursor at the NOP instruction, and click on the breakpoint icon 🖑. Set up display windows to view `in_buffer` and `out_buffer`. To ensure that the output data does not contain any previous values, we can fill the `out_buffer` memory with zeros by using **Edit** → **Memory** → **Fill** and specifying the address as `out_buffer` with a length of 300. Run the program, stop at the breakpoint, and update the display window for `out_buffer`.

The final step is to profile the IIR filter using assembly code. Set breakpoints at instructions MAC `*AR4+,*AR3-,A`, and STH `A,3,*AR2+`. Enable the clock, as described in the previous sections. Note that only 392 clock cycles are required to run the assembly code for IIR filtering, as compared to 582 cycles for the fixed-point C given in Section 7.7.4. The increase in cycle count in the C-compile-generated assembly code is due to the intensive use of stack for passing variables. In addition, it does not make full use of the powerful indirect addressing.

Implementation on the C55x

In this experiment, we modify the fixed-point C program `iir_fixpt_ccs.c`, given in Section 7.7.4, for an experiment. We replace the computation-intensive loop for IIR filtering with the subroutines written in C55x assembly language.

The new C program, `iir_fixpt_C55x.c`, calls two assembly routines for gain control and IIR filtering as follows:

```
gain_c55x(&out,&in_buffer[i],gain);
for (j=0; j<section; j++){
    iir_c55x(&out,&NUM[j][1],&DEN[j][1],&delay[1+(3*j)]);
}
*output++ = (int)(out*8);
```

The IIR function `iir_c55.asm` is written in C55x assembly language and is listed as follows:

```
.def  _iir_c55x
_iir_c55x
pshm  ST1_55                    ; save ST1, ST2, and ST3
pshm  ST2_55
pshm  ST3_55

or    #0x340,mmap(ST1_55)       ; set FRCT,SXMD,SATD
bset  SMUL
bset  SATA

mov   *AR0<<16,AC0              ; get the value of 'out'
masm  *AR2+,*AR3+,AC0           ; out-= DEN[j][1]*delay[1+(j*section)]
```

```
masm    *AR2,*AR3-,AC0          ; out-DEN[j][2]*delay[2+(j*section)]
mov     *AR3-,T0               ; save the value of delay[1+(j*section)]
mov     rnd(hi(AC0)),*AR3      ; delay[j*section]=
                              ; out-DEN[j][2]*delay[2+(j*section)]
mov     *AR3+,T1              ; save the value of delay[(j*section)]
mov     AC0,AC1
mov     #0,AC0
macm    *AR1+,*AR3+,AC0        ; delay[1+(j*section)]*NUM[j][1]
add     AC1,AC0              ; out=delay[(j*section)]+
                              ; delay[1+(j*section)]*NUM[j][1]
macm    *AR1,*AR3,AC0         ; out=out+delay[2+(j*section)]*NUM[j][2]
mov     T0,*AR3-             ; delay[2+(j*section)]=delay[1+
                              ; (j*section)]
mov     T1,*AR3             ; delay[1+(j*section)]=
                              ; delay[(j*section)]
mov     rnd(hi(AC0)),*AR0    ; save the value of 'out'
BCLR    FRCT
BCLR    SATD
BCLR    SXMD
BCLR    SATA
BCLR    SMUL
Popm    ST3_55              ; restore ST1, ST2, and ST3
Popm    ST2_55
Popm    ST1_55
ret
.end
```

It is important to note that this assembly IIR subroutine is different than the iircas51 function listed in Table 7.7. The reason is that the arrangement of IIR-filter coefficients in the coefficient header file iircoeff_ccs_16.h generated by FDATool is different than the IIR routines for the C54x and C55x in the DSPLIB. The user can also modify the coefficient header file to match the arrangement required by IIR routines given in the DSPLIB.

We can create a new project, similar to the procedure given in Section 6.6.6, to include the files iir_fixpt_c55x.c, gain_c55x.asm, iir_c55x.asm, c55x.cmd, and rts55.lib. The same C header file, iircoeff_ccs_16.h, is included in the main program. After irrcoeff_ccs_16.h successfully compiling and linking the program, we can load the executable file into a simulator for experiments. Following the identical procedure given in Section 6.6.6, use the probe point of CCS to connect in_int_ccs.dat for data file I/O, and use Graph to view the filter output waveform. By comparing the output waveform with the one obtained from the fixed-point C program in Section 7.7.4, we can verify that the assembly routine is correct.

To evaluate the efficiency of this mixed C-and-assembly routine, we set one breakpoint at the instruction

```
for (j=0; j<section; j++){
```

and the other at the instruction

```
*output++ = (int)(out*8);
```

Run the program to the first breakpoint by selecting **Debug** → **Run** from the menu. The execution stops at the first breakpoint. Double-click on **Clock Window** to clear the cycle count. Run the program to the second breakpoint, and **Clock Window** displays the number of cycles (Clock = 250) that the code took to execute IIR filtering on the C55x. Compared with Clock = 489 in the fixed-point C program given in Section 7.7.4, it clearly shows that the C55x assembly IIR routine is faster than the C code.

7.7.7 Implementation Using the C5000 Digital Signal Processing Library

As introduced in Section 7.4, Texas Instruments provides C54x and C55x DSPLIBS that contain C-callable, assembly-optimized functions. These routines are summarized in Table 7.6 and Table 7.7 for the C54x and C55x, respectively. In this experiment, we use a C program that calls the IIR-filtering routine iircas4 in the DSPLIB. This function implements the cascade of a direct-form II IIR filter with four coefficients per biquad, which can be called in C as follows:

```
oflag = short iircas4(DATA *x, DATA *h, DATA *r,
                      DATA *dbuffer, ushort nbiq, ushort nx)
```

The argument x is the pointer to the input-signal vector of nx elements, h is the pointer to the coefficient vector of size 4*nbiq (since there are four coefficients per biquad), r is the output-signal vector of nx elements, dbuffer is the delay buffers of 2*nbiq elements that must be initialized before this function is run, and oflag is the overflow flag.

The C program given in Section 7.7.4 can be modified to call the IIR-filter function. The new C code, iir_fixpt_dsplib.c, is listed as follows:

```
#include "math.h"
#include "stdio.h"
#include "tms320.h"
#include "dsplib.h"
#include "input1.h"        // input data file

short i;
short oflag;               // overflow flag
DATA gain=17;
DATA x1[300];
void main(void)
{
  for (i=0;i<NX;i++) r[i] =0;   // clear output buffer (optional)
  for (i=0; i<2*NBIQ; i++) dbuffer[i] = 0;     // clear delay buffer
                                               // (a must)
  for (i=0;i<NX;i++) x1[i]=(DATA)(((long)(DATA)x[i]*(long)
    (DATA)gain)>>15);
  oflag=iircas4(x1,h,r,&dp,NBIQ, NX);          // IIR filtering
  for (i=0;i<NX;i++) r[i] = r[i]*8;
  return;
}
```

Some important features of the preceding C program are discussed in Section 6.6.7. In the `iircas4` function, the coefficients are arranged in the format

```
h = a1k/2, a2k, b2k, b1k, …
```

where `k` is the biquad index. Note that the coefficients `a0k` and `b0k` are assumed to be 1, and `a1k` must be halved due to the algorithm used in the function.

As discussed in Section 6.6.7 for FIR filtering, IIR filtering also operates more efficiently in block mode. This fact can be verified by benchmarking the block-processing cycle count using different block sizes. The user can modify the C program to benchmark the code running at different block sizes of 300, 75, and 1. For example, the benchmark results of the C54x for different IIR-filtering functions and block sizes are given in Table 7.9. The user can verify the filtering results by modifying `iir_fixpt_dsplib.c` and using different input header files.

It is noted that the `iircas4` function is the fastest of the three, and all of the functions execute faster in block mode than in sample mode. However, using a longer block size only slightly reduces the cycle count, but it requires more memory to store the data samples.

If the stability of the IIR filter causes problems due to the use of Q.15 format, the double-precision IIR-filtering function `iir32` can be used. This function computes the direct-form II IIR filter with five coefficients per biquad using 32-bit coefficients and a 32-bit delay buffer. The input and output samples are still in 16-bit single precision. The user can refer to the file `iir_fixpt_dsplib32.c` for details on using the function `iir32`. The input data presented in Q.15 precision and the coefficients presented in Q.31 precision are stored in the file `input3a.h`. The benchmark is 69,389 clock cycles for `nx=300`, 69,656 cycles for `nx=75`, and 102,609 cycles for `nx=1`. These cycle counts are more than twice those listed in Table 7.9, which use single precision. Furthermore, the memory requirement for the coefficient and signal buffers is doubled.

Additional hands-on experiments using IIR filters are given in Appendix B. The additional applications include: (1) implement a reverberation circuit using an IIR filter, (2) implement a signal generator (sine, cosine, square, and triangular) using an IIR filter, and (3) implement a parametric equalizer using an IIR filter. The user can refer to the MATLAB and Simulink files available on the companion CD and Web for more details on these exercises and progressively work through these hands-on experiments to obtain the final solution using the CCS.

TABLE 7.9   Benchmark of Three C54x Biquad IIR Functions Using Different Block Sizes to Filter 300 Input Samples

Number of samples per block, nx	Cycle count (iircas4)	Cycle count (iircas5)	Cycle count (iircas51)
300	12,671	14,471	18,375
75	12,884	14,684	18,600
1 (sample mode)	40,509	42,309	47,409

PROBLEMS

PART A

Problem A–7–1

Design a narrow lowpass elliptic-IIR filter with a transition width of 20 Hz, a passband cutoff frequency at 100 Hz, and a sampling frequency of 20 kHz. The passband and stopband ripples can be selected as 0.5 dB and 50 dB, respectively. Use FDATool to examine the stability of the direct-form IIR filter for the following formats:

(a) single-precision floating-point
(b) double-precision floating-point
(c) Q.31 fixed-point

Problem A–7–2

Use FDATool to quantize filter coefficients, input, output, multiplicands, products, and sums in Section 7.3 to the single-precision floating-point format. Observe any significant changes of responses compared with the reference filter using the double-precision floating-point format.

Problem A–7–3

Design a lowpass IIR filter with the frequency specification Fpass = 4,000 Hz, Fstop = 4,100 Hz, Fsample = 48,000 Hz, passband ripple = 0.5 dB, and stopband ripple = 50 dB. Compare the designed IIR filter using Butterworth, Chebyshev Type I, Chebyshev Type II, and elliptic design methods in terms of the following:

(a) magnitude response
(b) phase response
(c) group delay
(d) pole-zero plot
(e) order of filter

Problem A–7–4

Transform the lowpass elliptic-IIR filter designed in Problem 3 to (a) another lowpass filter with Fpass = 5,000 Hz using **Transform Filter → Lowpass to lowpass** and (b) a highpass filter with Fpass = 8,000 Hz using **Transform Filter → Lowpass to highpass**. These transformation tools can be found in FDATool.

Problem A–7–5

In Example 7.2, what happens when pre-warping is not performed (i.e., $\Omega = 0.3\pi/T$)? Use MATLAB to compare the frequency plots of the digital IIR filters with and without pre-warping.

Problem A–7–6

DTMF tones are generated using second-order IIR filters, as shown in Fig. 7.24. Instead of using a unit-impulse function to excite the IIR filter, we can initialize the initial state of the IIR filter to make the filter self-oscillate without any excitation. Show how to initialize the IIR filter for this purpose.

Problem A–7–7

Instead of using Goertzel filters to detect DTMF tones, as explained in Section 7.6, we can use a set of highpass, lowpass, and bandpass filters, as shown in Fig. 7.33. The user can design the set of filters shown in the figure using FDATool based on the following specifications:

Filter	Fstop1	Fpass1	Fpass2	Fstop2
HP1: 680 Hz	650 Hz	680 Hz	-	-
HP2: 1180 Hz	1150 Hz	1180 Hz	-	-
LP1: 960 Hz	980 Hz	960 Hz	-	-
BP1: 697 Hz	687 Hz	692 Hz	702 Hz	707 Hz
BP2: 770 Hz	760 Hz	765 Hz	775 Hz	780 Hz
BP3: 852 Hz	842 Hz	847 Hz	857 Hz	862 Hz
BP4: 941 Hz	931 Hz	936 Hz	946 Hz	951 Hz
BP5: 1209 Hz	1199 Hz	1204 Hz	1214 Hz	1219 Hz
BP6: 1336 Hz	1326 Hz	1331 Hz	1341 Hz	1346 Hz
BP7: 1477 Hz	1467 Hz	1472 Hz	1482 Hz	1487 Hz
BP8: 1633 Hz	1623 Hz	1628 Hz	1638 Hz	1643 Hz

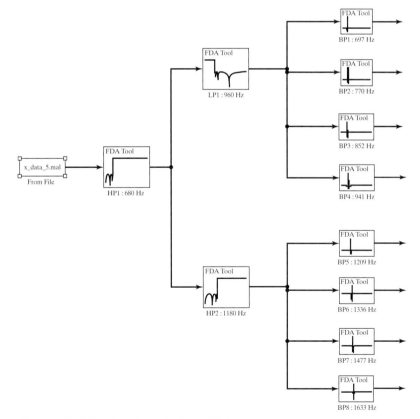

Figure 7.33 A filtering scheme for DTMF detection

The sampling frequency is 8,000 Hz. The passband and stopband ripples are set to 1 and 50 dB, respectively. Design the filters using the IIR filter-design method with the least order. Report the computational load required for implementing this filtering scheme, and compare it with the Goertzel algorithm.

Problem A–7–8

Write a MATLAB script to implement DTMF tone detection using the Goertzel algorithm. Determine a suitable threshold for the correct detection of dial tones. In addition, DTMF tone detection should include a means of rejecting a speech signal that has harmonics closed to DTMF tones. Send in the following test tones to verify the algorithm:

(a) a sequence of seven dial tones
(b) a sequence of seven dial tones corrupted with a speech signal (`timit1.asc` in Chapter 2) at a sampling frequency of 8 kHz
(c) a speech signal only
(d) silence with background noise

Problem A–7–9

We are given a sixth-order IIR filter with the following transfer function expressed in MATLAB:

```
Num=[0.06307331692864      0.04174696033121      0.14697805219203
     0.0885618516540       0.14697805219203      0.04174696033121
     0.06307331692864];
Den=[1.00000000000000     -2.18438961264166      3.60947038268932
    -3.44847072536905      2.43185706142163     -1.02517466489337
     0.24395366699689];
```

Use the MATLAB functions described in this chapter to factorize the transfer function into a series of second-order IIR filters. Group the zeros to the closest poles. The order of pole-zero pairs are in (1) up direction and (2) down direction. Perform the scaling based on the L_1, L_2, and Chebyshev norms, and determine the stability of the filter using Q.15 format.

Problem A–7–10

Use the MATLAB Filter Realization Wizard in the DSP Blockset (or in FDATool) to derive an IIR filter using (a) direct-form I, (b) direct-form II, (c) transposed-form II, and (d) cascade structure (using a transposed direct-form II biquad) for the following difference equation:

$$8y(n) + 6y(n-1) + y(n-2) - 4y(n-3) - y(n-4) =$$
$$x(n) - 3x(n-1) + 9x(n-2) - 13x(n-3) + 9x(n-4).$$

Determine whether the filter is stable. Is it possible to implement this IIR filter on a fixed-point DSP processor using Q.15 format? If so, how? Comment on the sign of the feedback coefficients when they are represented in difference-equation and polynomial forms.

Problem A–7–11

A desired transfer function (desired_tf.mat) is given. Use the invfreqz MATLAB function to derive a fifth-order IIR filter. In order to verify the accuracy of the derived IIR filter, we can use freqz to measure the difference between the desired transfer function and the derived transfer function. Also, compute the sum of the squared error between these two transfer functions.

Problem A–7–12

Following Problem 11, obtain an impulse response of the preceding desired transfer function. Apply the following time-domain IIR-filter-design MATLAB functions to obtain their respective fifth-order IIR filters:

(a) prony
(b) lpc
(c) stmcb

Comment on the accuracy of these methods.

Problem A–7–13

MATLAB provides a tool for designing an IIR filter using pole-zero placement. In SPTool, under the **Filter** column, click on the **New** icon. Under the **Algorithm** menu, choose **Pole/Zero Editor**, and **Delete all** to delete the default poles/zeros. Now, select the pole icon ▨ and the zero icon ▨ to place the poles and zeros on the z-plane. After placing these poles and zeros, the user can also move the poles and zeros with the mouse by clicking on the ▨ icon. Use this tool to design the following IIR filters:

(a) A second-order lowpass filter with complex poles and zeros and with a cutoff frequency of $\pi/3$.
(b) A second-order highpass filter with complex poles and zeros and with a cutoff frequency of $\pi/3$.
(c) A fourth-order bandpass filter with cutoff frequencies of $\pi/3$ and $\pi/2$. What happens when the poles or zeros are close to the unit circle?
(d) Place a complex pole-pair on the unit circle and observe its impulse response.
(e) Observe the magnitude response when six zeros are placed equally around the unit circle.
(f) What happens when the poles and zeros are overlapped?
(g) Design a second-order all-pass IIR filter by matching the poles with zeros located at the reciprocal conjugate of the poles.

Problem A–7–14

Use MATLAB to design a fourth-order Butterworth IIR lowpass filter with a passband from 0.2π to 0.3π and with passband and stopband ripples of 0.5 dB and 25 dB, respectively. Convert this filter into two biquads (direct-form II) in a cascade connection. Examine the output of the addition nodes of each section, and see whether overflow occurs. If it does, scale the filter coefficients in each section using L1 and L2 norms to avoid overflow. Examine the results after scaling.

Problem A–7–15

Explain why the result of the multiplication needs to be right-shifted by 15 bits, as shown in Section 7.7.4, when implementing fixed-point C for C5000 processors. Why is a right-shift of 16 bits used when programming in fixed-point C with intrinsic? Reprogram the fixed-point C (with intrinsic) code in Section 7.7.5 to cascade three biquads represented in direct-form I structure. Comment on the difference in coding, and profile the clock cycles.

PART B

Problem B–7–16

Compare and contrast IIR and FIR filters. When are the advantages of using an IIR filter?

Problem B–7–17

Use a second-order IIR filter to show how to convert the direct-form I structure to the direct-form II structure:

(a) Draw a signal-flow diagram of a second-order IIR filter (similar to that shown in Fig. 7.2).
(b) Exchange the feedforward section with the feedback section.
(c) Observe that both the buffers for the feedback and feedforward sections have the same signal samples. Combine these two buffers into one. Draw the final signal-flow diagram of the direct-form II structure shown in Fig. 7.3 for the second-order IIR filter.

Problem B–7–18

In Problem 17, assume the combined signal buffer for both the feedback and feedforward sections (similar to those shown in Fig. 7.5) is $\{d(n), d(n-1), d(n-2)\}$.

(a) Derive the I/O difference equation using the intermediate variables $d(n-i)$.
(b) Derive the transfer function of the second-order direct-form II IIR filter based on the I/O equation.

Problem B–7–19

We are given a second-order IIR filter

$$H(z) = \frac{0.7(1 - 0.36z^{-2})}{1 + 0.1z^{-1} - 0.72z^{-2}}.$$

(a) Draw the direct-form II structure.
(b) Realize the filter by cascading two first-order sections.
(c) Determine the stability of the IIR filter using the stability triangle.

Problem B–7–20

We are given a fourth-order IIR-filter transfer function

$$H(z) = \frac{(1 - 2z^{-1} + z^{-2})(1 + 1.414z^{-1} + z^{-2})}{(1 + 0.8z^{-1} + 0.64z^{-2})(1 - 1.0833z^{-1} + 0.25z^{-2})}.$$

(a) Find the poles and zeros of this filter, and determine its characteristics.
(b) Use the stability triangle to check for the stability of this filter.
(c) Draw a cascade realization of the direct-form II IIR filter using an ascending gain grouping.

Problem B–7–21

A first-order all-pass IIR filter has the transfer function

$$H(z) = \frac{z^{-1} - a}{1 - az^{-1}}.$$

(a) Sketch the pole and zero in the z-plane.
(b) Determine the magnitude response of the all-pass filter for all frequencies from 0 to π.
(c) Determine and sketch the phase response of the all-pass filter. How can the phase be controlled?

Problem B–7–22

Design a first-order lowpass filter, similar to the one shown in Example 7.2, with a cut-off frequency at 0.2π using a bilinear transformation.

Problem B–7–23

Find the z-transform of the signal $h(n) = \sin(\omega_k n)$ and $h(n) = \cos(\omega_k n)$. Instead of using two sinewave generators in the DTMF tone generator, implement two cosine-wave generators. Compare the differences between the two DTMF generators.

Problem B–7–24

A real-time audio application requires an analog signal to be sampled at a rate of 44.1 kHz. The digitized signal is then passed into an 80 MIPS DSP processor to perform digital equalization to the frequency range from 1 kHz to 4 kHz. Using a cascade IIR-filter routine in the C54x DSPLIB, find the maximum sampling rate that can be attained using an IIR filter of order 20 (for both the numerator and the denominator). What is the maximum sampling rate if a 100-tap FIR filter must be implemented using the FIR-filtering routine?

SUGGESTED READINGS

1 Parks, T. W. and C. S. Burrus. *Digital Filter Design*. New York, NY: Wiley, 1987.
2 Oppenheim, A. V., R. W. Schafer, and J. R. Buck. *Discrete-Time Signal Processing*. 2nd Ed. Upper Saddle River, NJ: Prentice Hall, 1999.
3 Proakis, J. G. and D. G. Manolakis. *Digital Signal Processing: Principles, Algorithms, and Applications*. 3rd Ed. Upper Saddle River, NJ: Prentice Hall, 1996.
4 Mitra, S. K. *Digital Signal Processing: A Computer-Based Approach*. 2nd Ed. New York, NY: McGraw-Hill, 2001.
5 Kuo, S. M. and B. H. Lee. *Real-Time Digital Signal Processing: Implementations, Applications and Experiments with the TMS320C55x*. New York, NY: Wiley, 2001.

   6 Kuo, S. M. and C. Chen. "Implementation of Adaptive Filters with the TMS320C25 or the TMS320C30." Chapter 7 in *Digital Signal Processing Applications with the TMS320 Family*, vol. 3, edited by P. E. Papamichalis. Upper Saddle River, NJ: Prentice Hall, 1990.

   7 Orfanidis, S. J. *Introduction to Signal Processing*. Upper Saddle River, NJ: Prentice Hall, 1996.

   8 Van de Vegte, J. *Fundamentals of Digital Signal Processing*. Upper Saddle River, NJ: Prentice Hall, 2002.

   9 Ahmed, N. and T. Natarajan. *Discrete-Time Signals and Systems*. Englewood Cliffs, NJ: Prentice Hall, 1983.

  10 The MathWorks. *Filter Design Toolbox, User's Guide*. Version *2.1*, 2000.

  11 The MathWorks. *Signal Processing Toolbox for Use with MATLAB*. Version 5, 2000.

  12 Texas Instruments. *Filter Library: Module User's Guide*. C24x Foundation Software, 2002.

  13 Texas Instruments. *TMS320C54x DSP Library Programmer's Reference*. SPRU518c, 2002.

  14 Texas Instruments. *TMS320C55x DSP Library Programmer's Reference*. SPRU422F, 2002.

  15 Texas Instruments. *TMS320C62x DSP Library Programmer's Reference*. SPRU402a, 2002.

  16 Texas Instruments. *TMS320C64x DSP Library Programmer's Reference*. SPRU565a, 2002.

  17 Texas Instruments. *TMS320C3x User's Guide*. SPRU031D, 1994.

8

Fast Fourier Transforms

As discussed in Section 2.4, the frequency analysis of digital signals and systems is the cornerstone of DSP. A time-domain digital signal can be transformed into its frequency-domain representation using the z-transform, discrete-time Fourier transform (DTFT), or discrete Fourier transform (DFT). The FFT is a family of fast computational algorithms for the DFT that has been widely used in many practical applications, including fast FIR filtering, spectral analysis and synthesis, and correlation. This chapter concentrates on the design, application, and implementation of the FFT on DSP processors. A detailed discussion on the theoretical aspects of these subjects can be found in [1–4].

8.1 INTRODUCTION TO THE DISCRETE FOURIER TRANSFORM

Expressing the signal in the z-transform, which was introduced in Section 2.2, and evaluating the spectrum on the unit circle can provide a frequency analysis of a digital signal. The DTFT of a sequence $x(n)$, discussed in Section 2.4.1, is defined as

$$X(\omega) = \sum_{n=-\infty}^{\infty} x(n)e^{-j\omega n}. \qquad (8.1.1)$$

The DTFT $X(\omega)$ is a function of a continuous-frequency variable ω, and the summation in Eq. (8.1.1) extends towards positive and negative infinity. Therefore, the DTFT is a theoretical Fourier transform of a digital signal, but it cannot be implemented for real applications.

Frequency analysis of a finite-length sequence, the DFT introduced in Section 2.4.2, is equal to the sampled version of the DTFT. In other words, the continuous-frequency variable ω is sampled at N equally spaced frequencies $\omega_k = 2\pi k/N$,

$k = 0, 1, \ldots, N - 1$ on the unit circle. These frequency samples (DFT coefficients) are expressed as

$$X(k) = X(\omega_k)\big|_{\omega_k = 2\pi k/N} = \sum_{n=0}^{N-1} x(n)e^{-j2\pi kn/N}$$

$$= \sum_{n=0}^{N-1} x(n)W_N^{kn}, k = 0, 1, \ldots, N - 1, \qquad (8.1.2)$$

where the twiddle factors are defined in Eq. (2.4.12) as

$$W_N^{kn} = e^{-j\left(\frac{2\pi}{N}\right)kn} = \cos\left(\frac{2\pi kn}{N}\right) - j\sin\left(\frac{2\pi kn}{N}\right). \qquad (8.1.3)$$

The DFT is based on the assumption that the signal $x(n)$ is periodic. Therefore, $X(k)$ for $k = 0, 1, \ldots, N - 1$ can uniquely represent a periodic sequence $x(n)$ of period N.

The inverse DFT (IDFT) is the reversed process of the DFT. It converts the frequency spectrum $X(k)$ back to the time-domain signal $x(n)$. As defined in Eq. (2.4.14), it can be expressed as

$$x(n) = \frac{1}{N}\sum_{k=0}^{N-1} X(k)e^{j2\pi kn/N} = \frac{1}{N}\sum_{k=0}^{N-1} X(k)W_N^{-kn}, n = 0, 1, \ldots, N - 1. \quad (8.1.4)$$

By comparing this equation with Eq. (8.1.2), it can be seen that the IDFT has a similar computational load in terms of multiply-and-add operations to the DFT. The differences between the DFT and IDFT are an additional scaling factor $1/N$ in the IDFT and a different exponent sign in the twiddle factors.

8.1.1 Computational Load of the Discrete Fourier Transform and the Inverse Discrete Fourier Transform

The computational load of the DFT can be studied by first expanding Eq. (8.1.2) into N equations (for $k = 0, 1, \ldots, N - 1$) expressed as

$$X(0) = x(0)W_N^0 + x(1)W_N^0 + \cdots + x(N - 1)W_N^0$$

$$X(1) = x(0)W_N^0 + x(1)W_N^1 + \cdots + x(N - 1)W_N^{(N-1)}$$

$$\cdots$$

$$X(N - 1) = x(0)W_N^0 + x(1)W_N^{(N-1)} + \cdots + x(N - 1)W_N^{(N-1)^2}.$$

These N equations can be expressed in a matrix form as

$$\mathbf{X} = \mathbf{W}\mathbf{x}, \qquad (8.1.5)$$

where the time-domain signal vector $\mathbf{x} = [x(0)x(1)\cdots x(N - 1)]^T$ is an $N \times 1$ column vector, the frequency-domain vector $\mathbf{X} = [X(0)X(1)\cdots X(N - 1)]^T$ is an

$N \times 1$ column vector, and the $N \times N$ twiddle-factor matrix is defined as

$$\mathbf{W} = \begin{bmatrix} W_N^0 & W_N^0 & \cdots & W_N^0 \\ W_N^0 & W_N^1 & \cdots & W_N^{N-1} \\ \vdots & \vdots & \ddots & \vdots \\ W_N^0 & W_N^{N-1} & \cdots & W_N^{(N-1)^2} \end{bmatrix}. \tag{8.1.6}$$

Computing one frequency sample $X(k)$ of an N-point DFT requires N complex multiplications and $(N-1)$ complex additions from the row-column multiplication (inner product) described by Eq. (8.1.5). Therefore, to compute a complete set of N DFT coefficients, $N \times N$ complex multiplications and $N \times (N-1)$ complex additions are required, which can be simply stated as a complexity of order $O(N^2)$.

Similarly, the IDFT given in Eq. (8.1.4) can be expressed in a matrix form as

$$\mathbf{x} = \frac{1}{N}\mathbf{W}^* \, \mathbf{X}, \tag{8.1.7}$$

where the scaling factor $(1/N)$ can be combined into the complex-conjugate twiddle-factor matrix $\mathbf{W}^*$ with elements $W_N^{-kn} = e^{j(2\pi/N)kn}$. Thus, a similar computational load is needed in computing the IDFT as the DFT described in Eq. (8.1.5).

Example 8.1

MATLAB provides the twiddle-factor function `dftmtx(N)` for computing the N-point DFT of a given data sequence of length N. In this example, we compute the DFT of an 8-point sequence $x(n) = \{1\,1\,1\,1\,1\,0\,0\,0\,0\}$. In addition, we verify the DFT results by computing the IDFT of the 8 DFT coefficients $X(k)$ using the following M-file (`example8_1.m`):

```
In_sig = [1 1 1 1 0 0 0 0];    % define x(n)
N = length(In_sig);            % length N
X_in = In_sig*dftmtx(N);       % performing DFT
In_recover = (1/N)*(X_in*conj(dftmtx(N)));   % performing IDFT
figure;
subplot(2,1,1); stem(In_sig); title('Original signal');
subplot(2,1,2); stem(real(In_recover)); title('DFT->IDFT');
```

As shown in the figures generated by the MATLAB script, the original square wave is identical to the recovered signal, which shows that the computation of the DFT and IDFT is correct.

8.1.2 Properties of Twiddle Factors

The twiddle factors defined in Eq. (8.1.3) are complex variables with unit amplitude and different phase angles. For example, an 8-point DFT uses the twiddle factors from W_N^0 to W_N^7, which are displayed in Fig. 8.1. Note that $W_N^0 = 1$ is the first twiddle factor, and the twiddle factor increases in a clockwise direction around the unit circle. The twiddle factor W_N^k is also the Nth root of one since $(W_N^k)^N = 1$ for

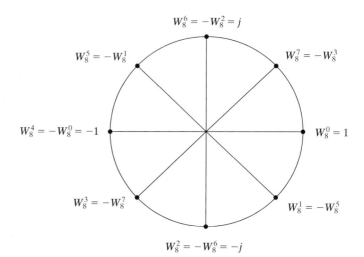

Figure 8.1 Twiddle factors for an 8-point DFT

all k. Some important properties that can be observed from Fig. 8.1 are summarized as follows:

1. The twiddle factor W_N^k repeats itself after every multiple of N. In the case shown in Fig. 8.1, $W_8^0 = W_8^8 = W_8^{16} \ldots, W_8^1 = W_8^9 = W_8^{17} \ldots$, etc. Therefore, W_N^k is a periodic function of period N, which can be expressed as

$$W_N^k = W_N^{k+iN}, \quad k = 0, 1, \ldots, N - 1, \qquad (8.1.8)$$

   where i is an integer.

2. Since all twiddle factors have identical magnitudes, the only difference is in the phase. In particular, there is a 180° phase shift between twiddle factors that are at opposite positions. For example, in an 8-point DFT, $W_8^0 = -W_8^4, W_8^1 = -W_8^5, \ldots$, and $W_8^7 = -W_8^3$. In general,

$$W_N^k = -W_N^{k+(N/2)}, \qquad (8.1.9)$$

   which is the symmetric property of the twiddle factors.

8.2 FAST FOURIER TRANSFORM ALGORITHMS

The FFT is a family of algorithms that efficiently implements the DFT. There are several types of FFT algorithms:

- Different radices, such as radix-2, radix-4, etc., and mixed-radix algorithms
- DIT and DIF algorithms
- Real and complex algorithms

In this chapter, we focus on the radix-2 algorithms that are commonly used for DSP processor implementations.

The basic principle of FFT algorithms is the "divide-and-conquer" approach, which decomposes an N-point DFT into progressively smaller DFTs, thus reducing the computational load as compared with that of the original N-point DFT. For example, dividing an N-point data sequence into two $N/2$-point sequences and performing the DFT on these two sequences individually results in the order of $2(N/2)^2 = N^2/2$ complexity, as compared with the original N^2 operations in an N-point DFT. Further computational reduction can be achieved by dividing these two sequences with $N/2$ samples into four subsequences with $N/4$ samples and performing an independent $N/4$-point DFT on these four shorter sequences. The process of dividing can be continued until a 2-point DFT is reached.

In addition to the preceding dividing process, we can exploit the periodic and symmetric properties of twiddle factors to reduce the computational load further. For example, some twiddle factors have real or imaginary parts equal to ± 1 (or 0), and these factors do not need multiplication. In the following sections, we show different ways of performing the divide-and-conquer approach in the FFT algorithms that were first introduced by Cooley and Tukey [5].

8.2.1 Decimation-in-Time Radix-2 Fast Fourier Transform

In this section, we examine the development of the DIT radix-2 FFT algorithm. In the DIT FFT, the N-point sequence $\{x(n), n = 0, 1, \ldots, N - 1\}$ is divided into the even sequence

$$x_e(n) = x(2m), m = 0, 1, \ldots, (N/2) - 1 \qquad (8.2.1)$$

and the odd sequence

$$x_o(n) = x(2m + 1), m = 0, 1, \ldots, (N/2) - 1. \qquad (8.2.2)$$

Therefore, the length of the radix-2 FFT is restricted to being a power of two (i.e., $N = 2^M$, where M is a positive integer).

The DFT defined in Eq. (8.1.2) can be rewritten as

$$X(k) = \sum_{n=0}^{N-1} x(n) W_N^{kn}$$

$$= \sum_{m=0}^{(N/2)-1} x(2m) W_N^{2mk} + \sum_{m=0}^{(N/2)-1} x(2m + 1) W_N^{(2m+1)k}. \qquad (8.2.3)$$

Since $W_N^{2mk} = W_{N/2}^{mk}$, Eq. (8.2.3) can be written as

$$X(k) = \sum_{m=0}^{(N/2)-1} x_e(m) W_{N/2}^{mk} + W_N^k \sum_{m=0}^{(N/2)-1} x_o(m) W_{N/2}^{mk}, \quad k = 0, 1, \ldots, N - 1. \qquad (8.2.4)$$

Using the symmetric property given in Eq. (8.1.9), Eq. (8.2.4) can be reduced to

$$X(k) = \begin{cases} X_e(k) + W_N^k X_o(k), & k = 0, 1, \ldots, (N/2) - 1 \\ X_e(k) - W_N^k X_o(k), & k = (N/2), \ldots, N - 1 \end{cases}, \qquad (8.2.5)$$

where $X_e(k)$ is the $N/2$-point DFT of $x_e(n)$ defined in Eq. (8.2.1), and $X_o(k)$ is the $N/2$-point DFT of $x_o(n)$ defined in Eq. (8.2.2).

The computation of $X_e(k)$ and $X_o(k)$ requires $2 \times (N/2)^2$ multiply-and-add operations. In addition, the computation of $W_N^k X_o(k)$ needs another $(N/2)$ multiply-and-add operations. The result is $(N^2 + N)/2$ operations, which is almost a 50% reduction over the N-point DFT if N is large. Figure 8.2 shows the flow graph of implementing Eq. (8.2.5), which decomposes an 8-point DFT into two 4-point DFTs and operates on the even and odd sequences independently. Note that the final outputs, $X(0), \ldots, X(7)$, are computed using four butterfly computation units. The DIT butterfly unit is clearly illustrated in Fig. 8.3, which contains one complex multiplication by the twiddle factor W_N^k, one complex addition, and one complex subtraction. Figure 8.2 is a single-stage decomposition of an 8-point DFT. The total operations required in Fig. 8.2 are 36 multiply and 40 add operations, which is almost a 50% reduction of the 8-point DFT that requires 64 multiply-and-add operations. Note that all multiply-and-add operations are computed using complex arithmetic.

The decomposition process can be further continued for implementing the 4-point DFT by using two 2-point DFTs. The 2-point DFT is the smallest computational unit in the radix-2 FFT algorithm. The complete decomposition of the 8-point

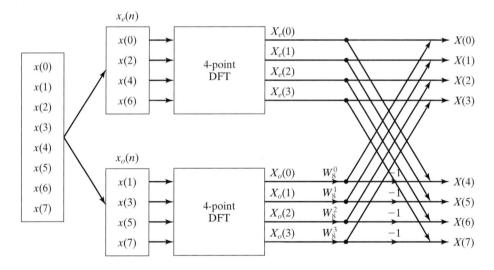

Figure 8.2   Decimation-in-time, 8-point DFT consists of two 4-point DFTs

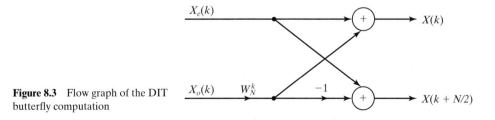

Figure 8.3   Flow graph of the DIT butterfly computation

DFT is shown in Fig. 8.4. The 2-point DFT can be easily derived from Fig. 8.3 by replacing W_N^k with $W_N^0 = 1$, which requires only one complex addition and one complex subtraction. Therefore, the total computational requirements of this 8-point FFT are only 8 complex multiplications and 24 complex additions. These requirements offer a substantial reduction in the computational load compared with the 8-point DFT that requires 64 complex multiplications and 64 complex additions. In addition, since $W_8^0 = 1$ in Fig. 8.4, only five complex multiplications are really needed in the 8-point FFT.

In general, a radix-2 FFT can be decomposed into M stages, where $M = \log_2 N$ since $2^M = N$. In each stage, $N/2$ complex multiplications by the twiddle factors and N complex additions (or subtractions) are required. Therefore, the total computational requirements are $(N \log_2 N)/2$ complex multiplications and $N \log_2 N$ complex additions. Note that these numbers are the upper limit since some multiplications by twiddle factors of value 1 can be skipped.

As shown in Fig. 8.4, the input sequence for the DIT radix-2 FFT must be scrambled in order to obtain the output in a natural order. To scramble the input sequence, the index of the data sample is represented in M-bit binary form, and its bits are reversed. For example, the data sample $x(3)$ has an index of 3 and is represented by a 3-bit binary number (011). This binary number is then bit reversed to 110; thus, the index becomes 6, which represents data sample $x(6)$. Table 8.1 shows the bit-reversal process for $N = 8$ using a 3-bit binary number as an index of sequence, where the rightmost column shows the bit-reversed order used in the input of Fig. 8.4. As discussed in Chapters 4 and 5, most modern DSP processors provide a bit-reversal addressing mode to support FFT algorithms. More details are

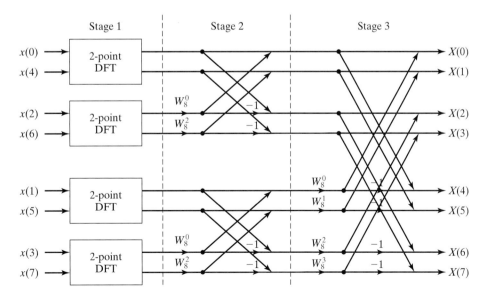

Figure 8.4   Flow graph of the final decomposition of an 8-point DIT FFT

TABLE 8.1   Bit-Reversal Process for $N = 8$

Decimal index in natural order	Binary representation	Bit-reversed representation	Decimal index in bit-reversed order
0	000	000	0
1	001	100	4
2	010	010	2
3	011	110	6
4	100	001	1
5	101	101	5
6	110	011	3
7	111	111	7

described in Sections 8.5 and 8.6 to illustrate the programming of the bit-reversal process in DSP processors.

In order to store the input and output of the N-point FFT, only one complex array of size N is required. The input sequence $x(n)$, the intermediate results from each stage of the FFT, and the final output $X(k)$ can be stored in the same buffer. This technique is called in-place computation, which is explained later.

8.2.2 Decimation-in-Frequency Radix-2 Fast Fourier Transform

Similar to the DIT algorithm, the DIF radix-2 FFT algorithm simply divides the time-domain sequence into two subsequences with $N/2$ samples: $\{x(0), x(1),\ldots, x(N/2 - 1)\}$ and $\{x(N/2), x(N/2 + 1),\ldots, x(N - 1)\}$. The DFT of $x(n)$ can be obtained by combining the DFT of these subsequences expressed as

$$X(k) = \sum_{n=0}^{(N/2)-1} x(n)W_N^{nk} + \sum_{n=N/2}^{N-1} x(n)W_N^{nk}$$

$$= \sum_{n=0}^{(N/2)-1} x(n)W_N^{nk} + \sum_{n=0}^{(N/2)-1} x\left(n + \frac{N}{2}\right)W_N^{nk}W_N^{(N/2)k}. \qquad (8.2.6)$$

Since $W_N^{(N/2)k} = (-1)^k$, Eq. (8.2.6) can be simplified to

$$X(k) = \sum_{n=0}^{(N/2)-1}\left[x(n) + (-1)^k x\left(n + \frac{N}{2}\right)\right]W_N^{nk}. \qquad (8.2.7)$$

This equation can be further expanded into two parts: one for even samples $X(2k)$ and the other for odd samples $X(2k + 1)$. In addition, using the relationships $W_N^{2kn} = W_{N/2}^{kn}$ and $W_N^{(2k+1)n} = W_N^n W_{N/2}^{kn}$, Eq. (8.2.7) can be partitioned as

$$X(2k) = \sum_{n=0}^{(N/2)-1}\left[x(n) + x\left(n + \frac{N}{2}\right)\right]W_{N/2}^{nk} = \sum_{n=0}^{(N/2)-1} x_1(n)W_{N/2}^{nk} \qquad (8.2.8)$$

and

$$X(2k + 1) = \sum_{n=0}^{(N/2)-1} \left[x(n) - x\left(n + \frac{N}{2}\right) \right] W_N^n W_{N/2}^{nk} = \sum_{n=0}^{(N/2)-1} x_2(n) W_N^n W_{N/2}^{nk}$$

(8.2.9)

for $k = 0, 1, \ldots, (N/2) - 1$. This first stage of the DIF FFT is illustrated in Fig. 8.5, and the detailed DIF butterfly computation is illustrated in Fig. 8.6. The output sequence $X(k)$ of the DIF FFT is bit reversed, while the input sequence $x(n)$ of the DIT FFT is bit reversed, as shown in Fig. 8.4. In addition, there is a slight difference in the butterfly computation. As shown in Fig. 8.3, complex multiplication is performed before complex addition (or subtraction) in the DIT FFT. In contrast, complex subtraction is performed before complex multiplication in the DIF FFT shown in Fig. 8.6. The subtraction before multiplication expressed in Eq. (8.2.9) has a close similarity with the symmetric FIR filter described in Section 6.1.3, which reduces the number of multiplications by half.

As shown in Fig. 8.7, the process of decomposition is continued until the last stage is reduced to the 2-point DFT. Since the frequency samples in the DIF FFT are bit reversed, the bit-reversal algorithm shown in Table 8.1 must be applied to these

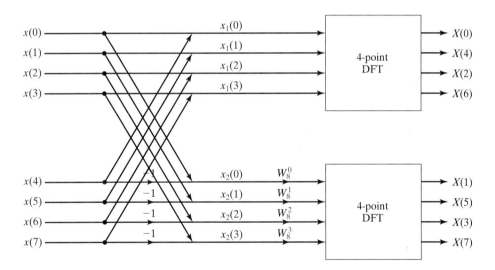

Figure 8.5   Decimation-in-frequency flow graph of an 8-point FFT

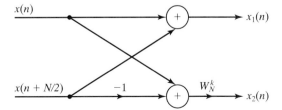

Figure 8.6   Flow graph of the DIF butterfly computation

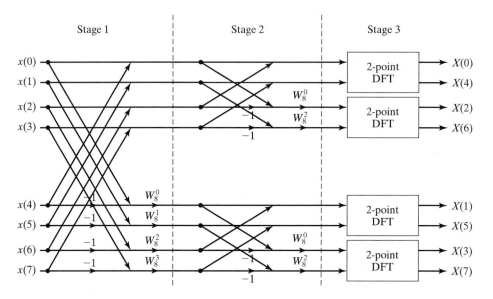

Figure 8.7   Final decomposition of an 8-point DIF FFT

frequency samples in order to obtain the natural order of frequency samples. Similarly, the DIF FFT algorithm also uses in-place computation.

8.2.3 Other Radix and Mixed-Radix Fast Fourier Transform Algorithms

The principle of decomposing the signal sequence into smaller subsequences can be extended to higher radices. An example of a commonly used higher radix is the radix-4 FFT algorithm, where the smallest decomposition consists of a 4-point DFT using a four-input and four-output butterfly computation. Similar to the DIF radix-2 FFT, the DIF radix-4 FFT first divides the input sequence into four subsequences, $\{x(0), \ldots, x(N/4 - 1)\}$, $\{x(N/4), \ldots, x(N/2 - 1)\}$, $\{x(N/2), \ldots, x(3N/4 - 1)\}$, and $\{x(3N/4), \ldots, x(N - 1)\}$, where the length N is a power of four. The $(N/4)$-point DFT is performed on these subsequences and is followed by simplification using the properties of the twiddle factors.

The output sequence is scrambled and needs to be bit reversed as in the radix-2 FFT. In the radix-4 FFT, base-4 digits $(0, 1, 2, 3)$ are used to represent the index of the data samples. For example, $X(12)$ is swapped with $X(3)$ since the index $(12)_{10} = (30)_4$, which is bit reversed to $(03)_4 = (03)_{10}$. In general, $M = \log_4 N$ stages are required to implement a radix-4 FFT algorithm.

The decomposition process in an FFT algorithm can also be applied to any nonprime number N. In this case, the length of the FFT is decomposed into $N = N_1 N_2$. This decomposition is useful when the length of the sequence is not a power of two for an implemention using a radix-2 FFT algorithm.

Although a higher radix or mixed-radix FFT can provide a further reduction in computation over the radix-2 FFT algorithm, it is often more complicated to

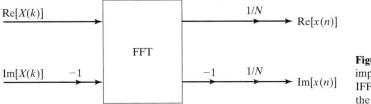

Figure 8.8 Efficient implementation of the IFFT using the FFT as the main building block

implement in programmable DSP processors. In the benchmark quoted in [18], the FFT speed is slower when implementing on radix-4 as compared with radix-2. Therefore, the radix-2 FFT is the most commonly used FFT algorithm, and we concentrate on its DSP implementations in later sections.

8.2.4 Inverse Fast Fourier Transform

The inverse FFT (IFFT) is the fast computation of the IDFT defined in Eq. (8.1.4), which is very similar to the DFT algorithm given in Eq. (8.1.2) except for the conjugate of twiddle factors and the scaling factor of $1/N$. Due to the close similarity with the FFT, IFFT can be easily computed from the FFT code without the need to have a different IFFT program.

By performing a complex conjugate on Eq. (8.1.4), we have

$$x^*(n) = \frac{1}{N} \sum_{k=0}^{N-1} X^*(k) W_N^{nk}, \quad n = 0, 1, \dots, N - 1. \tag{8.2.10}$$

This equation shows that an N-point IFFT can be treated as an N-point FFT of the data sequence $X^*(k)$ using the previously described FFT algorithm. The output data can be scaled by $1/N$ to obtain $x^*(n)$. The complex conjugate is then applied to $x^*(n)$ to obtain $x(n)$. This final complex conjugate may not be necessary if the time-domain signal is real. The implementation of IFFT is illustrated in Fig. 8.8, which only requires an additional $2N$ real multiplications for scaling by $1/N$.

8.3 ANALYSIS AND IMPLEMENTATION USING MATLAB AND C

It is important to understand the properties and implementations of the DFT for practical applications. In this section, we introduce some important characteristics of the DFT and its implementation using MATLAB and C.

8.3.1 Strengths and Weaknesses of the Discrete Fourier Transform and the Fast Fourier Transform

The DFT is able to compute any frequency component $X(k)$ for a given k using Eq. (8.1.2) without computing the entire spectrum. Therefore, we can zoom in to the frequency range of interest by using the DFT. This feature is not possible in the FFT, where N frequency samples $X(k)$ for $k = 0, 1, \dots, N - 1$ must be computed as a whole. Another advantage of the DFT is that it can operate on any length N

that best matches the period of the given signal. This feature is not possible in a radix-k FFT, which imposes a restriction on the length of the sequence to be $N = k^M$. For example, a radix-2 FFT must have a sequence length that is a power of two. Otherwise, zero padding or data truncation must be applied to satisfy this length constraint.

The weakness of the DFT is its computational load [i.e., $O(N^2)$ complex multiplications and additions are required]. One complex multiplication is equivalent to four real multiplications and two real additions, and one complex addition requires two real additions. Therefore, the total number of real multiplications and additions for an N-point DFT are $N(4N)$ and $N[2(N - 1) + 2N] = N(4N - 2)$, respectively. For example, a 512-point DFT requires approximately one million real multiplications and additions, which is a very demanding requirement for typical DSP processors. Thus, a faster computational algorithm such as the FFT is required. In addition, the higher computational load in DFT also introduces more quantization errors due to internal rounding after multiply-add operations. The FFT algorithm requires fewer computations and therefore reduces quantization errors. A more detailed analysis of finite-wordlength effects is given in Section 8.4.3.

The DFT (or FFT) is useful in transforming time-domain periodic signals, but is particularly poor in handling transient signals. A better transform for transient signals is the wavelet transform or joint time-frequency approaches, which are introduced briefly in Section 8.7.4.

8.3.2 Properties of the Discrete Fourier Transform

The DFT has several useful properties that can be exploited to achieve a more efficient implementation of the FFT for some applications. Therefore, it is important to understand and use these properties for solving practical problems.

Linearity

The DFT is a linear transformation that can be expressed as

$$\text{DFT}[ax_1(n) + bx_2(n)] = a\text{DFT}[x_1(n)] + b\text{DFT}[x_2(n)], \qquad (8.3.1)$$

where $x_1(n)$ and $x_2(n)$ are assumed to have the same length. If these sequences have different durations, the shorter sequence can be zero-padded to have the same length as the longer sequence. This property is illustrated in the MATLAB example property1.m.

Symmetry

The output of the DFT of a real-valued sequence is a conjugate-symmetric sequence. This property can be expressed as

$$X(k) = X^*(-k)_N = X^*(N - k)_N, \qquad (8.3.2)$$

where the symbol $(\ )_N$ is a modulo-N operator that restricts the frequency indices k from 0 to $N - 1$. The real part of the transform is an even sequence expressed as

$$\text{Re}[X(k)] = \text{Re}[X(N - k)_N]. \tag{8.3.3}$$

The imaginary part of the transform is an odd sequence expressed as

$$\text{Im}[X(k)] = -\text{Im}[X(N - k)_N]. \tag{8.3.4}$$

Equation (8.3.2) also implies that the magnitude and phase of $X(k)$ are even and odd functions, respectively. A simple example illustrating the symmetry property of the FFT for a real-valued sequence is given in the M-file `property2.m`. The results displayed in Fig. 8.9 show that the real part of $X(k)$ and the magnitude have even symmetry, and the imaginary part and the phase have odd symmetry. The first element of each plot is the DC coefficient, and the $(N/2)$-th element (index 4 in this example) is the Nyquist component of frequency, $f_s/2$. These two components are real because imaginary parts are zero. The rest of the frequency components are repeated about the Nyquist frequency $(N/2)$ in either even or odd symmetry. Therefore, memory can be saved by storing only the $(N/2) + 1$ elements from DC to Nyquist ($k = 0, 1, \ldots, N/2$) that correspond to the first half of the plot.

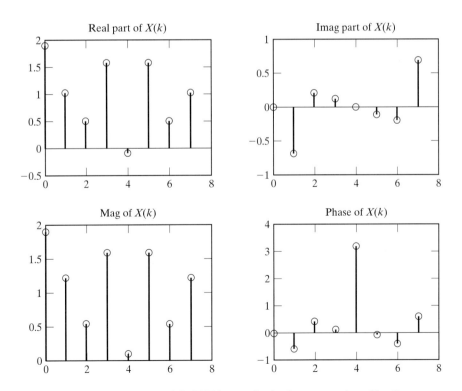

Figure 8.9 Symmetry property of the FFT for a real-valued sequence, where $N = 8$

Circular Shifting

The circular shifting of the data sequence by m units is equivalent to treating the sequence as a periodic signal with period N, performing the linear shift of m units, and extracting the data samples using an N-point rectangular window. Circular shifting is defined as $x(n - m) * w_N(n)$, where $x(n)$ is a periodic signal, m is the amount of shift, and $w_N(n)$ is the rectangular window function from $0 \le n < N$. Figure 8.10 shows that the data sequence of length $N = 8$ is shifted circularly by two samples; thus, the old $x(6)$ becomes the new $x(0)$, the old $x(7)$ becomes the new $x(1)$, and the rest of the samples are shifted to the right by two units. In contrast, linear shifting by two samples results in the insertion of two zeros at the beginning of the shifted data samples, and the last two samples are lost.

The FFT of the circular-shifted sequence is expressed as

$$\text{FFT}[x(n - m) * w_N(n)] = W_N^{km} X(k). \tag{8.3.5}$$

Thus, the spectrum $X(k)$ is multiplied by a twiddle factor with a phase shift of m units. Figure 8.11 shows that the magnitude response of the circular-shifted sequence remains unchanged, but the phase has been modified. An illustration of circular shifting is given in the MATLAB file `property3.m`.

The property of circular shifting can also be applied in the frequency domain, resulting in a time sequence that is multiplied by the complex twiddle factor expressed as

$$\text{IFFT}[X(k + m) * w_N(k)] = W_N^{nm} x(n). \tag{8.3.6}$$

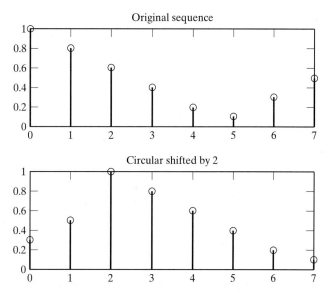

Figure 8.10 The original sequence (top) and the circular shift by two samples (bottom)

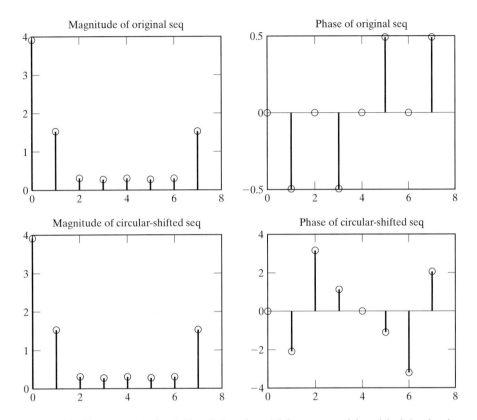

Figure 8.11 The magnitude (top left) and phase (top right) responses of the original signal and the magnitude (bottom left) and phase (bottom right) of the circular-shifted signal

Circular Convolution

The linear convolution of time-domain sequences $x(n)$ of length N_x and $h(n)$ of length N_h results in a sequence $y(n)$ of length $N_x + N_h - 1$. In the case of circular convolution, the sequences $x(n)$ and $h(n)$ are treated as periodic. Because circular convolution requires both sequences to have the same length, sequences with different lengths must be made equal by zero-padding the shorter sequence. The output of the circular convolution is expressed as

$$y(n) = h(n) \otimes x(n) = \sum_{k=0}^{N-1} x(k)h(n-k)_N, \qquad (8.3.7)$$

where the symbol $\otimes$ indicates circular convolution, and the output sequence length $N = \max(N_x, N_h)$. Circular shifting is used in circular convolution, and it results in an N-point output $y(n)$ instead of the $N_x + N_h - 1$ points in linear convolution. Circular convolution is the same as linear convolution if the length of the data sequences $x(n)$ and $h(n)$ are set to $N = N_x + N_h - 1$ using zero padding. Section 8.7.1 includes a more detailed discussion of using circular convolution for efficient FIR filtering of long data sequences.

The property of circular convolution is expressed as

$$\mathrm{DFT}[x_1(n) \circledast x_2(n)] = X_1(k)X_2(k). \qquad (8.3.8)$$

Therefore, circular convolution in the time domain is equivalent to multiplication in the frequency domain.

A duality of the time-domain circular convolution can be applied to the frequency domain. The DFT on the time-domain multiplication of $x_1(n)$ and $x_2(n)$ is equivalent to the circular convolution of $X_1(k)$ and $X_2(k)$ with a scaling factor $1/N$ expressed as

$$\mathrm{DFT}[x_1(n)x_2(n)] = \frac{1}{N}[X_1(k) \circledast X_2(k)]. \qquad (8.3.9)$$

The MATLAB script `property4.m` illustrates the steps used to perform circular convolution in both time and frequency domains.

Parseval's Theorem

The energy of the signal in time and frequency domains is expressed as

$$E = \sum_{n=0}^{N-1} |x(n)|^2 = \frac{1}{N}\sum_{k=0}^{N-1} |X(k)|^2. \qquad (8.3.10)$$

It is important to note that $|X(k)|^2$ is defined either as the energy spectrum when the signal $x(n)$ is of finite duration or the power spectrum if the signal $x(n)$ is a periodic sequence. An illustration of Parseval's theorem is given in the MATLAB file `property5.m`.

8.3.3 MATLAB Implementations

MATLAB provides the functions `fft` and `ifft` to compute FFT and IFFT, respectively. For example, the command

```
X = fft(x,N)
```

computes an N-point FFT (or DFT) of the time sequence $x(n)$ in vector x. The output vector X is a complex vector of length N. If the length of vector x is less than N, the vector is zero-padded to the length of N. On the other hand, the vector is truncated to N points if it is longer than N. The execution time of the `fft` function depends on the length of the transform. It is fastest if N is a power-of-two integer such that the radix-2 FFT algorithm can be applied. It is much slower for lengths that are prime or that have large prime factors. For example, if the length of x is 61, the command X = fft(x,64) is executed much faster than the command X = fft(x), which computes a 61-point DFT.

For some applications, it is more informative to view the DC component in the middle of the spectrum. MATLAB provides the function `fftshift(X)` to shift frequency components from the range of $f_s/2$ to f_s to the range of $-f_s/2$ to 0, thus displaying the frequency plots from the range of 0 to f_s to the range of $-f_s/2$ to $f_s/2$.

IFFT can be computed using the `ifft` function. To transform the frequency domain vector x into the time-domain signal vector x using the *N*-point IFFT, we can use the following command:

```
x = ifft(X,N)
```

The usage and efficiency of `ifft` is similar to `fft`.

Example 8.2

A signal contains two sinewaves at frequencies 20 Hz and 41 Hz, and a DC component of amplitude of 2 is expressed as

$$x(n) = \sin(2\pi(20)nT) + \sin(2\pi(41)nT) + 2, \quad n = 0, 1, \dots, 63,$$

where *T* is the sampling period of 1/128 seconds (or the sampling frequency of 128 Hz). A 64-point radix-2 FFT is used to compute its magnitude spectrum using the following MATLAB script (example8_2.m):

```
n = 0:1:63; T = 1/128;
x = sin(2*pi*20*n*T)+sin(2*pi*41*n*T)+2;    % generate composite signal
Xc = abs(fft(x,64));                         % magnitude spectrum
subplot(2,1,1),plot(Xc);title('spectrum plot');
subplot(2,1,2),plot(fftshift(Xc));title('shifted spectrum plot');
```

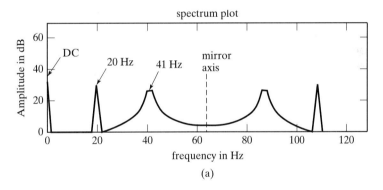

(a)

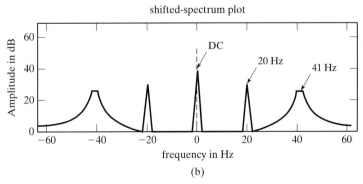

(b)

Figure 8.12   Magnitude spectra of real-valued signals: (a) frequency range from 0 to f_s and (b) frequency-shifted version from $-f_s/2$ to $f_s/2$

The amplitude spectrum of $x(n)$ and its shifted version are displayed in Fig. 8.12. Note that in Fig. 8.12(a), the spectrum is a mirror image about 64 Hz ($f_s/2$). The 64-element vector Xc contains the 1st element (index 0) at DC, the 2nd element at 2 Hz, etc., the 33rd element at half of the sampling frequency, and the last (64th) element at $(f_s - 2)$ Hz. In order to have a clear view of the DC component, the frequency-shifted plot is shown in Fig. 8.12(b). It clearly shows that when the input signal is a real sequence, only half of the spectrum from 0 to $f_s/2$ is needed because the other half is its mirror image and does not contain additional information.

Figure 8.12 also shows that the frequency components at 20 Hz and 41 Hz look different. The frequency component at 20 Hz is a spike (actually a line), while the component at 41 Hz is spread over to its neighboring frequencies and thus has a lower amplitude. The reason for this spreading is that the frequency resolution, $\Delta f = f_s/N = 2$ Hz, can represent frequencies only at 40 Hz and 42 Hz, and there is no frequency bin (or index) for representing 41 Hz. Therefore, the power of the signal at 41 Hz is spread across its neighboring frequency bins. This undesired phenomenon is called spectral leakage.

A simple solution for reducing spectral leakage is to increase the length of the FFT so that the frequency resolution is finer and thus able to represent the signal frequencies more closely. If the number of data samples cannot be increased, we can pad zeros at the end of the sequence to increase FFT size. Another way to reduce spectral leakage is to use the windowing approach, which is discussed further in Section 8.4.1.

In some applications, the DC value may be too large and may overwhelm other frequency components. We can eliminate the DC component using the detrending process expressed as

$$x'(n) = x(n) - \overline{x}, n = 0, 1, \dots, N - 1, \qquad (8.3.11)$$

where the DC component is calculated by

$$\overline{x} = \frac{1}{N} \sum_{n=0}^{N-1} x(n). \qquad (8.3.12)$$

This process removes the DC component before taking the FFT of the DC-free signal $x'(n)$.

8.3.4 C Implementation

This section presents a floating-point C program that performs a DIT radix-2 FFT. As discussed in Section 8.2.1, the sequence length N must be a power of two for the radix-2 FFT algorithm. The DIT FFT uses the bit_reversal.c function to arrange the input samples in bit-reversal order. We then apply this bit-reversed sequence to the radix-2 FFT function ditr2fft.c. These two functions are listed below:

```
#include "def_complex.h"      // floating-point header file
void bit_reversal(complex *x, unsigned int M)
```

```
{
  unsigned int i,j,k;
  unsigned int N = 1<<M;        // no. of FFT point N = 2^M
  unsigned int N2 = N>>1;       // N/2
  complex temp;                 // complex temporary storage
  for (j=0,i=1;i<N-1; i++){
    k = N2;
    while(k<=j){
      j-=k;
      k>=1;
    }
    j += k;
    if (i < j){            // swapping samples in bit-reversal order
      temp = x[j];
      x[j]=x[i];
      x[i]=temp;
    }
  }
}


#include "def_complex.h" // floating-point header file
#define pi 3.1415926535897

void ditr2fft(complex *x, unsigned int M, unsigned int SCALE)
{
  complex   W;              // store twiddle factor in complex array
  complex   temp;          // temporary storage for complex variables
  unsigned int iu,il;      // upper and lower point in butterfly
  unsigned int j;
  unsigned int N=1<<M;     // number of FFT points
  unsigned int m;          // FFT stage index
  unsigned int LEN;        // number of points in sub FFT at stage M
                           // and offset to next FFT stage
  unsigned int LEN1        // number of butterflies at stage m and
                           // offset to lower point at stage M

  float arg, w_real, w_imag
  float scale=0.5;         // scaling factor at intermediate stage
  if SCALE == 0            // testing of SCALE flag
    scale = 1.0;           // no scaling if SCALE = 0

  for (m=1; m<=M, m++)     // (c) for stage from m = 1 to M
  {
    LEN = 1<<m             // LEN = 2^m
    LEN1 = LEN>>1          // LEN1 = number of butterflies = LEN/2
    W.re = 1.0;            // starting from W^0=1+j0
    W.im = 0.0;

    for(j=0; j<LEN1; j++){ // (b) number of butterflies in each
                           // stage
      for(iu=j; iu<N; iu +=LEN){ // (a) perform butterflies
        il = iu+LEN1;            // determine spacing
```

```
        temp.re = (x(il).re*W.re-x(il).im*W.im)*scale;
        temp.im = (x(il).im*W.re+x(il).re*W.im)*scale;
        x(il).re = x(iu).re*scale-temp.re;
        x(il).im = x(iu).im*scale-temp.im;
        x(iu).re = x(iu).re*scale + temp.re;
        x(iu).im = x(iu).im*scale + temp.im;
      }
      // recursive compute higher order of W
      arg = pi/LEN1;
      w_real = cos(arg);
      w_imag = -sin(arg);
      W.re = W.re*w_real - W.im*w_imag;
      W.im = W.re*w_imag + W.im*w_real;
    }
  }
}
```

There are three loops in the FFT implementation. As marked in `ditr2fft.c`, mark (a) indicates the inner loop, mark (b) is the middle (or group) loop, and mark (c) is the outer loop. The inner loop performs the butterfly computation, and the group loop executes the group of independent butterflies that can be computed in parallel. For example, in the DIT radix-2 FFT shown in Fig. 8.4, stage 1 consists of four groups of one butterfly, stage 2 consists of two groups of two butterflies, and stage 3 consists of one group of four butterflies. Therefore, loops (a) and (b) complete all of the butterfly computations in each stage, while loop (c) indicates the number of stages to complete the FFT algorithm. Note that the total number of butterflies required in each stage is always $N/2$ for a radix-2 FFT.

The variables `temp.re` and `temp.im` are derived by multiplying the lower input `x(il)` with the twiddle factor `W` and the scaling factor `scale`. These terms are then subtracted and added to the upper point `x(iu)` and the lower point `x(il)`, respectively, to form the outputs of the butterfly. The in-place computation overwrites the input-memory locations with the results. A scaling factor of 0.5 is inserted to the real and imaginary paths of every stage when the flag `SCALE = 1` in order to prevent the result of the FFT from overflowing. MATLAB examples on the different scaling factors for the FFT are given in Section 8.4.4. In `ditr2fft.c`, the twiddle factor `W` is computed recursively in the main FFT routine. However, a better approach is to precompute these twiddle factors and store them in a look-up table. The table-lookup method is commonly used for the implementation of the FFT on DSP processors.

The user can modify the preceding routines so they become a DIF radix-2 FFT and a DIT radix-2 IFFT. A complete example of an FFT-based power-spectrum estimator using the preceding C functions for experiments is given in Section 8.8.2.

8.4 IMPLEMENTATION CONSIDERATIONS

In this section, we discuss several practical issues related to the implementation and application of FFT algorithms.

8.4.1 Frequency Resolution and Windowing

As discussed in Chapter 2, frequency representation varies in different literatures. In general, frequency can be represented by nature frequency f, radian frequency ω, normalized frequency F_1 with reference to the sampling frequency, normalized frequency F_2 with reference to the Nyquist (folding) frequency, or the frequency index k in the FFT. Figure 8.13 shows different representations of frequencies, units, ranges, and frequency resolutions.

Frequency representation can be converted from one form to another using the following relationship:

$$\frac{f}{f_s} = \frac{k}{N} = \frac{\omega}{2\pi}. \tag{8.4.1}$$

Note that two different normalized frequencies are defined in different books. F_2 (defined as F in Chapter 2) is used in this book and MATLAB. For example, assuming $f_s = 10{,}000$ Hz and $N = 200$, the natural frequency $f = 1{,}000$ Hz is equal to the frequency index $k = Nf/f_s = 20$, $\omega = 2\pi f/f_s = 0.2\pi$, $F_1 = f/f_s = 0.1$, and $F_2 = 0.2$. As discussed in Section 8.3.2, when computing an N-point FFT of a real-valued sequence, there are only $(N/2 + 1)$ unique samples $(k = 0, 1, \ldots, N/2)$ when N is even. The other $(N/2 - 1)$ samples are duplicated (complex conjugate) in the negative-frequency axis or from the frequency index $k = N/2 + 1$ to N.

Frequency Resolution

Frequency resolution is an important factor in determining the size of the FFT. In Fig. 8.13, the frequency axis is divided into N equally spaced intervals over the frequency

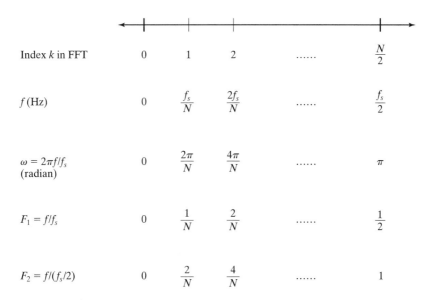

Figure 8.13 Summary of frequency units, resolutions, and ranges

range from 0 to f_s Hz, which results in different frequency resolutions and units expressed as

$$\Delta f = \frac{f_s}{N} \quad (\text{Hz}), \tag{8.4.2a}$$

$$\Delta k = 1 \quad (\text{frequency index}), \tag{8.4.2b}$$

$$\Delta \omega = \frac{2\pi}{N} \quad (\text{radians}), \tag{8.4.2c}$$

$$\Delta F_1 = \frac{1}{N} \quad (\text{normalized frequency vs. } f_s), \tag{8.4.2d}$$

or

$$\Delta F_2 = \frac{2}{N} \quad (\text{normalized frequency vs. } f_s/2). \tag{8.4.2e}$$

We can create a finer frequency resolution by increasing the FFT length, N, while keeping f_s constant. It is important to note that the frequency resolutions stated in Eq. (8.4.2) are based on the assumption that the number of available data L equals N. If the data length is less than the FFT length (i.e., $L < N$), N-point FFT can still be performed by zero padding with $(N - L)$ zeros, as explained in Section 8.3.3. However, the frequency resolution is no longer inversely proportional to N, as stated in Eq. (8.4.2); rather, it is computed based on the data length L. For example, when 1,000 data samples are padded by 24 zeros to compute a 1,024-point FFT, the frequency resolution, Δf, is not equal to $f_s/1{,}024$ Hz; rather, it is $f_s/1{,}000$ Hz. In the case when $L > N$, the N-point FFT is computed based on the truncated data sequence with the first N samples, and the frequency resolution is stated in Eq. (8.4.2). In summary, frequency resolution is defined as the ability to resolve two neighboring frequency components whose frequency separation is greater than or equal to the frequency interval provided by an N-point FFT. See the problems at the end of this chapter for more examples.

The FFT can also be viewed as a parallel bank of N bandpass filters (BPFs) centered at different frequencies, as shown in Fig. 8.14. When N increases, the bandwidth of these filters becomes narrower and thus able to separate closely spaced frequency components.

Windows

Windowing is the process of applying a window function $w(n)$ of length N to the signal sequence $x(n)$. The window output is $w(n)x(n)$ for $n = 0, 1, \ldots, N - 1$. Since the frequency resolution is inversely proportional to N, finer spectral information can be obtained by using a longer window. As the length of the window increases, the width of the mainlobe decreases. As a result, the transition width from the passband to the stopband decreases; thus, the finer frequency details are captured. However, longer windows cannot capture transient information and thus have a poorer time resolution. Therefore, choosing the window length N involves a time-frequency

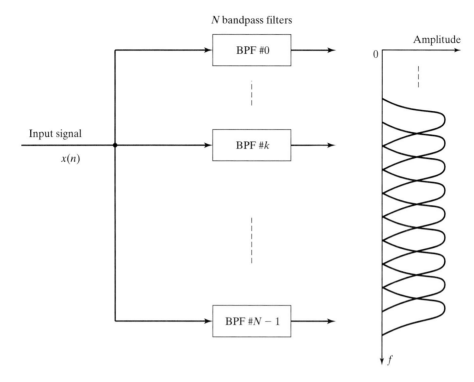

Figure 8.14 *N*-point FFT equivalent to *N* bandpass filters

tradeoff. A variable window length, such as the wavelet transform, can be used to capture fine details of both time and frequency information.

Since the FFT is a block-processing algorithm, a window must be applied to the original signal prior to the frequency analysis using the FFT. For example, a rectangular window is used to collect a block of input samples with unit weighting. However, the rectangular window introduces edge effects in the data, which leads to errors in frequency analysis. As shown in Fig. 8.14, an *N*-point FFT can be viewed as an *N*-band filter. Using the rectangular window, the frequency response has the sharpest mainlobe and smallest transition width and thus has the finest spectral resolution. Unfortunately, the tradeoff is that the rectangular window's sidelobe attenuation is poor (only −13 dB), and the sidelobe rolloff is only −6 dB/octave.

A better window function is the one that tapers to zero at the edges of the window. These tapering-window (or bell-shape) functions reduce edge effects of the signal and produce better sidelobe attenuation and faster rolloff compared with the rectangular window. Some commonly used windows including Hanning, Hamming, and Blackman are listed in Table 8.2. Note that there is always a tradeoff between the transition width (and the width of the mainlobe) and its sidelobe attenuation. Smooth tapering windows sacrifice their mainlobe bandwidth for a higher sidelobe attenuation.

MATLAB version 6.5 contains a window design and analysis tool (WDAT) in the Signal Processing Toolbox. WDAT allows the user to evaluate the characteristics

TABLE 8.2   Characteristics of Some Commonly Used Windows of Length N [2]

Window	Definition of $w(n)$ for $0 \leq n \leq N - 1$	Normalized transition width (F_2)	Sidelobe attenuation	Rolloff (dB/octave)
Rectangular	$w(n) = 1$	1.8/N	−13 dB	−6
Hamming	$w(n) = 0.54 - 0.46 \cos\left(\dfrac{2\pi n}{N - 1}\right)$	6.6/N	−43 dB	−6
Hanning	$w(n) = 0.5 - 0.5 \cos\left(\dfrac{2\pi n}{N - 1}\right)$	6.2/N	−31 dB	−18
Blackman	$w(n) = 0.42 - 0.5 \cos\left(\dfrac{2\pi n}{N - 1}\right)$ $+\ 0.08 \cos\left(\dfrac{4\pi n}{N - 1}\right)$	11/N	−58 dB	−18

of different windows. The user can open this tool by typing

```
wintool
```

in the MATLAB command window. A window appears (see Fig. 8.15), which displays both the time and frequency domains of the selected windows. The user can select the type of windows by providing a user-specified name, a window length, additional parameters for some windows (like Tukey), and periodic or symmetric sampling. Periodic sampling computes a length $N + 1$ window and returns the first N points, while symmetric sampling computes and returns N points specified in the **Length** field.

As shown in Fig. 8.15, the Blackman window has the lowest sidelobe attenuation and also has the widest mainlobe among the three windows. In the time domain, the Blackman window has the narrowest bell-shape window, which accounts for its reduced sidelobe attenuation and rolloff.

The problem with these tapering windows is that they reduce the magnitude of signal samples near the edges of the windows. Therefore, overlapping of data blocks is recommended to recover lost data, as shown in Fig. 8.16. Typically, 50% overlapping is used. This overlapping of windows also has the effect of reducing the variance of the spectrum estimation by allowing more FFT blocks to be averaged over a given timeframe. This issue is further explained in Section 8.7.2.

Another advantage of using overlapped segments is the ability to line up the central frequencies of the N bandpass filters to the frequencies of the signal. In the case of 50% overlap, each data sample is used twice, except for the first and the last $N/2$ samples. This overlapping increases the chance of fitting the period of the signal to match the window, which leads to a more accurate spectral analysis. However, the problems with overlapping segments are an increase in memory storage and slight increases in programming complexity.

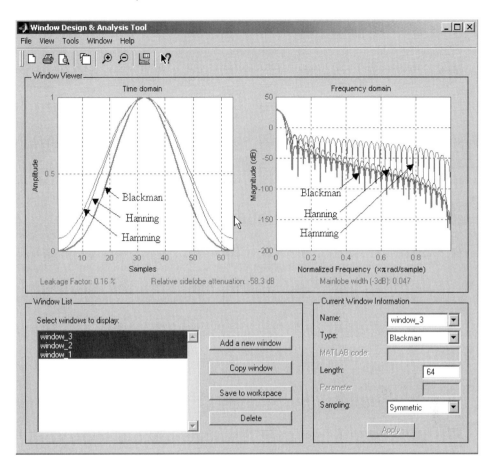

Figure 8.15   WDAT displays the time and frequency responses of Hamming, Hanning, and Blackman windows.

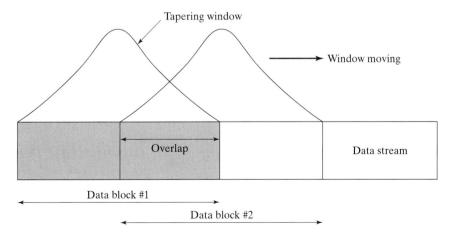

Figure 8.16   Overlapping of data windows

8.4.2 Computational Issues

FFT algorithms are used to reduce the computational load of the DFT. Further speed improvement in the FFT computation can be obtained by using the following methods:

1. Zero-padding or truncating the signal sequence to match the required length of the FFT based on the chosen radix.
2. Using a higher radix such as the radix-4 DIT FFT algorithm, which reduces the number of complex multiplications by 50% over the radix-2 DIT FFT algorithm.
3. Saving the twiddle factors in a look-up table instead of computing them in the program. However, doing so increases memory requirements.
4. Multiplying only when necessary since many twiddle factors are ±1. However, skipping redundant multiplications may result in an increased program size and execution time for handling these special conditions. Therefore, the reduction in the total number of multiply-and-add operations may not produce a faster processing program. A closer look at Sections 8.5 and 8.6 reveals an efficient programming style in coding the FFT for different DSP processors.
5. Using real-valued signals in applications including speech, audio, image, etc. The FFT routine usually operates on complex input signals; thus, the imaginary parts are set to zero for real input data. However, the FFT routine is not efficient in these cases. There are techniques that make use of the properties introduced in Section 8.3.2 to efficiently compute two N-point real-valued sequences using only one N-point FFT, or compute a $2N$-point FFT of a real-value sequence using an N-point FFT. These efficient implementations can reduce computations significantly.

We first introduce the technique of performing the FFT of two N-point real-valued sequences using only one N-point FFT. Instead of using an N-point FFT on the two sequences $x_1(n)$ and $x_2(n)$ separately, a complex N-point sequence $x(n)$ is first formed by combining these two real-valued sequences as $x(n) = x_1(n) + jx_2(n)$. Based on the linear property, an N-point FFT now can be applied to the complex sequence $x(n)$ to obtain

$$X(k) = \text{FFT}[x(n)] = X_1(k) + jX_2(k), \qquad (8.4.3)$$

where $X_1(k)$ is the FFT of the real sequence $x_1(n)$, and $X_2(k)$ is the FFT of the real sequence $x_2(n)$.

Using the symmetry property, the FFT of a real-valued sequence is an even symmetric

$$X(k) = X^*(N - k)_N \qquad (8.4.4)$$

and the FFT of an imaginary sequence is an odd symmetric

$$X(k) = -X^*(N - k)_N. \qquad (8.4.5)$$

We can derive the complex-conjugate of $X(k)$ by negating its imaginary terms to obtain $X^*(k)$ and then solving the following equations for $X_1(k)$ and $X_2(k)$:

$$X_1(k) = \frac{1}{2}[X(k) + X^*(N - k)_N]$$ (8.4.6a)

and

$$X_2(k) = -\frac{j}{2}[X(k) - X^*(N - k)_N].$$ (8.4.6b)

This technique requires only $(N/2)\log_2 N + N$ complex multiplications and $N \log_2 N + N - 2$ complex additions. This technique is in contrast to the use of two N-point, radix-2 FFTs to compute $X_1(k)$ and $X_2(k)$ independently, which requires a total of $N \log_2 N$ complex multiplications and $2N \log_2 N$ complex additions. Furthermore, calling two FFT routines requires an extra overhead for the subroutine calls. An example of this implementation is given in the MATLAB M-file combine_N.m.

The other efficient implementation is to perform the FFT of a real-valued $2N$-point sequence $y(n)$ using an N-point complex FFT. This packing algorithm can be implemented by forming two real-valued sequences of length N as even and odd sequences

$$x_1(n) = y(2n) \quad \text{and} \quad x_2(n) = y(2n + 1),$$ (8.4.7a)

for $0 \leq n \leq N - 1$ and by forming an N-point complex sequence as

$$x(n) = x_1(n) + jx_2(n).$$ (8.4.7b)

Compute the N-point complex FFT of $x(n)$ to obtain $X(k)$, derive its complex conjugates, and extract $X_1(k)$ and $X_2(k)$ using Eq. (8.4.6a) and Eq. (8.4.6b), respectively. By using the periodicity property of $X_1(k)$ and $X_2(k)$ expressed as

$$X_1(k) = X_1(k + N) \quad \text{and} \quad X_2(k) = X_2(k + N),$$ (8.4.8)

the final step is to link the $2N$-point FFT of $y(n)$ to the N-point FFT of $x(n)$. Recall that the DIT radix-2 FFT has a similar type of first-stage decomposition of using even and odd samples. We can use Eq. (8.2.4) to complete the last step as

$$Y(k) = X_1(k) + W_{2N}^k X_2(k), \quad 0 \leq k \leq 2N - 1.$$ (8.4.9)

The computational load of a $2N$-point FFT requires $N \log_2(2N)$ complex multiplications and $2N \log_2(2N)$ complex additions, and the modified method only requires $(N/2)\log_2 N + 2N$ complex multiplications and $N \log_2 N + 2N$ complex additions. Again, significant computational savings can be achieved with a higher order N. An example of using this technique is given in the MATLAB M-file combine_2N.m.

It is important to consider the practical implementation of FFT algorithms on a specific DSP processor. The DSP architecture, internal memory, data buses, and special instructions that handle bit reversal, parallelism, and looping affect efficient implementation. Therefore, we study the efficient implementation of FFT algorithms on different DSP processors in Sections 8.5 and 8.6.

8.4.3 Finite-Wordlength Effects

As explained previously, an FFT algorithm generates fewer quantization errors because it reduces arithmetic operations as compared with a direct computation of the DFT. However, overflow and roundoff errors still require special considerations, especially for implementing the FFT using fixed-point DSP processors.

Since multiplication with a twiddle factor results in quantization errors, we can re-examine the flow graph of the commonly used DIT radix-2 FFT shown in Fig. 8.4 in terms of the number of multiplications. Note that some butterfly multiplications by ± 1 (such as W_N^0 and $W_N^{N/2}$) do not require actual multiplication, thus avoiding roundoff errors. A single butterfly with a nontrivial twiddle factor requires one complex multiplication (or four real multiplications), which can be treated as four roundoff-noise sources. As discussed in Chapter 3, the noise is an uncorrelated zero-mean white noise with a variance of $2^{-2B}/12$. Thus, the total noise variance produced by one butterfly is

$$\sigma_e^2 = 4\frac{2^{-2B}}{12} = \frac{2^{-2B}}{3}, \tag{8.4.10}$$

where B is the total number of bits (or the wordlength).

In order to determine the roundoff noise at the outputs of the FFT, we need to determine the number of butterflies connected to the individual output sample. As shown in Fig. 8.4, there are $N/2$ butterflies in the first stage, $N/4$ butterflies in the second stage, and so on until a single butterfly is attached to the output node. Therefore, for an N-point FFT, $(N - 1)$ butterflies are connected to each node, thus resulting in a total roundoff-error variance of

$$\sigma_e^2 = \frac{(N - 1)2^{-2B}}{3}. \tag{8.4.11}$$

This equation shows that the total roundoff error is proportional to the length of the FFT N. Note that this is the upper bound since there is no need to perform multiplication for some twiddle factors with a value of ± 1.

Another finite-precision error in the FFT computation is the overflow problem. Methods such as switching to saturation mode and using guard bits of the accumulator were explained in Chapter 3. A more efficient technique for preventing overflow in FFT computations is to scale signals at different stages. The scaling of the input can be derived by examining the DFT operations given in Eq. (8.1.2). Since the twiddle factors have a magnitude of 1, we only need to scale the input signals. That is, $|x(n)| < (1/N)$ for all n to guarantee that the output $|X(k)| < 1$ for all k. The scaling by $1/N$ in a radix-2 FFT is the same as shifting right $x(n)$ by M bits, where $M = \log_2 N$. For example, performing a 512-point FFT requires the input to be shifted right by 9 bits, which reduces the precision of the arithmetic substantially.

A more practical approach is to scale the signal at each stage by 0.5. The scaling by 0.5 is equivalent to right-shifting 1 bit at the input of each stage. The next section examines the effect of quantizing the FFT and the effect of scaling in MATLAB.

8.4.4 Evaluation of a Quantized Fast Fourier Transform Using MATLAB

In this section, FDATool is used to design, analyze, and implement a quantized FFT algorithm. The following MATLAB command constructs a quantized FFT object:

```
F = qfft
```

The quantized FFT object is created and displayed in the main window as follows:

```
F =
                  Radix = 2
                 Length = 16
      CoefficientFormat = quantizer('fixed', 'round', 'saturate',
                          [16  15])
            InputFormat = quantizer('fixed', 'floor', 'saturate',
                          [16  15])
           OutputFormat = quantizer('fixed', 'floor', 'saturate',
                          [16  15])
     MultiplicandFormat = quantizer('fixed', 'floor', 'saturate',
                          [16  15])
          ProductFormat = quantizer('fixed', 'floor', 'saturate',
                          [32  30])
              SumFormat = quantizer('fixed', 'floor', 'saturate',
                          [32  30])
       NumberOfSections = 4
            ScaleValues = [1]
```

The preceding code shows that the default value for the quantized FFT is a 16-point, radix-2 FFTs for the fixed-point representations and arithmetic. The NumberOfSect-ions (M) is 4 for the four stages of the 16-point radix-2 FFT, and no scaling is applied to the input at each stage by setting ScaleValues = [1].

The user can change the default settings by specifying new values with the following syntax:

```
F = qfft('Property1',Value1, 'Property2',Value2, ...)
```

For example, we can change the 16-point, radix-2 FFT to a 64-point, radix-4 FFT by using:

```
F = qfft('Radix',4, 'Length',64)
```

The user can also specify the data formats used in computing the FFT algorithm with the following quantizer commands:

```
F.CoefficientFormat = quantizer('fixed', 'round', 'saturate',[16 15])
F.InputFormat = quantizer('fixed', 'round', 'saturate',[16 15])
```

```
F.MultiplicandFormat = quantizer('fixed', 'round', 'saturate',[16 15])
F.OutputFormat = quantizer('fixed', 'round', 'saturate',[16 15])
F.ProductFormat = quantizer('fixed', 'round', 'saturate',[32 31])
F.SumFormat = quantizer('fixed', 'round', 'saturate',[32 31])
```

These commands set the coefficients, input, output, and multiplicands to the Q.15 format and set the product and sum to the Q.31 format.

Finally, the user can set the different scaling values to different stages of the quantized FFT. For example, in a 16-point, radix-2 FFT, there are four stages of butterfly computation. The elements in the scale vector [0.5 0.5 0.5 0.5] can be used to scale the input at the corresponding stage by setting ScaleValues = [0.5 0.5 0.5 0.5]. We can use a simple example to illustrate several important arithmetic properties of the quantized FFT.

Example 8.3

A sinewave is generated and the FFT is computed using the following MATLAB script (example8_3.m):

```
M= 128;                      % length of FFT
x =sin(2*pi*(0:127)*0.1);    % generate a sinewave
X = fft(x);                  % compute FFT
magX=abs(X);                 % magnitude response
phaseX=phase(X);             % phase response
figure(1);
plot(magX); title('Magnitude plot using double precision');
figure(2);
plot(phaseX); title('Phase plot using double precision');
```

The default double-precision floating-point arithmetic is used by MATLAB.

We apply the Q.15 format to the coefficient, input, output, and multiplicand, and we apply the Q.31 format for the product and sum. The scale value 1 (without scaling) is used. The new quantized FFT parameters are set as follows:

```
Fq15=qfft('length',M,'radix',2);
Fq15.CoefficientFormat=quantizer('fixed', 'round', 'saturate',
                                 [16 15]);
Fq15.InputFormat=quantizer('fixed', 'round', 'saturate',[16 15]);
Fq15.MultiplicandFormat=quantizer('fixed', 'round', 'saturate',
                                  [16 15]);
Fq15.OutputFormat=quantizer('fixed', 'round', 'saturate',[16 15]);
Fq15.ProductFormat=quantizer('fixed', 'round', 'saturate',[32 31]);
Fq15.SumFormat=quantizer('fixed', 'round', 'saturate',[32 31]);
Fq15.ScaleValues=1;    % no scaling at each stage
Xq15=fft(Fq15,x);      % compute fixed-point FFT
magXq15=abs(Xq15);     % fixed-point magnitude response
phaseXq15=phase(Xq15); % fixed-point phase response
figure(1), hold on;
plot(magXq15,'r');
figure(2), hold on;
plot(phaseXq15,'r');
```

A fixed-point FFT is computed, and the result is compared with the result using the double-precision floating-point format. Figure 8.17(a) shows the magnitude responses and Fig. 8.17(b) shows the phase responses of double-precision and Q.15 implementations. It is obvious that the FFT using Q.15 format without scaling is not satisfactory.

A warning report generated by the `qreport` command for the Q.15 quantized FFT is listed as follows:

	Max	Min	NOverflows	NUnderflows
Coefficient	1	-1	7	6
Input	0.9511	-0.9511	0	25
Output	1	-1	32	1
Multiplicand	1	-1	146	10
Product	1	-1	12	0
Sum	2	-2	399	0

Magnitude plot using double precision vs. Q15

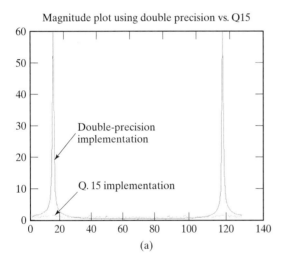

(a)

Phase plot using double precision vs. Q15

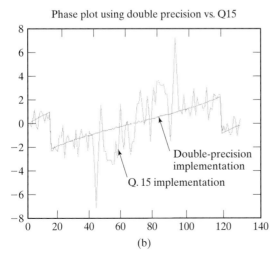

(b)

Figure 8.17   Magnitude and phase responses of a sinewave using the double-precision, floating-point and Q.15 formats: (a) Magnitude responses and (b) phase responses

This report states the number of overflows and underflows that occurred in different parts of the FFT computation. In addition, the maximum and minimum values of the parameters are also listed.

Modify the setting to Q.31 format for the coefficient, input, output, and multiplicand, and use the Q.63 format for the product and sum. The results are very similar to those using Q.15 format.

We continue the experiment by using the single-precision floating-point format. The following MATLAB commands are used for the experiment:

```
Fsingle=qfft('length',M,'radix',2);
Fsingle.quantizer = {'single'};
Xsingle = fft(Fsingle,x);
magXsingle=abs(Xsingle);
phaseXsingle=phase(Xsingle);
figure(1), hold on;
plot(magXsingle,'c');
figure(2), hold on;
plot(phaseXsingle,'c');
qreport(Fsingle);
```

A report generated by the qreport command is listed as follows:

	Max	Min	NOverflows	NUnderflows
Coefficient	1	-1	0	0
Input	0.9511	-0.9511	0	0
Output	48.76	-48.76	0	0
Multiplicand	30.03	-30.03	0	0
Product	17.89	-24.12	0	0
Sum	48.76	-48.76	0	0

There is no overflow and underflow using single-precision format, and the magnitude and phase match those of the double-precision format.

We can obtain accurate FFT results if we use proper scaling factors at each stage of the fixed-point (Q.15) implementation. It is interesting to examine the effects of using different scaling factors of [0.5 0.5 0.5 0.5 0.5 0.5 0.5], [1/128 1 1 1 1 1 1], and [1 1 1 1 1 1 1/128]. We can also evaluate the performance of the quantized IFFT using single-precision floating-point and Q.15 fixed-point formats.

8.5 FIXED-POINT IMPLEMENTATIONS

Most modern DSP processors support efficient implementation of numerically intensive algorithms such as the FFT. The powerful indirect-addressing indexing scheme facilitates the access of FFT butterfly legs with different spans. The repeat-block instruction reduces the looping overhead in FFT algorithms that heavily use program loops. In addition, bit-reversal addressing mode is specially designed to support FFT algorithms.

In general, the input sequence of the FFT routine is assumed to be complex. As discussed in Section 8.4.2, symmetric properties can be employed to compute the

FFT efficiently when the input is a real-valued signal. In this case, the original N-point real sequence can be packed into an $N/2$-point complex sequence, as shown in Eq. (8.4.7). An $N/2$-point FFT is performed on the packed complex sequence, and the resulting $N/2$-point complex output is unpacked into the N-point complex sequence, which corresponds to the FFT of the original N-point real-valued input signal. By using this technique, the FFT size can be reduced by half, and the speed becomes almost twice as fast as the general FFT algorithm. The principle of this technique is introduced in Section 8.4.2, and the detailed implementation using TMS320C54x processors is given in the next section.

8.5.1 Implementation Using the TMS320C54x

The C54x processor has special features to implement FFT efficiently. These features include a fast multiply-add computation unit, bit-reversal and circular-buffer addressing, block-repeat and parallel instructions, and a large internal memory. When implementing a DIT radix-2 FFT, a bit-reversal routine scrambles the N input samples to the bit-reversal order so that the output of the FFT is in natural order. The bit-reversal routine is listed as follows:

```
Bit_rev:
    SSBX    FRCT                    ; turn on fractional mode
    MVDK    data_input_addr,AR3     ; AR3 -> address of input array
    MVMM    data_bitrev_addr,AR2    ; AR2 -> address of bit-reversed
                                    ; array
    STM     #FFT_size-1,BRC         ; block repeat counter = FFT size
    STM     #FFT_size,AR0           ; AR0 = half of circular buffer
    RPTB    bit_rev_end-1
    MVDD    *AR3+,*AR2+
    MVDD    *AR3-,*AR2+
    MAR     *AR3+0B                 ; bit-reversal addressing
Bit_rev_end:
    RET
```

The code rearranges the input buffer pointed at by AR3 in bit-reversal order. Another buffer pointed at by AR2 stores the bit-reversed data. For example, to implement the FFT of 16 samples, the constant FFT_size must be set to half the data samples (i.e., $N/2 = 8$) and stored in AR0. After every access of the data, AR3 is incremented by $N/2$ using the MAR *AR3+0B instruction. Any carry from the addition of AR3 with AR0 is propagated to the right instead of to the left. This bit-reversed carry results in the generation of the bit-reversal address.

As mentioned earlier, if the input signal of length N is real, it can be packed into a complex vector with length $N/2$ before performing the bit reversal for the $N/2$ complex FFT routine. An example for the case of $N = 16$ is illustrated in Fig. 8.18. As shown in Eq. (8.4.7), the even-indexed real inputs $\{y(0), y(8),\ y(4), y(12),\dots\}$ form the real part of the $x(n)$ inputs $\{x.re(0), x.re(4), x.re(2), x.re(6),\dots\}$, and the odd-indexed real inputs $\{y(1), y(9), y(5), y(13),\dots\}$ form the imaginary part of the $x(n)$ inputs $\{x.im(0), x.im(4), x.im(2), x.im(6),\dots\}$. After this packing process, an 8-point complex FFT is performed on the packed complex sequence

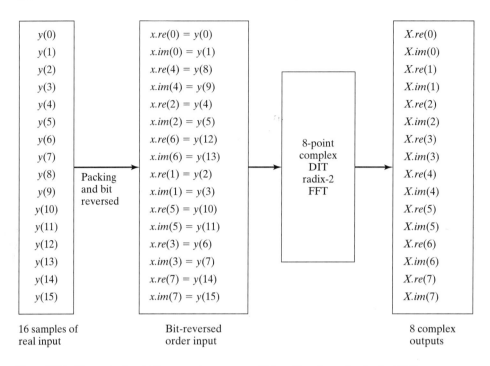

Figure 8.18   The packed and bit-reversed sequences of 16 real input samples for the FFT

$x(n)$. The DIT FFT outputs are $N/2$ complex samples in natural order. Since the FFT algorithm uses in-place computation, the resulting sequence $X(k)$ uses the same data buffer. Further steps to derive the FFT of $y(n)$ are described in Section 8.4.2.

The core of the FFT algorithm is the butterfly computation. The C54x assembly code used in implementing the butterfly computation is listed as follows:

```
Butterfly:
      LD    *WR,T            ; WR is the real part of twiddle
                             ; factor = T
      MPY   *low+,A          ; A = low_re*WR
      MACR  *WI+0%,*low-,A   ; A=(low_re*WR+low_im*WI) || ;low-> low_re
      ADD   *up,16,A,B       ; B=(low_re*WR+low_im*WI)+up_re
      ST    B,*up            ; replaced up_re
||    SUB   *up+,B           ; B=up_re-(low_re*WR+low_im*WI)||up->up_im

      LD    *WI,T            ; WI is the imaginary part of twiddle
                             ; factor
      ST    B,*low           ; replace low_re
||    MPY   *low+,A          ; A=low_re*WI || low->low_im

      MASR  *low,*WR+0%,A    ; A=(low_re*WI-low_im*WR)
      ADD   *up,16,A,B       ; B =(low_re*WI-low_im*WR)+up_im
      ST    B,*low+          ; replace low_im ||low->low_re
||    SUB   *up,B            ; B = up_im-(low_re*WI-low_im*WR)
```

```
     LD    *WR,T
     ST    B,*low+           ; replace up_im ||up->up_re
Butterfly_end:
     RET
```

Note that the preceding code only implements the butterfly routine shown in Fig. 8.3, which consists of multiplying the lower inputs of the butterfly (low_re and low_im) with the twiddle factors (WR and WI) and performing addition and subtraction with the upper inputs of the butterfly (up_re and up_im). The computed complex results replace the memory locations occupied by the inputs for the in-place computation. Additional code is required to implement different groups and stages of the FFT algorithm, as well as to unpack output elements. The complete assembly code can be obtained from [11].

The C54x is supported by the TMS320C54x DSPLIB [12], which contains a suite of comprehensive and optimized FFT and IFFT assembly routines (cfft, cifft, rfft, and rifft) for implementing 16-bit, radix-2 DIT FFTs and IFFTs for both complex and real inputs and also supports 32-bit, complex, radix-2 DIT FFT and IFFT functions (cfft32 and cifft32), as listed in Table 8.3. This library also

TABLE 8.3   Benchmark of *N*-Point FFT/IFFT Routines in the TMS320C54x DSP Library [12]

FFT/IFFT function	Description	Cycle count	Code size .text section	Data size .sintab section
cfft	Complex radix-2 DIT FFT with optional scale factor 1/N	8,542[N = 256] 19,049[N = 512] 42,098[N = 1,024]	343[N = 256] 391[N = 512] 439[N = 1,024]	367[N = 256] 750[N = 512] 1,517[N = 1,024]
cifft	Complex radix-2 DIT IFFT with optional scale factor 1/N	Same as cfft	Same as cfft	Same as cfft
cfft32	32-bit complex radix-2 DIT FFT (similar to cfft)	30,059[N = 256] 144,205[N = 512] 371,312[N = 1,024]	498[N = 256] 521[N = 512] 544[N = 1,024]	764[N = 256] 1,514[N = 512] 3,050[N = 1,024]
cifft32	32-bit complex radix-2 DIT IFFT (similar to cifft)	29,446[N = 256] 142,724[N = 512] 361,469[N = 1,024]	498[N = 256] 521[N = 512] 544[N = 1,024]	764[N = 256] 1,514[N = 512] 3,050[N = 1,024]
rfft	Real radix-2 DIT FFT (in place) with optional scale factor 1/N	5,470[N = 256] 11,881[N = 512] 25,716[N = 1,024]	357[N = 256] 405[N = 512] 453[N = 1,024]	367[N = 256] 750[N = 512] 1,517[N = 1,024]
rifft	Real IFFT (in place)	Same as rfft	Same as rfft	Same as rfft
cbrev	Obtain bit-reversal order array	3N + 23 (off place), 13N − 5 (in place)	50	2N (off place), N (in place)

includes the bit-reversal routine cbrev, which rearranges the input complex vector into an output vector in bit-reversed order. The input sequence needs to be bit-reversed using the cbrev routine before using the cfft or cfft32 function. The cbrev routine is also used before applying the rfft routine, which is equivalent to implementing a complex FFT cfft of size $N/2$. The rfft routine achieves a similar computational savings as the one described in Section 8.4.2 on implementing the FFT of an N-point real-value sequence using an $N/2$-point complex FFT algorithm.

In Table 8.3, N is the size of the FFT for the complex FFT or is half the FFT size for the real FFT. The benchmark shows that a 1,024-point complex FFT (cfft) needs approximately 0.55 msec to execute on a 100 MHz C54x processor, whereas, a 1,024-point real FFT (rfft) only requires 0.32 msec. These results are obtained by adding the cycles of cbrev (in-place) and cfft (or rfft) and then multiplying the result by the processor cycle time of 10 nsec. From the benchmark, a substantial increase in execution time (by more than four times) occurs when using 32-bit precision routines.

Table 8.3 also shows the benchmark for code size and data size. The code size increases marginally when moving from 16-bit complex FFT/IFFT routines to 32-bit complex FFT/IFFT routines. Data memory is used to store the twiddle factors. Data memory requirement is the same in the 16-bit complex and real FFT/IFFT cases, but doubles when using 32-bit precision complex FFT/IFFT routines. It is important to note that in-place computation reduces the data-buffer size from $2N$ to N, but at the cost of increasing the cycle count from $3N$ to $13N$.

Scaling can be carried out in both FFT and IFFT routines. When the SCALE option is set to zero, no scaling is performed. Otherwise, the scaling factor is set to N, which is implemented as scaling by 0.5 at each FFT stage. The reason for implementing progressive scaling is discussed in Section 8.4.4. For example, if a 256-point FFT is scaled at the input by 1/256, it immediately loses 8 bits of precision. A better approach is to perform progressive scaling by 0.5 at each stage. When using the 32-bit version of the FFT/IFFT routines cfft32 and cifft32 with 32-bit extended-precision input, it is possible to scale the input by N and still maintain good 24-bit precision.

8.5.2 Implementation Using the TMS320C2000

As discussed in Section 4.2, C2000 processors provide a bit-reversal-addressing mode for implementing radix-2 FFT algorithms. As with the C54x, this addressing mode supports incrementing or decrementing by an index amount using a reversed carry bit for implementing bit-reversal addressing. The value of the index amount must be stored in AR0. After the instruction uses the contents of the current auxiliary register as the data-memory address, the contents are incremented or decremented by the index amount. The addition or subtraction process is accomplished with the carry-bit propagation to the right instead of to the left as in normal arithmetic.

Bit-reversal addressing efficiently rearranges the data samples from the natural order to the bit-reversal order for a radix-2 FFT. For example,

```
LT  *BR0+   ; increment by AR0, adding with reversed carry
LT  *BR0-   ; decrement by AR0, subtracting with reversed carry
```

These instructions load TREG with the contents of the data memory pointed at by the current AR and then add (or subtract) the contents of AR0 to the contents of the current AR. The direction of the carry propagation in the ARAU is reversed. A typical use of this addressing mode requires AR0 initially to be set to $N/2$ and the current AR value to be set to the base address of the first sample in the buffer.

The C2000 FFT Library [13] contains 128-, 256-, and 512-point real/complex FFT programs, which are summarized in Table 8.4. The input sequence is assumed to be complex. In many real-world applications, the data sequences to be processed are real valued. Even though the data is real, complex-valued DFT algorithms can still be used. One simple approach for creating the complex sequence from the real sequence is to use the real data for the real parts of the complex sequence and zeros for all the imaginary parts. Thus, the complex FFT can be applied directly to this complex sequence.

However, this method is not efficient because it consumes $2N$ memory locations (real and imaginary parts) for an N-point sequence. As mentioned earlier, one optimized FFT algorithm for a $2N$-point real-data sequence is the packing algorithm. In this method, the original $2N$-point real sequence is packed as an N-point complex sequence, and an N-point complex FFT is performed on the packed complex sequence. The resulting N-point complex output is unpacked into another $(N + 1)$-point complex sequence, which corresponds to frequency index k from 0 to N. Spectral bins from 0 to N are sufficient because the remaining bins from $(N + 1)$ to $(2N - 1)$ are complex conjugates of the corresponding spectral bins from $(N - 1)$ to 1. Therefore, the real FFT algorithm computes the FFT of the real sequence almost twice as fast as does the complex FFT algorithm.

8.5.3 Implementation Using the TMS320C55x

Similar to the C54x DSPLIB, the C55x DSPLIB [14] also provides a set of FFT/IFFT functions and a bit-reversal routine for both real and complex FFTs. Table 8.5 shows the benchmark for the cycle count and code size (bytes). Compared with the benchmark of the C54x DSP, the C55x complex FFT and IFFT routines

TABLE 8.4   List of FFT Assembly Programs for the C2000 [13]

Module name	Description
cfft64a	64-point complex FFT
cfft128a	128-point complex FFT
cfft256a	256-point complex FFT
cfft512a	512-point complex FFT
rfft128sa	128-point real FFT
rfft256sa	256-point real FFT
rfft512sa	512-point real FFT

TABLE 8.5   Benchmark of FFT/IFFT Routines in the TMS320C55x DSP Library [14]

FFT/IFFT function	Description	Cycle count	Code size .text section
cfft	Complex radix-2 DIT FFT with optional scale factor; in-place computation; required `twiddle.inc`, which contains the twiddle factors	With scaling: $6,183[N = 256]$ $13,638[N = 512]$ $30,043[N = 1,024]$ Without scaling: $5,563[N = 256]$ $12,434[N = 512]$ $27,689[N = 1,024]$	With scaling: 490 Without scaling: 361
cifft	Radix-2 DIT IFFT with optional scale factor	Very close to cfft	Very close to cfft
rfft	Radix-2 real DIT FFT, equivalent to $N/2$ complex FFT (cfft)		
rifft	Radix-2 real DIT IFFT, equivalent to $N/2$ complex IFFT (cifft)		
cbrev	Bit reversal for both complex and real FFT.	$2N$ (off place) $4N + 6$ (in place)	81

generally need fewer cycles and less code size. The real FFT function, rfft, is a macro (defined in dsplib.h) that calls cfft and cbrev as follows:

```
#define rfft(x,nx,type) \
    (\
        cfft_##type(x,nx/2),\
        cbrev(x,x,nx/2),\
        unpack(x,nx)\
    )
```

An example of using rfft in a fixed-point C program for computing FFT for experiments is given in Section 8.8.6.

Bit-reversal addressing is a special type of indirect addressing. It uses one of the auxiliary registers (AR0-AR7) as a base pointer and uses the temporary register T0 as an index register. The syntaxes for each of the two bit-reversal-addressing modes supported by the C55x are

```
*(ARn+T0B)    ; address = ARn, ARn = (ARn+T0)
*(ARn-T0B)    ; address = ARn, ARn = (ARn-T0)
```

When T0 is added to (or subtracted from) the auxiliary register ARn using bit-reversal addressing, the address is generated in a bit-reversed fashion, with the carry propagating from left to right instead of from right to left.

Bit-reversal addressing enhances execution speed for FFT algorithms. Given a complex FFT of size N, we can arrange the input vector (or output vector for DIF algorithms) in bit-reversed order by executing the following steps:

Step 1. Write 0 to the ARMS bit of status register 2 to select DSP mode for AR indirect addressing, and then use the `.arms_off` directive to notify the assembler of this selection.

Step 2. Determine the base pointer of the input vector that must be aligned to match the given vector format. If the real and imaginary parts of each complex element are stored at consecutive memory locations as Re-Im-Re-Im- ..., the $M + 1$ LSBs of the vector base address must be 0, where $M = \log_2 N$. If the real and imaginary data are stored in separated arrays, M bits must be 0.

Step 3. Load the index register T0 properly. If the real and imaginary parts of each complex element are stored at consecutive memory locations, $T0 = 2^M$; otherwise, $T0 = 2^{M-1}$.

Step 4. Ensure that the entire array fits within a 64K boundary.

We can have a bit-reversed array write over the original array (in place) or write over the array in a separate place in memory (off place). An example of an off-place bit-reversal operation is given as follows:

```
BCLR    ARMS        ; reset ARMS bit to allow bit-reversal addressing
.arms_off           ; notify assembler of ARMS bit = 0
RPTBLOCAL{
    MOV    dbl(*AR0+),AC0          ; AR0 points to input array
    MOV    AC0,dbl(*(AR1+T0B))     ; AR1 points to output array
}
```

In this example, an input array of N complex elements pointed at by AR0 is bit-reversed into an output array pointed at by AR1. If the input array is assumed to be a Re-Im-Re-Im arrangement, each complex element (Re-Im) of the input array is loaded into AC0. This element is then transferred to the output array using bit-reversal addressing. Although it requires twice the memory, off-place bit reversal (2 cycles per complex data sample) is faster than in-place bit reversal, which needs 4 cycles per complex data sample.

8.5.4 Implementation Using the TMS320C62x

As described in Chapter 4, the C62x processor contains two data paths with eight functional units that can be accessed in one cycle. The C62x processor has special SIMD instructions, such as ADD2 and SUB2, that allow dual 16-bit additions and subtractions to be performed in a single cycle to speed up the FFT and IFFT. However, the C62x processor does not support bit-reversal-addressing mode.

An efficient FFT algorithm for real sequences can be developed entirely in C. However, to achieve higher performance, some tasks and routines must be coded in linear assembly. The assembly optimizer assigns functional units and registers to

TABLE 8.6   Benchmark for the FFT/IFFT Routines in the TMS320C62x DSP Library [15]

FFT/IFFT function	Description	Cycle count	Code size
DSP_bitrev_cplx	Complex bit-reversal routine	$7(N/4 + 2) + 18$	352
DSP_radix2	Complex radix-2 DIF FFT	$\log_2 N \times (4N/2 + 7) + 34 + N/4$	800
DSP_r4fft	Complex radix-4 DIF FFT; input may be scaled by $1/N$	$\log_4 N \times (10 \times N/4 + 29) + 36 + N/4$	736
DSP_fft16x16r	Complex mixed radix FFT; scaled by 0.5 at each stage	$2.5N \times \text{ceil}[\log_4 N] - N/2 + 164$	1,344

each instruction and arranges the assembly instructions in parallel. Detailed C and assembly programs are available in [15].

Table 8.6 shows the complex FFT and IFFT functions that are available in the C62x DSPLIB. A potential problem for the C62x implementation is memory-bank hits. Therefore, the signal vector and the twiddle-factor coefficient vector should be in different data sections or memory spaces. No in-place FFT/IFFT algorithm is available, as shown in Table 8.6.

8.5.5 Implementation Using the TMS320C64x

As explained in Chapter 4, the C64x processor is upward (object-code) compatible with the C62x processor. In addition, the C64x processor has a wider memory interface for handling up to 128 bits and has more multiplier and arithmetic hardware. Furthermore, it provides advanced SIMD instructions that allow more data words to be executed simultaneously. The dual 16-bit arithmetic supported by six of the eight functional units is equipped with the bit-reversal instruction (BITR) that improves the speed of the bit-reversal operation in the FFT.

The C64x DSPLIB [16] also provides a set of FFT/IFFT routines (DSP_bitrev_cplx, DSP_radix2, and DSP_r4fft) that are backward compatible with the C62x library. In addition, extra functions are created in the C64x DSPLIB to take advantage of the added precision of the C64x processor. Table 8.7 shows the benchmark of using these FFT and IFFT routines. A speed improvement of 2-3 times over the C62x can be achieved using the optimized FFT and IFFT routines.

8.6 FLOATING-POINT IMPLEMENTATIONS

As evaluated in Section 8.4.4 using MATLAB, 32-bit, single-precision floating-point arithmetic is good for implementing FFT algorithms. In this section, we introduce the implementation of FFT algorithms on the TMS320C3x and TMS320C67x floating-point processors.

TABLE 8.7   Benchmark of the FFT/IFFT Routines in the TMS320C64x DSP Library [16]

FFT/IFFT function	Description	Cycle count	Code size (bytes)
DSP_bitrev_cplx	Backward compatible with C62x library		
DSP_radix2	Backward compatible with C62x library		
DSP_r4fft	Backward compatible with C62x library		
DSP_fft	Complex radix-4 DIF FFT; off place computation	$1.25N \times \log_4 N - 0.5N + 23 \log_4 N - 1$	984
DSP_fft16x16r	Complex mixed-radix 16x16-bit FFT with scaling; off place computation	$\text{Ceil}(\log_4 N - 1) \times (5N/4 + 25) + 5N/4 + 26$	868
DSP_fft16x16t	Complex mixed-radix 16x16-bit FFT with scaling; off place computation	$(10N/8 + 19) \times \text{ceil}(\log_4 N - 1) + 7(N/8 + 2) + 28 + N/8$	1,004
DSP_fft16x32 (similar to SP_fft16x16t)	Complex mixed-radix 16x32-bit FFT	$(13N/8 + 24) \times \text{ceil}[\log_4 N - 1] + 1.5(N + 8) + 27$	1,068
DSP_fft32x32 (similar to DSP_fft16x16t)	Complex mixed-radix 32x32-bit FFT	$[10(N/4 + 1) + 10] \times \text{ceil}[\log_4 N - 1] + 6(N/4 + 2) + 27$	932
DSP_fft32x32s (similar to DSP_fft16x16t)	Complex mixed-radix 32x32-bit FFT with scaling	$[10(N/4 + 1) + 10] \times \text{ceil}[\log_4 N - 1] + 6(N/4 + 2) + 27$	932
DSP_ifft16x32 (similar to DSP_fft16x16t)	Complex mixed-radix 16x32-bit FFT; off place computation	$[13N/8 + 25] \times \text{ceil}[\log_4 N - 1] + 1.5(N + 8) + 30$	1,064
DSP_ifft32x23 (similar to DSP_fft16x16t)	Complex mixed-radix 32x32-bit FFT; off place computation	$[10(N/4 + 1) + 10] \times \text{ceil}[\log_4 N - 1] + 6(N/4 + 2) + 27$	932

8.6.1 Implementation Using the TMS320C3x

As introduced in Section 5.2, there are eight extended-precision registers (R0-R7) that can be used as accumulators or general-purpose registers. For many applications, these registers are sufficient for the temporary storage of values, and there is no need to use memory locations. Arithmetic using these registers greatly increases programming efficiency. In addition, the repeat block implemented by the RPTB instruction reduces the looping overhead in FFT algorithms that use loops intensively. For example, the butterfly computation can be implemented in a block-repeat form, thereby saving execution time and program space.

The TMS320C3x provides powerful indirect addressing for accessing butterfly legs with different spans. The two index registers, IR0 and IR1, are used for indexing the contents of the auxiliary registers (AR0-AR7), thus easing the access of the butterfly legs and the twiddle factors. In addition, bit-reversal-addressing mode eliminates the need for swapping memory locations.

To implement a radix-2 FFT algorithm efficiently on the C3x, the base address of the signal buffer in memory must be on a 2^M ($M = \log_2 N$) boundary. That is, the first memory address has zeros for the last M bits. For example, in the case of a 1,024-point FFT, $M = 10$ such that the last 10 bits of the memory address are zeros.

An auxiliary register (ARn, n = 0, 1, ..., 7) is used to point at the physical location of data in the scrambled buffer. The index register IR0 contains a number equal to $N/2$. After every access of the data, ARn is incremented by IR0 using the operand *ARn++(IR0)B. This bit-reversal-addressing mode causes the contents of ARn to be incremented by the contents of IR0; however, if there is a carry in this addition, the carry propagates to the right instead of to the left. The result is the generation of addresses in bit-reversed order.

In implementing the complex FFT, the output is complex even when the input is real. The preceding description of bit-reversed-addressing mode assumes that the real and the imaginary parts are stored in separate arrays in memory. In this case, each of the real and imaginary arrays can be accessed as described. However, in most cases the real and imaginary data alternate in the same array, as shown in the second column of Fig. 8.18. In this arrangement, we have to set IR0 equal to N instead of to $N/2$. At every access, ARn points to the real part of the data. The imaginary part is located in the next higher location; thus, it can be easily accessed.

TABLE 8.8   TMS320C3x FFT Timing Benchmarks in Clock Cycles

N	Complex, radix-2	Complex, radix-4	Real, radix-2	Real, radix-2 IFFT
64	2,770	2,050	810	1,070
128	6,170		1,760	2,370
256	13,600	10,400	3,940	5,290
512	29,740		8,860	11,740
1,024	64,570	50,670	19,820	25,900

The complete C3x assembly programs for a (1) complex, radix-2 DIF FFT, (2) complex, radix-4 DIF FFT, (3) real, radix-2 FFT, and (4) real, radix-2 inverse FFT are given in the *TMS320C3x User's Guide* [17]. The C3x can execute FFT lengths of up to 1,024 (complex) or 2,048 (real) samples in on-chip memory. FFT timing benchmarks are summarized in Table 8.8. For the complex FFT, the radix-4 algorithm reduces the execution time by about 20-25% compared with the execution time of the radix-2 algorithm, depending on the FFT size N. A detailed discussion of the implementation of the FFT on the TMS320C30 and a listing of programs are available in [18].

8.6.2 Implementation Using the TMS320C67x

As introduced in Chapter 5, the C67x processor can execute C62x object code without modification. The weakness in executing floating-point arithmetic in the C67x processor are the longer latency (four cycles for single-precision multiplications and additions) and the need for more than one cycle of throughput for double precision as compared with fixed-point arithmetic. As with the C62x device, there is no bit-reversal addressing in the C67x instruction set.

DSP functions, such as the FFT routines described in Table 8.6, can be used for the C67x processor. In addition, the C6x compiler provides a set of intrinsics for performing special functions in assembly code. As explained in previous chapters, intrinsics can help to improve C-code efficiency. The C62x and C67x share a set of common intrinsics that perform various types of tasks [19].

8.7 APPLICATIONS

As mentioned earlier, the FFT is an important tool used in DSP applications. In this section, we introduce some commonly used FFT applications that include fast convolution, power-spectrum estimation, and STFT.

8.7.1 Fast Convolution

The concept of circular convolution and its relation to linear convolution is introduced briefly in Section 8.3.2. In this section, we focus on the implementation of linear convolution using the FFT. To perform linear convolution of the filter-impulse response $h(n)$ (of length N_h) with the N_x samples of signal sequence $x(n)$, the following steps are needed:

Step 1. Append $(N - N_x)$ and $(N - N_h)$ zeros to the sequences $x(n)$ and $h(n)$, respectively. The new sequences have the same N, where $N \geq N_x + N_h - 1$, and N is a power-of-two integer if a radix-2 FFT is used.

Step 2. Perform N-point FFT on $x(n)$ and $h(n)$ to obtain $X(k)$ and $H(k)$, respectively.

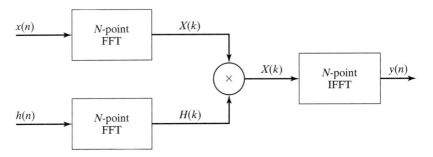

Figure 8.19   Fast convolution using the FFT and IFFT

Step 3. Multiply $X(k)$ and $H(k)$ in the frequency domain to obtain $Y(k) = X(k)H(k)$.

Step 4. Perform the IFFT of $Y(k)$ to form the output $y(n)$.

The preceding process, called fast convolution, is summarized in Fig. 8.19.

Fast convolution uses two FFTs, one IFFT, and N complex multiplications to produce N output samples. Therefore, the total number of real multiplications required is $4 \times ([3N/2]\log_2 N + N)$. When compared with computing $N = N_x + N_h - 1$ output samples using linear convolution that requires $N_x \times N_h$ real multiplications, FFT-based fast convolution requires fewer computations if

$$6N \log_2 N + 4N < N_x N_h. \tag{8.7.1}$$

In general, the computational saving is significant when the lengths of the sequences to be convoluted are long.

Example 8.4

An input sequence $x(n)$ of length 1,024 is convoluted with a 64-tap FIR filter of impulse response $h(n)$ using both time-domain and frequency-domain techniques in the following MATLAB script (example8_4.m):

```
x = randn(1,1024);
h = hanning(64)';
total_length = length(x)+length(h)-1;
y_t = conv(x,h);             % time domain linear convolution
X = fft(x,total_length);     % compute X(k)
H = fft(h,total_length);     % compute H(k)
Y = X.*H;                    % frequency domain multiplicaiton
y_f = ifft(Y,total_length);  % compute y(n) using IFFT

figure;
subplot(2,1,1), stem(y_t); title('Time domain linear convolution');
subplot(2,1,2),stem(real(y_f));title('Frequency domain fast
  convolution');
```

The results are shown in Fig. 8.20.

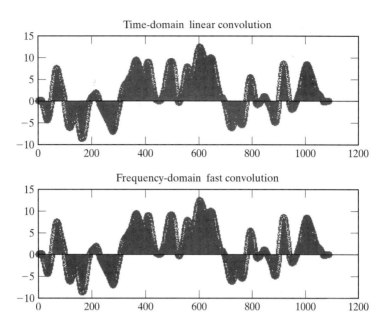

Figure 8.20   Results of time-domain linear convolution and frequency-domain fast convolution

The disadvantage of using FFT-based fast convolution is the increased memory locations for storing the signal sequence for block processing, whereas time-domain FIR filtering operates on each input sample as it arrives. In addition, the FFT approach also introduces a processing delay.

In some applications, the input signal $x(n)$ is too long for computing the FFT. A solution to this problem is to break down the input sequence into smaller blocks of size N, where N is a power-of-two integer, and each block is convoluted with the impulse response of the filter. This process produces a set of output sections that must be assembled to form the output sequence. Unfortunately, the process of segmentation, convolution, and recombination are not straightforward due to the edge effect of each block. The most commonly used techniques for the segmentation and recombination of the data are the overlap-save and overlap-add algorithms.

Overlap-Save Technique

The overlap-save technique is based on the fact that when circular convolution is performed, aliasing is introduced in certain portions of the output sequence. For example, when circular convolution is performed between the input block of length N and the filter-impulse response of length N_h (when $N > N_h$), an output block of length N is obtained. However, the first $(N_h - 1)$ samples are wrong and only the remaining $(N - N_h + 1)$ samples are the correct output samples.

To solve this problem, the overlap-save technique overlaps $(N_h - 1)$ input samples on each consecutive block. The output segments derived from the circular convolution between the overlapped input block with the filter-impulse response are truncated to avoid overlapping, followed by concatenation of these truncated blocks, as shown in Fig. 8.21. The steps for implementing the overlap-save technique are summarized as follows:

Step 1. Segment the long input sequence $x(n)$ into overlap-data blocks. Each block has N samples with the overlapping of $N_h - 1$ samples, as shown in Fig. 8.21. Transform these data blocks $x_i(n)$ to obtain $X_i(k)$ using N-point FFT.

Step 2. Perform an N-point FFT to obtain $H(k)$ for $k = 0, 1, \ldots, N - 1$. (The impulse response of the filter (length N_h) is zero-padded to form a coefficient vector of length N.)

Step 3. Perform $Y_i(k) = X_i(k)H(k)$ for $k = 0, 1, \ldots, N - 1$.

Step 4. Perform N-point IFFT on each $Y_i(k)$ to obtain $y_i(n)$ for $n = 0, 1, \ldots, N - 1$.

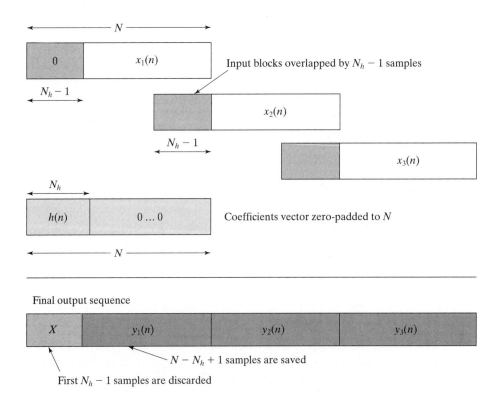

Figure 8.21   Overlap-save technique

Step 5. Discard the first $(N_h - 1)$ samples of every output block $y_i(n)$ and concatenate these blocks to form the final output sequence, as shown in the bottom row of Fig. 8.21.

An illustration of the overlap-save technique is given in the M-file `overlap_save.m`.

Overlap-Add Technique

As shown in Fig. 8.22, the overlap-add technique divides the input sequence $x(n)$ into nonoverlapping segments of length N_{xb}. These nonoverlap segments are expressed as

$$x_i(n) = x(n + iN_{xb}), \quad n = 0, 1, \ldots, N_{xb} - 1, \tag{8.7.2}$$

where $i = 1, 2, \ldots$ is the block index. In order to perform linear convolution, both the filter-impulse response and the data blocks are zero-padded to the length

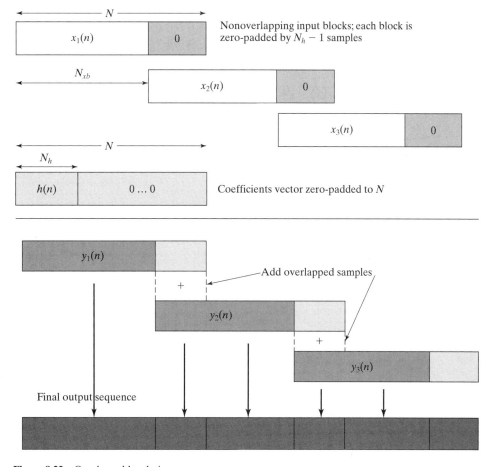

Figure 8.22 Overlap-add technique

$N (\geq N_{xb} + N_h - 1)$, where N is a power-of-two integer. This zero padding is followed by applying the N-point FFT of $h(n)$ and $x_i(n)$ to form $H(k)$ and $X_i(k)$, respectively. We multiply $X_i(k)$ from each block with $H(k)$ to generate $Y_i(k)$ and use N-point IFFT to obtain the output sequences $y_i(n)$. The output segment, $y_i(n)$, overlaps with the following segment, $y_{i+1}(n)$, as shown in Fig. 8.22. These overlapping samples are added to form the final output sequence, $y(n)$.

An illustration of the overlap-add technique is given in the MATLAB program `overlap_add.m`. MATLAB also provides the function `fftfilt` for performing efficient FIR filtering using the overlap-add technique. This function can be used as follows:

```
y = fftfilt(h,x)      % MATLAB selects an efficient order N
```

or

```
y = fftfilt(h,x,N)    % user specify the order N
```

We observe that the overlap-save method discards $(N_h - 1)$ samples from each block. However, the overlap-add method needs to perform extra additions on the overlapped sections.

8.7.2 Power-Spectrum Estimation

Simple power-spectrum estimation uses a single block of data and performs FFT to analyze the frequency components of signals. Power-spectrum (or periodogram) estimation is defined as

$$P(k) = \frac{1}{N}|X(k)|^2 = \frac{1}{N}X(k)X^*(k), \qquad (8.7.3)$$

where $X(k)$ is the FFT of the input sequence $x(n)$. This periodogram is adequate for spectral analysis only if the signal does not change its characteristics. For time-varying signals, single block FFT-based power-spectrum estimation is not suitable for extracting or displaying time events.

Example 8.5

A chirp signal sweeps linearly from 0 Hz to 100 Hz in 1 second. The chirp signal is sampled at 1,000 Hz, and the power-spectrum estimation of this signal can be obtained using the following MATLAB code (`example8_5.m`):

```
t=0:0.001:3;                  % 3 seconds at 1kHz sampling rate
x = chirp(t,0,1,100);         % chirp signal sweep linearly
N = 1024; Fs=1000;            % using 1024-point FFT & Fs=1000Hz
pse = abs(fft(x,N)).^2/N;     % power spectrum estimation
pse_dB = 10*log10(pse);       % power spectrum in dB
figure; plot(0:1000/N:500,pse_dB(1:513));
title('Power Spectrum Estimation of x');
xlabel('Frequency axis(Hz)'); ylabel('Power (dB)');
```

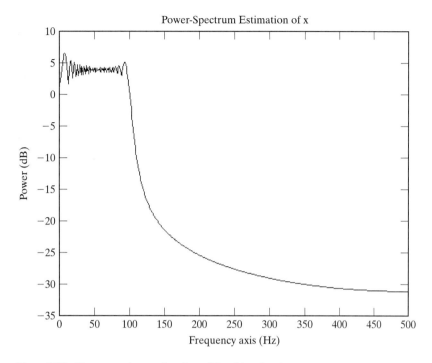

Figure 8.23   Power-spectrum estimation of the chirp signal

Figure 8.23 shows the magnitude spectrum of the chirp signal. The power spectrum only displays the frequency contents from 0 to around 100 Hz.

An alternate method for estimating the power spectrum involves the use of the following MATLAB function:

```
psd(x,N,Fs,window,noverlap)
```

The variable `x` is the input vector, `N` is the length of the FFT, `Fs` is the sampling frequency, `window` indicates the type of window to be used, and `noverlap` denotes the number of samples that is overlapped between the adjacent segments. The `psd` function estimates the power-spectrum density (or power-density spectrum) using Welch's modified (averaged) periodogram method. Figure 8.24 shows the process of segmenting the long data sequence into smaller blocks and applies a bell-shaped window $w(n)$ (as described in Table 8.2) to the time-domain segments to reduce spectral leakage. Overlapping the data segments by M samples increases the total number of segments, resulting in a more accurate averaged spectrum for a given sequence. Typically, 50% $(M = N/2)$ overlap between successive segments is used in most applications. Therefore, the maximum number of segments I that can fit into a data sequence of length L is given as

$$I_{max} = \text{int}\left(\frac{L - N}{M} + 1\right),$$
(8.7.4)

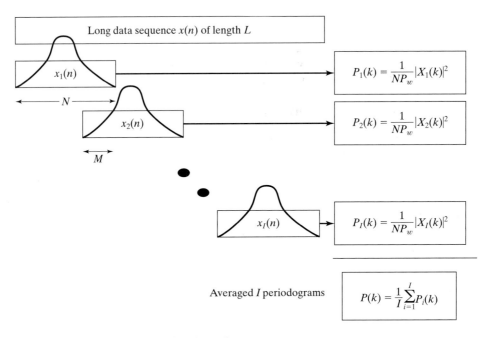

Figure 8.24   Welch algorithm for a long input-data sequence

where int(.) is the integer operator. An N-point FFT is then applied to each windowed segment, and the ith power-spectrum estimation (or periodogram) of each segment is described on the right-hand side of Fig. 8.24.

The contribution of the window function in estimating the power spectrum at each segment must be compensated. The power of the window function can be expressed as

$$P_w = \frac{1}{N}\sum_{n=0}^{N-1} w^2(n).\tag{8.7.5}$$

This power can be used to normalize the power spectrum expressed as

$$P(k) = \frac{1}{NIP_w}\sum_{i=1}^{I}|X_i(k)|^2, k = 0, 1, \ldots, N - 1.\tag{8.7.6}$$

This Welch algorithm results in a smoother power-spectrum estimation when compared with the periodogram that uses only one segment of data and also provides a more accurate estimation. However, the frequency resolution of the Welch method is poorer compared with the periodogram due to the usage of smaller data segments for the FFT and the effect of windowing. Therefore, a suitable segment length must be chosen to achieve good resolution and spectrum averaging.

The preceding power-spectrum estimation methods assume that the signal is stationary. However, in most real-world signals such as speech and audio, this assumption is not true. This fact can be seen in Fig. 8.23, where only the frequency of the chirp signal can be observed and where there is no time information on exactly when the frequency components are changed. The FFT only extracts frequency information and is suitable for analyzing stationary signals. For nonstationary (or time-varying) signals, we compute a short-time FFT-based spectrum that measures the contents over short time intervals.

8.7.3 Short-Time Fourier Transform

The STFT is similar to Welch's averaged periodogram. The input data is windowed into smaller overlapping segments of length N, and the FFT spectra of these short-time segments $x_i(n)$ are computed. Instead of averaging these spectra as in Welch's periodogram, the STFT displays the magnitude of the spectrum $|X_i(k)|$ as a function of the frequency index k and the block (time) index i in a three-dimensional plot. This three-dimensional plot is called the spectrogram, which normally uses a color bar to indicate the strength of the magnitude at different frequencies and times. For example, we can perform the spectrogram of the chirp signal x given in Example 8.5 using the following MATLAB function:

```
specgram(x,[],Fs); colorbar;
```

This function displays the spectrogram with a color bar that uses a darker color to represent the higher magnitude and a lighter color to represent the lower magnitude of the spectrum, as shown in Fig. 8.25. The default overlapping between segments is 50%. Note that the x-axis is the time-axis, which increases linearly from left to right, while the y-axis represents the frequency, which ranges from 0 Hz (bottom) to $f_s/2 = 500$ Hz (top). Figure 8.25 shows that a maximum-magnitude (dark red) line starts from the bottom-left corner and moves toward the top-right corner of the plot. This signal is interpreted as having an increasing frequency as time increases, which corresponds to the characteristic of the chirp signal.

The default length of the time window is 256 samples or the length of x if the length is shorter. A longer window produces a coarser time resolution Δt, but a finer frequency resolution Δf. A shorter window produces a finer time resolution, but a poorer frequency resolution. This time-frequency tradeoff, known as the Heisenberg uncertainty principle, is expressed as

$$\Delta t \Delta f \geq 1/4\pi, \tag{8.7.7}$$

which states that the time-bandwidth product is a constant. Therefore, a practical challenge is to determine the correct window size to use for a given application. For example, a typical window of 30 msec (or 240 samples when the sampling rate is 8,000 Hz) may be used for speech applications.

8.7.4 Wavelet Transform

A better solution to this time-frequency resolution tradeoff is to use the wavelet transform [10], which has the flexibility to vary the window size for the best resolution in both time and frequency. In the wavelet transform, a shorter window is used for analyzing high frequency components. This short window results in good time

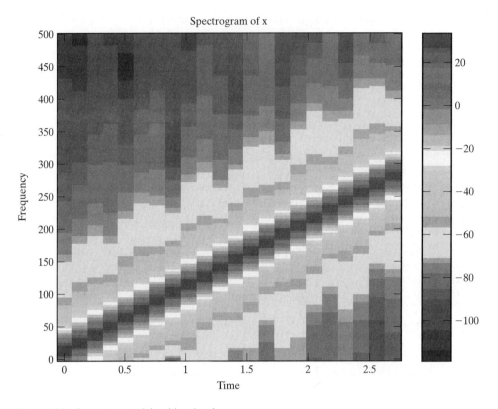

Figure 8.25   Spectrogram of the chirp signal

resolution, but poor frequency resolution, which is valid since high-frequency components usually vary rapidly. When analyzing low-frequency components, a longer window is used, which results in poor time resolution, but provides better frequency resolution. This tradeoff is acceptable because low-frequency components normally change slowly with time. When comparing the STFT and wavelet transform in a two-dimensional plot, as shown in Fig. 8.26, it is clear that STFT uses a constant

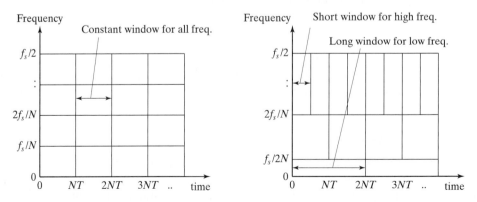

Figure 8.26   Time-frequency resolution for the STFT (left) and wavelet transform (right)

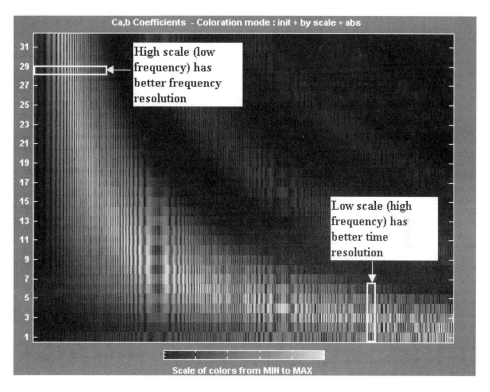

Figure 8.27   Wavelet in terms of a scale-time plot

window for both time and frequency, but the wavelet transform uses a shorter window for higher frequency ranges.

In wavelet transform, the frequency axis is replaced by the scale axis (w), which is inversely related to the frequency f as

$$w = \frac{\alpha}{f}, \tag{8.7.8}$$

where α is a constant. Therefore, the wavelet transform is presented in the time-scale plot shown in Fig. 8.27. The y-axis indicates the scale axis, where a higher scale indicates a lower frequency, and vice versa. The x-axis indicates the time axis from left to right. Notice that different time-scale resolutions are used when the chirp signal increases its frequency over time. Time resolution is good when the scale is small (high frequency), whereas frequency resolution is good when the scale is high (low frequency). Figure 8.27 shows the wavelet transform of a chirp signal. This wavelet transform is plotted using the MATLAB Wavelet Toolbox, which can be opened by typing wavemenu in the MATLAB command window. A GUI menu is displayed that allows the user to plot the 1-D continuous-wavelet transform using different wavelet families. A detailed description of the Wavelet Toolbox is given in [20], and a more detailed discussion on the theoretical aspects of this advanced topic can be found in [10]. See the problems at the end of this chapter for a comparison between the STFT and the wavelet transform.

8.8 EXPERIMENTS AND PROBLEMS

In this section, we examine the complete design and implementation processes for an FFT-based power-spectrum estimation using all of the tools described previously. We start with high-level MATLAB tools and end by using optimized assembly code for fixed-point C54x and C55x processors.

8.8.1 Spectrum Estimation Using MATLAB

We first investigate the design of power-spectrum estimation using MATLAB with double-precision floating-point arithmetic. In this experiment, we generate and mix three sinusoidal signals at 1 kHz, 1.2 kHz, and 1.4 kHz using the MATLAB program sine3.m and save the mixed signal in the file in.dat. We can use MATLAB functions in the Signal Processing Toolbox for computing the periodogram and the averaged periodogram, such as the Bartlett and Welch periodograms introduced in Section 8.7.2. The MATLAB program psd_881.m is listed as follows:

```
load in.dat -ascii;     % load the signal in in.dat data file
NFFT = 2^(nextpow2(length(in)));
NFFT1 = 256;    % window of 256-sample is used for averaged
                % periodogram
Fs = 10000;
figure;
% PSD plot (1): Periodogram
[Pxx1,F1] = periodogram(in,[],[],Fs);
subplot(3,1,1); plot(F1,10*log10(Pxx1));
title('Power spectrum of sine waves (Periodogram)');
grid on; axis([0 5000 -100 50]);

% PSD plot (2): Bartlett using rectangular window and no overlapping
[Pxx2,F2] = pwelch(in,boxcar(NFFT1),0,NFFT1,Fs);
subplot(3,1,2); plot(F2,10*log10(Pxx2));
title('Power spectrum of sine waves (Bartlett)');
grid on; axis([0 5000 -100 50]);

% PSD plot (3): Welch using Hamming window and 50% overlapping
[Pxx3,F3] = pwelch(in,hamming(NFFT1),NFFT1/2,NFFT1,Fs);
subplot(3,1,3); plot(F3,10*log10(Pxx3));
title('Power spectrum of sine waves (Welch)');
grid on; axis([0 5000 -100 50]);
```

As shown in Fig. 8.28, the periodogram (top plot) results in the best frequency resolution since a 512-point FFT is computed over the entire 500 samples of data. No window (or the rectangular window) is used in this case. The periodogram provides adequate results for estimating the spectrum of narrowband signal, but becomes very noisy for broadband signal. A more reliable power spectrum estimator is to use the averaging periodogram methods explained in Section 8.7.2.

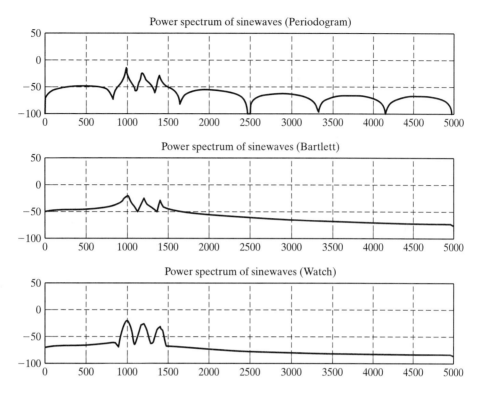

Figure 8.28   Power-spectrum plots using periodogram method (top), Bartlett method (middle), and Welch method (bottom)

The middle plot of Fig. 8.28 shows the result from the Bartlett-averaging periodogram, which computes the spectrum by averaging the spectra of two nonoverlapping segments of 256 samples each. No window and no overlapping between segments are used in the Bartlett method. The spectrum obtained from the Welch periodogram is shown in the bottom plot of Fig. 8.28. In this algorithm, the spectrum is computed by averaging the spectra of four overlapping (50% overlapped) segments of 256 samples each. In addition, the Hamming window is used to reduce sidelobe effects. The Welch method provides a smoother spectrum and has larger sidelobe attenuations compared with the Bartlett method. Therefore, the averaging method is commonly used for providing a better spectrum estimation over the periodogram.

The preceding power spectrum can also be implemented using the basic fft function in MATLAB. The user can refer to the M-file psd_ex.m to see a detailed implementation of the power-spectrum estimation using several basic functions.

We can simulate the 16-bit (Q.15) implementation further and analyze the quantization error of the FFT-based power-spectrum estimation using fixed-point arithmetic, as explained in Section 8.4.4. This program is given in the M-file psd_ex_quant.m, and the user can experiment with different wordlengths for evaluating the performance of quantized FFT algorithms.

8.8.2 Implementation Using Floating-Point C

Section 8.3.4 introduces the C code for bit reversal (bit_reversal.c) and DIT radix-2 FFT (ditr2fft.c). Based on these C functions, we can implement the power-spectrum estimation using the floating-point C program psd_floatpt.c, which is listed as follows:

```c
#include <stdio.h>
#include <stdlib.h>
#include <math.h>
#include "def_complex.h"  // complex.h header file

/*     external functions     */
extern void ditr2fft(complex *, unsigned int, complex *, unsigned int);
extern void bit_reversal(complex *, unsigned int);

/*     global variables and constants     */
#define N 512              // FFT size
#define EXP 9              // EXP=log2(N)
#define pi 3.1415926535897
complex X[N];              // declare input array
complex W[EXP];            // twiddle factor e^(-j2pi/N)
complex temp;
float spectrum[N];
float xn;

void main()
{
  unsigned int i,j,LE,LE1;
  FILE *xn_in;                     // file pointer of x(n)
  FILE *yn_out;                    // file pointer of y(n)
  xn_in = fopen("in.dat","r");   // open file for input x(n)
  yn_out = fopen("out.dat","w");// open file for output y(n)

  /*     Step 1: create a twiddle factor table     */
  for (j=1; j<=EXP; j++){         // create twiddle factor table
    LE=1<<j;                       // LE=2^L=points of sub FFT
    LE1=LE>>1;                     // number of butterflies in sub-FFT
    W[j-1].re = cos(pi/LE1);      // real part
    W[j-1].im = -sin(pi/LE1);     // imaginary part
  }
  /*     Step 2: enter input data to signal buffer     */
  for(i=0; i<N; i++){             // input signal
    fscanf(xn_in,"%f",&xn);{
      X[i].re = xn;
      X[i].im = 0.0;               // set imaginary part to 0
    }
  }
  /*     Step 3: perform bit reversal, follow by FFT     */
  bit_reversal(X,EXP); // arrange X[] in bit-reversal order
  ditr2fft(X,EXP,W,1); // FFT with scaling 0.5 at each stage
```

```
/*    Step 4: perform magnitude-square    */
for (i=0; i<N; i++){ // compute magnitude spectrum
  temp.re = X[i].re*X[i].re;
  temp.im = X[i].im*X[i].im;
  spectrum[i] = (temp.re+temp.im);
  fprintf(yn_out,"%f\n",spectrum[i]);
}
fcloseall();
}
```

The input signal to be analyzed is read from the data file in.dat. The output data file out.dat contains the power-spectrum samples, which can be loaded into MATLAB for evaluation by comparing them with the results obtained in the previous section. The floating-point C program can also be modified to include the averaged periodogram using the Bartlett and Welch methods. These modifications can be found in the C files psd_bar_floatpt.c and psd_wel_floatpt.c.

8.8.3 Implementation Using Fixed-Point C

In Section 8.8.1, we verified that the fixed-point implementation of the power-spectrum estimation results in satisfactory performance. In this section, we implement the algorithm using a fixed-point C program. Some modifications are required: (1) the header file is changed to def_complex_fixpt.h to include the complex-integer data type for Q.15 and Q.31 formats, and (2) the floating-point C functions ditr2fft.c and bit_reversal.c are replaced by the fixed-point functions ditr2fft_fixpt.c and ibit_reversal.c, respectively.

More work is needed to handle multiplication in Q.15 format, as shown in the following partial listing of ditr2fft_fixpt.c:

```
for (j=0; j<LE1;j++)
{
  for(i=j; i<N; i+=LE) // Do the butterflies
  {
    id=i+LE1;
    ltemp.re = ((((long)(int)X[id].re*(long)(int)U.re)>>15)-
              (((long)(int)X[id].im*(long)(int)U.im)>>15))>>scale;
    ltemp.im = ((((long)(int)X[id].im*(long)(int)U.re)>>15)+
              (((long)(int)X[id].re*(long)(int)U.im)>>15))>>scale;
    X[id].re = (X[i].re>>scale) - (int)ltemp.re;
    X[id].im = (X[i].im>>scale) - (int)ltemp.im;
    X[i].re = (X[i].re>>scale) + (int)ltemp.re;
    X[i].im = (X[i].im>>scale) + (int)ltemp.im;
  }
  /*    recursive compute W^k as U*W^(k-1)    */
  ltemp.re = ((((long)(int)U.re*(long)(int)W[L-1].re)>>15)-
            (((long)(int)U.im*(long)(int)W[L-1].im)>>15));
  U.im = (int)((((long)(int)U.re*(long)(int)W[L-1].im)>>15)+
          (((long)(int)U.im*(long)(int)W[L-1].re)>>15));
  U.re = (int)ltemp.re;
}
```

In the code, every multiplication is first upgraded to long multiplication, and the result is shifted right 15 bits and then added to (or subtracted from) the other product. Scaling by 0.5 at each stage is done to maintain the numerical range for computing the FFT.

The main program `psd_fixpt.c` calls the integer bit-reversal function `ibit_reversal.c` and the DIT radix-2 FFT function `ditr2fft_fixpt.c`. The arithmetic operations inside this program are modified to handle fixed-point computation. In addition, the input signal to the main program is converted to a 16-bit integer using `sine3.m`, and the result is saved in the file `in_int.dat`. It is important to note that we must scale down the final result `spectrum[i]` in order to achieve a dynamic range similar to that in the floating-point C program.

8.8.4 Implementation Using Fixed-Point C for C5000 Processors

The fixed-point C code developed in the previous experiment can be modified for C5000 processors. The modified main program `psd_fixpt_ccs.c` is included in the project with the modified functions `ditr2fft_fixpt_ccs.c` and `ibit_reversal_ccs.c`. The project also includes `vectorsc54.asm` (for the C54x only), `cmdc54.cmd` (or `C55x.cmd` for the C55x), and `rts.lib` (or `rts55.lib` for the C55x). After compiling and linking the program successfully, we load the executable file into the C5000 simulator. In addition, we also modify `in_int.dat` to `in_int_ccs.dat` by adding the CCS data header.

The only difference between the fixed-point C code for the C5000 and the previous fixed-point code for the general-purpose computer is that the internal variables `U.re` and `U.im` in the function `ditr2fft_fixpt_ccs.c` are set to `lcomplex` instead of `complex`. The input file `in_int_ccs.dat` is read by the main program `psd_fixpt_ccs.c` using the `dataIO` function. As with the previous FIR and IIR filtering experiments, we use the probe point to transfer data from `in_int_ccs.dat` to the memory array `in_buffer`.

Probe points and breakpoints are inserted into the program by placing the cursor at the line `dataIO()` and clicking on the probe-point button 🖑 and the breakpoint button 🖐. Choose **File → File I/O**, and then click on **Add File**. Select `in_int_ccs.dat`, and type `in_buffer` and `512` in the **Address** and **Length** boxes, respectively. Click on **Add Probe Point** and highlight the line. In the **Connect To** field, click on the down arrow, select the `in_int_ccs.dat` file, and finally click on **Replace** to complete the link between the data file and the probe point. In addition, insert the second breakpoint at the end of the power-spectrum loop. A block of data is read from the file and collected before being passed to the bit-reversal (`ibit_reversal_ccs.c`) and FFT (`ditr2fft_fixpt_ccs.c`) routines. The FFT results are converted to the power spectrum in dB and saved in the buffer `spectrum_db`.

The time-domain plots of `in_buffer` and `spectrum_db` can be displayed by choosing **View → Graph → Time/Frequency** and setting the display-property dialog for a 256-point display. Click on the run button to run the program to the first breakpoint and then click on the animate button 🏃, which executes the program until it reaches the second breakpoint. The program halts, and the display windows are

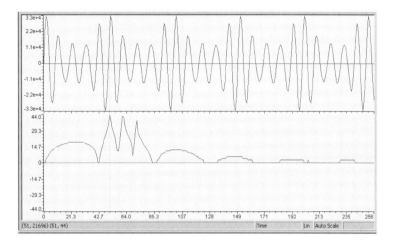

Figure 8.29   Time-domain signal `inp` (upper) and its spectrum `spectrum_db` in dB (bottom)

updated, (from left to right) as shown in Fig. 8.29. Stop the program by clicking on the halt button after 256 points are displayed.

The 512-point FFT-based spectrum estimator (periodogram) code can be profiled in terms of the cycle count. Select **Profiler** → **Enable Clock** and **Profiler** → **View Clock** from the menu. We can set one breakpoint at `ibit_reversal_ccs( )` and the other at `ditr2fft_fixpt_ccs( )` to get the cycle count for running the bit-reversal function. Run the program until the execution stops at the first breakpoint, and then double-click on the **Clock Window** to clear the cycle count. Run the program to the second breakpoint, and the **Clock Window** displays the number of cycles (`Clock = 44896`) that the C code took to execute the code `ibit_reversal_ccs`. Similar results can be observed in the C55x experiment, which shows `Clock = 34349` cycles.

We can set another breakpoint at the line immediately below `ditr2fft_-fixpt_ccs( )` and repeat the same steps to benchmark the cycle count for the FFT function `ditr2fft_fixpt_cc`. We obtain `Clock = 1073577` for the C54x and `Clock = 704649` for the C55x. In addition, the user can also generate a map file to examine the memory usage of the fixed-point function and tabulate these results for comparison with the implementations described in the following sections.

8.8.5 Implementation Using Fixed-Point C with Intrinsics

The preceding C code is portable between different DSP processors. In order to improve code efficiency, the user can use specific DSP intrinsics, which act like function calls and which can be mapped by the compiler directly onto the target processor's instructions. The intrinsic function is a higher-level instruction, which lets the compiler handle register allocation and instruction scheduling.

The intrinsics used in this experiment are the same for both C54x and C55x processors. The C file `ditr2fft_fixpt_intr.c` replaces the C statements with its equivalent intrinsics, as shown in the following partial listing:

```
for (j=0; j<LE1;j++)
{
  for(i=j; i<N; i+=LE) // do the butterflies
  {
    id=i+LE1;
    ltemp.re = _lsmpy(X[id].re, U.re);
    temp.re = (_smas(ltemp.re, X[id].im, U.im)>>16);
    temp.re = _sadd(temp.re, 1)>>scale; // rounding & scale
    ltemp.im = _lsmpy(X[id].im, U.re);
    temp.im = (_smac(ltemp.im, X[id].re, U.im)>>16);
    temp.im = _sadd(temp.im, 1)>>scale; // rounding & scale
    X[id].re = _ssub(X[i].re>>scale, temp.re);
    X[id].im = _ssub(X[i].im>>scale, temp.im);
    X[i].re = _sadd(X[i].re>>scale, temp.re);
    X[i].im = _sadd(X[i].im>>scale, temp.im);
  }
  /*    recursive compute W^k as U*W^(k-1)    */
  ltemp.re = _lsmpy(U.re, W[L-1].re);
  ltemp.re = _smas(ltemp.re, U.im, W[L-1].im);
  ltemp.im = _lsmpy(U.re, W[L-1].im);
  ltemp.im = _smac(ltemp.im, U.im, W[L-1].re);
  U.re = ltemp.re>>16;
  U.im = ltemp.im>>16;
}
```

Similar to the reason stated in previous chapters, intrinsic functions such as `_lsmpy`, `_sadd`, and `_ssub` are used to complete FFT operations. Besides applying intrinsics to the FFT routine, the user can also extend the usage of intrinsics in the spectrum computation within the `psd_fixpt_intr.c` file.

The time required to perform these intrinsics within the `ditr2fft_fixpt_ccs.c` file is 653,604 cycles for the C54x (or 395,771 cycles for the C55x), an almost 40% reduction in cycle count compared to that of the fixed-point C program in Section 8.8.4. Similar improvement can be observed in the power-spectrum computation using intrinsics.

8.8.6 Implementation Using the C5000 Digital Signal Processing Library

The final experiment uses mixed C-and-assembly code. The bit-reversal and FFT routines are replaced with the optimized assembly code from Texas Instruments given in the TMS320C54x and TMS320C55x DSPLIBs [12, 14]. Unlike the intrinsics, the same routine name in the C54x and C55x DSPLIBs may have different functions.

The C program calls these assembly routines inside the main function. In this typical example of mixing C with assembly code, time-critical sections are hand-optimized using assembly routines. Housekeeping tasks such as data allocation, handling, and simple arithmetic are coded in C. For example, the main program `psd_fixpt_c54x.c` for the C54x experiments is listed below:

```
#include "math.h"                    // C header files
#include "tms320.h"
#include "dsplib.h"
#include "in_int_asm.dat"            // data file
#include "def_complex_fixpt.h"       // complex.h header file
#pragma DATA_SECTION(spectrum, "fft_out");

lcomplex ltemp;                      // variables and constants
DATA spectrum[N/2+1];

void main()
{
  unsigned int i;
  /* perform bit reversal, follow by FFT */
  cbrev(inp,inp,N/2);      // arrange X[] in bit-reverse order
  rfft(inp,N,1);           // perform FFT with scaling 0.5 in
                           // each stage
  /* perform magnitude-square and display results */
  spectrum[0]= (int)(((long)inp[0]*(long)inp[0])>>15)+1;    // DC
  spectrum[N/2]= (int)(((long)inp[1]*(long)inp[1])>>15)+1; // Fs/2
  for (i=1; i<N/2; i++){    // compute magnitude spectrum
    ltemp.re = (((long)inp[2*i]*(long)inp[2*i])>>15);
    ltemp.im = (((long)inp[(2*i)+1]*(long)inp[(2*i)+1])>>15);
    spectrum[i] = (int)(ltemp.re+ltemp.im)+1;
  }
  return;
}
```

Note that the bit-reversal operation is included in the function `rfft()` for the C55x; thus, the function `cbrev()` is not needed in the C55x program `psd_fixpt_c55x.c`.

Create a new project file that consists of the main program `psd_fixpt_c54x.c` (or `psd_fixpt_c55x.c` for the C55x), which includes the header files `dsplib.h`, `tms320.h` and `math.h`. These header files contain the function definitions and data types used in the library routines. The user also needs to add the C54x (or C55x) object library `54xdsp.lib` (or `55xdsp.lib` for the C55x) into the project. In addition, the included search path has to be set up in the CCS to locate the DSPLIB. This setup can be done by clicking on **Project → Build Options → Compiler → Preprocessors** and adding the search path `c:\ti\c5400\dsplib\include\` for the C54x (or `c:\ti\c5500\dsplib\include\` for the C55x).

The input-data array `inp` (which is defined in `in_int_asm.dat`) is included in the main program, which shows another way that the input file can be loaded into the main program instead of using the `dataiIO()` function described in Chapters 6 and 7. The DSP functions `cbrev.asm` (complex bit reversal, used for the C54x only) and `rfft.asm` (in-place real FFT computation, which also includes a bit-reversal operation for the C55x) complete the bit reversal and DIT radix-2 FFT operations. Since the real FFT computation is in place, the variable `inp` contains the FFT results. The results are arranged in a real-imaginary format, as explained in Section 8.5. Finally, the last section of the code computes the power spectrum.

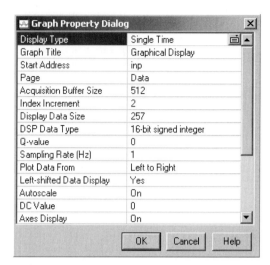

Figure 8.30   Graph Property Dialog window

Perform a complete build and load the executable code into the simulator. Set up two display windows to view the variables inp and spectrum. Since the outputs of the FFT are arranged in real-imaginary format, we can display the real part of the FFT output in inp by setting the properties of the display window, as shown in Fig. 8.30. It is important to note that **index increment** is set as 2 to pick up only the real parts. Please refer to the manuals [12, 14] for a complete description of data placement.

Run the program, and the program pointer stops with an updated display for inp and spectrum, as shown in Fig. 8.31.

The final step is to profile the DIT radix-2 FFT using the optimized assembly code. For the C54x program, set a breakpoint at the instruction cbrev(inp,inp,N/2), another at the instruction rfft(inp,N,1), and a third at the line immediately below it. For the C55x code, set a breakpoint at instruction rfft(inp,N,1) and another at the line immediately below it. Enable the clock and view the cycle counts. The cycle counts in running different functions for the C54x are 4,068 cycles for cbrev and 21,891 cycles for rfft. The C55x needs 11,424 cycles for rfft, which also includes a bit-reversal operation.

Significant cycle-count reduction is observed in using the optimized assembly FFT routines as compared with the previous fixed-point C and C with intrinsic functions. The obtained benchmarks for in-place bit reversal and forward-real FFT functions also match the reported benchmarks in [12] and [14].

Some additional applications and experiments using FFT are introduced in Appendix B. The user can refer to the MATLAB or Simulink files for more details on these exercises and progressively work through these hands-on experiments to obtain the final solution using the CCS. These additional hands-on experiments include implementation of an FFT-based convolution using overlap-add and overlap-save methods, implementation of a sliding FFT algorithm, and implementation of a downsampling FFT analyzer.

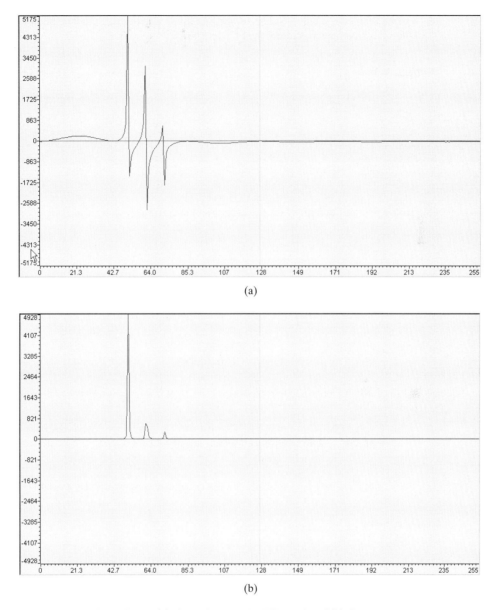

Figure 8.31 Updated display: (a) The real part of the FFT result and (b) the power spectrum

PROBLEMS

PART A

Problem A–8–1

Compute the time taken (using the `etime` function) to perform a DFT using `dftmtx` and an FFT using `fft` in MATLAB on a 2,048-point data sequence. Repeat the same experiment using 1,024-, 512-, 128-, 64-, 32-, 16-, and 8-point data sequences. Find the minimum N such that it is advantageous to use the FFT function.

Problem A–8–2

Use the MATLAB functions `fft` and `ifft` to verify that an IFFT can be implemented by using a FFT.

Problem A–8–3

A signal consists of three tonal components that have an amplitude of 1V at 1 kHz, 2 kHz, and 3 kHz. It also contains a DC component of 5V. This signal is sampled at 8 kHz, and 200 samples are acquired. Perform a radix-2 FFT on the signal using MATLAB, and view and comment on the magnitude spectrum of this signal. Since we are only interested in detecting the three tones of the signal, how can you improve the spectrum plot for this purpose?

Problem A–8–4

A data sequence of 256 samples is obtained by sampling an analog signal that consists of two sinewave components with a sampling frequency $f_s = 1,024$ Hz. The digital sequence is expressed as

$$x(n) = \sqrt{2}\sin(2\pi(200)n/f_s) + \sin(2\pi(200 + \Delta f_1)n/f_s), \quad n = 0,1,\dots 255,$$

where Δf_1 is the frequency separation between the two sinewaves. Use MATLAB to answer the following questions:

(a) Can a 1,024-point radix-2 FFT be used to resolve the two frequency components with $\Delta f_1 = 2$ Hz? If not, can a 2,048-point FFT be used?
(b) What is the smallest frequency separation that can be resolved?
(c) What is the frequency resolution if a 512-point FFT is used?
(d) What are two ways of resolving the frequency separation of $\Delta f_1 \leq 2$ Hz?

Problem A–8–5

Write a simple MATLAB script to determine the number of unique points in a real FFT computation. This routine should also compute the next higher power-of-two order to perform the radix-2 FFT. Since the MATLAB function `fft` does not include any magnitude scaling, the user must also include magnitude normalization. Test the MATLAB script using a simple sinewave, and determine whether the correct magnitude has been obtained.

Problem A–8–6

Use MATLAB functions to implement (a) `freqz` for the DTFT in Section 2.4.1, (b) `dftmtx` for the DFT in Section 8.1.1, and (c) `fft` for the FFT in Section 8.3.3 on a time-domain sequence $x(n) = \{1, 2, 3, 4\}$. Zero-pad the data to an 8-point FFT, and comment on the magnitude spectrum.

Problem A–8–7

Zero-padding is commonly used in the time-domain sequence to interpolate the frequency spectrum. What is the effect of zero-padding the frequency sequence? Derive a way of adding zeros in the real and imaginary frequency responses, and observe the effect of this frequency zero-padding in its time response. Use MATLAB to illustrate your findings.

Problem A–8–8

Use the power-spectrum density-estimate function psd to estimate the spectrum of the speech file timit1.asc. Will a time-reversed speech signal result in the same power-spectrum density estimation? Why or why not?

Problem A–8–9

Use the computational trick in computing the FFT of two real sequences $x(n)$ and $h(n)$ using only one FFT. Obtain the efficiency improvement using the FFT-based linear convolution over the time-domain linear convolution. Write a MATLAB script for implementation.

Problem A–8–10

Modify the C program ditr2fft.c into a DIF radix-2 FFT (difr2fft.c), and compute the power spectrum of a 1,000-point input sinewave $x(n)$ generated by MATLAB. After the FFT has been performed, a DIT IFFT can be used to recover the time-domain signal $x(n)$. Why is this approach (a DIF FFT followed by a DIT IFFT) of recovering the signal useful?

Problem A–8–11

Extend the C and assembly codes in Section 8.8 to analyze a speech signal timit1.asc corrupted by the three-tonal noise. The power-spectrum estimation of the combined signal can be analyzed using the averaging periodogram. Select a suitable size of the FFT, a suitable averaging technique using a Bartlett or Welch periodogram, and percentage of overlapping for the analyzing window.

Problem A–8–12

Write mixed C-and-assembly code using the C5000 DSPLIB to implement fast convolution between a long sequence in_int_asm.dat and a 10-tap lowpass filter with an impulse response of $h(n) = \{0.0021\ 0.0580\ 0.1234\ 0.1814\ 0.2155\ 0.2155\ 0.1814\ 0.1234\ 0.0580\ 0.0021\}$. The user can use either the overlap-add or overlap-save method. Observe the spectrum of the output from the fast convolution. Profile this code, and compare it to that obtained using linear convolution (i.e., the FIR filtering given in Chapter 6).

Problem A–8–13

A linear-sweep chirp signal, as described in Example 8.5, has been distorted by a sinewave that only occurs within a short interval. This distorted chirp signal is saved as distorted_chirp.mat. Use the MATLAB spectrogram function and the Wavelet Tool-box (using a Daubechies order-8 wavelet) to determine the time interval when the sinewave distortion occurs. Determine which method yields the best result.

PART B

Problem B–8–14

Compute the DFT of the finite sequence

$$x(n) = a^n, \quad n = 0, 1, \ldots, N - 1.$$

Problem B–8–15

If $x(n), n = 0, 1, \ldots, N - 1$ is a real sequence and N is an even number, show that $X(0)$ and $X(N/2)$ are real numbers.

Problem B–8–16

Find the DFT of the sequence $x(n) = \{1, 2, 3, 4\}$.

Problem B–8–17

Derive the general I/O equation for the 2-point DFT, and draw the butterfly diagram for it.

Problem B–8–18

Compute the circular convolution of the following sequences:

$$x_1(n) = \{1, 2, 3, 4\} \text{ and } x_2(n) = \{1, 2, 3, 4\}.$$

Problem B–8–19

The TMS320C54x DSPLIB contains a matrix multiplication function mmul that allows the user to perform matrix-matrix multiplication. The dimensions of these two matrices are defined as [row1 x col1] and [row2 x col2]. We can implement a 512-point DFT using this function. The instruction cycles needed for executing this DFT function is row1×(7+(11+(6×col1))×col2)+71 instruction cycles, and the code size is 65 (16-bit) words. If the C54x processor is running at 100 MIPS, determine the maximum sampling frequency allowed for implementing this 512-point DFT. What is the total memory required for storing both the program and data for this routine?

Problem B–8–20

A 15-second audio signal is sampled at 48 kHz. An FFT is performed using a window of 30 msec. Answer the following questions:

(a) How many data samples are collected?
(b) What is the frequency resolution of the FFT?
(c) How many blocks of the FFT can be performed within the 15 seconds of audio signal for (1) no overlapping, (2) 50% overlapping, and (3) 75% overlapping?

Problem B–8–21

An analog signal that consists of two sinusoidal components at 2 kHz and 8 kHz is sampled at 20 kHz to form the digital signal $y_1(n)$. A second analog signal consisting of two sinewaves at 12 kHz and 18 kHz is also sampled at 20 kHz to form the digital signal $y_2(n)$. The FFT is used to observe the magnitude spectrum of these two signals. Is there any difference in the frequency spectrum of $y_1(n)$ and $y_2(n)$? If not, is there any method for differentiating between these two signals?

Problem B–8–22

What is the consequence of performing scaling of $(1/N)$ in the FFT algorithm instead of in the IFFT algorithm? When it is suitable to perform scaling in the FFT algorithm?

Problem B–8–23

Besides using the efficient implementation of the FFT on real sequences, as explained in Section 8.4.2, the following fast Hartley transform can also implement the FFT efficiently:

$$H(f) = \frac{1}{N} \sum_{n=0}^{N-1} x(n) \left[\cos\left(\frac{2\pi f n}{N}\right) + \sin\left(\frac{2\pi f n}{N}\right) \right].$$

Show that its relationship with the FFT is expressed as

$$F_{real}(f) = H(f) + H(N - f)_N$$
$$F_{imag}(f) = H(f) - H(N - f)_N.$$

Problem B–8–24

A bandlimited continuous-time signal is sampled at 40 kHz over a time interval of 50 seconds. The power spectrum of the signal is estimated using a nonoverlap periodogram and a Welch periodogram with 50% overlapping. Answer the following questions:

(a) What is the length of the data record?
(b) If a radix-2 FFT is used to compute both periodograms, what is the minimum length N of the segment if the desired frequency resolution is restricted to less than 10 Hz?
(c) What is the number of segments for the nonoverlap and Welch periodograms?
(d) What is the computational load in performing these spectral estimations?
(e) What are two methods of further improving the statistical stability of the Welch's periodogram while maintaining the frequency resolution of 10 Hz?

Problem B–8–25

In speech-processing applications, a speech signal is sampled at 8 kHz. The algorithm collects 20 msec of speech samples and computes an N-point DFT, N-point spectral weighting, and N-point IDFT to get the processed time-domain signal. Answer the following questions:

(a) What is a suitable N for this application?

(b) If it takes 20 nsec to perform a real multiply-add operation, how much time remains after the DFT, spectral weighting, and IDFT are computed?

(c) Can the preceding transformation be implemented in real time using the preceding DSP processor?

(d) What are the differences between using the radix-2 FFT and IFFT?

(e) If additional time-domain windowing is required, what is the additional cost?

Problem B–8–26

A bandlimited continuous-time signal is sampled at 20 kHz over a time interval of 60 seconds. The power spectrum of the signal is estimated using the Bartlett method and the Welch method with 50% overlapping. Answer the following questions:

(a) What is the length (number of samples) of the data record? If a radix-2 FFT is used to compute the average periodogram, what is the minimum length N of the FFT if the desired frequency resolution of the power spectrum cannot be greater than 10 Hz? What is the number of segments K that fit in the data record?

(b) Assume that the number of segments L for a Welch periodogram (with 50% overlap) is twice that of a Bartlett periodogram. What is the computational load in performing these spectral estimations?

(c) The instruction cycles needed to carry out an N-point radix-2 FFT, magnitude square, and windowing are given as $4N \log_2 N$, $15N$, and $2N$, respectively. What is the total number of cycles required to compute one segment of the periodogram? What is a suitable MIPS rating for this DSP processor? Describe a relevant buffering technique for performing the data acquisition and block processing of the signal.

Problem B–8–27

A bandlimited analog signal $x(t)$ is sampled for 10 seconds and generates a sequence of 10,240 samples. The power spectrum of the signal is estimated using the Bartlett and Welch (50% overlapping) periodograms. Answer the following questions:

(a) What is the highest frequency in $x(t)$ if it is sampled without aliasing? If a 1,024-point radix-2 FFT of the sampled signal is computed, what is the frequency resolution in Hz?

(b) A DFT can be used to compute the frequency spectrum within a frequency range from 100 Hz to 200 Hz. What are the computational complexities in evaluating the preceding frequency range using the DFT and the DIT radix-2 FFT? Can selective frequency-spectrum analysis be computed using the FFT?

(c) A DSP processor calculates the 1,024-point Bartlett periodogram and displays the spectrum in 0.3 sec. How much time does the DSP processor need to wait until the acquisition of the next block of input is completed? How long does it take to calculate the average of 10 blocks for the Bartlett periodogram? Answer the preceding questions for the Welch periodogram with 50% overlapping.

SUGGESTED READINGS

1 The MathWorks. *Filter Design Toolbox, User's Guide.* Version 2.1, 2000.

2 Mitra, S. K. *Digital Signal Processing: A Computer-Based Approach.* 2nd Ed. New York, NY: McGraw-Hill, 2001.

3 Kuo, S. M. and B. H. Lee. *Real-Time Digital Signal Processing.* New York, NY: Wiley, 2001.

4 Smith, W. W. and J. W. Smith. *Handbook of Real-Time Fast Fourier Transform.* New York, NY: IEEE Press, 1995.

5 Cooley, J. W. and J. W. Tukey. "An Algorithm for the Machine Computation of Complex Fourier Series." *Mathematical Computations* 19 (1965): 297–301.

6 Bracewell, R. N. "The Fast Hartley Transform." *Proceedings of the IEEE* 72 (1984): 1010–1017.

7 Embree, P. M. *C Algorithms for Real-Time DSP.* Upper Saddle River, NJ: Prentice Hall, 1995

8 The MathWorks. *Signal Processing Toolbox for Use with MATLAB.* Version 5, 2000.

9 Oppenheim, A. V., Schafer, R. W., and J. R. Buck. *Discrete-Time Signal Processing.* 2nd Ed. Englewood Cliffs, NJ: Prentice Hall, 1999.

10 Strang, G. and T. Nguyen. *Wavelets and Filter Banks.* Wellesley, MA: Wesley-Cambridge Press, 1996.

11 Texas Instruments. *Implementing FFT Algorithms of Real-Valued Sequences with the TMS320 DSP Family.* SPRU291, 1997.

12 Texas Instruments. *TMS320C54x DSP Library Programmer's Reference.* SPRU518c, 2002.

13 Texas Instruments. *FFT Library, Module User's Guide, C24x Foundation Software.* May 2002.

14 Texas Instruments. *TMS320C55x DSP Library Programmer's Reference.* SPRU422e, 2002.

15 Texas Instruments. *TMS320C62x DSP Library Programmer's Reference.* SPRU402a, 2002.

16 Texas Instruments. *TMS320C64x DSP Library Programmer's Reference.* SPRU565a, 2002.

17 Texas Instruments. *TMS320C3x User's Guide.* SPRU031D, 1994.

18 Papamichalis, P. "An Implementation of FFT, DCT, and Other Transforms on the TMS320C30." Chapter 4 in *Digital Signal Processing Applications with the TMS320 Family,* vol. 3. Texas Instruments, 1990.

19 Texas Instruments. *TMS320C62x/C67x: Programmer's Guide.* SPRU198C, 1999.

20 The MathWorks. *Wavelet Toolbox for Use with MATLAB.* Version 2, 2002.

9

Adaptive Filtering

The coefficients of adaptive filters continuously and automatically adapt to given signals in order to achieve the desired response. Adaptive filtering can be used in system identification (e.g., echo cancellation in telecommunications), inverse system modeling (e.g., channel equalization in modems), noise cancellation (e.g., active noise-control systems), signal prediction (e.g., speech coding), and many others. In this chapter, we introduce some widely used adaptive-filter structures and adaptive algorithms, as well as their basic characteristics and important applications. We also demonstrate the design, analysis, and quantization of adaptive filters using MATLAB and Simulink and then implement the designed filters using C programs and assembly programs on both fixed-point and floating-point DSP processors. Detailed discussion on the theoretical aspects of adaptive filters can be found in textbooks [1–3].

9.1 INTRODUCTION TO ADAPTIVE FILTERS

As discussed in Chapters 6 and 7, the coefficients of the digital filter determine the filter's characteristics. Based on a given application, we can design digital (FIR and IIR) filters with fixed coefficients using filter-design software packages such as MATLAB. However, in many practical applications, filter specifications are unknown at design time and/or change with time. In addition, it may be necessary to attenuate noise that has spectral overlap with the desired signals. For these applications, we have to use adaptive filters with time-varying coefficients updated by adaptive algorithms to track the unknown and/or changing environments. Therefore, designing an adaptive filter does not require the same frequency specification as designing fixed-coefficient FIR and IIR filters, as discussed in Chapters 6 and 7.

As illustrated in Fig. 9.1, an adaptive filter consists of two main functional units: a digital filter with time-varying coefficients to perform desired filtering and an adaptive algorithm to adapt the coefficients of the filter in order to improve its performance. The filter structure is determined and fixed at the design stage, but the adaptive algorithm continuously adjusts its coefficients.

The generic adaptive filtering given in Fig. 9.1 shows that the input signal, $x(n)$, is filtered by the digital filter to produce the output signal, $y(n)$. The adaptive algorithm adjusts the coefficients of the filter contained in the vector $\mathbf{w}(n)$ to minimize the error signal, $e(n)$, which is the difference between the desired signal, $d(n)$, and the filter output, $y(n)$. Therefore, the adaptive-filter design itself automatically is based on the characteristics of the input signal $x(n)$ and the desired signal $d(n)$. In this way, the adaptive filter adapts to the environment prescribed by these signals. When the environment changes, the filter adapts to the new characteristics by generating a new set of coefficients. Note that for certain applications such as channel equalization, the desired signal, $d(n)$, may not be available for the adaptive filter. In addition, for active noise control applications [4], the error signal $e(n)$ is measured by a sensor (e.g., a microphone) instead of being the computed difference between $d(n)$ and $y(n)$.

The digital filter shown in Fig. 9.1 may be an FIR, symmetric FIR, IIR, or other structure. The coefficients of the adaptive filter are time functions because the adaptive algorithm continuously updates them either sample by sample or block by block. To design the adaptive filter, we have to select the correct filter structure. For the selected filter structure, many adaptive algorithms can be used to update the filter coefficients. The complexity of an adaptive filter is usually measured in terms of its multiplication and memory requirements. Many computationally efficient and high-performance algorithms for adaptive filtering have been developed over the past 30 years. In this chapter, we focus on the least-mean-square (LMS) algorithm.

Because of the increasing speed and flexibility of DSP processors, real-time adaptive filtering is becoming an enabling technology for communication, network, audio, and control systems. To apply adaptive filter to practical applications,

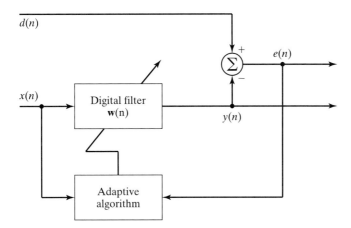

Figure 9.1 A generic block diagram of an adaptive filter

engineers must understand the basic characteristics of adaptive filtering, which include stability, convergence speed, steady-state performance, and numerical effects. The design of an adaptive filter involves the selection of a filter structure and an adaptive algorithm, determines the optimum values of the parameters, and minimizes the finite-wordlength effects, all of which require a general knowledge of the system environment and a deep understanding of the particular application.

The filter structure used for filtering and the adaptive algorithm used for updating filter coefficients are introduced in Section 9.2. In Section 9.3, we discuss characteristics of adaptive filters that are critical for determining filter parameters, such as the length of the filter and the convergence factor. Adaptive filters can be used in various applications with different input and output configurations. Some widely used applications are introduced in Section 9.4, and additional applications on acoustic echo cancellation and active noise control are given in Appendix B. The design and implementation of adaptive filtering using MATLAB and C programs are presented in Section 9.5. In addition, we introduce the practical implementation of adaptive filters using fixed-point and floating-point DSP processors in Sections 9.6 and 9.7, respectively. Finally, Section 9.8 guides the user through a series of hands-on experiments for designing and implementing adaptive filters on fixed-point DSP processors.

9.2 ADAPTIVE-FILTER STRUCTURES AND ALGORITHMS

As shown in Fig. 9.1, an adaptive filter updates its coefficients using the adaptive algorithm to minimize the error signal, $e(n)$, based on a given criterion. Most reported developments and applications use the FIR filter with the LMS algorithm because it is relatively simple to design and implement. It is also well understood and very robust and thus is best suited for many real world embedded applications. In this chapter, we focus on the adaptive FIR filter with the LMS algorithm. However, different filter structures and algorithms for some specific considerations are also briefly introduced. These structures and algorithms generally trade increased complexity for improved performance.

9.2.1 Filter Structures

Several filter structures can be used in the design of adaptive systems. As discussed in Chapter 7, IIR filters have both poles and zeros and thus are able to offer the same performance as FIR filters with lower complexity. However, the major problem with the adaptive IIR filter is the potential instability if the poles move outside the unit circle during the adaptation process. This problem is a very difficult one because filter coefficients continuously change, and the order of the filter is usually high for most applications. To guarantee filter stability, most practical applications use FIR filters.

The adaptive FIR filter based on the fixed-coefficient FIR filter shown in Fig. 6.2, is illustrated in Fig. 9.2, in which the filter coefficients change with time. The I/O

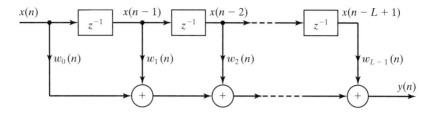

Figure 9.2   Signal-flow diagram of an adaptive FIR filter

equation of the adaptive FIR filter is expressed as

$$y(n) = \sum_{i=0}^{L-1} w_i(n)x(n-i) = \mathbf{w}^T(n)\mathbf{x}(n), \qquad (9.2.1)$$

where $w_i(n)$ is the ith adjustable coefficient (or weight) at time n. The coefficient vector $\mathbf{w}(n)$ consists of L coefficients expressed as

$$\mathbf{w}(n) = [w_0(n)\, w_1(n)\ldots w_{L-1}(n)]^T, \qquad (9.2.2)$$

and the signal vector $\mathbf{x}(n)$ at time n is given as

$$\mathbf{x}(n) = [x(n)x(n-1)\ldots x(n-L+1)]^T. \qquad (9.2.3)$$

As discussed in Chapter 6, an FIR filter that has a symmetric impulse response (coefficients) has a linear phase. In applications such as audio and data processing, linear-phase filters are preferred to avoid phase distortion. The adaptive-symmetric FIR filter structure, similar to the one shown in Fig. 6.4, is illustrated in Fig. 9.3. In

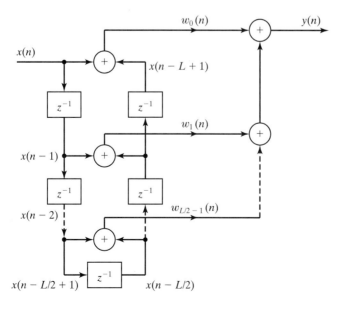

Figure 9.3   Signal-flow diagram of an adaptive symmetric FIR filter (Type 2)

Fig. 9.3, L is assumed to be an even number. This filter is actually an FIR filter with coefficients that are symmetric about the center coefficient $w_{L/2-1}(n)$. The output of the filter is computed as

$$y(n) = \sum_{i=0}^{L/2-1} w_i(n)[x(n - i) + x(n - L + 1 + i)]. \qquad (9.2.4)$$

Other filter structures such as IIR, lattice, and frequency-domain filters can also be applied for adaptive filtering. These advanced structures are introduced in adaptive signal-processing books [1–3].

9.2.2 Adaptive Algorithms

Many adaptive algorithms can be viewed as approximations of the Wiener filter. As shown in Fig. 9.1, the adaptive algorithm uses the error signal

$$e(n) = d(n) - y(n) \qquad (9.2.5)$$

to update the filter coefficients in order to minimize a predetermined criterion. The most widely used mean-square-error (MSE) criterion is defined as

$$\xi = E[e^2(n)], \qquad (9.2.6)$$

where $E[.]$ denotes the expectation operator.

Substituting Eqs. (9.2.1) and (9.2.5) into Eq. (9.2.6), we obtain

$$\xi = E[d^2(n)] + \mathbf{w}^T(n)\mathbf{R}\mathbf{w}(n) - 2\mathbf{w}^T(n)\mathbf{p}, \qquad (9.2.7)$$

where

$$\mathbf{R} = E[\mathbf{x}(n)\mathbf{x}^T(n)] = \begin{bmatrix} r_{xx}(0) & r_{xx}(1) & \cdots & r_{xx}(L-1) \\ r_{xx}(1) & r_{xx}(0) & & \\ \vdots & & \ddots & r_{xx}(1) \\ r_{xx}(L-1) & & r_{xx}(1) & r_{xx}(0) \end{bmatrix} \qquad (9.2.8)$$

is the $L \times L$ autocorrelation matrix, and

$$\mathbf{p} = E[d(n)\mathbf{x}(n)] = [r_{dx}(0) \quad r_{dx}(1) \quad \cdots \quad r_{dx}(L-1)]^T \qquad (9.2.9)$$

is the $L \times 1$ crosscorrelation vector. The element of the $\mathbf{R}$ matrix, $r_{xx}(k)$, $k = 0$, $1, \ldots, L - 1$, is the autocorrelation function defined in Eq. (2.5.7). Similarly, the crosscorrelation function is defined as $r_{dx}(k) = E[d(n)x(n - k)]$, $k = 0, 1, \ldots$, $L - 1$.

The optimum coefficient vector, which minimizes the MSE, can be derived by solving the following partial-differential equation

$$\frac{\partial \xi}{\partial \mathbf{w}(n)} = \mathbf{0}. \qquad (9.2.10)$$

From the error function defined in Eq. (9.2.7), we obtain the optimum coefficient vector expressed as

$$\mathbf{w}^{\text{o}} = \mathbf{R}^{-1}\mathbf{p}, \tag{9.2.11}$$

where the inverse matrix exists because of the unique features of the $\mathbf{R}$ matrix shown in Eq. (9.2.8). Substituting Eq. (9.2.11) into Eq. (9.2.7), we obtain the minimum MSE, $\xi_{\min}$. Equation (9.2.11) is known as the Wiener–Hopf solution, and the filter with optimum coefficients given in Eq. (9.2.11) is called the Wiener filter.

Several methods can be used to calculate the optimum solution given in Eq. (9.2.11) for the optimum filter. In block-by-block processing such as linear predictive coding of speech, the input signal is divided into short segments. The $\mathbf{R}$ matrix and $\mathbf{p}$ vector are estimated, and the optimum coefficient vector corresponding to each segment is computed using an efficient matrix-inversion algorithm. Block-processing algorithms assume the signal is stationary at least inside each block and may have problems in maintaining continuity of the filter coefficients from one block to another. They also introduce a block-processing delay and requires intensive computations for matrix operations.

The more efficient method is to adjust the filter coefficients recursively on a sample-by-sample basis when the new input sample is received. In this chapter, we focus on recursive, sample-by-sample processing algorithms. In particular, we consider the simple LMS algorithm, which is based on a gradient search for learning the unknown environment and also tracking time-varying signal characteristics.

Least-Mean-Square Algorithm

A plot of the MSE defined in Eq. (9.2.7) against the filter coefficients is an $(L + 1)$-dimensional bowl-shaped surface with a unique bottom (minimum MSE) at the optimum vector $\mathbf{w}^{\text{o}}$. This quadratic performance surface is always positive and thus is concave upward. The recursive-adaptive algorithm is the process of seeking the minimum point on the performance surface. In the steepest descent algorithm, the next coefficient vector $\mathbf{w}(n + 1)$ is updated by an amount proportional to the negative gradient of the MSE function at time n. The LMS algorithm is based on the steepest descent method that adapts the coefficients sample by sample toward the optimum vector on the performance surface. That is,

$$\mathbf{w}(n + 1) = \mathbf{w}(n) - \mu\nabla(n), \tag{9.2.12}$$

where μ is the step size (or convergence factor) that determines the stability and convergence rate (speed) of the algorithm.

The computation of the true gradient, $\nabla(n)$, requires the exact MSE function given in Eq. (9.2.6), which assumes the input signal is stationary. The estimation of the exact gradient requires intensive computation and memory and is often inaccurate for only a finite number of available data samples. The LMS algorithm developed by Widrow uses the instantaneous squared error, $\hat{\xi} = e^2(n)$, as an estimate of the MSE. Thus, the gradient estimate can be simplified to

$$\hat{\nabla}(n) = \frac{\partial e^2(n)}{\partial \mathbf{w}(n)} = -2e(n)\mathbf{x}(n). \tag{9.2.13}$$

By substituting this gradient estimate into the steepest-descent algorithm defined in Eq. (9.2.12), we obtain the LMS algorithm expressed as

$$\mathbf{w}(n + 1) = \mathbf{w}(n) + 2\mu e(n)\mathbf{x}(n). \tag{9.2.14}$$

This algorithm is also called the stochastic-gradient algorithm. In practical implementation, the constant 2μ is usually replaced by μ. Thus, the vector equation (9.2.14) can be expressed in scalar form for each coefficient as follows:

$$w_i(n + 1) = w_i(n) + \mu e(n)x(n - i), \quad i = 0, 1, \ldots, L - 1. \tag{9.2.15}$$

Adaptive filtering using an FIR filter with the LMS algorithm is described by Eqs. (9.2.1), (9.2.5), and (9.2.15). The LMS algorithm requires only $2L$ multiplications and additions and is the most efficient adaptive algorithm in terms of computation and storage requirements. The complexity is much lower than that of other adaptive algorithms such as Kalman and recursive-least-square (RLS) algorithms. Furthermore, it does not suffer from the numerical-instability problem inherent in the other two advanced algorithms. Thus, the LMS algorithm is the most widely used adaptive algorithm for practical applications, especially for implementation on fixed-point DSP processors.

In the following subsections, we present a set of LMS-type algorithms obtained by the modification of the LMS algorithm. The motivation for each is a practical consideration such as faster convergence, simplicity of implementation, or robustness of operation. These algorithms are described in the context of the FIR filter. However, they can also be applied to symmetric FIR, IIR, lattice, and frequency-domain structures.

Normalized Least-Mean-Square Algorithm

The maximum step size, μ, to guarantee stability of the LMS algorithm is inversely proportional to the filter length, L, and the power of the input signal, $x(n)$. One important technique for optimizing the speed of convergence while maintaining the independence of the signal power is known as the normalized LMS algorithm. The normalized LMS algorithm is expressed as

$$\mathbf{w}(n + 1) = \mathbf{w}(n) + \mu(n)e(n)\mathbf{x}(n), \tag{9.2.16}$$

where $\mu(n)$ is the time-varying step size normalized by L and the power of the signal $x(n)$. This dynamic step size is computed as

$$\mu(n) = \frac{\alpha}{L\hat{P}_x(n)}, \tag{9.2.17}$$

where α is a constant and $\hat{P}_x(n)$ is the power estimate of $x(n)$ at time n.

As introduced in Chapter 2, we can use the simple recursive method for estimating the power of a signal sample by sample, which is expressed as

$$\hat{P}_x(n) = (1 - \beta)\hat{P}_x(n - 1) + \beta x^2(n), \tag{9.2.18}$$

where $\beta < 1$ is a smoothing parameter. As shown in Eq. (2.5.10), $\beta = 1/N$ is the reciprocal of the length of a moving-average filter. Since it is not desirable that the power estimate $\hat{P}_x(n)$ be zero or very small such that the normalized step size, $\mu(n)$, in Eq. (9.2.17) becomes infinity or very large, a software constraint is required for practical implementations. A simple method of software constraint for the implementation of the normalized LMS algorithm in C is introduced in Section 9.5.3.

Leaky Least-Mean-Square Algorithm

Insufficient spectral excitation of the LMS algorithm may result in divergence of the adaptive algorithms. In addition, finite-precision effects (discussed in Section 9.3.4) can cause unconstrained filter coefficients to grow without bound, resulting in over-flow during the coefficient update process. These long-term stability problems are undesirable for real-time processing.

Divergence may be avoided by using a leaking mechanism during the coefficient adaptation process. The leaky LMS algorithm is expressed as

$$\mathbf{w}(n + 1) = \nu\mathbf{w}(n) + \mu e(n)\mathbf{x}(n), \tag{9.2.19}$$

where ν is the leaky factor with range $0 \ll \nu < 1$. We have shown that leakage is equivalent to adding low-level white noise into filter coefficients. This approach results in some degradation of the adaptive-filter performance.

In general, the value of the leaky factor is determined by the designer on an experimental basis as a compromise between the robustness of the adaptive algorithm and the loss of performance of the adaptive filter. For a fixed-point implementation, the multiplication of each coefficient with the value ν can introduce additional roundoff noise. Therefore, leakage effects must be incorporated into the design procedure in order to determine the required wordlength for filter coefficients and internal data.

Signed Least-Mean-Square Algorithms

Three simplified versions of the LMS algorithm further reduce the number of multiplications required and therefore extend the real-time bandwidth for some applications. However, the convergence rates of these signed LMS algorithms are slower than that of the LMS algorithm, which is shown in simulations given in Section 9.5.

In a previous subsection, the instantaneous squared error, $e^2(n)$, was used to estimate the MSE given in Eq. (9.2.6). For the sign-error algorithm, an instantaneous absolute error is used as the performance criterion. That is,

$$\hat{\xi} = |e(n)|. \tag{9.2.20}$$

Following the similar derivation in an earlier subsection, the sign-error LMS algorithm can be derived as

$$\mathbf{w}(n + 1) = \mathbf{w}(n) + \mu\mathbf{x}(n)\text{sgn}[e(n)], \tag{9.2.21}$$

where

$$
\text{sgn}[e(n)] \equiv
\begin{cases}
1, & e(n) > 0 \\
0, & e(n) = 0. \\
-1, & e(n) < 0
\end{cases}
\tag{9.2.22}
$$

This simple sign operation is equivalent to a very harsh quantization of $e(n)$.

Note that if μ is chosen to be a negative power of two, $\mu\mathbf{x}(n)$ can be computed with a right shift in the binary representation of $x(n)$. Therefore, no multiplication is required, but one IF-ELSE decision is needed per iteration. For a constant step size μ, the reduced computation of the sign-error LMS algorithm comes at the expense of a slower convergence rate.

Similar to Eq. (9.2.21), the sign operation can be performed on data $x(n)$ instead of error $e(n)$. Thus, the sign-data LMS algorithm is expressed as

$$
\mathbf{w}(n+1) = \mathbf{w}(n) + \mu e(n)\text{sgn}[\mathbf{x}(n)].
\tag{9.2.23}
$$

The computation of $\mu e(n)$ can be implemented as a right shift of the binary representation of $e(n)$. In DSP implementations, conditional tests require more instruction cycles than multiplications in the LMS algorithm. Since L branch (IF-ELSE) instructions are required inside the adaptation loop to determine the sign of $x(n-i)$, $i = 0, 1, \ldots, L-1$, slower throughput than the sign-error LMS algorithm is expected. The convergence rate is also slower compared with that of the LMS algorithm.

Finally, the sign operation is applied to both $e(n)$ and $x(n)$, which results in the sign-sign LMS algorithm expressed as

$$
\mathbf{w}(n+1) = \mathbf{w}(n) + \mu\,\text{sgn}[e(n)]\,\text{sgn}[\mathbf{x}(n)].
\tag{9.2.24}
$$

This equation shows that the sign-sign LMS algorithm requires no multiplication. This simplified LMS algorithm is designed for a VLSI or ASIC implementation to save multiplications. It is used in the adaptive-differential pulse-code modulation (ADPCM) for speech compression. However, when this algorithm is implemented on a DSP processor with a pipeline architecture and parallel hardware multipliers, the throughput is slower than the standard LMS algorithm because the determination of signs can break the instruction pipeline and therefore severely reduce the execution speed.

Complex Least-Mean-Square Algorithm

The LMS algorithm was developed to process real-valued signals. However, some adaptive-filtering applications and frequency-domain adaptive filtering require complex operations. For example, in the adaptive filtering of data signals in modems, the data is translated to complex baseband signals for maintaining their phase relationships.

Similar to Eqs. (9.2.2) and (9.2.3), the complex adaptive filter uses the complex input vector $\mathbf{x}(n)$ and complex coefficient vector $\mathbf{w}(n)$ expressed as

$$
\mathbf{x}(n) = \text{Re}[\mathbf{x}(n)] + j\,\text{Im}[\mathbf{x}(n)] = \mathbf{x}_R(n) + j\mathbf{x}_I(n)
\tag{9.2.25}
$$

and

$$\mathbf{w}(n) = \text{Re}[\mathbf{w}(n)] + j \,\text{Im}[\mathbf{w}(n)] = \mathbf{w}_R(n) + j\mathbf{w}_I(n). \qquad (9.2.26)$$

The complex output $y(n)$ is computed as

$$y(n) = \mathbf{w}^T(n)\mathbf{x}(n), \qquad (9.2.27)$$

where all multiplications and additions are complex operations.

The complex LMS algorithm adapts the real and imaginary parts of $\mathbf{w}(n)$ simultaneously and is expressed as

$$\mathbf{w}(n + 1) = \mathbf{w}(n) + \mu e(n)\mathbf{x}^*(n), \qquad (9.2.28)$$

where * denotes a complex conjugate such that $\mathbf{x}^*(n) = \mathbf{x}_R(n) - j\mathbf{x}_I(n)$.

Delayed Least-Mean-Square Algorithm

The LMS algorithm is commonly used in adaptive FIR filtering. As described by Eqs. (9.2.1), (9.2.5), and (9.2.14), the filter output, $y(n)$, is computed and subtracted from the desired signal, $d(n)$. The error signal, $e(n)$, is then used to update the coefficients for the next iteration. In some practical applications, the desired signal, and thus the error signal, is not available until several sampling intervals later. In addition, in the implementation of adaptive filters using a pipeline architecture, the computational delay is an inherent problem. Therefore, there is a delay in the LMS algorithm for those applications.

The delayed LMS algorithm can be expressed as follows:

$$y(n - \Delta) = \mathbf{w}^T(n - \Delta)\mathbf{x}(n - \Delta), \qquad (9.2.29)$$
$$e(n - \Delta) = d(n - \Delta) - y(n - \Delta), \qquad (9.2.30)$$
$$\mathbf{w}(n + 1) = \mathbf{w}(n) + \mu e(n - \Delta)\mathbf{x}(n - \Delta). \qquad (9.2.31)$$

The delay in the coefficient adaptation has only a slight influence on the steady-state behavior of the LMS algorithm. The delayed LMS algorithm with delay $\Delta = 1$ is widely used in implementing adaptive FIR filtering on DSP processors with a pipeline architecture. This issue is discussed in Sections 9.6.2 and 9.6.3.

After the LMS algorithm has reached its minimum MSE, the adaptive process is completed, and the filter coefficients have converged to the optimum solution defined in Eq. (9.2.11). Therefore, the output from the adaptive filter closely matches the desired signal. When the signal characteristics change because of a time-varying environment, the filter tracks the new environment by generating a new set of coefficients. Some important properties of this widely used LMS algorithm are discussed next.

9.3 PROPERTIES OF ADAPTIVE FILTERS

In order to design and implement adaptive filters for a given application, we have to determine the values of parameters such as the step size, μ, the filter length, L, and the initial coefficient vector, $\mathbf{w}(0)$. To properly select these parameters, we have to understand important properties of adaptive algorithms. In this section, we summarize the properties of the LMS algorithm. Derivation of these results can be found in references [1–3].

9.3.1 Stability Conditions

As discussed in Section 9.2.1, the adaptive FIR filter has only adjustable zeros. However, we should not conclude that adaptive FIR filters are always as stable as the fixed-coefficient FIR filters introduced in Chapter 6. In fact, the stability of the filter depends on the algorithm that adjusts its coefficients. Starting from an arbitrary initial coefficient vector $\mathbf{w}(0)$, the vector $\mathbf{w}(n)$ gradually converges to the optimum vector defined in Eq. (9.2.11) if and only if the step size, μ, satisfies the stability condition

$$0 < \mu < \frac{1}{\lambda_{\max}}, \tag{9.3.1}$$

where $\lambda_{\max}$ is the largest eigenvalue of the matrix $\mathbf{R}$ defined in Eq. (9.2.8).

It is difficult to estimate $\lambda_{\max}$ during the algorithm-development time without real data. This problem may be solved by finding the upper bound of $\lambda_{\max}$. Since $\mathbf{R}$ is an autocorrelation matrix with nonnegative eigenvalues, the upper bound of $\lambda_{\max}$ becomes the trace of the matrix $\mathbf{R}$, which is LP_x, where P_x is the power of the input signal, $x(n)$. Therefore, an easy-to-use stability condition for the LMS algorithm is

$$0 < \mu < \frac{1}{LP_x}. \tag{9.3.2}$$

This equation shows that the step size, μ, is inversely proportional to the filter length, L, and the power of input signal, P_x, which can be estimated recursively using Eq. (9.2.18). Therefore, the normalization of the step size using the power estimate results in the more robust normalized LMS algorithm described in Eq. (9.2.16).

9.3.2 Convergence Rate

In applications with slowly changing signal statistics, the performance function drifts in time. Adaptation is the process of tracking the signals and environments. Thus, for many adaptive signal-processing applications, the most important consideration is the speed of convergence.

Algorithm convergence is attained when the MSE is reduced to its minimum value. The average time needed for the algorithm to converge is approximated by

$$\tau_{MSE} = \frac{1}{\mu \lambda_{\min}}, \tag{9.3.3}$$

where $\lambda_{\min}$ is the smallest eigenvalue of the matrix $\mathbf{R}$ defined in Eq. (9.2.8). Equation (9.3.3) shows that the required time for algorithm convergence is inversely proportional to the step size, μ. However, Eq. (9.3.1) shows that we cannot use arbitrary large step sizes to speed up convergence because of the stability constraint.

Substituting the upper bound of the step size given in Eq. (9.3.1) into Eq. (9.3.3), we obtain

$$\tau_{MSE} > \frac{\lambda_{\max}}{\lambda_{\min}}, \tag{9.3.4}$$

where $\lambda_{\max}/\lambda_{\min}$ is called the eigenvalue spread of the **R** matrix. Therefore, the convergence rate depends on the characteristics of the input signal. The eigenvalue spread can be approximated by the ratio of maximum-to-minimum spectrum magnitudes of the input signal. For signals that have a relatively flat spectrum, such as white noise, faster convergence can be expected. However, for other signals with signal components that have large power differences, such as speech, the convergence rate is slower.

The LMS algorithm is able to track slow time-varying signal characteristics. In this case, the LMS algorithm follows the moving minimum MSE, but may be lagging behind due to its slow convergence speed. This lag error decreases with an increase in the step size, μ, for faster tracking.

The time-domain LMS algorithm can be constructed in a frequency domain for computational savings. These fast implementations are called frequency-domain (or block) adaptive filters. They are capable of saving computational burdens with the same convergence properties as time-domain algorithms.

9.3.3 Steady-State Performance

With a true gradient and under noise-free conditions, the adaptive algorithm converges to the minimum MSE and remains there because the gradient is zero at the optimum solution. However, the derivation of the LMS algorithm given in Section 9.2.2 uses the gradient estimate in Eq. (9.2.13) instead of the true gradient. This noisy gradient estimate may not be zero at the minimum MSE, thus causing the coefficients to be updated randomly around the optimum values. This additive algorithm noise generates extra noise at the output of the adaptive filter in a steady state. Therefore, the steady-state performance of the algorithm is measured by the average of extra noise called excess MSE, which is the average increase in the MSE over the minimum MSE after convergence.

The excess MSE has been shown to be approximately

$$excess\ \text{MSE} = \mu L P_x \xi_{\min}, \tag{9.3.5}$$

where $\xi_{\min}$ is the minimum MSE at the optimum solution in a steady state. A normalization of the excess MSE with the minimum MSE is called the misadjustment. Equation (9.3.5) shows that the algorithm noise is proportional to the step size, μ. Equations (9.3.2) and (9.3.5) also clearly indicate that the step size, μ, must be reduced in proportion to the adaptive filter length, L. Therefore, using a longer filter length than necessary not only requires higher cost, but also introduces more misadjustment noise. When the environment is time invariant, the goal of the adaptive filter is to converge to the optimum coefficient vector that gives the best performance. To obtain a higher steady-state performance, a smaller value of μ is required. Unfortunately, a small μ results in slower convergence, as indicated by Eq. (9.3.3). When the environment is time varying, the adaptive filter should adjust its coefficients continuously using a larger μ to track environmental changes.

9.3.4 Finite-Precision Effects

Finite-wordlength effects of DSP systems and methods of preventing overflow are discussed in Chapter 3. For adaptive filters, the dynamic range of the filter output is determined by the time-varying filter coefficients, which are unknown at the design stage. In addition, the feedback of $e(n)$ makes signal scaling (to avoid overflow) more complicated.

In general, decreasing the magnitude of $d(n)$ reduces the gain demand on the filter, therefore reducing the coefficient values. Scaling the desired signal, $d(n)$, is equivalent to scaling the filter output and coefficients. Usually, the required value of the scaling factor α (<1) is not expected to be very small. Since it only scales the desired signal, $d(n)$, it does not affect the speed of convergence, which depends on the input signal, $x(n)$. An alternate method for preventing coefficient overflow is to use the leaky LMS algorithm introduced in Section 9.2.2.

The finite-precision LMS algorithm can be expressed as follows:

$$y(n) = Q\left[\sum_{i=0}^{L-1} w_i(n)x(n-i)\right], \qquad (9.3.6)$$

$$e(n) = Q[\alpha d(n)] - y(n), \qquad (9.3.7)$$

$$w_i(n+1) = w_i(n) + Q[\mu e(n)x(n-i)], \quad i = 0, 1, \ldots, L-1, \qquad (9.3.8)$$

where $Q[x]$ denotes the quantization (rounding) of the value x. When calculating the FIR filter output in Eq. (9.3.6) using a multiplier with a double-precision accumulator such as the fixed-point DSP processors introduced in Chapter 4, intermediate roundoff noise can be avoided. Therefore, roundoff error is only introduced when the final sum of products is rounded to single precision and transferred to a memory location. When updating coefficients according to Eq. (9.3.8), the computation of the update term $\mu e(n)x(n-i)$ produces a double-precision number, which is rounded and then added to the old coefficient value $w_i(n)$ to form the updated value $w_i(n+1)$.

The total output MSE can be derived as

$$\xi = \xi_{min} + \mu L P_x \xi_{min} + \frac{1}{\alpha^2}[\|\mathbf{w}^o\|^2 + k]\sigma_e^2 + \frac{L\sigma_e^2}{\alpha^2\mu}, \qquad (9.3.9)$$

where σ_e^2 is the variance of the quantization noise defined in Eq. (3.3.2). The second term represents excess MSE, as discussed in Section 9.3.3, which is proportional to the step size, μ. The third term in Eq. (9.3.9) arises because of quantization errors in the quantized input vector and filter output, $y(n)$, where k is the number of rounding operations in Eq. (9.3.6). In most fixed-point DSP processors, $k = 1$ since a double-precision accumulator is used.

For fixed-point arithmetic, the finite-precision error given in Eq. (9.3.9) is dominated by the last term, which reflects the error in the quantized coefficient vector and is inversely proportional to the step size, μ. Whereas the excess MSE is proportional to μ, the power of the roundoff noise is inversely proportional to μ. Although a small value of μ reduces the excess MSE, it may result in a large quantization error. Therefore, an optimum step size achieves a compromise between these two competing goals.

There is another factor to be considered in the selection of the step size, μ. From Eq. (9.3.8), the LMS algorithm modifies the current coefficients by adding an update term $Q[\mu e(n)x(n - i)]$. As stated in Section 9.2, the adaptive algorithm is aimed at minimizing the error signal, $e(n)$. That is, as the algorithm converges, the error signal decreases. At some point, the quantized update term, $Q[\mu e(n)x(n - i)]$, is rounded to zero because $e(n)$ is getting smaller and smaller. At this point, the adaptation of the filter virtually stops. Thus, the roundoff of update terms precludes the coefficients reaching the optimum values. This phenomenon is known as "stalling" or "lockup," which may be resolved by increasing the step size.

In order to stabilize the digital implementation of the LMS algorithm, we may use the leaky LMS algorithm to reduce numerical errors accumulated in filter coefficients. The leaky LMS algorithm prevents overflow in a finite-precision implementation by providing a compromise between minimizing the MSE and constraining the values of the adaptive-filter coefficients. It also minimizes the stalling effect by adding a small amount of random noise to the coefficient vector.

9.4 APPLICATIONS

The most important feature of an adaptive filter is its capability to operate effectively in an unknown environment and to track time-varying characteristics of input signals. Figure 9.1 shows a general form of adaptive filter that operates on input signals $x(n)$ and $d(n)$, and produces output signals $y(n)$ and $e(n)$. The adaptive filter can be used in different applications with different input/output configurations. In some applications, the desired signal, $d(n)$, is not available. For others, the error signal, $e(n)$, is measured by an error sensor. Adaptive filters have been widely used in communication, control, and many other systems. In this section, we introduce four basic applications that include adaptive system identification, adaptive inverse modeling, adaptive noise cancellation, and adaptive prediction.

9.4.1 Adaptive System Identification

Adaptive system identification is illustrated in Fig. 9.4. The adaptive filter is connected in parallel with the unknown system (or plant) to be modeled. The modeling signal $x(n)$ excites both the unknown system and the adaptive filter. The objective of the adaptive filter is to adapt to the unknown plant; thus, the adaptive filter output, $y(n)$, closely matches the unknown system output, $d(n)$. This is achieved by minimizing the error signal, $e(n)$, which is the difference between the physical response $d(n)$ and the model response $y(n)$. If the excitation signal, $x(n)$, is rich in

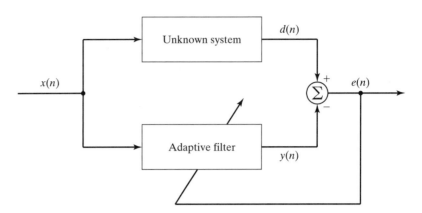

Figure 9.4   System identification using adaptive filters

frequency contents such as white noise and the internal plant noise is small, the adaptive filter converges to the unknown system from the I/O viewpoint.

In some real-world applications, the plant being modeled may change slowly with time. Furthermore, measurement noise is usually present at the sensor output, which corrupts filter coefficients. Therefore, the adaptive filter must identify and track the time-varying characteristics of the plant in the presence of plant and measurement noises.

Adaptive system identification can be applied to many practical applications such as control systems, geophysics, and communications. For example, in adaptive network (line) echo cancellation, the unknown system to be identified is a hybrid circuit located in the central office. The input signal $x(n)$ is far-end speech. Because of an impedance mismatch, part of the speech leaks back to the far end as an echo $d(n)$. The adaptive filter uses the far-end speech $x(n)$ to model the unknown system in order to generate the echo mimic $y(n)$, which approximates and cancels the undesired echo $d(n)$. The residual echo $e(n)$ is minimized and is sent back to the far end. Adaptive echo cancellers are widely used in long-distance voice communications, full-duplex data modems, and high-performance hands-free telephones.

Example 9.1

Adaptive system identification can be simulated using the Simulink program `sysident.mdl`, as shown in Fig. 9.5, which provides an example of adaptive echo cancellation. This simulation uses a 32-tap adaptive FIR filter with the LMS algorithm to model an unknown system, which is a lowpass FIR filter of length 32. A white noise from **Random Source** is used as input to excite both the unknown system and the adaptive filter. An additional noise source is introduced as plant noise at the output of the unknown system before subtracting the output of the adaptive filter to compute the error signal.

The initial values of the adaptive filter's coefficients are set to zero. The filter-output signal, $y(n)$, is shown in Fig. 9.6 (top plot), while the squared error is shown in the bottom plot. It shows that the magnitude of error signal decreases with time, which means that the adaptive filter gradually converges to the unknown system.

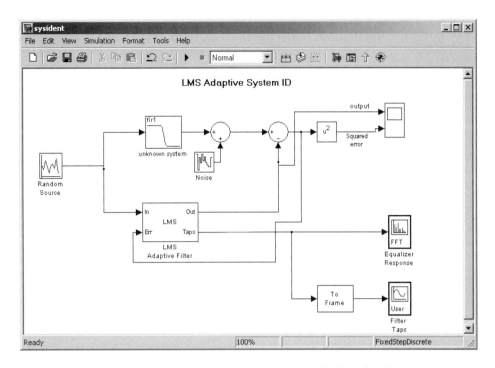

Figure 9.5   System identification using an adaptive FIR filter with the LMS algorithm

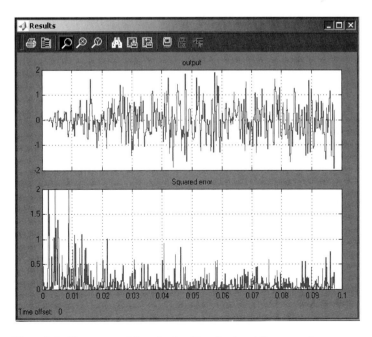

Figure 9.6   The output of the adaptive filter (top) and the squared error (bottom)

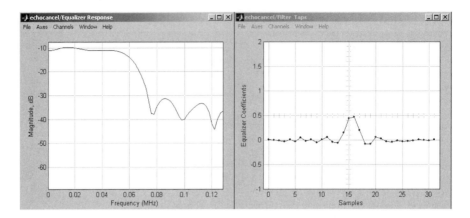

Figure 9.7   Magnitude (left) and impulse (right) responses of an adaptive filter

The magnitude (left) and impulse (right) responses of the adaptive filter are shown in Fig. 9.7. The magnitude response shows that the adaptive filter converged to the unknown system (lowpass filter). We can conduct further experiments using different step sizes and implement the normalized LMS algorithm by double-clicking on the LMS block and checking the **Use normalization** box. We can compare the differences in terms of convergence speed and steady-state performance by examining the squared-error signal.

9.4.2 Adaptive Inverse Modeling

In adaptive inverse-modeling applications, an unknown system is cascaded with an adaptive filter, as illustrated in Fig. 9.8. The desired signal, $d(n)$, is derived by delaying the signal $s(n)$ using a delay unit $z^{-\Delta}$, where $\Delta \approx L/2$. The purpose of the delay is to compensate for the propagation delay through the unknown system and the adaptive filter. The delay allows the adaptive filter to converge to a causal filter, which is the inverse of the unknown system. The adaptive filter equalizes the unknown system (or channel), thus recovering the delayed version of the signal, $s(n)$, at the output of the adaptive filter, $y(n)$. Similar to the forward modeling shown in Fig. 9.4, if $s(n)$ has a flat spectrum and the plant noise is small, the adaptive filter can be adapted to an accurate inverse model of the unknown system.

Inverse modeling is widely used in adaptive control, deconvolution, and channel equalization. For example, the adaptive equalizer in modems compensates for the distortion caused by transmission over telephone and radio channels. In adaptive channel equalization, $s(n)$ is the original data at the transmitter which is transmitted to the receiver through the communication channel (unknown system). At the receiver, the received data $x(n)$ is distorted by the unknown channel because each data symbol transmitted over a time-dispersive channel extends beyond the time interval used to represent that symbol, thus causing an overlay of received symbols. Since most channels are time varying and unknown in advance, the adaptive filter is required to converge to the inverse of the unknown system so that $y(n)$ approximates the delayed version of $s(n)$.

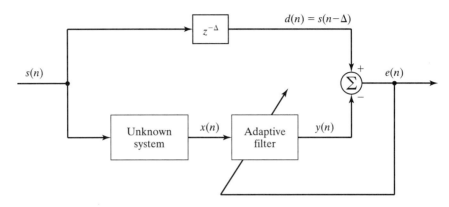

Figure 9.8   Inverse system modeling using an adaptive filter (training mode)

After a short training period using random numbers known to both the transmitter and receiver in modems, the transmitter begins to transmit the modulated data signal $s(n)$. To track the possible changes in the channel, the adaptive equalizer coefficients must be adjusted continuously while receiving the data sequence. However, the desired signal is no longer available to the adaptive equalizer located in the receiver at the other end of the channel. This problem can be solved by treating the output of the decision device at the receiver as correct and using it as the desired signal, $d(n)$, to compute the error signal. This technique is called the decision-feedback equalizer, which works well when decision errors occur infrequently.

Example 9.2

Adaptive inverse modeling can be simulated using the Simulink program `adeq.mdl` shown in Fig. 9.9. This model is modified from the Simulink demo file on solving a channel-equalization problem. The simulation starts in a training mode that transmits a known binary sequence (generating by the bandlimited white-noise block and the sign block) across a dispersive channel, which is further corrupted by a white noise. The received signal is used as input for the LMS block (adaptive FIR filter with 11 taps and a step size of 0.05), which adapts to the inverse of the dispersive channel. The error signal is derived by subtracting the output of the adaptive filter from the training signal, which is obtained by delaying the original binary signal by half the length of the adaptive filter.

The results of the adaptation are shown in Fig. 9.10, where the squared error (bottom plot) decreases below a level of 0.5 in 1.2 msec. Figure 9.11 shows that the adaptive filter converges to a highpass filter, which is the inverse of the lowpass channel used for this simulation.

After the adaptive filter has converged to a steady state, the adaptive equalizer can be switched to the decision feedback mode by clicking on the Manual Switch. This mode quantizes the output of the adaptive filter using a simple slicer. The slicer output is used as the desired signal and is compared with the output of the adaptive filter to generate an error signal for continuously adjusting the coefficients of the

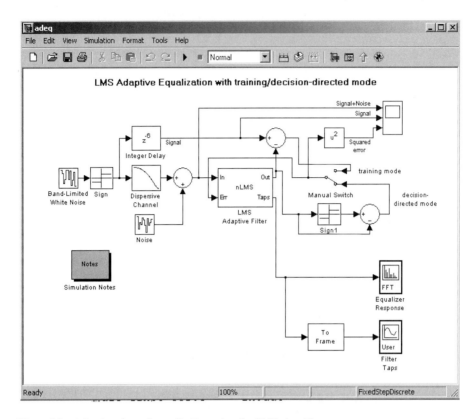

Figure 9.9   Adaptive channel equalization using the LMS algorithm

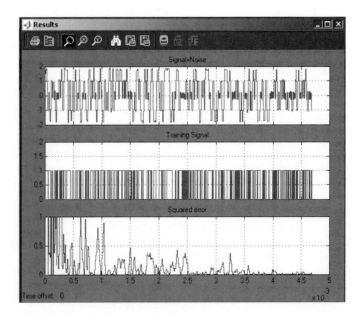

Figure 9.10 Performance of
an adaptive channel equalization (the top plot shows the
received signal, the middle plot
shows the binary sequence, and
the bottom plot shows the
squared error)

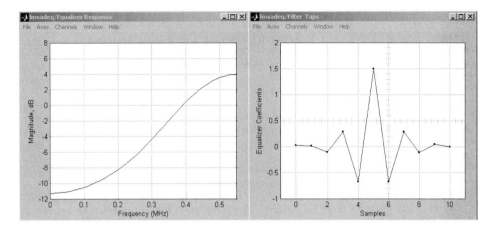

Figure 9.11 The magnitude (left) and impulse (right) responses of the converged adaptive filter to the inverse of a dispersive channel

adaptive filter. Similar to Example 9.1, further experiments on using the normalized LMS algorithm and different step sizes can be conducted for observing different convergence rates and steady-state performances. In addition, a more accurate performance measurement of the adaptive equalization system can be obtained by computing the bit-error rate.

9.4.3 Adaptive Noise Cancellation

Adaptive filters can be used to cancel noise components that overlap with an uncorrelated signal over the same frequency range. As illustrated in Fig. 9.12, the primary signal, $d(n)$, picked up by the primary sensor contains the desired signal, $s(n)$, and the undesired noise, $v(n)$. In order to remove noise from the primary signal, the

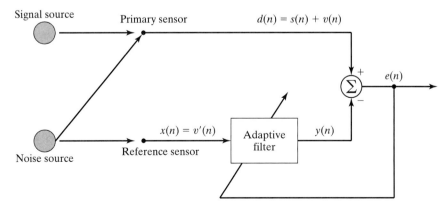

Figure 9.12 Adaptive noise cancellation

reference sensor is used to pick up the correlated noise, $v'(n)$, which is used as input $x(n)$ to the adaptive filter. Because $v'(n)$ and $s(n)$ are uncorrelated, the adaptive filter can only adjust the reference noise $v'(n)$ to produce $y(n)$ that approximates the noise $v(n)$. Therefore, the noise component $v(n)$ in the primary signal is canceled by $y(n)$; thus, the difference signal, $e(n)$, gradually converges to an approximation of the clean signal, $s(n)$.

As shown in Fig. 9.12, the primary sensor is placed close to the signal source, and the reference sensor is located close to the noise source. These two sensors should be positioned carefully such that the noise components $v(n)$ and $v'(n)$ are highly correlated, but that the signal $s(n)$ is not sensed by the reference sensor. There are tradeoffs between these two conflicting goals. For example, a simple way of increasing correlation between $v(n)$ and $v'(n)$ is to place these two sensors close together. Unfortunately, this arrangement also increases the amount of signal leakage to the reference sensor. Applications of the adaptive noise canceller include the cancellation of various forms of interference in electrocardiography, noise in fighter cockpit environments, antenna sidelobe interferences, and 60 Hz hums.

Example 9.3

A Simulink program `anc.mdl` is modified from the Simulink demo program `lmsdemo.mdl`. As shown in Fig. 9.13, a speech signal `timit1.wav` (sampled at 8,000 Hz) is corrupted by a white noise of a zero mean and unit variance and is picked up by the primary sensor. A 32-tap FIR lowpass filter is used to simulate the transfer function between the reference-noise source and the primary sensor. Therefore, the two noises are highly correlated. The reference noise picked up by the reference sensor is used as input to the adaptive FIR filter with the LMS algorithm.

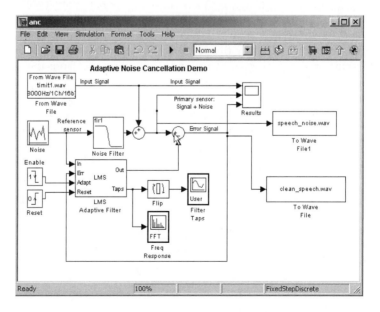

Figure 9.13   Adaptive noise cancellation using the LMS algorithm

As shown in the bottom plot of Fig. 9.14, the error signal gradually converges to a cleaner speech signal. After the simulation, both the clean signal (`clean_speech.wav`) and the corrupted signal (`speech_noise.wav`) can be played back to evaluate the performance of the adaptive noise canceller. The adaptive filter also converges to a low-pass filter, as shown in Fig. 9.15. Care must be taken in selecting proper step sizes. Select different step sizes that will not result in instability, and check the **Use normalization** box in the LMS block to implement the normalized LMS algorithm. The user can also experiment on different plants, input signals, and noises to examine the effect of correlation between the noise and signal. In addition, signal leakage to the reference sensor can also be simulated, and its effects on the convergence of the adaptive filter can be observed (see the problems at the end of this chapter for details).

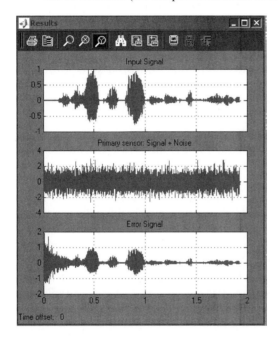

Figure 9.14   Results of the adaptive noise cancellation using the LMS algorithm (the original speech signal is shown in the top plot, the noisy signal is displayed in the middle plot, and the output signal is shown in the bottom plot)

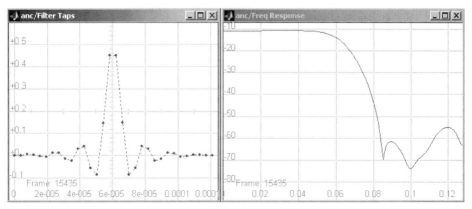

Figure 9.15   Coefficients vector (left) and magnitude response (right) of an adaptive filter

9.4.4 Adaptive Prediction

Adaptive prediction is illustrated in Fig. 9.16. The desired signal, $d(n)$, is delayed by Δ samples to form the input signal $x(n) = d(n - \Delta)$ for the adaptive filter, which adapts its coefficients to minimize the error signal, $e(n)$. The input signal $d(n)$ is assumed to have predictable signal components. The adaptive filter predicts the current value of the desired signal based on the past values. A major application of adaptive prediction is the waveform coding of speech such as ADPCM. In this case, the adaptive filter is designed to exploit the correlation between adjacent samples of the speech so that the value of the prediction error is much smaller than the original signal on average. This prediction error is then quantized using fewer bits for transmission, resulting in a lower bit rate. Other applications of adaptive prediction include spectral estimation and signal whitening.

The adaptive predictor can also be used to enhance the narrowband signal corrupted by broadband noise or to remove the narrowband interference from the corrupted broadband signal, depending on whether $y(n)$ or $e(n)$ is the desired output signal. For example, in adaptive line enhancement, $d(n)$ consists of desired narrowband components corrupted by broadband noises. The decorrelation delay, Δ, is sufficiently large so that the broadband noise components are uncorrelated, but the narrowband components still correlate in $d(n)$ and $x(n)$. The adaptive filter compensates for the delay in order to cancel the narrowband components in $d(n)$. Therefore, the filter output $y(n)$ consists of narrowband components only. This structure enhances the narrowband signal and is called the adaptive line enhancer (ALE). Usually, $\Delta = 1$ is adequate for white noises, and $\Delta > 1$ is used for color noises.

The adaptive predictor shown in Fig. 9.16 can also be used to attenuate undesired narrowband noise in a broadband signal. In this case, $d(n)$ consists of desired broadband components corrupted by narrowband interference. The delay is used to decorrelate the broadband components such that the adaptive predictor only predicts the narrowband components. The predictor output $y(n)$ is subtracted from $d(n)$ to cancel the narrowband noise, thus yielding an interference-free broadband signal at the output $e(n)$.

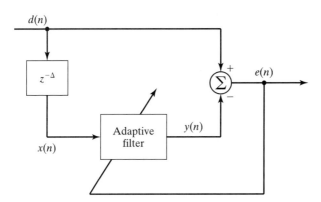

Figure 9.16   Block diagram of an adaptive predictor

Example 9.4

The Simulink program `ale.mdl` (modified from the Simulink demo file `lmsadlp.mdl`) shown in Fig. 9.17 illustrates the concept of an adaptive predictor in the case of enhancing a narrowband signal. In this ALE simulation, a 200 Hz sinewave is corrupted by a Gaussian white noise with a zero mean and a variance of 0.1. The sampling frequency is 8,000 Hz. A 16-tap adaptive FIR filter with the LMS algorithm is used to separate the narrowband sinewave from the broadband white noise.

The results are displayed in Fig. 9.18, which shows that the output of the adaptive filter quickly converges to a cleaner sinewave (bottom plot) as compared with the original noisy signal (top plot). The prediction error (middle plot) also converges to small values.

Similar to previous examples, we can perform further experiments by using different step sizes. Also, we can enable the **Use normalization** box to use the normalized LMS algorithm and observe the performance improvements.

The Simulink model shown in Fig. 9.17 can also be modified to attenuate the undesired narrowband noise from a broadband signal such as speech. In this case, the desired output is the prediction error. The user can use the Simulink file `adapred.mdl` to demonstrate the operation of the adaptive filter in this application. Note that the speech signal `mtlb.mat` must be loaded into the MATLAB workspace before this Simulink simulation can be run.

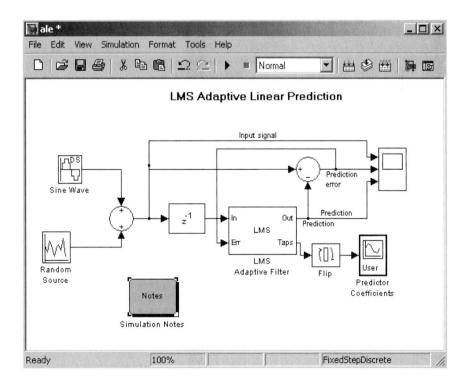

Figure 9.17   Adaptive line enhancer using the LMS algorithm

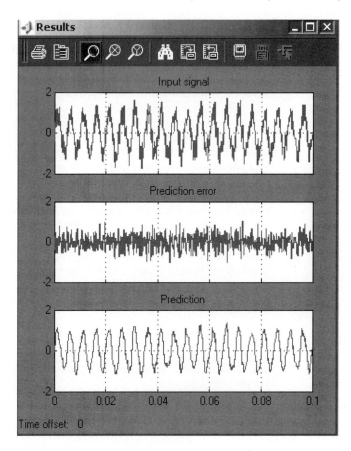

Figure 9.18 Results of the adaptive line enhancer (the noisy signal is shown in the top plot, the error signal is displayed in the middle plot, and the enhanced sinewave is shown in the bottom plot)

9.5 DESIGN AND IMPLEMENTATION USING MATLAB AND C

With many adaptive structures and algorithms introduced in Section 9.2, selecting the best one for the given application requires careful consideration. Some important issues such as filter performance, finite-precision effects, and algorithm complexity determine the suitability of adaptive filters. The design of an adaptive filter requires intensive computer simulations based on real data. MATLAB provides several adaptive algorithm functions to simplify the evaluation of the adaptive algorithms using different wordlengths. In this section, we use MATLAB to design, analyze, quantize, and implement adaptive FIR filters. Then we use C programs to implement the designed adaptive filters.

The Filter Design Toolbox contains several functions for performing adaptive filtering. As shown in Table 9.1, MATLAB supports the LMS, normalized LMS, signed LMS, RLS, and Kalman algorithms for updating filter coefficients. These functions offer the first step in studying the performance of adaptive filters. In addition, the toolbox also supports finite-wordlength analysis and implementation using fixed-point arithmetic.

TABLE 9.1 Adaptive Filtering Functions
Available in MATLAB

Function	Algorithm
adaptlms	LMS
adaptnlms	Normalized LMS
adaptsd	Sign-data LMS
adaptse	Sign-error LMS
adaptss	Sign-sign LMS
adaptrls	RLS
adaptkalman	Kalman

9.5.1 Design Examples

In this section, we study the performance of the adaptive filter with the LMS algorithm. The following command performs adaptive FIR filtering with the LMS algorithm:

```
[y,e,s] = adaptlms(x,d,s);
```

In the function, the input vectors x and d contain the input signal, $x(n)$, and desired signal, $d(n)$, respectively. The output vectors y and e contain the output signal, $y(n)$, and error signal, $e(n)$, respectively. The adaptive filtering function uses the structure s to set up parameters such as step size, initial coefficient vector, etc. These parameters are summarized in Table 9.2. The function initlms sets up the initial values of the structure elements.

TABLE 9.2 Structure Elements for the LMS Algorithm

Structure element	Description	initlms element
s.coeffs	FIR-filter coefficients. Initialized with coefficient vector w0. Updated coefficients are returned in s.coeffs when s is used as an output argument	w0
s.steps	Step size μ. Initialized with mu, it can be changed during the adaptation process	mu
s.states	States of the FIR filter after adaptation. It can be used to specify initial filter states	zi
s.leakage	Leakage factor ν for the leaky LMS algorithm. The default value is 1	lf
s.iter	Total number of iterations	

Example 9.5

We use simple adaptive system identification as an example to illustrate the usage of MATLAB functions and structures for adaptive filtering. The MATLAB script (exmp9_5.m) is listed as follows:

```
x   = 0.1*randn(1,640); % normal distributed random numbers as input
                        % x(n)
b   = fir1(31,0.3);     % an FIR filter as unknown system to be
                        % identified
d   = filter(b,1,x);    % generate desired signal d(n)
w0  = zeros(1,32);      % initialize filter coefficient vector w(0)
                        % to 0
mu  = 0.75;             % step size mu
s   = initlms(w0,mu);   % initialize w(0) and the step size mu
[y,e,s] = adaptlms(x,d,s); % adaptive filtering with the LMS algorithm
p1=stem(b,'r^'); hold on;  % impulse response of unknown system
p2=stem(s.coeffs);         % impulse response of adaptive FIR model
legend([p1,p2],'Actual','Estimated');
title('System identification of an FIR filter');grid on;
```

In the preceding MATLAB code, the input signal, $x(n)$, stored in vector x is the zero-mean random noise of variance 0.01. The unknown system to be modeled is an FIR lowpass filter with a normalized cutoff frequency of 0.3. The coefficients of the FIR filter are given in vector b. The desired signal, $d(n)$, is obtained by filtering the input signal, $x(n)$, through the unknown system described by vector b and is saved in vector d. The LMS algorithm adapts the coefficients of the adaptive FIR filter to identify the unknown system. A step size of $\mu = 0.75$ is used, which falls within the stability bound ($1/LP_x = 1/0.32 = 3.125$) stated in Eq. (9.3.2). The function initlms initializes the step size and coefficients for the adaptive filter before running the function adaptlms. The initial coefficients are normally set to zero without prior information about the optimum solution. As shown in Fig. 9.19, the filter coefficients (impulse response) converge to match the coefficients of the unknown system after 640 iterations.

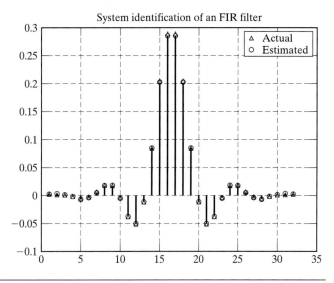

Figure 9.19 Coefficients of the unknown system (marked by a triangle) and the adaptive filter (marked by a circle)

The convergence of the adaptive filter is usually evaluated by examining the error signal. The plot of the MSE, $E[e^2(n)]$, versus time n is called the learning curve, which shows that the MSE gradually converges (or decreases) to a constant value greater than or equal to zero. In the MATLAB script `error_plot.m`, we simply plot the error signal, $e(n)$, vs. n. As shown in Fig. 9.20, the error signal, $e(n)$, gradually converges to zero, which indicates that the filter output, $y(n)$, approximates the unknown system output, $d(n)$. Thus, the adaptive filter identifies the unknown system.

As shown in Table 9.1, several adaptive algorithms can be used to update the coefficients of the adaptive filter. We repeat the adaptive system-identification simulation given in Example 9.5 using the MATLAB script `algorithms.m` for six different adaptive algorithms and compare their performance by plotting the corresponding error signals. Figure 9.21 shows the error signals obtained by running the different adaptive algorithms for 640 iterations. It shows that the RLS algorithm has the fastest convergence, while the sign-sign LMS algorithm has the slowest convergence and largest excess MSE. Note that the scale used for the sign-sign LMS algorithm is five times larger than that of the other algorithms, and the step size used in the signed LMS algorithms is reduced to 0.1, as compared to 0.75 used in the LMS algorithm. The use of smaller step sizes prevents instability in sign-LMS algorithms, which must satisfy a tighter step size bound when compared with the LMS algorithm.

In general, signed LMS algorithms quantize input and/or error signals. Therefore, these algorithms must use a smaller step size to ensure stability, which in turn leads to slower convergence. Signed LMS algorithms were originally developed to reduce multiplications in hardware implementation. Today, these algorithms are seldom used due to the increased processing power of DSP processors, which can

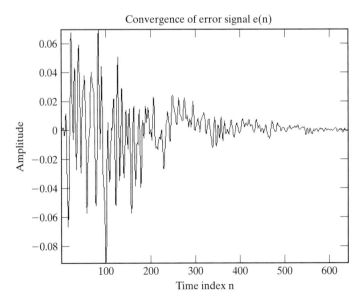

Figure 9.20

Convergence of an error signal toward zero

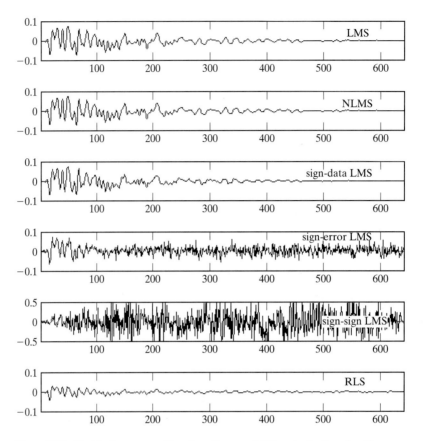

Figure 9.21   Residual errors resulting from the use of different adaptive algorithms (from top down): LMS, normalized LMS, sign-data LMS, sign-error LMS, sign–sign LMS, and RLS

perform very fast multiply-add operations, as discussed in Chapters 4 and 5. RLS and Kalman algorithms have the advantage of having faster convergence, but the disadvantage is the higher complexity for DSP implementation in terms of computation and memory requirements. In addition, these algorithms may suffer a numerical-instability problem.

9.5.2 Quantization of Adaptive Filters

In this section, we investigate the effects of quantizing adaptive filters. In fixed-point implementations, filter coefficients, step size, and signals are quantized to fixed-point numbers. The Filter Design Toolbox provides the functions `quantizer` and `quantize` to perform necessary quantization of the filter parameters with different wordlengths. Using the same adaptive system-identification program given in Example 9.5, we quantize the coefficient vector `w0`, input signal `x`, desired signal `d`, internal states of the adaptive filter, and step size mu to Q.15 format using the following script:

```
q = quantizer('mode','fixed','roundmode','floor',
    'overflowmode','saturate','format',[16,15]);
w0_q = quantize(q,w0); % quantize coefficient vector w0
mu_q = quantize(q,mu); % quantize step size mu
states_q = quantize(q,zeros(1,length(b)-1)');
s_q = initlms(w0_q,mu_q,states_q);
x_q = quantize(q,x);   % quantize the input signal x
d_q = quantize(q,d);   % quantize the desired signal d
```

Adaptive filtering with the LMS algorithm using quantized parameters is implemented as follows:

```
for n = 1:length(x)
    [y(n),e(n),s_q] = adaptlms(x_q(n),d_q(n),s_q);
    s_q.coeffs=quantize(q,s_q.coeffs);
    s_q.states = quantize(q,s_q.states);
end
```

We can repeat this adaptive system-identification simulation using Q.7 format by changing the quantizer parameter from [16, 15] to [8, 7].

We can observe the quantized parameters in the MATLAB command window and compare the difference between double-precision floating-point and Q.15 representations. It is necessary to change the display format to long by clicking on **Files → Preference → Command Window**. From **Command Window Preferences**, change **Numeric format** from short to long in order to view the numbers in long format.

The complete MATLAB script for simulating the quantized adaptive filter in a system identification application is given in the file quant_af.m. The converged adaptive-filter coefficients and the FIR-filter coefficients (of the unknown system to be modeled) are shown in Fig. 9.22. Figure 9.22(a) shows that the performance of the adaptive filter with Q.15 format is close to that of the adaptive filter using double-precision, floating-point format, as shown in Fig. 9.22(b). Thus, the fixed-point processor can be used for practical applications. Figure 9.22(c) shows that Q.7 format is not suitable since the filter coefficients converge to the wrong values. Therefore, fixed-point implementation of adaptive filters may affect the performance of the adaptive algorithm. Recall that signed LMS algorithms are a very crude form of quantization, thus resulting in poor performance. Some adaptive algorithms are more sensitive to quantization errors than others.

The Filter Design Toolbox is very useful in evaluating the performance of adaptive filters, analyzing their finite-wordlength effects, and providing possible remedies before implementation of algorithms using DSP processors. In Sections 9.6 and 9.7, we introduce the actual fixed-point and floating-point implementation of adaptive filters.

System identification of an FIR filter using LMS with Q.15 quantization

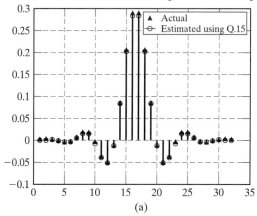

(a)

System identification of an FIR filter using LMS with double precision quantization

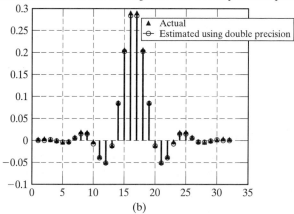

(b)

System identification of an FIR filter using LMS with Q.7 quantization

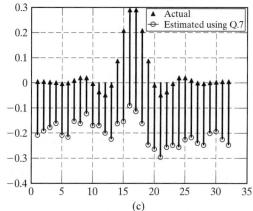

(c)

Figure 9.22 Performance of adaptive filters using different formats: (a) Q.15,
(b) double-precision, floating-point, and (c) Q.7

9.5.3 Adaptive Filtering in C

In this section, we implement the adaptive FIR filter with the LMS algorithm in C. The complete algorithm consists of three equations: Eq. (9.2.1) for obtaining the filter output, $y(n)$, Eq. (9.2.5) for computing the error signal, $e(n)$, and Eq. (9.2.15) for updating the filter coefficients for the next iteration at time $n + 1$. We also implement the normalized, leaky, and signed LMS algorithms in C.

As shown in Fig. 9.2, the FIR filter generates its output, $y(n)$, by performing the convolution operations defined in Eq. (9.2.1). The implementation of Eq. (9.2.1) is written in C as follows:

```
y[n] = 0.;                    // clear output buffer
for (i=0; i<L; i++){
   y[n] += wn[i]*xn[i];    // FIR filtering
}
```

In the code, wn[i] represents $w_i(n)$, and xn[i] represents $x(n - i)$. It is a good practice to initialize all coefficients to 0; however, the best initial coefficient vector is $\mathbf{w}(0) = \mathbf{w}^o$ if $\mathbf{w}^o$ can be estimated from off-line training operations. For example, startup training for the adaptive channel equalizer using white noise is a common practice used in modems.

The LMS algorithm is described in Eq. (9.2.15). Since $\mu e(n)$ is a constant for updating all of the L coefficients, the error signal $e(n)$, is first multiplied by μ to get $\mu e(n)$. This value is stored in memory location uen and then is multiplied by $x(n - i)$ to update the corresponding $w_i(n)$ for $i = 0, 1, \ldots, L - 1$. A C implementation of the LMS algorithm given in Eq. (9.2.15) is written as follows:

```
uen = mu*en;                  // compute μe(n)
for (i=0; i<L; i++){
   wn[i] += uen * xn[i];    // LMS update
}
```

In the code, en is the variable that represents $e(n)$, and mu is the variable that represents the step size, μ.

The normalized LMS algorithm described in Eqs. (9.2.16)-(9.2.18) can be implemented as follows:

```
powerXn = beta1*powerXn + beta*xn[0]*xn[0];   // power estimate
temp = alpha*en/(powerXn+saveMargin);   // normalize step size
for (i=0; i<L; i++){
   wn[i] += temp * xn[i];               // normalized LMS update
}
```

In the code, powerXn represents $\hat{P}_x(n)$, which is estimated using Eq. (9.2.18). The variable beta1 represents $(1 - \beta)$, which is precomputed during program development time, and the variable beta represents β. The temporary storage temp is computed using Eq. (9.2.17) with a small constant saveMargin for preventing the division of a very small number if the power estimate, $\hat{P}_x(n)$, is too small. This software constraint is very important for practical applications.

The leaky LMS algorithm described in Eq. (9.2.19) can be implemented as follows:

```
uen = mu*en;                        // compute μe(n)
for (i=0; i<L; i++){
   wn[i] = leaky*wn[i] + uen*xn[i]; // leaky LMS update
}
```

In the code, the leaky factor ν is represented by the variable `leaky`. The value of the leaky factor is slightly less than one.

The C implementation of the sign-error LMS algorithm given in Eq. (9.2.21) is listed as follows:

```
tu = mu;                    // assume e(n) is positive
if (en < 0.0){
   tu = -mu; }              // e(n) is negative
for (i=0; i<L; i++){
   wn[i] += tu * xn[i];     // sign-error LMS update
}
```

The sign-data LMS algorithm given in Eq. (9.2.23) is implemented in C as follows:

```
uen = mu*en;                // compute μe(n)
for (i=0; i<L; i++){
   if(xn[i] >= 0.0)         // x(n) is positive
      wn[i] += uen;         // sign-data LMS update
   else                     // x(n) is negative
      wn[i] -= uen;         // sign-data LMS update
}
```

Finally, the sign-sign LMS algorithm given in Eq. (9.2.24) is implemented in C as follows:

```
for (i=0; i<L; i++){
   if (en >= 0.0){          // if e(n) is positive
      if (xn[i] >= 0.0)     // if x(n) is positive
         wn[i] += mu;
      else                  // x(n) is negative
         wn[i] -= mu;}
   else{                    // e(n) is negative
      if (xn[i] >= 0.0)     // x(n) is positive
         wn[i] -= mu;
      else                  // x(n) is negative
         wn[i] += mu;}
}
```

The C implementation of the sign-sign LMS algorithm does not require multiplication. However, with the pipeline architecture and the powerful hardware parallel multiplier available on DSP processors, the LMS algorithm given in Eq. (9.2.15) is more efficient than the signed LMS algorithm.

As discussed in Chapter 4, the architecture of DSP processors has been optimized for the convolution operations given in Eq. (9.2.1) to compute the filter output, $y(n)$, in L cycles. However, the coefficient-update operations described in Eq. (9.2.15) cannot take advantage of this special architecture because each coefficient update involves loading the old coefficient value into the accumulator, performing a multiply-add operation, and storing the new coefficient back into the same memory location. The memory-transfer operation needs one clock cycle as the multiply-add, thus resulting in a dominant computational burden involved in updating filter coefficients.

As discussed in Chapter 3, real-time processing requires the processing time to be less than the sampling period. Thus, if the algorithm is too complicated, we need to reduce the complexity of the computation. Because the output convolution is more important and can be implemented very efficiently using DSP processors, Eq. (9.2.1) is executed for all coefficients, but only part of the coefficients in Eq. (9.2.15) are updated. The disadvantage of updating only a fraction of the coefficients at each sampling period is a slower convergence rate. In a worst-case scenario, we might update only one coefficient at each sampling period. The next coefficient would be updated for the next sampling period, and so on. When the computing power permits, a group of coefficients can be updated during each sampling period.

9.6 FIXED-POINT IMPLEMENTATIONS

Adaptive filters are widely used in communication systems and devices such as network echo cancellers, modems, and cellular phones. These DSP implementations are usually based on fixed-point processors due to cost considerations. In this section, we implement the adaptive FIR filter with the LMS algorithm on TMS320C2000, C5000, C62x and C64x processors.

9.6.1 Implementation Using the TMS320C2000

The implementation of several adaptive filter structures and adaptive algorithms on the C2000 processor can be found in an earlier publication [5].

In order to produce the fastest possible adaptive-filtering routine, all signal and coefficient buffers are stored in fast on-chip DARAM. As the memory map shows in Fig. 4.9, the TMS320C2000 processor has 544 words of on-chip DARAM, which are divided into three blocks: B0 (256 words), B1 (256 words), and B2 (32 words). For adaptive filtering, we generally use B0 for the coefficient buffer and B1 for the data buffer. In this case, we can configure B0 as program memory for performing FIR filtering and configure it as data memory for updating coefficients.

After the execution of the instruction

```
SETC   CNF      ; configure block B0 as program memory
```

the filter coefficient $w_i(n)$ from B0 (via the program bus) and the signal $x(n-i)$ from B1 (via the data bus) are available simultaneously for the parallel multiplier.

The MACD instruction enables complete multiply-accumulate, data-move, and pointer-update operations in a single cycle. The implementation of the inner product in Eq. (9.2.1) can be made even more efficient with the repeat instruction, RPT. Thus, an FIR filter of length L can be implemented as follows:

```
LAR    AR1,#LASTAP   ; AR1 point to x(n–L+1)
RPT    #L-1          ; repeat next instruction L times
MACD   #COEFFP,*-    ; multiply/accumulate and pointer update
```

In the code, AR1 is an address register that initially points to $x(n - L + 1)$, and the PC points to the last coefficient $w_{L-1}(n)$. When the MACD instruction is repeated, the coefficient address contained in the PC is incremented by one during its operation. The elements of the coefficient vector $\mathbf{w}(n)$ are stored in B0, as shown in Fig. 4.8. The MACD instruction in the repeat mode also copies data in B1 pointed at by AR1 to the next (higher) DARAM location, which is also depicted in Fig. 4.8.

The C2000 processor provides two instructions for performing the LMS update given in Eq. (9.2.15). The ZALR instruction loads a data-memory value into the high-order half of the accumulator while simulating the effect of rounding by setting bit 15 of the accumulator to 1 and setting bits 0-14 of the accumulator to 0. Since the updated coefficient is truncated to single precision before being stored, adding 1 to bit 15 is equivalent to rounding. The MPYA instruction adds the previous product in the PREG register (32 bits) to the accumulator, multiplies the operand with the data in the TREG register (16 bits), and stores the result in the PREG register.

In order to update the filter coefficients stored in B0, we have to configure block B0 as data memory using the following instruction:

```
CLRC   CNF       ; configure block B0 as data memory
```

Assuming that $\mu e(n)$ is already computed and stored in the TREG register and that the current address pointer is AR3, the coefficient update is implemented by the following instructions:

```
     LAR    AR1,#L-1       ; load loop counter AR1=L-1
     LAR    AR2,#COEFFD    ; AR2 points to wL-1(n)
     LAR    AR3,#LASTAP+1  ; AR3 points to x(n-L+1)
     MPY    *-,AR2         ; PREG=(mu)*e(n)*x(n-L+1
ADP  ZALR   *,AR3          ; load wi(n) and add 1/2 LSB
     MPYA   *-,AR2         ; ACC=PREG+wi(n), PREG=mu*e(n)*x(n-i)
     SACH   *+,0,AR1       ; store wi(n+1)
     BANZ   ADP,*-,AR2     ; loop if counter AR1>0, decrement AR1
```

For each iteration, L cycles are needed to perform the FIR filtering given in Eq. (9.2.1), and $6L$ cycles are needed to perform the coefficient updates defined in Eq. (9.2.15). Thus, the total number of cycles needed is $7L$. Note that BANZ needs three cycles for execution, which can be avoided by using straight-line code.

The leaky LMS algorithm is given in Eq. (9.2.19), where ν is slightly less than 1. Assuming $\nu = 1 - c$ (where $c \ll 1$), the leaky LMS algorithm given in Eq. (9.2.19) can be implemented efficiently as

$$w_i(n + 1) = w_i(n) - cw_i(n) + \mu e(n)x(n - i). \qquad (9.6.1)$$

Assuming that the constant c is selected as

$$c = 2^{-M}, \qquad (9.6.2)$$

the second term $cw_i(n)$ can be calculated using the barrel shifter to shift $w_i(n)$ M bits to the right. Since the accumulator is 32 bits and since the high word (bits 31-16) is used for updating $w_i(n)$, shifting $w_i(n)$ M bits to the right can be implemented by loading $w_i(n)$ and shifting it $(16-M)$ bits to the left. The C2000 implementation of the leaky LMS algorithm given in Eq. (9.6.1) is listed as follows:

```
        LAR   AR1,#L-1        ; load loop counter AR1=L-1
        LAR   AR2,#COEFFD     ; AR2 points to wL-1(n)
        LAR   AR3,#LASTAP+1   ; AR3 points to x(n-L+1)
        MPY   *-,AR2          ; PREG=mu*e(n)*x(n-L+1)
AD      ZALR  *,AR3           ; load wi(n) and add 1/2 LSB
        MPYA  *-,AR2          ; ACC=PREG+wi(n), PREG=mu*e(n)*x(n-i)
        SUB   *,#LEAKY        ; LEAKY=16-M
        SACH  *+,0,AR1        ; store wi(n)
        BANZ  AD,*-,AR2       ; loop if counter AR1>0 and decrement
                                AR1
```

For each iteration, $7L$ cycles are needed to perform the adaptation process ($6L$ cycles for the LMS algorithm). Therefore, the total number of cycles needed is $8L + 34$.

9.6.2 Implementation Using the TMS320C54x

Due to the importance of adaptive filtering in practical applications, the C54x processor provides a special instruction LMS that computes the FIR filtering and coefficient update in parallel.

The syntax for the LMS instruction is

```
LMS    Xmem x Ymem
```

where Xmem and Ymem point at the coefficient and signal elements, respectively. This instruction performs the FIR filtering, saves the product in ACC B, and updates the old coefficient in ACC A in parallel as follows:

```
    B = B + Xmem × Ymem
 || A = rnd(error+Xmem)
```

This instruction can be interpreted as follows:

```
      MAC    Xmem+,Ymem+,B   ; B = B+(*coeff+ * *input+)
   ||{ADD    Xmem,16,A       ; A = (*coeff+A) << 16
      RND    A}              ; A = round(A)
```

The first instruction performs the multiply-add operation of the FIR filtering. To avoid the conflict of accessing the same signal-coefficient pair, the multiply-add operation operates on the first coefficient-signal pair and places the result in ACC B. At the same time, the second coefficient element is added to ACC A. The result is left-shifted by 16 bits and rounded before being placed into ACC A. In this way, the filtering and the adaptation of the coefficient are carried out in ACC B and ACC A, respectively.

It is important to note that by using the LMS instruction, it is not possible to update the coefficient based on the current error signal, $e(n)$. Instead, a delayed LMS algorithm is implemented, as shown in Eq. (9.2.31). As mentioned in Section 9.2.2, the delayed LMS algorithm with delay $\Delta = 1$ is the most efficient adaptive algorithm for DSP processors with a pipeline architecture such as the TMS320C5000. A delay-update by one sample is implemented as

$$w_i(n + 1) = w_i(n) - \mu x(n - i - 1)e(n - 1), \qquad (9.6.3)$$

where adaptation using the previous error is given as

$$e(n - 1) = d(n - 1) - y(n - 1). \qquad (9.6.4)$$

A consequence of using the delayed LMS algorithm is a slightly slower convergence when compared with that of the original LMS algorithm.

A partial code [12] for the implementation of an adaptive filter with the LMS algorithm on the C54x processor is listed as follows:

```
      .asg    AR3,coeffs      ; assign AR3 as coeffs
      .asg    AR4,input       ; assign AR4 as input
      .asg    AR1,result      ; assign AR1 as result
      STM     #a,coeffs       ; coeffs is pointing to coefficients
      STM     #x,input        ; input is pointing to signal
      STM     #y,result       ; result is pointing to output
      STM     #1,AR0          ; setup AR0=1
      STM     #taps,BK        ; setup circular buffer size
      LD      update,T        ; T = μ*e
      LD      #0,B            ; initialize B to zero
      STM     #taps-2,BRC     ; block loop counter = taps-2
      RPTBD   lms_end-1       ; start of block repeat loop
      MPY     *input,A        ; A = T * input (not inside block loop)
      LMS     *coeffs,input+  ; B = w0*x0 || A = e0+w0 (not inside
                              ; block loop)
      ST      A,*coeffs+0%    ; update w0
   || MPY     *input,A        ; A = next updating term
      LMS     *coeffs,*input+; B=accumulate filtering || A=e+w
lms_end:
      STH     A,*coeffs+0%    ; store last updated w
      STH     B,*result+      ; store the result
```

As discussed in Section 9.5.3, $\mu \times e(n)$ is computed and the result is stored in the temporary register, T. This value is multiplied with each input sample $x(n - i)$

to form the update term and then added to the corresponding old coefficient, $w_i(n)$, to obtain the new coefficient, $w_i(n + 1)$. The delayed block-repeat instruction, RPTBD, allows the instructions MPY *input,A and LMS *coeff,input+ to be fetched and executed before the RPTBD instruction, thus maximizing pipeline operations. The parallel instructions ST||MPY within the block-repeat loop multiply the signal sample by the constant $\mu e(n)$ stored in the T register. The LMS instruction updates the coefficient and accumulates the filtered output.

The TMS320C54x DSPLIB [8] provides three adaptive-filtering functions that use the LMS instruction for adaptive filtering. These functions and their benchmarks are summarized in Table 9.3. All three functions are based on the delayed LMS algorithm. The normalized-delayed LMS filter routines ndlms and nblms normalize the step size to improve convergence. The block LMS function nblms partitions the L coefficients into nb blocks, with each block consisting of the bs coefficient (thus, $L = nb \times bs$), and only updates a block of coefficients for each iteration.

9.6.3 Implementation Using the TMS320C55x

Similar to the TMS320C54x, the C55x also supports the LMS instruction for implementing the delayed LMS algorithm. The syntax of this instruction is expressed as

```
LMS    Xmem,Ymem,ACx,ACy
```

This instruction performs the following two parallel operations in one cycle:

```
ACy = ACy + (Xmem × Ymem)
:: ACx = rnd(ACx +(Xmem << #16))
```

The first operation performs a multiplication and an accumulation in the DU, while the second operation adds the accumulator contents and the contents of data memory operand Xmem shifted left by 16 bits.

The LMS instruction implements the FIR filtering given in Eq. (9.2.1) and part of the LMS adaptation given in Eq. (9.2.15). For example, if AR0 points at the coefficient buffer and AR1 points at the signal buffer, the instruction

```
LMS    *AR0,*AR1,AC0,AC1
```

performs two parallel operations. The first operation multiplies the coefficient $w_i(n)$ addressed by AR0 with the signal $x(n - i)$ pointed at by AR1, and the

TABLE 9.3   Benchmark for the Adaptive Routines in the TMS320C54x DSP Library [8]

Function	Algorithm	Cycle count	Code size
dlms	Delayed LMS	$N \times (14 + 2L) + 45$	62
ndlms	Normalized-delayed LMS	$N \times [63 + 2(L - 1)] + 52$	144
nblms	Normalized block-delayed LMS	$N \times [85 + bs + L +$ $(18 + bs) \times nb] + 88$	144

product is added to AC1. The second operation shifts left the data $w_i(n)$ (pointed at by AR0) by 16 bits and added to AC0, which contains the update term $\mu e(n-1)x(n-i-1)$. It is important to note that before the completion of Eq. (9.2.1), we do not have the current output, $y(n)$, and thus we cannot compute the current error signal, $e(n)$. Therefore, we have to use the previous error signal $e(n-1)$ and the delayed update term $\mu e(n-1)x(n-i-1)$ instead of the current update term $\mu e(n)x(n-i)$ given in Eq. (9.2.15).

Similar to the C54x processor, the C55x also provides the delayed LMS routine dlms in the TMS320C55x DSPLIB [9]. This routine needs a code size of 122 bytes, and the cycle count is $N \times (5 + 2L) + 26$. A special requirement imposed in the C55x dlms routine is that it requires a minimum of two input samples and two coefficients. The delayed LMS algorithm is implemented in DLMS.asm using the TMS320C55x assembly language. This assembly subroutine can be called from the C program. The kernel of code is listed as follows:

```
StartSample:
     MOV    #0,AC1                          ; clear AC1 for initial error
                                            ; signal
  || RPTBLOCAL Outer_End-1                  ; while starting outer loop
     MOV    HI(AC1),T3                       ; place error signal in T3
     MOV    *ar_input+,*ar_data+            ; copy input -> x(n)
     MPYM   *ar_data+,T3,AC0                ; place first update term in
                                            ; AC0
  || MOV    #0,AC1                          ; while clearing FIR value
     LMS    *ar_coef,*ar_data,AC0,AC1       ; AC0 = update coefficient
                                            ; AC1 = start of FIR output
  || RPTBLOCAL  Inner_End-1                 ; while starting inner loop
     MOV    HI(AC0),*ar_coef+               ; store updated coefficient
  || MPYM   *ar_data+,T3,AC0                ; while calculating next
                                            ; update term
     LMS    *ar_coef,*ar_data,AC0,AC1       ; AC0 = updated coefficient
Inner_End:
                                            ; AC1 = update of FIR output
     MOV    HI(AC0),*ar_coef+               ; store updated coefficient
  ||MOV     rnd(HI(AC1)),*ar_output+        ; and store FIR output
     SUB    AC1,*ar_des+<<#16,AC2           ; AC2 is error amount
  || AMAR   *ar_data+                       ; point to oldest data sample
     MPYR   T_step,AC2,AC1                  ; place updated mu_error term
                                            ; in AC1
Outer_End:
```

In the code, it is assumed that XAR0 through XAR4 already are loaded with pointers to the input vector, coefficient vector, output vector, expected output array, and delay-buffer structure, respectively. The registers T0 and T1 are loaded with the values of the step size, μ, and filter length, L, respectively. The filter coefficients are stored in a circular buffer. The delay buffer is a structure composed of an index register and a circular buffer of length $L + 1$. The index register is pointing at the oldest signal sample in the circular buffer.

In addition to dlms.asm, the C55x DSPLIB also provides the function dlmsfast.asm for fast implementation of the delayed LMS algorithm. Unlike the dlms function, which uses the LMS instruction, this fast LMS algorithm separates coefficient updating and filtering to get a better cycle count. In this implementation, two input samples are processed as a pair. The filtering operation uses a dual MAC to process two signal samples, and two coefficients are updated in correspondence with these two samples. The cycle count of this function is $(N/2) \times (26 + 3L) + 71$, and the code size is 322 bytes. Compared with dlms.asm, this function is good for a large number of input samples and requires a minimum of 10 coefficients.

9.6.4 Implementation Using the TMS320C62x

As explained in Section 6.3.4, FIR filtering can be implemented efficiently using the C6000 compiler-intrinsic function _mpy. This intrinsic can also be used to implement the adaptive FIR filter with the LMS algorithm. For example, the coefficient and signal vectors w and x are declared as short (16 bits); the output yout is defined as int (32 bits); and the error signal err, the step size mu, and the desired signal d all are declared as short. The C code using three _mpy intrinsics for adaptive filtering with the LMS algorithm can be written as:

```
for (i=0; i<L, i++)
   yout += ((_mpy(x[i],w[i]))<<1);   // FIR filtering
err = d - (int)(yout>>16);        // shift right by 16 bits for
                                  // alignment
mu_err = (int)((_mpy(mu,err))>>15); // shift right by 15
for (i=L-1; i>=0, i--)            // start from last input &
                                  // coefficient
{
   w[i]+=((_mpy(mu_err,x[i])>>15);  // coefficient updates
   x[i]=x[i-1];                      // perform data shifting
}
```

Note that the multiplications are carried out in Q.15 format and must be left-shifted by 1 bit to get rid of the extra sign bit, as explained in Chapter 3. Therefore, we need to right-shift the product by 15 bits to align it for 16-bit addition.

The TMS320C62x DSPLIB [10] provides an adaptive FIR filtering function DSP_firlms2 using the delayed LMS algorithm. This function uses the error signal derived from the previous sample. The number of coefficients in the FIR filter is restricted to multiples of two. The code size of this routine is 256 bytes and the cycle count is $3L/2 + 26$.

9.6.5 Implementation Using the TMS320C64x

The C64x processor is an extension of the C62x processor. The new features of the C64x include wider data paths, a larger register file, and new instructions that support packed-data processing. For example, it has the capability of processing four 16-bit $\times$ 16-bit multiplications in a single cycle.

As introduced in Section 6.3.5, we can use the _dotp2 intrinsic (which uses the DOTP2 instruction) to implement two 16-bit × 16-bit multiply-add operations in one cycle and thus reduce the loop count by half.

Similarly, _smpy2 and _add2 intrinsics can be applied to perform two 16-bit × 16-bit multiplications and two 16-bit additions in one cycle. The implementation of the LMS algorithm using these two intrinsics is shown below:

```
for (i=0; i<L/2, i++)
  yout += _dotp2(x[i],w[i]);      // FIR filtering
err = d - (int)(yout>>16);        // shift right by 16 bits for
                                  // alignment
mu_err = _mpy(mu,err);            // mu*e(n)
mu_err_v = _spack2(mu_err,mu_err); // packing
for (i=L/2; i>0, i--)             // start from the last input &
                                  // coefficient
{
  temp =((_smpy2(mu_err_v,x[i])));  // compute update term
  temp_hi = _hi(temp);              // unpacking
  temp_lo = _lo(temp);
  temp_com = _spack(temp_hi,temp_lo)
  w[i] = _add2(temp_com,w[i]);      // LMS update
}
```

The variables x, w, mu_err, and yout in the code are defined as 32-bit int.

Similar to the C62x processor, the C64x version of the adaptive-filtering function DSP_firlms2 is included in the C64x DSPLIB [11]. It uses 16-bit signal and coefficient vectors and a 16-bit error signal, but the output is in long (32-bit) format. Note that the number of coefficients must be a multiple of four. The code size is 148 bytes, and the cycle count is $3L/4 + 17$.

9.7 FLOATING-POINT IMPLEMENTATIONS

In this section, we implement adaptive filters using floating-point TMS320C3x and TMS320C67x processors.

9.7.1 Implementation Using the TMS320C3x

As introduced in Chapter 5, the architecture of the C3x allows two data memory addresses to be generated at the same time. Thus, parallel data store, load, or one data store with one data load can be performed simultaneously. In addition, the hardware multiplier and the ALU of the C3x are separated. The processor can do one multiplication and one addition (or subtraction) at the same time with proper operand arrangement. In combining these two features, the C3x can execute several parallel instructions.

Using single repeat instruction, RPTS, the FIR filtering operations given in Eq. (9.2.1) can be implemented as follows [5]:

```
    MPYF3   *AR0++(1)%,*AR1++(1)%,R1    ; w0(n)*x(n)
    RPTS    L-2                         ; repeat L-1 times
    MPYF3   *AR0++(1)%,*AR1++(1)%,R1    ; wi(n)*x(n-i)
  ||ADDF3   R1,R2,R2                    ; y(n)=y(n)+wi(n)*x(n-i)
    ADDF3   R1,R2,R2                    ; add the last product
```

The auxiliary registers AR0 and AR1 point to $w_i(n)$ and $x(n - i)$ arrays, respectively. The first product of the first instruction prior to the repeat instruction initializes R1. The addition in the parallel instruction sums the previous values of R1 and R2. Note that the program uses circular buffers for the $x(n - i)$ and $w_i(n)$ arrays.

Although the C3x does not provide any special instructions for updating filter coefficients with the LMS algorithm, it still can update each weight using two instructions due to its architecture. As discussed in Chapter 5, the C3x has a repeat-block instruction, RPTB, which allows a block of instructions to be repeated without loop overhead. Assuming that the step size, μ, is stored in R5 and the error signal, $e(n)$, is stored in R7, the adaptation of filter coefficients is listed as follows [5]:

```
    MPYF    R5,R4           ; mu*e(n) -> R4
    LDI     @ADDW,AR0       ; AR0 points to w^L-1(n)
    LDI     @ADDXN,AR1      ; AR1 points to x(n-L+1)
    LDI     ORDERW-2,RC     ; L-2 -> repeat counter
    LDI     ORDERW,BK       ; circular buffer size = L
    CALL    LMS             ; call LMS routine to update W(z)
```

The subroutine LMS, which performs the LMS algorithm, is listed as follows:

```
LMS     MPYF3   *AR1++(1)%,R4,R1    ; mu*e(n)*x(n-L+1) -> R1
        RPTB    LOOP1               ; repeat following instructions
                                    ; L-1 times through label LOOP1

        MPYF3   *AR1++(1)%,R4,R1    ; mu*e(n)*x(n-i) -> R1
      ||ADDF3   R1,*AR0,R2          ; wi(n)+R1 -> R2
LOOP1   STF     R2,*AR0++(1)        ; store wi(n+1)
        NOP
        ADDF3   R1,*AR0,R2          ; add last product
        STF     R2,*AR0             ; store w0(n+1)
        RETS                        ; return to main program
```

Performing the LMS algorithm for updating an adaptive FIR filter of length L on the C3x processor requires $2L$ cycles.

9.7.2 Implementation Using the TMS320C67x

The C67x processor is the floating-point version of the C62x processor. However, unlike C62x and C64x processors, the C67x does not provide a DSPLIB. The C67x contains floating-point instructions, which operate on data in either single-precision or double-precision format. Floating-point formats and instructions are described in Chapters 3 and 5.

In this section, we explore the implementation of the adaptive FIR filter using single-precision, floating-point arithmetic on the C67x processor. The data variables x and w are specified as single precision (32 bits). These variables are loaded into registers and used by single-precision multiply (MPSYSP) and add (ADDSP) instructions, which is an extension of the FIR-filtering code described in Section 6.4.2. The first loop is repeated until all of the coefficients in the FIR filter have been multiplied by the corresponding signal samples, and the second loop is used to update the coefficients with the LMS algorithm. This kernel code is listed as follows:

```
fir_loop       LDW          *x_pointer++,x
               LDW          *w_pointer++,w
               MPYSP        x,w,yout
               ADDSP        yout,prod,yout
               SUB          loop_count,1,loop_count
[loop_count]B               fir_loop
               SUBSP        des,yout,err
               MPYSP        mu,err,muerr
adap_loop      MPYSP        muerr,x,update
               ADDSP        w,update,w
               SUB          loop_count,1,loop_count
[loop_count]B               adap_loop
```

We can improve the performance further by using the 64-bit data access available in the C67x processor. Instead of loading the word using the LDW instruction, the C67x processor can load a double word of two 32-bit data samples (x1:x0) and two 32-bit coefficients (w1:w0) using the LDDW instruction. This instruction is followed by two MPYSP instructions to perform single-precision multiplications (x1 × w1 and x0 × w0) independently. These partial products are accumulated in the yout1 and yout0 variables for even and odd elements using two single-precision ADDSP instructions, as shown in Section 6.4.2. Finally, the MPYDP and ADDDP instructions perform double-precision (64-bit) FIR filtering.

9.8 EXPERIMENTS AND PROBLEMS

In this section, we design an adaptive filter for the adaptive line enhancement shown in Fig. 9.16 to recover the sinewave corrupted by white noise. The ALE was first tested in Example 9.4 using Simulink. The adaptive filter used in the simulation is the FIR filter with the LMS algorithm. The filter length is 16, and the step size is 0.01.

9.8.1 Design an Adaptive Filter Using MATLAB

In this experiment, we implement the ALE application shown in Fig. 9.16 using a MATLAB script. In the program, we use the adaptlms function to perform adaptive FIR filtering with the LMS algorithm. The M-file ale_lms.m that implements ALE is listed as follows:

```
clear all, close all;
Fs = 8000;                          % sampling rate
```

```
n = sqrt(0.1)*randn(1,500);          % noise with var = 0.1
s = sin(2*pi*200*(1:1:500)/Fs);      % signal with power = 0.5
d = s+n;                             % desired signal, d(n)
x = [0 d(1:length(d)-1)];            % input to AF, delayed version
                                     % of d(n)
L = 16;                              % length of FIR filter
w0 = zeros(1,L);                     % initialize filter coefficients
                                     % to 0
mu_lms = 0.01;                       % step size

% Adaptive filtering with the LMS algorithm
Slms = initlms(w0,mu_lms);
[ylms,elms,Slms] = adaptlms(x,d,Slms);
figure(1)
subplot(311); plot(d); title('Sinewave corrupted by noise');
subplot(312); plot(ylms);
subplot(313); plot(elms);
figure(2)
subplot(211);psd(d); title('PSD before ALE'); grid on;
subplot(212);psd(ylms);title('PSD after ALE (LMS)');grid on;
```

The desired signal d contains the sinewave corrupted by white noise, as shown at the top of Fig. 9.23. The spectrum of the noisy sinewave is shown at the top of Fig. 9.24. The input signal x to the adaptive filter is the signal d delayed by one sample. As discussed in Section 9.4.4, one sample delay works well to decorrelate the white noise. The enhanced output is displayed in the middle of Fig. 9.23, which shows that the output gradually converges to a cleaner sinewave. The spectrum of

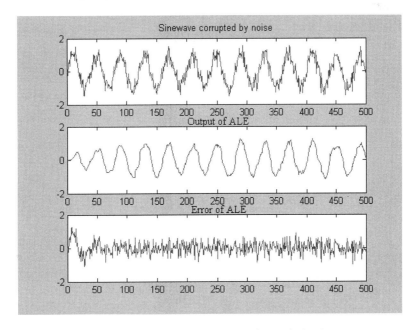

Figure 9.23   Input (top), output (middle), and error (bottom) signals

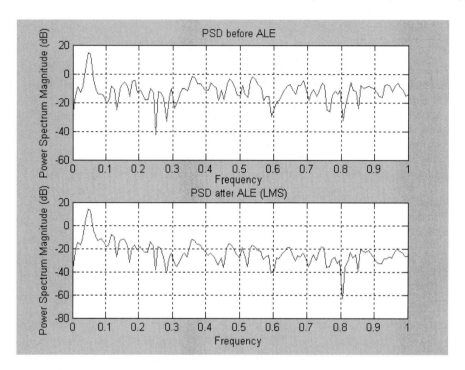

Figure 9.24   Spectra of ALE input (top) and output (bottom) signals

the ALE output is displayed at the bottom of Fig. 9.24, which shows that the broadband noise is attenuated by the adaptive filter. The error signal is reduced to the white noise, as shown at the bottom of Fig. 9.23.

The performance of the adaptive filter depends on the selection of the step size, filter length, and algorithms. The user can modify the M-file ale_lms.m as follows:

Step 1. Change the step size from 0.01 to 0.2, 0.1, 0.01, 0.005, and 0.001 for the LMS algorithm, and observe the different results.

Step 2. Increase the filter length, L, of the adaptive filter from 16 to 32, 64, and 128, and examine the convergence time and performance using the same step size of 0.01.

Step 3. Use the normalized LMS and RLS algorithms, and compare the convergence rate and computational cost.

Step 4. Plot the magnitude spectrum of the impulse response (filter coefficients) after convergence.

Step 5. Generate a new input signal d that consists of two sinewaves embedded in white noise. Observe the performance of the ALE related to the frequency difference between these two sinewaves.

The M-file for further simulations of ALE is written in adaptive_ale.m.

9.8.2 Implementation Using Floating-Point C

The ALE developed in Section 9.8.1 can be implemented using floating-point C on general-purpose computers. First, we use `singen9_8.m` to generate the same noisy sinewave used in `ale_lms.m` and then save it in the data file `in.dat`. The C program reads the input signal $d(n)$ from this data file. The floating-point C program `ale_floatpt.c` is listed as follows:

```c
#include <stdio.h>
#include <stdlib.h>
#include <math.h>

void main()
{
  #define NL 16                 // length of FIR filter
  float yn = 0.0;
  float dn, err, mu_err;
  float xnbuf[NL];
  float w[NL];
  int i,j,k;
  float mu = 0.01;
  float xn = 0;

  FILE *xn_in;                  // file pointer of x(n)
  FILE *yn_out;                 // file pointer of y(n)
  xn_in = fopen("in.dat","r");  // open file for input x(n)
  yn_out = fopen("out.dat","w"); // open file for output y(n)

  /*    Start of the main program    */
  for(k=0;k<NL;k++){
    xnbuf[k] = 0.0;             // clear signal buffer
    w[k] = 0.0;                 // clear coefficient buffer
  }
  while ((fscanf(xn_in,"%f",&dn)) != EOF)
  { /*  read in d(n) from data file and processing it  */
    for (i=NL-1; i>0; i--){
      xnbuf[i] = xnbuf[i-1]; // refresh signal buffer
    }
    xnbuf[0] = xn;             // insert new sample into buffer
    xn = dn;                   // delayed d(n) by 1 to get x(n)
    yn = 0.0;
    for (j=0; j<NL; j++){
      yn += xnbuf[j] * w[j]; // FIR filtering
    }
    err = dn-yn;               // e(n) = d(n) - y(n)
    mu_err = mu*err;           // ue(n) = mu*e(n)
    for (k=0; k<NL; k++){      // update coefficients
      w[k]=w[k]+mu_err*xnbuf[k];  // LMS update
    }
    fprintf(yn_out,"%f\n",yn);
  }
```

```
    printf("Finish");
    fcloseall();                    // close all opened files
}
```

The output data file `out.dat` contains the enhanced sinewave after processing the corrupted signal in `in.dat` by the ALE. The error signal is computed by subtracting the output of the adaptive filter from the desired signal. The filter coefficients are updated by adding a small correction term `mu*err*xnbuf[k]`. The user can adjust the filter length `NL` and the step size `mu` to observe the change of convergence rate by examining the output signal, $y(n)$, and the error signal, $e(n)$.

9.8.3 Implementation Using Fixed-Point C

The floating-point C program can be modified to become a fixed-point C program that runs on a general-purpose computer. First, the input signal must be converted and stored in 16-bit integer format, which is done by `singen9_8.m` and is saved in the file `in_int.dat`. Details on the data conversion are given in Section 6.6.3. A partial listing of the fixed-point C code `ale_fixpt.c` is shown as follows:

```
yn = 0;
for (j=0; j< NL; j++){ //    FIR filtering
  yn += ((long)(int)xnbuf[j]*(long)(int)w[j])>>15;
}
err = dn-yn; // e(n) = d(n) - y(n)
mu_err = (int)(((long)(int)mu*(long)(int)err)>>15);

for (k=0; k< NL; k++){ //    update filter coefficients
  w[k]=w[k]+(int)(((long)(int)mu_err*(long)(int)xnbuf[k])>>15);
}
```

In fixed-point C code, only the output variable `yn` is declared as a `long` variable, whereas the other variables, such as `xnbuf[]`, `w[]`, `err`, `dn`, and `mu_err`, are assigned as `int`. The 16-bit fixed-point implementation needs to scale back the product after multiplication in order to prevent the sum of products from overflow. This process is carried out by right-shifting the product 15 bits after multiplication. The reason for performing a 15-bit right shift instead of 16-bit right shift is to account for the extra sign bit, as explained in Chapter 3.

The fixed-point C program `ale_fixpt.c` saves the output signal, $y(n)$, the first coefficient $w_0(n)$ trajectory, and the error signal, $e(n)$, in the data files `out_int.dat`, `w_int.dat`, and `err_int.dat`, respectively. These files can be loaded into MATLAB for analyzing the performance of the adaptive filter.

9.8.4 Implementation Using Fixed-Point C for C5000 Processors

We can modify the fixed-point C program developed in the previous section to run on the C54x and C55x processors. The CCS for the C5000 is used for testing this fixed-point C program.

In this experiment, we use the FILE I/O capability of CCS to read the signal from a data file for running the C code on C54x and C55x simulators. The C code reads the signal samples via a probe point, performs ALE operations, and displays the enhanced output. The main function `ale_fixpt_ccs.c` enters an infinite loop that runs the `dataIO` and `aleproc` functions. The function `aleproc` is listed as follows:

```
static int aleproc(int *input, int *output, int *error)
{
  #define NL 16              // length of FIR filter
  long yn = 0;
  int xnbuf[16];             // signal buffer
  int w[16];                 // coefficient buffer
  int i,j,k;
  int mu = 33;               // 33(0.001); 338(0.01); 655(0.02)
  int mu_err, err;
  int xn = 0;
  int size = 500;
  for(k=0;k<NL;k++){
    xnbuf[k]=0;              // clear signal buffer
    w[k] = 0;               // clear coefficient buffer
  }
  while(size--){
    for (i=NL-1;i>0;i--){
      xnbuf[i]=xnbuf[i-1]; // refresh signal buffer
    }
    xnbuf[0]=xn;            // inject new sample
    xn = *input++;         // delayed d(n) by 1
    yn = 0;
    for (j=0; j< NL; j++){ // FIR filtering
      yn += ((long)(int)xnbuf[j]*(long)(int)w[j])>>15;
    }
    err = xn-yn;            // compute e(n)
    mu_err = (int)(((long)(int)mu*(long)(int)err)>>15);
    for (k=0; k< NL; k++){ // update filter coefficients
      w[k]=w[k]+(int)(((long)(int)mu_err*(long)(int)xnbuf[k])>>15);
    }
    *output++ = (int)(yn);
    *error++ = (int)(err);
  }
  return(TRUE);
}
```

The C program `ale_fixpt_ccs.c` is included in the project with the files `vectorsc54.asm` (for the C54x only), `cmdc54.cmd` (or `c55x.cmd` for the C55x), and `rts.lib` (or `rts55.lib` for the C55x). After successfully compiling and linking the program, we load the executable file into the CCS simulator. A probe point is set to transfer data from the data file `in_int_ccs.dat` to the memory `in_buffer`. Note that `in_int_ccs.dat` is obtained by adding the CCS header to the data file `in_int.dat`, which is used in the previous section.

A probe point and breakpoint are inserted into the program by placing the cursor at the line `dataIO()` and clicking on ✳ and 🖑. Select **File → File I/O** from the menu, and click on **Add File** in the **File Input** tab. Key in `in_int_ccs.dat`, and fill in in_buffer and 500 in the **Address** and **Length** boxes, respectively. Click on **Add Probe point** and select the in_int_ccs.dat file to complete the link between the data file and probe point. Insert another breakpoint at the line *error++=(int)(err) for updating the display plots. The time-domain plots of out_buffer and err_buffer can be displayed by choosing **View → Graph → Time/Frequency**. In this experiment, we set up the dual plots, as shown in Fig. 9.25, to display both $y(n)$ and $e(n)$.

Once the preceding steps are completed, click on the animate button 🐾, which executes the program until it reaches the breakpoint. The program halts, and the graphic display windows are updated. Figure 9.26 shows the time-domain plots obtained by displaying the out_buffer (upper) and err_buffer (bottom). Notice that the output signal is the enhanced version of the input sinusoidal and matches the results in the previous C implementation on a general-purpose computer. The error

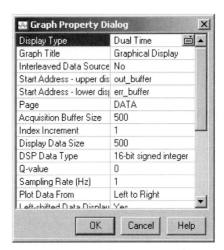

Figure 9.25 Configuring the graphic window for the dual plots of $y(n)$ and $e(n)$

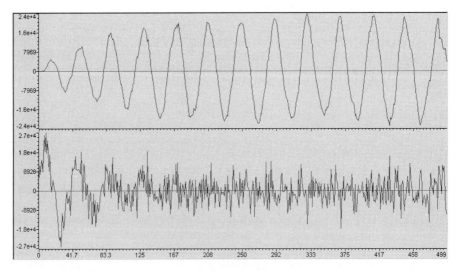

Figure 9.26 Time-domain plots of out_buffer (top) and err_buffer (bottom)

plot indicates the rate of convergence and the residual error after convergence. The user can change the time-domain plots to show the spectra of $y(n)$ and $e(n)$.

The ALE code can be profiled in terms of the number of clock cycles. Select **Profiler** → **Enable Clock** and **Profiler** → **View Clock** from the menu. Set one break-point at the instruction `for(i = NL - 1; i > 0; i--)` and the other at the instruction `*output++ = (int)(yn)`. Run the program to the first breakpoint, and double-click on the **Clock Window** to clear the cycle count. Run the program to the second breakpoint, and the **Clock Window** displays the number of cycles (`Clock = 1800` for the C54x and `Clock = 1422` for the C55x) that the C code need-ed to execute the adaptive filtering. The user can also compile the C code with the −k option to keep the C-compiler-generated assembly code.

9.8.5 Implementation Using Fixed-Point C with Intrinsics

In this experiment, the fixed-point C code for C5000 processors is modified to include intrinsics for both C54x and C55x processors. For a reason similar to that stated in Chapter 6, we use separate intrinsics (`_lsmpy` for multiplication and `_lsadd` for addition) to implement the multiply-add operation of FIR filtering. Note that right-shifting by 16 bits is required to prevent arithmetic overflow.

The partial listing of the fixed-point C program `ale_fixpt_intr.c`, which uses intrinsics for adaptive filtering, is shown as follows:

```
while(size--){
  for (i=NL-1;i>0;i--){
    xnbuf[i] = xnbuf[i-1]; // refresh signal buffer
  }
  xnbuf[0] = xn;
  xn = *input++;           // delayed version of xn

  yn = 0;
  for (j=0; j< NL; j++){   // FIR filtering
    temp = _lsmpy(xnbuf[j],w[j])>>16;
    yn = _lsadd(yn,temp);
  }
  err = xn-yn;             // e(n) = d(n) - y(n)
  mu_err = (int)(_lsmpy(mu,err)>>16);

  for (k=0; k< NL; k++){   // update filter coefficient
    temp = (int)(_lsmpy(mu_err,xnbuf[k])>>16);
    w[k] = _lsadd(w[k],temp);
  }
  *output++ = (int)(yn);   // save output signal y(n)
  *error++ = (int)(err);   // save error signal e(n)
```

This C program can be tested in CCS using the procedure given in Section 9.8.4.

9.8.6 Implementation Using Assembly for C5000 Processors

The development of assembly code to implement the ALE on C5000 processors is given in this experiment.

Implementation on the C54x

The C54x assembly code `ale_fixpt_c54x.asm` implements the ALE for experiments. The input data is copied physically into data memory using the `.copy` directive, while the coefficients are loaded into the section labeled as `coeff_table`. The uninitialized sections `out`, `error`, and `coeff` are reserved to store 500 samples of output and error; and 16 filter coefficients in the internal memory of the C54x. The coefficients of the FIR filter are initialized to zero before adaptation. The linker-command file `c54.cmd` is used to link these sections. The complete assembly code is listed as follows:

```
        .def      start
out_buffer  .usect    "out",500     ; reserve 500 locations for y
err_buffer  .usect    "error",500   ; reserve 500 locations for
                                    ; error
coeff       .usect    "coeff",16    ; reserve 16 locations for
                                    ; coefficients

            .sect     "coeff_table"
init_coeff  .int      0,0,0,0,0,0,0,0,0,0,0,0,0,0,0,0

            .sect     "indata"
in_buffer   .copy     "in1_int.dat" ; copy in1_int.dat to memory
                                    ; in_buffer

            .sect     "adapt"
two_mu      .int      676   ; mu = 33(0.001); 338(0.01); 655(0.02)
two_mu_err  .int      125   ; initialize to the e0
    .mmregs
    .text                   ; create code section
start:
    SSBX    FRCT            ; turn on fractional mode
    SSBX    SXM             ; turn on sign extension mode
    RSBX    OVM             ; turn on overflow mode

* Copy data to vector x using indirect addressing mode
copy:
    STM     #coeff,AR2      ; AR1 pointing to coeff
    RPT     #15             ; repeat next instruction 16 times
    MVPD    init_coeff,*AR2+ ; copy the init_coeff to coeff

* Setup the pointers to point to the coeff and data sample
    STM     #coeff, AR2     ; setup the pointer for coeff
    STM     #in_buffer,AR3  ; setup the pointer for in_buffer
    STM     #out_buffer,AR4 ; setup the pointer for out_buffer
    STM     #err_buffer,AR1 ; setup the pointer for err_buffer
    STM     #1,AR0
    STM     #499,AR6
    STM     #16,BK          ; circular buffer for coeff at 16-taps
    LD      #two_mu_err,DP
    LD      @two_mu_err,T   ; init T=two_mu_err
loop: LD    #0,B            ; clear Acc B for filter acc. output
```

```
      STM    #14,BRC                 ; loop (16-2)=14 times
      RPTBD  adapt_end-1
      MPY    *AR3,A                  ; A = mu*err*input
      LMS    *AR2,*AR3+              ; B = a0*x0+0 || A=e0+A (A = a0)
      ; RPTB starts here
      ST     A,*AR2+0%               ; updates coeff and
      ||MPY  *AR3,A                  ; A = x1*T(2*mu*err) in parallel
      LMS    *AR2,*AR3+              ; B = filtered output || A=e+A(A =ai)
adapt_end:
      STH    A,*AR2+0%               ; store final coeff
      STH    B,*AR4                  ; store final result
      LD     *AR3,A
      SUB    *AR4+,A                 ; obtain the error signal
      STL    A,*AR1                  ; store error signal
      LD     @two_mu,T
      MPY    *AR1+,A
      STH    A,@two_mu_err
      LD     @two_mu_err,T
      MAR    *+AR3(#-15)             ; reset input vector back by 15 time
      BANZ   loop,*AR6-
complete:    NOP
```

In the main ALE program, the coefficients located in the external memory coeff_table are transferred into the internal memory coeff. After this transfer operation, pointers AR1, AR2, AR3, and AR4 are set to point at the variables error_buffer, coeff, in_buffer, and out_buffer, respectively. The ALE program consists of three parts: (1) setting up the pointers, (2) performing a block repeat for adaptive filtering using the LMS instruction (explained in Section 9.6.2), and (3) computing the error signal, updating out_buffer and err_buffer, and resetting the pointers to coeff and in_buffer. It is important to note that the LMS instruction implements the delayed LMS algorithm. The user can modify the code to implement the LMS algorithm without using the LMS instruction (see the problems at the end of this chapter).

We create a new project to include files ale_fixpt_c54x.asm, c54.cmd, and vectors.asm, which is similar to the procedure given in Section 9.8.4. Perform a complete build, and load the executable program into memory. Use **GEL →** **C5402_Configuration → CPU_Reset** to reset the program pointer to the start of the program. Set a breakpoint at the end of the program by placing the cursor at the NOP instruction and clicking on the breakpoint icon 🖑 . Set up a display window to view out_buffer and err_buffer as in the previous experiments. In order to ensure that the output data does not contain any previous values, we can fill the out_buffer memory with zeros. Run the program and display out_buffer and err_buffer.

The final step is to profile the ALE using complete assembly code. Set breakpoints at the instructions LD #0,B and BANZ loop,*AR6−. Enable and view the clock, as described in the previous sections. Note that only 384 cycles are required to run the assembly-coded adaptive FIR filtering as compared to 1,800 cycles in fixed-point C. The user can open the assembly file ale_fixpt_ccs.asm generated (in Section 9.8.4) by the C compiler and compare it with the assembly program `ale_fixpt_c54x.asm`. The user can also modify the C54x assembly code to implement the normalized LMS and other LMS-type algorithms.

Implementation on the C55x

In this experiment, we modify the fixed-point C program `ale_fixpt_ccs.c` given in Section 9.8.4 for experiments. We replace the main computation loop for adaptive filtering with the subroutine written in C55x assembly language.

The new C program `ale_fixpt_c55x.c` calls two assembly routines for refreshing the signal buffer (`shift_c55x.asm`) and FIR filtering with the LMS updating (`adapt_c55x.asm`) as follows:

```
shift_c55x(xnbuf,NL/2,dn);          // refresh signal buffer
dn = *input++;                       // delay by 1
err = dn-yn;                         // compute e(n-1)
adapt_c55x(&err,xnbuf,w,NL,mu,&yn);  // FIR filtering with LMS
                                     // update
*output++ = yn;
*error++  = err;
```

Adaptive FIR filtering with the LMS algorithm is implemented using C55x assembly code `adapt_c55x.asm` and is listed as follows:

```
    .def _adapt_c55x        ; adaptive FIR filter with LMS algorithm
    .text
_adapt_c55x
    sub   #2,T0
    mov   T0,BRC0            ; set up repeat loop counter
    mov   T1,hi(AC1)
    mpym  *AR0,AC1,AC2       ; mu_err = mu * err
    mpym  *AR1,AC2,AC0       ; mu_err*xnbuf[i]
    mov   #0,AC1
    LMS   *AR2,*AR1,AC0,AC1  ; FIR filtering and update coefficients
    rptblocal   loop_end-1
    mov   hi(AC0),*AR2+
||  mpym  *AR1+,AC2,AC0
    LMS   *AR2,*AR1,AC0,AC1
loop_end
    mov   hi(AC0),*AR2       ; store the last updated coefficient
    mov   rnd(hi(AC1)),*AR3  ; store the output of FIR filtering
    ret
    .end
```

We create a new project to include `ale_fixpt_c55x.c`, `shift_c55x.asm`, `adapt_c55x.asm`, `c55x.cmd`, and `rts55.lib`, which is similar to the procedure given in Section 9.8.4. After successfully compiling and linking the programs, we load the executable file into a simulator for experiments. Following the identical procedure given in Section 9.8.4, use the probe point of CCS to connect `in_int_ccs.dat` for data file I/O, and use **Graph** to view the filter output and error signals. By comparing the output waveform with the one obtained from the fixed-point C program in Section 9.8.4, we can verify that the assembly routine is correct.

To evaluate the efficiency of this mixed C-and-assembly routine, we set one breakpoint at the instruction

```
shift_c55x(xnbuf,NL/2,dn);
```

and the other at the instruction

```
*output++ = yn;
```

Run the program to the first breakpoint, double-click on the **Clock Window** to clear the cycle count, and run the program again to the second breakpoint. The **Clock Window** displays the number of cycles (`clock = 164`) for the code to execute adaptive filtering on the C55x. Compared with `clock = 1422` in the fixed-point C program given in Section 9.8.4, it clearly shows that the mixed C-and-C55x assembly routines is the most efficient and effective technique for software development.

9.8.7 Implementation Using the C5000 Digital Signal Processing Library

As introduced in Section 9.6, Texas Instruments provides C54x and C55x DSPLIBs that contain C-callable, assembly-optimized functions. In this experiment, we introduce the C code that calls the adaptive-filtering routine in the DSPLIB. As discussed in Section 9.6.2, three adaptive filtering routines are available in the C54x DSPLIB (`dlms.asm`, `ndlms.asm`, and `nblms.asm`). For example, the adaptive FIR filter with the delayed LMS algorithm can be called as follows:

```
oflag = short dlms(DATA *x, DATA *h, DATA *r, DATA *d,
        DATA *des, DATA step, ushort nh, ushort nx)
```

The argument `x` is the pointer to the input-signal vector of `nx` elements, `h` is the pointer to the filter-coefficient vector of size `nh`, `r` is the output-signal vector of `nx` elements, `d` is the pointer to a memory location containing the address of the delay buffer, `des` is the desired signal vector, `step` is the step size, μ, and `oflag` is the overflow-error flag.

As discussed in Section 9.6.3, the C55x DSPLIB provides `dlms.asm` and `dlmsfast.asm` for supporting adaptive filtering. For example, the assembly function `dlms.asm` can be called in C as follows:

```
short oflag = dlms(DATA *h, DATA *x, DATA *r, DATA *des,
            DATA *dbuffer, DATA step, ushort nh, ushort nx)
```

It is important to note that the arguments are almost identical to those in the C54x routine, except that `dbuffer[nh + 2]` is used in the C55x for the delay buffer, and the order of arguments (`h` and `x`) is different.

In this experiment, we modify the C code in Section 9.8.4 to call the `dlms` function for experiments. The user can also load the C54x file `ale_fixpt_dsplib.c` into

the CCS environment to compile, build, and run the code. The new C program is listed
as follows:

```
#include <math.h>
#include <tms320.h>
#include <dsplib.h>

#include "newtest.h"      // data file used for testing
DATA err[NX];
short i;
short oflag;
#define STEP 1310         // 66(2*0.001); 676(2*0.01); 1310(2*0.02)

void main(void)
{
  asm(" STM           #0, SWWSR"); // setting simulator to operate in
                                   // 0ws
  for (i=0; i<NH; i++) h[i] =0;   // clear coefficient buffer
                                   // (optional)
  for (i=0; i<NX; i++) r[i] =0;   // clear output buffer (optional)
  for (i=0; i<NH; i++) dbuffer[i] = 0; // clear delay buffer (must)

  oflag = dlms(x,h,r,&dp,des,STEP,NH,NX);// different order for
                                         // C55x
  for (i=0;i<NX;i++){
    err[i]=des[i]-r[i];
  }
  return;
}
```

Some important features of the C program are discussed in Section 6.6.7. Since
the dlms function uses the Q.15 data type DATA, the coefficients and the input signals
are modified to this data type. The file newtest.h contains the input and desired signals. It also contains the initial coefficients, which are all set to zero. The user can
insert an instruction asm("STM #0,SWWSR") at the beginning of the program for C54x
experiments, which sets the simulator to operate at a zero-wait state.

As with FIR filtering and IIR filtering, the adaptive filter with the delayed
LMS algorithm operates more efficiently in block mode, which can be verified by
benchmarking the block-processing cycle count using different block sizes. The user
can modify the C program and benchmark the code to run in a block of 500, 125, and
2 samples. The benchmark results for using the C54x dlms function with different
block lengths are summarized in Table 9.4.

We can compare the filter output with the previous fixed-point C program by
displaying the output vector r. The user can also check whether any overflow occurs
in the adaptive filtering by examining the variable oflag. In addition, the user can
refer to the *TMS320C54x DSP Library Programmer's Reference* [8] for complete
information on using the function nblms for the normalized-block LMS filter and
the function ndlms for the normalized-delayed LMS filter. Finally, the user can modify the C program ale_fixpt_dsplib.c for C55x experiments using different block
sizes.

TABLE 9.4 Benchmark of the C54x `dlms` Function Using Different Block Sizes to Complete 500 Input Samples

Number of samples per block, nx	Cycle count
500	23,333
125 (4 blocks to complete 500 samples)	23,543
2 (2 samples per block)	91,020

Some additional applications and experiments on adaptive filters are given in Appendix B. The user can refer to the MATLAB, Simulink, or C files for more details of these exercises and can progressively work through these hands-on experiments to obtain the final solution using CCS. These additional hands-on experiments include implementation of an acoustic echo canceller and an active noise-control system.

PROBLEMS

PART A

Problem A–9–1

Design an adaptive filter to cancel the noise component in the primary signal $d(n) = s(n) + v(n)$, as shown in Fig. 9.12. The desired signal $s(n)$ is a 100 Hz sinewave with a sampling frequency of 8 kHz. The noise $v(n)$ is random noise with normal distribution, a zero mean, and a variance of 0.1. A highly correlated noise signal $v'(n)$ is derived by delaying the $v(n)$ by 2 samples [i.e., $v'(n) = v(n - 2)$]. Use the Filter Design Toolbox to implement the adaptive filter using the LMS and normalized LMS algorithms. Compare the error signals with the original sinewave to evaluate the performance of these algorithms.

Problem A–9–2

Redo Problem 1 using Q.15 format for signal samples, parameters, and filter coefficients.

Problem A–9–3

Modify the Simulink program `anc.mdl` to include a speech-signal leakage of 10% to the reference sensor. Run the simulation, and observe the rate of convergence of the adaptive filter. Try to increase the percentage of signal leakage until the adaptive noise canceller fails to converge.

Problem A–9–4

Implement an adaptive filter similar to that shown in Fig. 9.8 to equalize the given channel (unknown system), which is a 5-tap FIR filter with a coefficient vector of [0.3 0.5 1 0.5 0.3]. The adaptive filter models the inverse of the channel. Select a suitable delay in deriving the desired signal $d(n)$, and choose a proper filter length to compensate for the unknown channel. Again, use the Filter Design Toolbox to implement the adaptive algorithms using the LMS and normalized LMS algorithms, and compare their performances.

Problem A–9–5

Redo Problem 4 using Q.15 format for signal samples, parameters, and filter coefficients.

Problem A–9–6

Repeat the simulation in Problem 4 by experimenting with the tracking ability of the adaptive filter. After the adaptive filter in Problem 4 has converged to a steady state, it is switched to decision-feedback mode. During this time, the center tap in the aforementioned 5-tap channel starts to drift, as given in this new channel's transfer function $H(z) = 0.3 + 0.5z^{-1} + (1 + e^{-k/1000})z^{-2} + 0.5z^{-3} + 0.3z^{-4}$, where k is the iteration index, which only occurs for the first 500 iterations after switching to decision-feedback mode. Observe the coefficients of the adaptive filter during this period. Test the effects of the step size on its tracking ability.

Problem A–9–7

Replace the unknown system in the Simulink program `sysident.mdl` with an IIR filter defined as $H(z) = (0.2 + 0.6z^{-1} + 0.2z^{-2}/1 - 0.8z^{-1} + 0.8z^{-2})$. Experiment on a suitable filter length L for the FIR adaptive filter so that the adaptive system identification converges to a small error. What happens when a speech signal `timit1.wav` is used as an excitation signal instead of using white noise?

Problem A–9–8

Implement an ALE, as shown in Fig. 9.16, using single-precision, floating-point format. The input signal $x(n)$ contains the narrowband signal (sinewave) corrupted by the broadband white noise and is delayed by one sample unit before feeding into the adaptive filter. The adaptive filter predicts the narrowband component using the delayed signal $x(n - 1)$. Therefore, the adaptive-filter output $y(n)$ contains the narrowband component, while the error signal $e(n)$ contains the broadband noise component.

Problem A–9–9

Redo Problem 8 using Q.15 format for signal samples, parameters, and filter coefficients.

Problem A–9–10

Investigate the effects of representing the adaptive-filter parameters in Q.15 and Q.7 formats. Different adaptive algorithms such as the LMS and normalized LMS algorithms are used to adapt the coefficients for system identification, inverse system modeling, and noise-cancellation applications.

Problem A–9–11

Modify the ALE assembly programs listed in Section 9.8.6 to implement the adaptive filter with the LMS algorithm without using the LMS instruction. Observe any change in the performance of the implemented ALE over the program given in Section 9.8.6. Profile and compare the cycle counts for executing these two ALE programs.

Problem A–9–12

Modify the preceding ALE experiments to process an input signal that is speech plus sinusoidal noise. Show that the error signal contains an enhanced speech signal and that the output of the adaptive filter contains the sinusoidal noise after the adaptive filter has converged. Note that the delay must be larger than 1 to decorrelate the broadband speech.

Problem A–9–13

Redo Problem 12 using Q.15 format for signal samples, parameters, and filter coefficients.

Problem A–9–14

Modify the Simulink file `ale.mdl` to include three sinewaves at normalized frequencies with reference to the Nyquist frequencies of 0.2, 0.22, and 0.5. Their respective powers are 10, 0.1, and 0.5. In addition, change the Gaussian white-noise block to a zero mean and a variance of 0.5. Use a 128-tap adaptive FIR filter with the LMS algorithm, run the simulation, and observe the spectra of the adaptive-filter output and error signals. Comment on the ability of the ALE to resolve these three sinewaves and the Gaussian white noise. How does this method compare with the FFT-based periodogram in terms of resolving the weaker signal?

PART B

Problem B–9–15

Find the autocorrelation matrix $\mathbf{R}$ defined in Eq. (9.2.8) for the following input signals:

(a) $x(n) = v(n)$, where $v(n)$ is a zero-mean white noise with a variance of σ_v^2
(b) $x(n) = A\cos(\omega_0 n) + v(n)$

Problem B–9–16

Given the adaptive filter shown in Fig. 9.1, where the digital filter is an FIR filter with $L = 2$ and $x(n) = \sin(\omega_0 n)$ and $d(n) = 2\cos(\omega_0 n)$, find the following:

(a) The $\mathbf{R}$ matrix and $\mathbf{p}$ vector
(b) The MSE function defined in Eq. (9.2.7)
(c) The optimum coefficient vector
(d) The minimum MSE

Problem B–9–17

Discuss the effects of choosing the filter L for an adaptive FIR filter with the LMS algorithm in terms of the stability, convergence rate, and excess MSE of the algorithm.

Problem B–9–18

Discuss the effects of choosing the step size, μ, for an adaptive FIR filter with the LMS algorithm in terms of the stability, convergence rate, excess MSE, and finite-precision effects of the algorithm.

Problem B–9–19

As shown in Fig. 9.4, if the unknown system is given as $P(z) = 0.75 + 0.5z^{-1} - 0.6z^{-2}$, the adaptive filter is $W(z) = w_0 + w_1 z^{-1} + w_2 z^{-2}$, and the excitation signal $x(n)$ is a zero-mean white noise with variance 1, find the following:

(a) The **R** matrix and **p** vector
(b) The optimum solution of adaptive filter $\mathbf{w}^\circ$
(c) The minimum MSE after the convergence of the adaptive filter

Problem B–9–20

Compare and summarize the complexity (in terms of number of multiplications, number of additions, and memory locations) of the LMS, normalized LMS, and leaky LMS, as will as the sign-error, sign-data, sign-sign, and complex LMS algorithms.

Problem B–9–21

Derive the practical bound for the step size given in Eq. (9.3.2) from its theoretical bound given in Eq. (9.3.1).

Problem B–9–22

Discuss the methods that can be used to prevent the coefficients of an adaptive filter from overflowing during the adaptation when the LMS algorithm is implemented on fixed-point, 16-bit DSP processors. In addition, how can the LMS algorithm be prevented from being stalled or locked up?

SUGGESTED READINGS

1 Widrow, B. and S. D. Stearns. *Adaptive Signal Processing*. Englewood Cliffs, NJ: Prentice Hall, 1985.
2 Haykin, S. *Adaptive Filter Theory*. 4th Ed. Upper Saddle River, NJ: Prentice Hall, 2002.
3 Treichler, J. R., C. R. Johnson, Jr., and M. G. Larimore. *Theory and Design of Adaptive Filters*. Upper Saddle River, NJ: Prentice Hall, 2001.
4 Kuo, S. M. and D. R. Morgan. *Active Noise Control Systems*. New York, NY: John Wiley, 1996.
5 Kuo, S. M. and C. Chen. "An Implementation of Adaptive Filters with the TMS320C25 or the TMS320C30." Chapter 7 in *Digital Signal Processing Applications*, vol. 3, Ed. P. E. Papamichalis. Englewood Cliffs, NJ: Prentice Hall, 1990. Also available from Texas Instruments, SPRA017, 1990.
6 The MathWorks. *Filter Design Toolbox User's Guide*. Version 2.1, 2000.
7 The MathWorks. *Signal Processing Toolbox User's Guide*. Version 5, 2000.
8 Texas Instruments. *TMS320C54x DSP Library Programmer's Reference*. SPRU518c, 2002
9 Texas Instruments. *TMS320C55x DSP Library Programmer's Reference*. SPRU422e, 2002
10 Texas Instruments. *TMS320C62x DSP Library Programmer's Reference*. SPRU402a, 2002
11 Texas Instruments. *TMS320C64x DSP Library Programmer's Reference*. SPRU565a, 2002
12 Texas Instruments. *TMS320C54x DSP Design Workshop*. Oct 2000.

Answers to Selected Problems

13. $X(z) = \dfrac{z}{z - a}$, ROC is $|z| < |a|$.

14. $X(z) = \dfrac{1}{z^{N-1}} \dfrac{z^N - a^N}{z - a}$.

15. (a) $\delta(n - k) \leftrightarrow z^{-k}$.

    (b) $u(n - k) \leftrightarrow \dfrac{z^{-(k-1)}}{z - 1}$.

16. $y(n) = \dfrac{1 - a^{n-1}}{1 - a} u(n)$.

17. $y(n) = \{1, 2, 3, 3, 2, 1\}$.

18. $X(\omega) = e^{j\omega(N-1)/2} \dfrac{\sin(\omega N/2)}{\sin(\omega/2)}$.

19. (b) $H(\omega) = 2e^{-j\omega/2}\cos(\omega/2)$, $|\omega| \le \pi$.

   (c) $|H(\omega)| = 2\cos(\omega/2)$, $\phi(\omega) = -\omega/2$.

20. (a) $H(z) = \dfrac{z^{-1} + z^{-2}}{2 + z^{-1} + 0.9z^{-2}}$.

   (b) $H(z) = \dfrac{z^{-1} + 0.75z^{-2}}{1 - 0.5z^{-1} + 0.8z^{-2}}$.

21. (a) $y(n) = x(n) + 0.25x(n-1) - 0.5x(n-2) - 0.5y(n-1)$.

   (b) $y(n) = \dfrac{1}{8}[x(n-1) + 6y(n-1) - y(n-2)]$.

22. (b) $H(\omega) = \dfrac{1}{1 - ae^{-j\omega}}$, $|a < 1|$.

   (c) $|H(\omega)| = \dfrac{1}{\sqrt{1 + a^2 - 2a\cos\omega}}$.

23. (a) $H(\omega) = \dfrac{1}{3}e^{-j\omega}\dfrac{\sin(3\omega/2)}{\sin(\omega/2)}$.

   (b) $|H(\omega)| = \dfrac{1}{3}\left|\dfrac{\sin(3\omega/2)}{\sin(\omega/2)}\right|$. $\phi(\omega) = \begin{cases} -\omega, & if\ \dfrac{\sin(3\omega/2)}{\sin(\omega/2)} \ge 0 \\ -\omega + \pi, & if\ \dfrac{\sin(3\omega/2)}{\sin(\omega/2)} < 0 \end{cases}$.

24. $b = -a$.

25. $H(z) = \dfrac{4 + (4/3)z^{-1}}{1 + (2/3)z^{-1} - (1/3)z^{-2}}$.

   $y(n) = -(2/3)y(n-1) + (1/3)y(n-2) + 4\delta(n) + (4/3)\delta(n-1)$.

27. $m_x = a + b$. $\sigma_x^2 = b^2$.

29. $F_0 = 0.2$. $\omega_0 = F_0\pi = 0.2\pi$.

30. $\Delta f = \dfrac{f_s}{N} = 10$ Hz. $k = f_k/\Delta f = 10$.

CHAPTER 3

2. 3fd0000000000000h,
   c024e66666666666h,
   409b800000000000h.

5. 256-tap FIR filter.

9. (a) 0.01110 and 1.01110.
   (b) 0.01110 and 1.10001.
   (c) 0.01110 and 1.10010.

10. (a) -0.375 (1.101).
    (b) 0.375 (10.011).

11. Q.7 representation: $26, 56, -92, 105$.

12. Q.5 representation: $3, 7, -12\ 13$.

13. Q8.7 equivalent: (a) 36.40625, (b) 0, (c) -256, (d) -221.7265, (e) -168.3984.

14. (a) 0.
    (b) 0.015625.

15. Fractional multiplication error is $0.21875 - 0.125 = 0.09375$.
    Integer multiplication error is $42 - 2 = 40$.

17. Smallest error $= 1.175 \times 10^{-38}$, largest error $= 3.403 \times 10^{-38}$.

19. (a) 0.4143.
    (b) 13/32 (for 1), 9/32 (for 0.707).
    (c) Quantization errors are 8.05×10^{-3} and 0.0175.
    (d) FL and FR is $20 \times \log(13/32) = -7.8$ dB;
        $FC = SL = SR = 20 \times \log(9/32) = -11$ dB.

21. $L_{max} < 2{,}225$ for sample mode and $L_{max} < 2{,}261$ for block mode
    (32 samples/block).

23. 12 audio channels.

CHAPTER 4

CHAPTER 5

11. Single precision: 1.18×10^{-38} to 3.4×10^{38} (or ~1,530 dB),
    Double precision: 2^{-1022} to 2^{1024} (or ~12,282 dB).

12. Total number of bits $= 488$ bits. 23.75% reduction.

16. $+5.8774717 \times 10^{-39}$ and $+3.4028234 \times 10^{+38}$.

CHAPTER 6

14. (a) $B(z) = 1 + z^{-L} = \dfrac{z^L + 1}{z^L}$, L poles at the origin.
    (b) $H(w) = 1 + e^{-jwL}$.
    (c) $L = 10$.

16. $B(\omega + 2\pi) = \displaystyle\sum_{n=0}^{N-1} h(n)e^{-j(\omega+2\pi)n} = \sum_{n=0}^{N-1} h(n)e^{-j\omega n}e^{-j2\pi n} = B(\omega).$

17. (a) $|B(\omega)| = |b_M + 2\displaystyle\sum_{i=1}^{M} b_{M-i}\cos(\omega i)|.$

   (b) $|B(\omega)| = |2\displaystyle\sum_{i=1}^{L/2} b_{L/2-i}\cos[\omega(i - 0.5)]|.$

19. Type 1: $y(n) = b_M x(n - M) + \displaystyle\sum_{i=0}^{M-1} b_i[x(n - i) + x(n - L + 1 + i)].$

   Type 3: $y(n) = \displaystyle\sum_{i=0}^{M-1} b_i[x(n - i) - x(n - L + 1 + i)].$

   Type 4: $y(n) = \displaystyle\sum_{i=0}^{L/2-1} b_i[x(n - i) - x(n - L + 1 + i)].$

20. Maximum bandwidth: $\sim$32 kHz, maximum length of FIR filter: $L = 415$ taps.

21. 48 MIPS, $L_{\max} = 347$ taps, total of 3.3 K words of internal memory.

CHAPTER 7

18. (a) $y(n) = b_0 d(n) + b_1 d(n - 1) + b_2 d(n - 2)$

   $d(n) = x(n) - a_1 d(n - 1) - a_2 d(n - 2)$

   (b) $H(z) = \dfrac{b_0 + b_1 z^{-1} + b_2 z^{-2}}{1 + a_1 z^{-1} + a_s z^{-2}}$

19. (b) $H_1(z) = \dfrac{1 + 0.6z^{-1}}{1 - 0.8z^{-1}}, \quad H_2(z) = \dfrac{1 - 0.6z^{-1}}{1 + 0.9z^{-1}}$, and $H(z) = 0.7H_1(z)H_2(z).$

   (c) $|a_2| = 0.72 < 1, |a_1| = 0.1 < 1 + 0.72,$

20. (a) Two real zeros at $z = 1$, two complex zeros at $z = -0.707 \pm j0.7072$. Two complex poles at $z = -0.4 \pm j0.6928$, two real poles at $z = 0.7499$ and $z = 0.3334.$

   (b) All-pole section 1, $\dfrac{1}{1 + 0.8z^{-1} + 0.64z^{-2}}$ is stable

   All-pole section 2, $\dfrac{1}{1 - 1.0833z^{-1} + 0.25z^{-2}}$ is stable

   (c) $H(z) = \dfrac{(1 - 2z^{-1} + z^{-2})}{(1 - 1.0833z^{-1} + 0.25z^{-2})} \cdot \dfrac{(1 + 1.414z^{-1} + z^{-2})}{(1 + 0.8z^{-1} + 0.64z^{-2})}.$

21. (b) Magnitude response is flat from 0 to π.

   (c) $\theta_{ap} = -\omega - 2\tan^{-1}\left(\dfrac{a \sin \omega}{1 - a \cos \omega}\right)$.

22. $H(z) = 0.2452\dfrac{1 + z^{-1}}{1 - 0.5096z^{-1}}$.

23. $\sin(\omega_k n) \leftrightarrow \dfrac{\sin \omega_k z^{-1}}{1 - 2 \cos \omega_k z^{-1} + z^{-2}}$ and $\cos(\omega_k n) \leftrightarrow \dfrac{1 - \cos \omega_k z^{-1}}{1 - 2\cos \omega_k z^{-1} + z^{-2}}$.

24. 729 kHz for IIR filter and 704 kHz for FIR filter.

CHAPTER 8

14. $X(k) = \dfrac{1 - a^N}{1 - ae^{-j\frac{2\pi k}{N}}}$.

15. $X(0) = \displaystyle\sum_{n=0}^{N-1} x(n)e^{-j\frac{2\pi 0}{N}n} = \sum_{n=0}^{N-1} x(n)$ is real since $x(n)$ is real.

   $X(N/2) = \displaystyle\sum_{n=0}^{N-1} x(n)e^{-j\frac{2\pi N/2}{N}n} \sum_{n=0}^{N-1}(-1)^n x(n)$ is also real.

16. $X(0) = 10; X(1) = -2 + j2; X(2) = -2; X(3) = -2 - j2.$

18. $y(n) = \{26, 28, 26, 20\}.$

20. (a) 720,000 samples
   (b) 33.33 Hz.
   (c) 500 blocks if no overlap, 999 blocks for 50% overlap, 1,997 blocks for 75% overlap.

24. (a) 2×10^6 samples.
   (b) $N = 4,096$ data samples/segment.
   (c) 488 segments and 976 segments.
   (d) $\sim 100 \times 10^6$ real multiply/add (excluding magnitude-square operations).

25. (a) 160 samples.
   (b) 15.9 msec.
   (c) Yes.
   (d) 0.348 msec.
   (e) Another 260 real multiplications.

26. (a) 1.2×10^6 samples, 2048-point FFT, and 600 segments (2,000 sample/block) for Bartlett.

    (b) 6.76×10^6 complex multiply/add for Bartlett.

    (c) 120,832 cycles (for Bartlett), and 10 MIPS processor.

27. (a) 512 Hz, 1Hz.

    (b) 103,424 complex multiply-add operations.

    (c) 0.7 sec, 10 sec for Bartlett.

CHAPTER 9

15. (a) $\mathbf{R} = \sigma_v^2 \begin{bmatrix} 1 & 0 \\ 0 & 1 \end{bmatrix}$.

    (b) $r_{xx}(k) = \dfrac{A^2}{2}[\cos(2\omega_0 n + \omega_0 k) + \cos(\omega_0 k)] + \sigma_v^2 \delta(k)$.

16. (a) $\mathbf{R} = \begin{bmatrix} 0.5 & 0.5\cos\omega_0 \\ 0.5\cos\omega_0 & 0.5 \end{bmatrix}$, $\mathbf{p} = [r_{dx}(0)\ r_{dx}(1)]^T = [0\ -\sin\omega_0]^T$.

    (b) $\xi = 2 + 0.5(w_0^2 + w_1^2) + w_0 w_1 \cos\omega_0 + 2w_1 \sin\omega_0$.

    (c) $\mathbf{w}^o = [2\cot\omega_0\ -2\csc\omega_0]^T$.

    (d) $\xi_{\min} = 0$.

19. (a) $\mathbf{R} = \begin{bmatrix} 1 & 0 & 0 \\ 0 & 1 & 0 \\ 0 & 0 & 1 \end{bmatrix}$, $\mathbf{p} = \begin{bmatrix} 0.75 \\ 0.5 \\ -0.6 \end{bmatrix}$.

    (b) $\mathbf{w}^o = [0.75\ \ 0.5\ \ -0.6]^T$.

    (c) $\xi_{\min} = 0$.

Index